FARADAY DISCUSSIONS
NO. 117   2000

Excited States at Surfaces

The Faraday Division
The Royal Society of Chemistry
London

Organising Committee
Professor S. Holloway (*Chairman*)
Dr G. R. Darling
Professor E. Hasselbrink
Professor R. G. Jones
Dr K. W. Kolasinski
Dr D. Lennon
Dr M. R. S. McCoustra

ISBN: 0-85404-971-1

ISSN: 0301-7249

Typeset by Santype International Ltd., Netherhampton Road, Salisbury, Wiltshire and printed and bound in Great Britain by Black Bear Press, Cambridge, UK.

A General Discussion

on

Excited States at Surfaces

4th, 5th and 6th September, 2000

A General Discussion on Excited States at Surfaces was held at the University of Nottingham, UK on 4th, 5th and 6th September, 2000.

Contents

1   Introductory Lecture: Excited states at surfaces: Fano profiles in STM spectroscopy of adsorbates
    J. W. Gadzuk and **M. Plihal**

15 Long-lived adsorbate states on metal surfaces
    Jean-Pierre Gauyacq, Andreï G. Borisov, Georges Raşeev and **Andrey K. Kazansky**

27 Effect of the projected band gap on the formation of negative ions in grazing collisions from Cu surfaces
    T. Hecht, H. Winter, A. G. Borisov, J. P. Gauyacq and **A. K. Kazansky**

41 Charge-transfer reactions in atom scattering from ionic surfaces: A time-dependent wave-packet approach
    G. R. Darling, Y. Zeiri and **R. Kosloff**

55 General Discussion

65 Direct and indirect DIET and DIMET from semiconductor and metal surfaces: What can we learn from 'toy models'?
    Peter Saalfrank, Gisela Boendgen, Cécile Corriol and **Tohru Nakajima**

85 Photoinduced charge-transfer reaction at surface. Part I. $(HCl)_m \cdot Na_n/LiF(001) + h\nu$ (640 nm) $\rightarrow (HCl)_{m-1} Cl^- Na_n^+/LiF(001) + H(g)$
    Javier B. Giorgi, Tae Geol Lee, Fedor Y. Naumkin, John C. Polanyi, Sergei A. Raspopov and **Jiaxi Wang**

99 Semiclassical treatment of reactions at surfaces with electronic transitions
    Christian Bach and **Axel Groß**

109 Quantum dynamics of the dissociation of H_2 on Cu(100): Dependence of the site-reactivity on initial rovibrational state
    Drew A. McCormack, Geert-Jan Kroes, Roar A. Olsen, Jeroen A. Groeneveld, Joost N. P. van Stralen, Evert Jan Baerends and **Richard C. Mowrey**

133 Energy disposal during desorption of D_2 from the surface and subsurface region of Ni(111)
    S. Wright, J. F. Skelly and **A. Hodgson**

147 The role of rotational excitation in the activated dissociative chemisorption of vibrationally excited methane on Ni(100)
    Ludo B. F. Juurlink, Richard R. Smith and **Arthur L. Utz**

161 General Discussion

191 Structure and dynamics of excited electronic states at the adsorbate/metal interface: C_6F_6/Cu(111)
    C. Gahl, K. Ishioka, Q. Zhong, A. Hotzel and **M. Wolf**

203 Excitation mechanisms and photochemistry of adsorbates with spherical symmetry
    Kazuya Watanabe and **Yoshiyasu Matsumoto**

213	Controlling organic reactions on silicon surfaces with a scanning tunneling microscope: Theoretical and experimental studies of resonance-mediated desorption **Saman Alavi, Roger Rousseau, Gregory P. Lopinski, Robert A. Wolkow** and **Tamar Seideman**
231	Electronic mechanism of STM-induced diffusion of hydrogen on Si(100) **K. Stokbro, U. J. Quaade, R. Lin, C. Thirstrup** and **F. Grey**
241	Inelastic interactions of tunnel electrons with surfaces **Andrew J. Mayne, Franck Rose** and **Gérald Dujardin**
249	The initiation and characterization of single bimolecular reactions with a scanning tunneling microscope **Lincoln J. Lauhon** and **Wilson Ho**
257	General Discussion
277	Theoretical aspects of tunneling-current-induced bond excitation and breaking at surfaces **Nicolas Lorente** and **Mats Persson**
291	Dynamics of charge transfer states on metal surfaces: The competition between reactivity and quenching ■ **Ronnie Kosloff, Gil Katz** and **Yehuda Zeiri**
303	Self-trapped excitons at the quartz(0001) surface **J. Song, R. M. VanGinhoven, L. R. Corrales** and **H. Jónsson**
313	Abstractive chemisorption of O_2 on Al(111) **Marcello Binetti, Olaf Weiße, Eckart Hasselbrink, Andrew J. Komrowski** and **Andrew C. Kummel**
321	Chemical selectivity in the remote abstractive chemisorption of ICl on Al(111) **Kharissia A. Pettus, Peter R. Taylor** and **Andrew C. Kummel**
331	General Discussion
347	Concluding remarks **A. W. Kleyn**
351	Additions and corrections
353	List of Posters
355	List of Participants
357	Index of Contributors

■ Electronic supplementary information is available on http://www.rsc.org/esi See article for further information.

Introductory Lecture

Excited states at surfaces: Fano profiles in STM spectroscopy of adsorbates

J. W. Gadzuk and M. Plihal

National Institute of Standards and Technology, Gaithersburg, MD 20899, USA

Received 30th October 2000
First published as an Advance Article on the web

The Fano–Anderson model for a discrete state embedded within a continuum is revisited within the context of excitation and decay processes which lead to some manifestations of Fano lineshape profiles. The phenomenon of resonance tunneling between an STM tip and a metal surface upon which there are isolated adsorbed atoms is discussed and the relationship between the spectroscopic signature of such systems and that of the Fano profile is taken up. Recent experimental studies of Kondo systems of magnetic adsorbates such as Co and Ce adsorbed on noble metal ($\bar{1}11$) surfaces have motivated this work.

1 Introduction

A topic as non-specific as inferred from the title of this Introductory Lecture ('Excited states at surfaces') invites the kinds of generalities that usually go unread because they are of equal interest to everyone, which is to say, of insufficient immediate interest to anyone to merit a read. Thus, other than a brief nod to some issues of excited state dynamics in the final section, this paper will focus on one very well defined realization of the excited state, the seemingly ubiquitous Fano lineshape.[1,2] Briefly stated, Fano considered the observable consequences of interference effects occurring when an excited state of an atomic system is created or decays through competing alternative paths. As a student of Enrico Fermi, it was suggested to him by Fermi that he try to understand the highly asymmetric lineshapes observed by Beutler[3] in the photoabsorption spectra of the noble gases which were glaringly different from the symmetric Lorentzian or Breit–Wigner shape expected for excited states whose occupation probability decays exponentially with time.[4,5] The end result of this exercise spanning 25 years, many of them with Fano working at the National Bureau of Standards in Washington, was his 1961 *Physical Review* paper[2] in which the lineshape formula, in his chosen notation, takes the form $I(\varepsilon) \sim (q + \varepsilon)^2/(1 + \varepsilon^2)$ where $\varepsilon = 2(E - E_0)/\Gamma$ is the dimensionless energy measured from the line center at E_0, in units of Γ, roughly the inverse lifetime of the excited state, and q is a parameter to be discussed later that characterises the interference between pathways. With over 3200 citations in the scientific literature since 1973, this is the most frequently referenced manuscript by an author from NBS/NIST. It is noteworthy that significant numbers of these references came from the literature outside atomic physics, many from condensed matter physics. These observations have been made by C. W. Clark in his reflective vignette on the Fano papers which will appear in a commemorative volume "A Century of Excellence in Measurements, Standards, and Technology" to be published in the year 2001, the centennial year of NBS/NIST.[6] Since both Fano and subsequently, one of the present authors (JWG, who was significantly influenced by his persona and by his legacy) spent a large part of their respective scientific lives there dealing with problems involving the excited state in atoms and at surfaces, it seems timely, in light of the centennial, to focus in on this topic. This is particularly fitting since recent realizations of Fano lineshapes in scanning tunneling microscope (STM) spectroscopy studies[7-10] of magnetic atoms adsorbed upon non-magnetic surfaces have stimulated a wave of theoretical activity[11-14] which is relevant to many of the topics addressed at this Faraday Discussion. With both the historical reflections on NBS/NIST and also the current research interests of this Faraday Discussion in mind, it seems appropriate to examine the inter-related topics of Fano lineshapes, the excited state, and contemporary resonance tunneling spectroscopy studies using the STM. General aspects of excited state resonance processes involving interfering alternatives, as envisioned by Fano, will be taken up in Section 2. Section 3 looks into complimentary resonance tunneling processes occuring in STM spectroscopy of surfaces. The electron Green's function and special attributes of surface *vs.* bulk state electron transport at surfaces, as relevant to the resonance STM studies, are presented in Section 4. Specific applications of the resonance model within a tunneling and/or STM environment are dealt with in Section 5. The current studies[7-10] involving Kondo impurities[15,16] on the (111) face of noble metals are also discussed in Section 5. Some final thoughts are offered in Section 6.

2 Fano–Anderson model for a localized state in the continuum

The abstract problem of a discrete state coupled to a continuum appears in many branches of physics.[17-20] Mahan poses the problem this way: "*We shall solve exactly the Hamiltonian*

$$H = \varepsilon_0 b^+ b + \sum_{\vec{k}} \varepsilon_{\vec{k}} c_{\vec{k}}^+ c_{\vec{k}} + \sum_{\vec{k}} V_{a\vec{k}}(c_{\vec{k}}^+ b + b^+ c_{\vec{k}}). \qquad (1)$$

It describes a localised state of energy ε_0 and operators b and b^+. This localized state will be called the impurity, and we assume only one exists. There is a continuous set of states of energy ε_k with operators c_k and c_k^+. This set of states could have a finite bandwidth, as often occurs in tight binding models in solids, or else it could be a free-particle model. The last term of the Hamiltonian includes the mixing between these two kinds of states. It contains processes whereby the continuum particle hops onto the impurity ($b^+ c_k$) and where the particle hops off the impurity into the continuum ($c_k^+ b$). If the hopping particle could change its spin orientation, the problem would become harder [i.e. the Kondo problem].

Since the Hamiltonian is quadratic in operators, its solutions are equivalent to diagonalizing a matrix ... This model Hamiltonian was introduced simultaneously by Anderson (1961)[21] and Fano (1961).[2] Anderson applied it to solid-state physics, while Fano used it in atomic spectra. It tends to be called the Anderson model or the Fano model depending on whether the speaker is a solid-state or atomic physicist. We shall call it the Fano–Anderson model."[22] Although solid-state applications are of main interest here, the initial method-of-choice will be the Fano formulation in which diagonalizing the Hamiltonian matrix implied by eqn. (1) is considered in terms of the matrix elements

$$\langle \varphi_a | H | \varphi_a \rangle = E_\varphi, \qquad (2a)$$

$$\langle \psi_{E(k')} | H | \psi_{E(k'')} \rangle = E(k')\delta(E(k') - E(k'')), \qquad (2b)$$

$$\langle \psi_{E(k)} | H | \varphi_a \rangle = V_{ak}, \qquad (2c)$$

formed from φ_a, the single discrete wavefunction and $\{\psi_{E(k)}\}$, the set of unperturbed continuum wavefunctions. The corresponding eigenvector to be determined has the form

$$\Psi_E = a(E)\varphi_a + \int dE' b(E,E')\psi_{E'}, \qquad (3)$$

which is a coherent admixture of the original discrete state with the unperturbed continuum states. In the (single-center) atomic physics problems treated by Fano, the discrete states were imagined to be two-electron excited-but-bound state configurations that were degenerate with a continuum of ionization states, coupled by the inter-electron configuration interaction (CI). For these atomic applications, the matrix elements can be treated as pure real quantities up to an irrelevant common phase factor. Passage from the quasi-discrete to continuum states is known as autoionization. On the otherhand, for many of the (multi-center) solid-state and resonance tunneling realizations, the discrete state is often a single electron bound state on an impurity or adsorbate center, geometrically separated from $\{\psi_{E(k)}\}$, the continuum of conduction band states of the host with eigenvalues $\varepsilon_{\vec{k}}$ and with $\vec{k}$ the three-dimensional wave vector, in effect being now regarded as a so-called shape resonance. The off-diagonal CI matrix element is now labeled V_{ak} and is referred to as a transfer or hopping matrix element. In Fano's language (but omitting the actual derivations[2] which are also presented in enlightening alternative ways elsewhere[22-24]), the CI "dilutes" the discrete state into a band of stationary states characterised by the resonance profile

$$|a(E)|^2 = \frac{|V_E|^2}{(E - E_\varphi - \Lambda_a(E))^2 + \pi^2 |V_E|^4} \tag{4}$$

which is equivalent to the projected density of states preferred by the solid-state community,

$$\rho_a(E) = |\langle \Psi_E | \varphi_a \rangle|^2 = \frac{1}{\pi} \frac{\Delta_a(E)}{(E - E_a)^2 + \Delta_a^2(E)}, \tag{5}$$

where $\Lambda_a(E) = \text{Re } \Sigma_a(E)$ and $\Delta_a(E) = \text{Im } \Sigma_a(E)$, in terms of the self-energy

$$\Sigma_a(E) = \sum_{\vec{k}} \frac{|V_{ak}|^2}{E - \varepsilon_{\vec{k}}}$$

and $E_a = E_\varphi + \Lambda_a(E)$. In either representation, when the energy dependence of the functions $\Lambda_a(E)$ and $\Delta_a(E)$ is small within the interval $\sim E_a \pm \Delta_a$, setting them constant allows the discrete state to be regarded as broadened into a Lorentzian resonance profile characterized by a delocalization (=dephasing) or an autoionization (=decay) lifetime $\approx \hbar/\Delta_a$.

Having more-or-less established the attributes of the discrete-state/continuum model, one can move to the actual object of experimental importance, the probability for excitation of the stationary state Ψ_E, given the preparation of some unstable initial state $|i\rangle$. As Fano notes, "*Whatever the excitation mechanism is, this probability may be represented as the squared matrix element of a suitable transition operator T between the initial state $|i\rangle$ and the state Ψ_E.*" Using the fact that the matrix elements are not complex, he then demonstrates that the famed lineshape formula, given by the ratio of transition probabilities coupling the excited state to the perturbed and to the unperturbed continuum states, is indeed

$$Y_{\text{Fano}}(\varepsilon) \equiv \frac{|\langle \Psi_E | T | i \rangle|^2}{|\langle \psi_E | T | i \rangle|^2} = \frac{(q + \varepsilon)^2}{1 + \varepsilon^2} \tag{6}$$

where

$$q \equiv \frac{\langle \Phi_a | T | i \rangle}{\pi V_E \langle \psi_E | T | i \rangle} \tag{7}$$

is the ratio of transition amplitudes from the initial excited state to

$$\Phi_a \equiv \varphi_a + P \int dE' \frac{V_{E'} \psi_{E'}}{E - E'} \tag{8}$$

the original discrete "*state φ modified by an admixture of states of the continuum*", (with P denoting a principle part integral) divided by the amplitude for direct decay into the unperturbed continuum. The extra factor of V_E appearing in the definition of q arises from the procedure used by Fano, "energy normalization" of the continuum states, a procedure rarely appearing in the solid-state literature, since its role is taken over by the density of states function. If the various matrix elements and thus q are weak functions of energy, then the relative transition probabilities are given by the exquisitely simple eqn. (6) which produces a family of generally asymmetric curves characterized by the single parameter q which is a measure of the strength of the coupling of the initial state to the perturbed discrete state relative to its coupling to the unperturbed continuum. When $q = 0$, the lineshape is a symmetric dip with a minimum right on resonance, a point that will be returned to later. For large q, the lineshape is basically a Lorentzian (but not in the distant tails!) reflecting the fact that the dominant coupling is only between the originally discrete state resonance and a symmetric, structureless, and unbounded continuum. For intermediate q where coupling to both the discrete state and to the continuum are comparable, an asymmetric profile is produced with a dip on either the high or the low energy side of the maximum, depending upon the sign (hence phase of the interference channels) of q. If one were able to independently and continuously control the value of any of the components of the q-parameter, then systematic study of the lineshape would provide an additional source of detailed information on the excitation and decay of the coupled discrete state-continuum system.

3 Resonance tunneling and the STM

Resonance tunneling involving metal surfaces (the continuum), adsorbed atoms (the discrete state), and the vacuum or appropriately biased other electrode (the excited state) presented one of the early solid-state realizations of the Fano scenario in which the combination of direct tunneling, resonance tunneling through the adsorbate, and interference between the two channels yielded the first spectroscopic lineshapes (near-Lorentzian, large-q-limit) ever observed for atoms (single or in monolayers) adsorbed on metal surfaces.[25,26] Since this initial study carried out at NBS in the late 60's is also being revisited in the NBS/NIST centennial volume,[6] this current juxtaposition with the Fano work has served as an additional stimulus for this present contribution.

The geometry of the STM resonance tunneling experiment is shown in Fig. 1. The tip, in the standard Tersoff and Hamann model,[27] is approximated as a single atom s-wave electron source attached to a macroscopic wire or electrode with an identifiable chemical potential/Fermi level. The tip is a controllable distance Z_t from the surface under investigation and is displaced an amount $\vec{R}_\parallel$ in the plane of the surface from a possible adsorbate centered at $z = Z_0$, $\vec{R}_\parallel = 0$. The unperturbed continuum here is the manifold of electron conduction band states of the substrate, possibly including surface states if and when appropriate.[28–32] For instance, the near-Fermi-level surface states on the (111) face of the noble metals show many transport properties similar to those of a two-dimensional electron gas and it is believed that this may be fundamental to many of the recent STM spectroscopy studies involving electron communication/propagation along such surfaces.[7–10,33–36] In analogy with Fano's system of eqns. (2) and (3), the adsorbate (also known as the discrete state embedded within the continuum) is frequently thought of in terms of the Lorentzian projected density of states, eqn. (5), in which the level full width at half maximum = $2\varDelta_a \approx (2\pi/\hbar)\rho_s\langle|V_{\vec{a}\vec{k}}|^2\rangle$ with ρ_s the (assumed constant) substrate density of states and the brackets implying an averaging over k-directions. Typically values of $\varDelta_a \approx 0.1$–1.0 eV are observed in photoemission studies of simple atoms or molecules.[37] In addition, if the adsorbate–substrate combination forms a Kondo system,[15,16] then a very narrow (~ 10 meV) Fermi level structure known as the Kondo resonance is superimposed on the basic adatom density of states[13] and it is this additional feature observed in the STM spectra from both Co/Au(111)[7] and Ce/Ag(111)[8] Kondo systems that has triggered the recent theoretical activity.[11–14]

Let now the tunneling process from tip to surface, including the possible role of the adsorbate, be considered as some form of the Fano de-excitation outlined in Section 2. From this point of

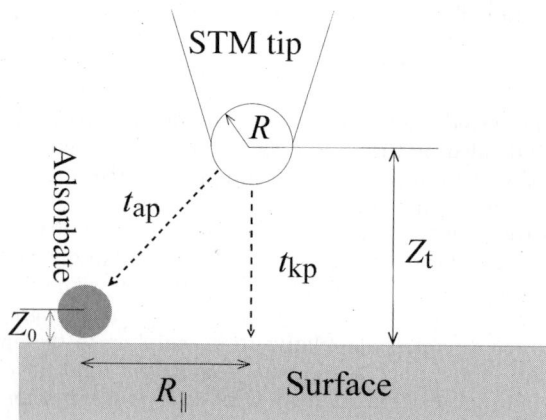

Fig. 1 Schematic picture of the STM probing a surface with an adsorbed atom upon it.

view, the initial excited state is the tunneling electron (or hole) state localised within the STM tip at an energy eV above the substrate Fermi level, where V is the applied voltage. The "de-excitation" matrix elements are proportional to tunneling amplitudes whose behaviour is dominated by a decaying exponential dependence on the product of the tip-to-"acceptor" separation multiplied by the square root of some average barrier height. Operationally this amounts to the substitutions:

$$\langle \psi_E | T | i \rangle \Rightarrow \text{tip-to-unperturbed surface} \equiv t_p(Z_t) \sim \exp(-\kappa Z_t), \quad (9a)$$

$$\langle \varphi_a | T | i \rangle \Rightarrow \text{tip-to-adsorbate} \equiv t_a \sim \exp(-\beta(R_\parallel^2 + (Z_t - Z_0)^2)^{1/2}), \quad (9b)$$

$$\langle \psi_E | T | \varphi_a \rangle \Rightarrow \text{adsorbate-to-surface} \equiv V_a \sim \exp(-\kappa_a Z_0), \quad (9c)$$

$$\langle \Psi_E | T | i \rangle \Rightarrow \text{tip-to-perturbed surface} \equiv \bar{t}_p(\vec{R}_\parallel, Z_t), \quad (9d)$$

where the decay constants $\approx (2m\varepsilon_{\text{barrier}}/\hbar^2)^{1/2}$ are all of the order of ~ 0.5–1.0 Å^{-1}. Both this aspect along with some other implications of three-dimensionality will be further elaborated upon.

The STM equivalent of the coupling of the excited state to the continuum perturbed by the discrete state, the matrix element in the numerator of eqn. (6), can be written down simply using a physically motivated picture, somewhat akin to ray tracing or path integral intuition. It should be apparent from Fig. 1 that there are several classes of interfering paths for the "excited" tip electron (or hole) to travel from tip to a target region within the surface, which for definiteness will be taken to be under the projected shadow of the tip. Based on past experience with enhanced resonance tunneling[25,26] the obvious paths include direct tunneling and through-adsorbate resonance tunneling. In addition, since the adsorbate perturbs the continuum conduction band states, there must be some contribution to the complete tunneling amplitude corresponding to direct tip-to-perturbed part of the continuum. The lowest order tunneling amplitude which incorporates all three of these paths can be written as

$$\bar{t}_p(\vec{R}_\parallel, Z_t) \approx t_p(Z_t) + G(\vec{R}_\parallel, 0; \omega) V_a G_a(\omega) t_a + G(\vec{R}_\parallel, 0; \omega) V_a G_a(\omega) V_a G(0, \vec{R}_\parallel; \omega) t_p \quad (10)$$

where $G_a(\omega)$ is the adsorbate/discrete state Green's function and $G(\vec{R}; \omega) \sim \Lambda(\vec{R}, \omega) + i\gamma(\vec{R}, \omega)$ is the unperturbed substrate electron Green's function that describes the electron propagation between points 0 and $\vec{R}$, for instance from adsorbate to tip shadow ($G(\vec{R}_\parallel, 0; \omega)$) and *vice versa* ($G(0, \vec{R}_\parallel; \omega)$) within the plane of the surface, and the pure real functions Λ and γ are proportional to the real and imaginary parts of $G(\vec{R}; \omega)$. In order to make contact with Fano's theory, the augmented matrix element for tip-to-adsorbate tunneling is defined as

$$\bar{t}_a = t_a + \pi \rho_s V_a \Lambda t_p \quad (11)$$

within which the second term on the right represents a coherent process of tip-to-surface tunneling, through-surface propagation *via* Λ, and surface-to-adsorbate hopping. The total amplitude described by eqn. (11) is completely analogous to the matrix element $\langle \Phi_a | T | i \rangle$ in eqn. (7) involving decay of the initial excited state to the perturbed discrete state, that one mixed with the continuum as given by eqn. (8). Note that only the real part of G has been incorporated into the definition of $\bar{t}_a$ in order to parallel Fano. In present notation the tip-position-dependent Fano q-parameter is defined as

$$q(\vec{R}_t, 0; \omega) \equiv \frac{\bar{t}_a(\vec{R}_t, 0; \omega)}{\pi \rho_s t_p(Z_t; \omega) V_a}. \quad (12)$$

In eqn. (12) it is explicitly noted that while $\bar{t}_a$ depends upon both the lateral ($\vec{R}_\parallel$) and the surface-normal ($Z_t \hat{i}_z$) position of the tip, within the Tersoff–Hamann s-wave model, t_p, the tunneling probability into the unperturbed continuum/surface, is only a function of Z_t, the tip–surface separation.[27]

It is a rather standard exercise to show that the equilibrium tunneling current between the tip and the clean, adsorbate-free (jellium) surface can be expressed as[14,38,39]

$$I_{eq}(Z_t, V) \approx \frac{2e}{h} \int_{-\infty}^{\infty} d\omega [f_t(\omega') - f_s(\omega)] \rho_t(\omega') \rho_s(\omega) |t_p(Z_t)|^2 \quad (13)$$

where f_t and f_s (ρ_t and ρ_s) are the tip and substrate Fermi function (density of states) and $\omega' = \omega - eV$. Assuming constant tip density of states, the differential conductance is given as

$$g_0(Z_t, V) = dI_{eq}/dV = \frac{2e}{h} \int_{-\infty}^{\infty} d\omega \left(\frac{\partial f_t(\omega')}{\partial \omega}\right) \rho_t \rho_s(\omega) |t_p(Z_t)|^2$$

which at zero temperature (where $\partial f_t/\partial \omega \Rightarrow \delta(\omega - eV)$) is simply

$$g_0(Z_t, V) = \frac{2e}{h} \rho_t \rho_s(eV) |t_p(Z_t, V)|^2. \tag{14}$$

The generalization to tunneling into a coupled adsorbate–surface system (akin to the Fano deexcitation to the system of discrete state embedded in a continuum) is straightforward, not surprisingly leading to

$$g_{ads}(\vec{R}_\parallel, Z_t, V) = \frac{2e}{h} \rho_t \rho_s(eV) |\bar{t}_p(\vec{R}_\parallel, Z_t, V)|^2 \tag{15}$$

where $\bar{t}_p$ is formally given by eqn. (10). With the use of eqns. (10)–(12), elsewhere[14] we have carried out the detailed algebraic rearrangements leading, in analogy with eqn. (6), to the expression for the ratio of the tunneling probabilities (hence differential conductivity) coupling "*the excited state [also known as the STM state] to the perturbed and unperturbed continuum states*" given by

$$Y_{STM}(\vec{R}_\parallel, Z_t, \omega) = \frac{|\bar{t}_p(\vec{R}_\parallel, Z_t)|^2}{|t_p(Z_t)|^2} = v(\vec{R}, \omega) + \frac{\Gamma_{as}}{2} \{(\gamma^2 - q^2) \text{Im } G_a + 2q\gamma \text{ Re } G_a\} \tag{16}$$

where $v(\vec{R}, \omega) \equiv \rho_s(\vec{R}, \omega)/(|t_p(Z_t)|^2 \rho_s(\omega))$ is a dimensionless, normalized local density of substrate states at position $\vec{R}$ above the surface, $\Gamma_{as}/2 = \pi \rho_s V_a^2$, and both γ and q depend upon the transverse tip position through $G(\vec{R}, \vec{R}'; \omega)$. An additional tip-position-dependence for $q(\vec{R}, \omega)$ appears through $t_a(\vec{R}_t)$. As written in eqn. (16), the adatom Green's function is quite general. For a nonmagnetic adsorbate described by eqn. (1), the ($U = 0$) Anderson model,[21] the adsorbate Green's function is

$$G_a(\omega) = [\omega - \varepsilon_\varphi - \Lambda_a(\omega) + i\Gamma_{as}/2]^{-1} \tag{17}$$

where Λ_a and $\Gamma_{as}/2 = \Delta_a$ were defined after eqn. (5). Taking Λ_a and Γ_{as} independent of ω and introducing the dimensionless energy variable $\varepsilon = 2(\omega - \varepsilon_\phi - \Lambda_a)/\Gamma_{as}$, the generic adatom Green's function is $G_a(\varepsilon) = [2(\varepsilon - i)/\Gamma_{as}]/(1 + \varepsilon^2)$, which allows eqn. (16), the differential conductance ratio in lowest order in t_p and t_a, to be expressed simply as

$$Y_{STM}(\vec{R}_t, \varepsilon) = v + \frac{q^2 - \gamma^2 + 2\varepsilon\gamma q}{1 + \varepsilon^2}, \tag{18}$$

the similar-but-different STM equivalent of eqn. (6), the Fano lineshape for atomic systems. To proceed from here, specific surface Green's functions must be introduced so that γ and Λ, hence $\bar{t}_a$ and thus q can be obtained for use in eqn. (18).

4 Electron propagators

The unperturbed substrate electron Green's function is

$$G(\vec{r}, \vec{r}'; \omega) = \sum_{\vec{k}} \frac{\psi_{\vec{k}}(\vec{r}) \psi_{\vec{k}}^*(\vec{r}')}{\omega - \varepsilon_{\vec{k}} + i\eta} \tag{19}$$

which requires specification of details in order to be useful. In the region outside the surface that is relevant to tunneling, the substrate wavefunctions are reasonably taken as

$$\psi_{\vec{k}}(\vec{r}) \sim e^{-\kappa_s z} e^{i\vec{k} \cdot \vec{\rho}_\parallel} \tag{20}$$

for both the tails of the three-dimensional propagating conduction band states as well as for the tails of the bandgap surface states[28–32,37,40] displaying two-dimensional $\vec{k}_\parallel$ propagation within the plane of the surface. The decay constant of the exponential tail is

$$\kappa_s = ((2m/\hbar^2)(\varphi + \varepsilon_{Fermi} - \varepsilon_z))^{1/2}$$

where φ is the workfunction, ε_{Fermi} the substrate Fermi energy, and $\varepsilon_z = \varepsilon_{Tot} - \hbar^2 k_\parallel^2/2m$ the so-called normal energy, expressed here as the difference between the total energy of the tunneling electron and the energy associated with the conserved transverse motion. It is not unreasonable to let $\varepsilon_{Tot} \simeq \varepsilon_{Fermi}$ and then retain only the leading terms in an expansion of the square root, in which case

$$\kappa_s \approx \lambda^{-1} + \lambda k_\parallel^2/2 \tag{21}$$

with $\lambda^{-1} = (2m\varphi/\hbar^2)^{1/2}$. Eqns. (20) and (21) contain the important message that tunneling processes involving states with non-vanishing $k_\parallel$ components are limited in amplitude by the Gaussian distribution of $k_\parallel$ maximized at $k_\parallel = 0$. The obvious consequence is that an interesting surface transport process requiring that the involved electron has some $k_\parallel$ will have a much lower likelihood, the larger is the required $k_\parallel$, if the source of the $k_\parallel$ is due to the original spatial localization of the tunneling electron within the STM tip,[27,41–43] as is the case here. This is just restating some consequences for an initial spherical s-wave tip state, but phrased in Cartesian coordinates. In our comprehensive study,[14] two-dimensionless and rescaled functions related to the real and imaginary parts of the Green's functions were introduced:

$$\Lambda(\vec{R},\omega) = e^{z/\lambda} \frac{\text{Re } G(\vec{R},0;\omega)}{\pi \rho_s(\omega)} \tag{22}$$

and

$$\gamma(\vec{R},\omega) = -e^{z/\lambda} \frac{\text{Im } G(\vec{R},0;\omega)}{\pi \rho_s(\omega)}. \tag{23}$$

Both functions carry information about the nature of the spatial extent of an electron introduced at the origin as felt at a point $\vec{R}$, (or *vice versa* when $G(R,0)$ is replaced by $G(0,R)$). The extra multiplicative factors in eqns. (22) and (23) have been inserted for later notational compactness. It is also convenient to take advantage of the Tersoff–Hamann result that $t_p \sim \psi(\vec{R}_t)$, that is the tip-to-surface tunneling amplitude is proportional to the value of the surface wavefunction at the position of the tip, thereby permitting substitution of the product $G(0,\vec{R}_\parallel;\omega)t_p(Z_t)$ with the function $G(0,\vec{R}_t;\omega)$. Combining eqns. (19)–(23), the propagator functions to be calculated are given by

$$\Lambda(\vec{R}_t,\omega) + i\gamma(\vec{R}_t,\omega) = \frac{1}{\pi\rho_s} \sum_{\vec{k}} \frac{e^{-\lambda Z_t k_\parallel^2/2} e^{-i\vec{k}_\parallel \cdot \vec{R}_\parallel}}{\omega - \varepsilon_{\vec{k}} + i\eta} |\psi_{\vec{k}}(0)|^2. \tag{24}$$

A crucial issue that must be addressed here is the character of the electron states which are conveying the message between $\vec{R}_\parallel$ and 0; are they the tails of the three-dimensional conduction band states or are they a two-dimensional band of surface states? Much of the most pictorially dramatic STM work has been carried out on the (111) surfaces of Cu, Ag or Au[7–10,31,32] which are well known to have a band of surface states falling within the band gap of the projected 2D band structure.[30,40] Operationally, since the noble metals are free-electron-like near the Fermi level (aside from the (111) neck in the Fermi surface), $\varepsilon_{\vec{k}} \simeq \hbar^2 k^2/2m$ for band states. The (111) surface states also show parabolic dispersion in $k_\parallel$, with $\varepsilon_{ss}(k_\parallel = 0) \sim 0.5$ eV below the Fermi level and with $\varepsilon_{ss}(k_\parallel) = \varepsilon_{Fermi}$ requiring that $k_\parallel(\varepsilon_{Fermi}) \simeq 0.15$–$0.20$ k_{Fermi}.[30,40] This also requires that the effective mass of such a surface state electron is given by $m_{ss}/m_{band} \sim 0.4$.

Finally, the propagator functions are explicitly obtained from eqn. (24) by setting $\varepsilon_{\vec{k}} = \hbar^2 k_\parallel^2/2m^*$ where m^* is the appropriate electron effective mass, by realizing that

$$\frac{1}{\omega - \varepsilon_{\vec{k}} + i\eta} \Rightarrow P\left(\frac{1}{\omega - \varepsilon_{\vec{k}}}\right) + i\pi\delta(\omega - \varepsilon_{\vec{k}}),$$

and by replacing the $\vec{k}$-sum by either $\int d^2\vec{k}_\parallel/(2\pi)^2$ or by $\int d^3\vec{k}/(2\pi)^3$ for surface state or band contributions. The resulting integrals are straightforward leading to

$$\gamma_{2D}(\vec{R}_t,\omega) = J_0(k_\omega R_\parallel) e^{-\lambda_t Z_t k_\omega^2/2}, \tag{25}$$

$$\gamma_{3D}(\vec{R}_t,\omega) = \int_0^1 dx J_0(k_\omega R_\parallel (1-x^2)^{1/2}) e^{-\lambda_t Z_t k_\omega^2(1-x^2)/2}, \tag{26}$$

and

$$\Lambda_{nD}(\vec{R}_t,\omega) = \frac{1}{\pi\rho_s} P \int d\varepsilon \rho_s(\varepsilon) \frac{\gamma_{nD}(\vec{R}_t,\varepsilon)}{\omega - \varepsilon} \tag{27}$$

for the surface state (2D) and band (3D) contributions respectively where J_0 is the zeroth order Bessel function and k_ω is the wavevector of the appropriate substrate state at energy ω. Eqn. (27) states that in both cases, Λ is just the Hilbert transform of the relevant γ.

5 Results

The spatial variations of several derived/calculated quantities that are components in the STM/Fano-like lineshape function, eqn. (7), are shown in Fig. 2. The effectiveness of t_a, the direct tip-to-adsorbate tunneling given by eqn. (9b) is expected to die very quickly with distance due to the contracted nature of the adsorbate orbitals responsible for the narrow resonance. For a reasonable value of $\beta \approx 1.3$ Å^{-1}, the short range function t_a (normalized to one at $R_\parallel = 0$, $Z_t = Z_0$) is shown as the bold solid line in Figs. 2(A) and (B), where it is apparent that tunneling processes involving direct tunneling through the adsorbate should not be very important for transverse tip positions more than a few lattice constants away from the adsorbate.

The conduction band surface propagator $\gamma_{3D} \sim \text{Im } G(\vec{R}_\parallel,Z_t;\omega)$ given by eqn. (26) has been evaluated at normal tip-to-surface positions $Z_t = 0$, 2.5 and 5 Å for two choices of $k_\omega = 1.2$ and 0.6 Å^{-1} in Figs. 2(A) and (B) respectively. Note the different $R_\parallel$ scales in the left and right side panels. For comparison of the range of band vs. surface state propagation, consider $J_0(k_\omega R_\parallel) = \gamma_{2D}(R_\parallel,0;\omega)$, according to eqn. (25). This is shown as the light dotted curve in Figs. 2(A) and (B) with the following caveat. In this form the comparison between γ_{3D} and γ_{2D} is being made using a common value of k_ω. For actual data interpretation involving (111) noble metal surface states, the surface state Fermi wave vector is roughly an order of magnitude smaller so the range of the oscillatory structure is an order of magnitude longer. Thus when considering γ_{2D} as inferred from Fig. 2, the x-axis should be rescaled by the multiplicative factor $k_\omega/k_{\text{Fermi, ss}}$.

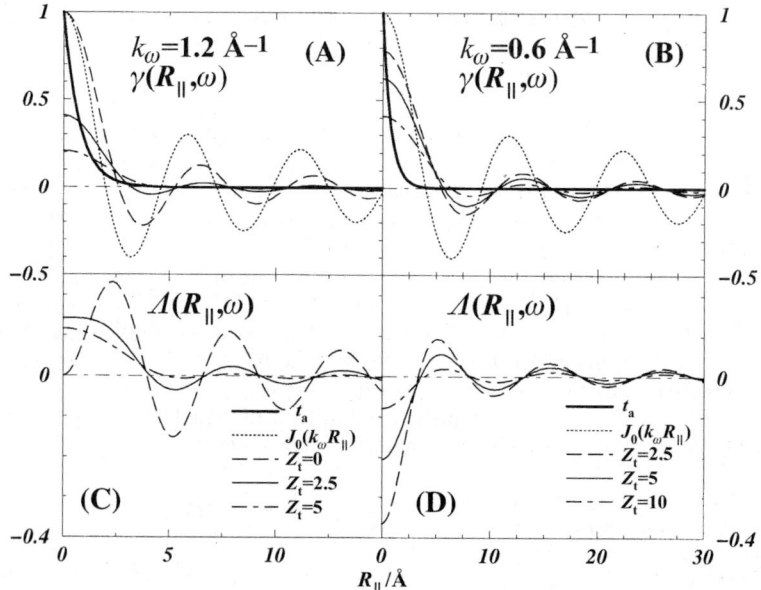

Fig. 2 Various spatially-dependent functions controlling electron transport on the surface. (A) and (B) show the spatial variation of $t_a \sim \exp(-\beta R_\parallel)$ with ($\beta \approx 1.3$) (bold curve) and of γ_{3D} obtained from eqn. (26), evaluated at $k_\omega = 1.2$ Å^{-1} (left panels) and $k_\omega = 0.6$ Å^{-1} (right panels) with different values of Z_t, as labeled. Also shown as the dotted curve is $J_0(k_\omega R_\parallel) = \gamma_{2D}(R_\parallel,0;\omega)$ for the same k_ω. The lower panels (C) and (D) show the complimentary propagators, $\Lambda_{3D}(\vec{R}_\parallel,Z_t;\omega)$.

In any event, both γ_{3D} and γ_{2D} decay much more slowly than does t_a, showing long range oscillatory behavior with relatively slow amplitude decay. The farther removed is the tip from the surface, the less important are the consequences of any of the processes involving large $k_\parallel$ because they get disproportionately killed by the Z_t-dependent exponential factors in eqns. (25) and (26).

The real part of the conduction band function, $\Lambda_{3D}(\vec{R}_\parallel, Z_t; \omega)$ given by eqn. (27) is shown for both values of k_ω in Figs. 2(C) and (D), showing a behavior complimentary to γ_{3D}. However being a Hilbert transform of the imaginary part, the nodes in Λ appear at the positions of local extrema of γ and *vice versa*. For these calculations, the conduction band/continuum principle-part integration in eqn. (27) was taken over a half filled, assumed bounded band ~ 11 eV wide with an elliptic density of states function, in which case $k_\omega \approx 1.2$ Å^{-1} corresponds to a near-Fermi level energy in the middle of the band. Therefore $\Lambda(\vec{R}_t = 0) = 0$ at this energy but is negative for smaller energies ($k_\omega = 0.6$ Å^{-1}) since in this case there are more high energy continuum states repelling the discrete state downward than low energy states pushing it up. Also note that since Λ enters the expression for $\bar{t}_a$ and thus q, the Fano parameter could also be oscillatory in sign and the asymmetry of the resonance profile could be reversed, particularly for $R_\parallel \gtrsim 4$–5 Å^{-1} where t_a becomes negligible.

Before looking at specific examples it is worthwhile to consider some generic characteristics of both the pure Fano lineshape, eqn. (6), and the STM version, eqn. (18). It is immediately obvious that the pure Fano expression is not appropriate for the STM application, when one thinks about the $R_\parallel$-dependence of q. Since for large $R_\parallel$, both t_a and Λ go to zero, $\bar{t}_a$ and thus $q = 0$ in this limit. The resulting lineshape implied by eqn. (6) is $Y_{Fano} = \varepsilon^2/(1 + \varepsilon^2)$, showing an energy-dependent signature due to the discrete state (the symmetric dip around $\varepsilon = 0$) rather than a constant value of unity for all energies. In contrast, Y_{STM} displays the physically correct limiting behavior. For the jellium surface, $v(\vec{R}, \omega) = 1$ outside the surface. Therefore since both q and γ go to zero at large $\vec{R}_t$, $Y_{STM} \to 1$ for all energies, as it must. In the other extreme limit in which $R_\parallel$, Z_t and Z_0 all go to zero, in effect restoring the single center atomic system envisioned in the original Fano treatment, $\gamma \to 1$ as is apparent from eqns. (25) and (26) and also Fig. 2(A). In this limit but only in this limit, $Y_{STM}(\varepsilon) = Y_{Fano}(\varepsilon)$. While the pure Fano profile has been invoked to account for experimentally observed lineshapes[7,8] obtained at arbitrary tip positions, it must be realized that this is at best justified only at a qualitative level since the pure Fano profile does not have the correct asymptotic behavior demanded of the STM configuration. On the otherhand, the lineshape profile $Y_{STM}(\vec{R}_t, \varepsilon)$, expressed by eqn. (18) and first presented in our extended paper,[14] gives the generically correct form in both the large and small separation limits, as well as for intermediate tip-to-adsorbate spacing.

Now consider two specific examples of the STM spectroscopic signature expected to be observed on the basis of the present theory. In Fig. 3 the normalized differential conductance ($\Delta g_{eq} \approx g_{ads} - g_0$, from eqns. (14) and (15)) as a function of energy (or applied voltage) is displayed for various transverse tip positions, as labeled, with $Z_t = 0$ Å and $t_a = 0$ (dark curves) in order to illustrate the dramatic changes in lineshape that are possible, cycling between anti-resonances and right or left sided resonances as $R_\parallel$ is varied, all this due to the oscillations in Λ and γ. The possibility for such anti-resonances had been raised in the work of Kawasaka *et al.*[11] and Schiller and Herschfield[12] although such behavior was not observed in the experimental studies.[7,8] In fact, on the basis of our theory such striking behavior would not be expected due to the detrimental consequences that follow tip placement at the operational distances from the surface required in real experiments. It is apparent from eqns. (25)–(27) that as Z_t increases, only the smallest $k_\parallel$ (longest wavelength) oscillations survive and then only with significantly reduced amplitude. This is also shown in Fig. 3 where the light line profiles are obtained with $Z_t = 5$ Å. The shape or asymmetry does not change much within the range of $R_\parallel$ considered here. The major observable consequence is just the decrease in the relative magnitude of the signal.

Finally, although this paper has concentrated mainly on global issues regarding STM-observable spectral lineshapes from simple adsorption systems that initially might be expected to be pure Fano profiles, the theory, in its full expression,[14] also handles STM resonance tunneling involving Kondo systems. Some spectral results for differential conductivity were obtained in an attempt to understand the systems Co/Au(111)[7] and Ce/Ag(111)[8] first studied experimentally. For the purposes of STM spectroscopy, the defining features of a Kondo system are shown in the inset of Fig. 4. The solid curve is the spectral function or imaginary part of the adsorbate Green's function, similar to that which follows from eqns. (5) and/or (17), here drawn for an adsorbate with

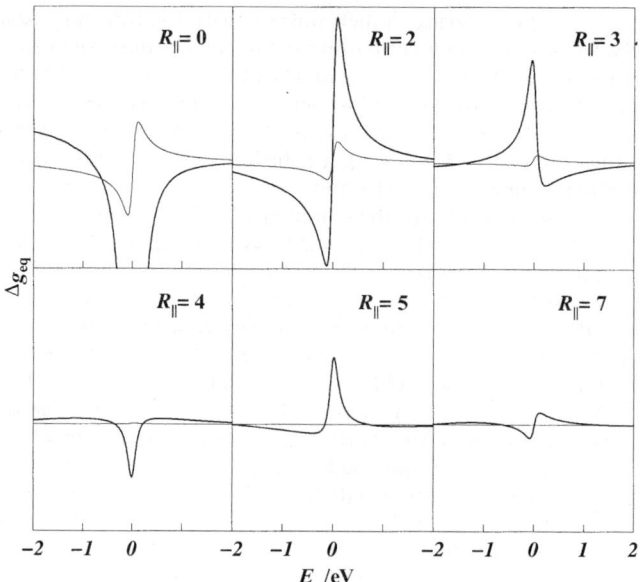

Fig. 3 Normalized differential conductance (Δg_{eq} vs. energy (or voltage), for various choices of lateral tip position $R_\parallel$ with $t_a = 0$, $Z_t = 0$ (heavy curves), and $Z_t = 5$ Å (light curves).

a broad (~1 eV) resonance centered ~0.75 eV below the Fermi level. The prominent narrow feature at the Fermi level is the Kondo resonance due to substrate spin compensation of the magnetic moment on the adsorbate.[13-16] The Kondo resonance width is given by kT_K, with the surface Kondo temperature $T_K \simeq 70$ K. The real part of the Kondo–atom Green's function is shown as the dotted curve within the inset.

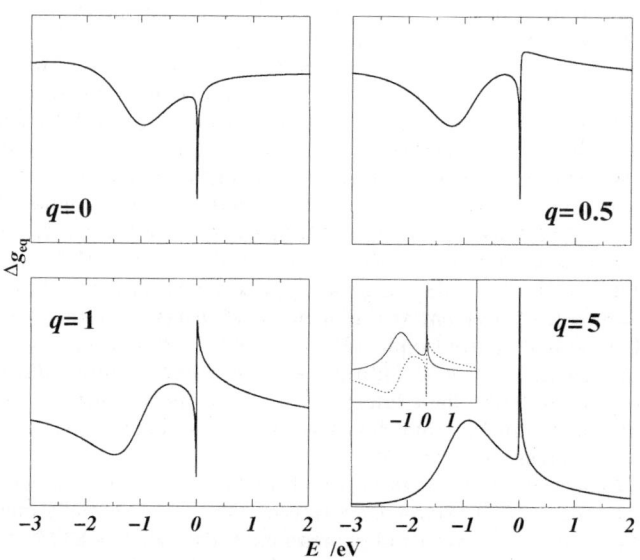

Fig. 4 Normalized differential conductance Δg_{eq} vs. energy or (voltage), for a model Kondo system that includes a broad resonance ~0.75 eV below and a sharp Kondo resonance right at the Fermi level. This is seen directly in the inset of the $q = 5$ panel where the full curve is the adsorbate spectral function and the dotted curve is the real part of the Kondo adsorbate Green's function. Each panel corresponds to a different value for q at the Fermi level.

Fig. 4 shows Δg_{eq}, the STM-observable spectral properties of the system, as calculated from eqn. (16) with an adsorbate Green's function obtained using the non-crossing approximation.[16,44] The differential conductivity is displayed on a large-scale energy range comparable to the conduction band width, for different values of q at the Fermi level in order to emphasize the qualitative range of variations in the spectral profile of the Kondo resonance, as q varies over a reasonable and realistic range that could be attained by repositioning the tip with respect to the Kondo adsorbate.

The resonance lineshapes reported for both Co/Au(111) and Ce/Ag(111) correspond to small values of q. Madhavan et al.[7] fit the observed resonance to a Fano profile with $q \sim 0.7$ when the tip was directly over the adsorbate, decreasing to a value of $q \sim 0.1$ when the tip was displaced ~ 4 Å transversely. Of course the meaning of this type of analysis must be re-examined in the light of the general discussion earlier in this section concerning the differences between $Y_{Fano}(\varepsilon)$, the simple Fano lineshape given by eqn. (6), and $Y_{STM}(\vec{R}_t, \varepsilon)$, the anticipated STM lineshape that follows from eqns. (16) or (18). At least for the spectrum obtained at $\vec{R}_\parallel = 0$, our best fit requires approximately the same q value as implied by the measurements. In the case of Ce/Ag(111), the observed profile is an almost symmetric antiresonance corresponding to $q \sim 0$, likely due to the extreme compactness of the Ce 4f electron orbital. The larger value of q in the Co/Au(111) system is expected because the participating 3d orbital in Co is not as tightly bound. The interested reader is urged to look at the full report for extensive treatment of STM spectroscopy on both the non-magnetic and the Kondo adsorption systems.[14]

6 Conclusions

The topic of *excited states at surfaces* has been "introduced" in this lecture/paper mainly by looking at some of the fundamental issues associated with the classic abstract problem of decay of an initially excited state into a continuum within which a quasi-discrete state is embedded. This is the inherent process addressed by Fano[1,2] within an atomic physics context. He found that interference effects between alternative decay/de-excitation/autoionization paths resulted in asymmetric spectral lineshapes possibly showing not only resonance but also anti-resonance behavior. As a tribute to its fundamental significance, the Fano lineshape has become an icon in almost all areas of spectroscopic inquiry on quantum systems.

The main body of the present paper has demonstrated that the basic ideas behind the Fano profile, when suitably modified, can be used to construct a theory applicable to STM spectroscopy of atoms adsorbed on surfaces.[14] For this to be realized, the state of the tunneling electron within an STM tip that is biased with respect to the surface is regarded as the initially excited state. The atom on the surface plays the role of Fano's discrete state embedded in the continuum. The normalized differential conductivity associated with the resonance, the STM equivalent of a spectral lineshape, is described by an expression (eqn. (16)/(18)) that is related to, but somewhat more complicated than the Fano formula (eqn. (6)). The differences follow since the atomic physics problems considered by Fano involve single-center processes in which the relative phase between states and/or matrix elements is easily dealt with. In contrast, the STM resonant tunneling processes involve multiple atomic centers (the tip, adsorbate, suitably grained substrate, *etc.*) with variable and controllable separations. The coherence or phase acquired in electron transport between centers is an essential ingredient in the STM lineshape and it is here where the considerations with regards surface Green's functions enter. This is the crucial distinction between the atomic physics and STM processes and it is for this reason that $Y_{STM}(\vec{R}_t, \varepsilon)$ rather than $Y_{Fano}(\varepsilon)$ is required to properly account for the lineshapes observed in STM spectroscopy of adsorbates. Two specific examples drawn from our comprehensive paper[14] were presented which illustrate some generic lineshape dependences on controllable parameters.

Concerning the general but non-specific issue of *excited states at surfaces*, the first question that one might ask is "how excited?" Since this Faraday Discussion is concerned mainly with issues on a chemical energy scale, for present purposes this suggests that core level excitations can be put aside. This is only partially right since low energy shakeup excitation often accompanies such excitation.[45] In these final lines, some highly subjective but hopefully useful suggestions will be offered on how to gain entry into the vast literature. For starters, the Introductory Lecture drew heavily on tutorial material that appears in refs. 45–47. Other useful and informative entry points that have survived the time test are the Feature Article by Avouris and Persson[48] on *excited states at metal surfaces* and the comprehensive volume edited by Langreth and Suhl.[49] There exist a series of eight DIET workshop proceedings (Springer books and special issues of Surface Science) that contain a wealth of *excited states at surfaces* information. A sampling of some interesting solid-state/chemical physics realizations of Fano lineshapes can be found in refs. 50–54. A number of informative overviews concerning optical/laser excited states at surfaces are also heartily recommended.[55–59] Excitation and chemical dynamics occuring in molecular beam scattering including the vast field of H_2/Cu is accessible through refs. 60–64. The current state of STM spectroscopy is well stated in a special issue of the *Journal of Electron Spectroscopy* entitled *Scanning Tunneling Spectroscopy*.[65] A Tully overview provides both insights and an excellent survey of the contemporary literature in surface dynamics.[66] Finally, in addition to previous Faraday Discussions (in particular No. 96, "Dynamics at the Gas/Solid Interface" (1993)[67] and No. 110, "Chemical Reaction Theory" (1998)[68]), the articles and references therein appearing in this Faraday Discussion volume should be the ultimate roadmap for those interested in negotiating the current world of *excited states at surfaces*.

References

1 U. Fano, *Nuovo Cimento*, 1935, **12**, 154.
2 U. Fano, *Phys. Rev.*, 1961, **124**, 1866.
3 H. Beutler, *Z. Phys.*, 1935, **93**, 177.
4 V. Weisskopf and E. Wigner, *Z. Phys.*, 1930, **63**, 54.
5 J. M. Blatt and V. F. Weisskopf, *Theoretical Nuclear Physics*, Wiley, New York, 1952, p. 401.
6 *A Century of Excellence in Measurements, Standards, and Technology; Selected Publications of NBS/NIST, 1901–2000*, ed. D. R. Lide, NIST Special Publication 958, 2001.
7 V. Madhavan, W. Chen, T. Jamneala, M. F. Crommie and N. S. Wingreen, *Science*, 1998, **280**, 567.
8 J. Li, W-D Schneider, R. Berndt and B. Delley, *Phys. Rev. Lett.*, 1998, **80**, 2893.
9 H. C. Manoharan, C. P. Lutz and D. M. Eigler, *Nature*, 2000, **403**, 512.
10 M. F. Crommie, *J. Electron Spectrosc. Relat. Phenom.*, 2000, **109**, 1.
11 T. Kawasaka, H. Kasai, W. A. Dino and A. Okiji, *J. Appl. Phys.*, 1999, **86**, 6970.
12 A. Schiller and S. Hershfield, *Phys. Rev. B*, 2000, **61**, 9036.
13 O. Újsághy, J. Kroha, L. Szunyogh and A. Zawadowski, *Phys. Rev. Lett.*, 2000, **85**, 2557.
14 M. Plihal and J. W. Gadzuk, *Phys. Rev. B*, 2001, in the press.
15 J. Kondo, *Solid State Phys.*, 1969, **23**, 183.
16 A. C. Hewson, *The Kondo Problem to Heavy Fermions*, Cambridge University Press, Cambridge, 1993.
17 G. F. Koster and J. C. Slater, *Phys. Rev.*, 1954, **96**, 1208.
18 C. O. Almbladh and P. Minnhagen, *Phys. Rev. B*, 1978, **17**, 929.
19 J. W. Gadzuk, *Phys. Rev. B*, 1981, **24**, 1651.
20 M. Cini and A. d'Andrea, *J. Phys. C*, 1988, **21**, 193.
21 P. W. Anderson, *Phys. Rev.*, 1961, **124**, 41.
22 G. D. Mahan, *Many-Particle Physics*, Plenum, New York, 1990, p. 272.
23 A. Böhm, *Quantum Mechanics*, Springer-Verlag, New York, 1979, p. 399.
24 C. Cohen-Tannoudji, J. Dupont-Roc and G. Grynberg, *Atom–Photon Interactions*, John Wiley, New York, 1992, p. 49.
25 E. W. Plummer, J. W. Gadzuk and R. D. Young, *Solid State Commun.*, 1969, **7**, 487.
26 J. W. Gadzuk and E. W. Plummer, *Rev. Mod. Phys.*, 1973, **45**, 487.
27 J. Tersoff and D. R. Hamann, *Phys. Rev. B*, 1985, **31**, 805.
28 F. Forstmann and V. Heine, *Phys. Rev. Lett.*, 1970, **24**, 1419.
29 J. W. Gadzuk, *J. Vac. Sci. Technol.*, 1972, **9**, 591.
30 S. G. Davison and M. Stęślicka, *Basic Theory of Surface States*, Clarendon Press, Oxford, 1992.
31 N. Memmel, *Surf. Sci. Rep.*, 1998, **32**, 91.
32 R. M. Osgood, Jr. and X. Wang, *Solid State Phys.*, 1998, **51**, 1.
33 E. J. Heller, M. F. Crommie, C. P. Lutz and D. M. Eigler, *Nature*, 1994, **369**, 464.
34 W-D. Schneider and R. Berndt, *J. Electron Spectrosc. Relat. Phenom.*, 2000, **109**, 19.
35 S. Crampin, *J. Electron. Spectrosc. Relat. Phenom.*, 2000, **109**, 51.
36 P. Hyldgaard and M. Persson, *J. Phys.: Condens. Matter*, 2000, **12**, L13.
37 E. W. Plummer and W. Eberhardt, *Adv. Chem. Phys.*, 1982, **49**, 533.
38 *Scanning Tunneling Microscopy*, in *Methods of Experimental Physics*, ed. J. A. Stroscio and W. J. Kaiser, Academic Press, San Diego, 1993, vol. 27.
39 *Scanning Tunneling Microscopy III: Theory of STM and Related Scanning Probe Methods*, ed. R. Wiesendanger and H.-J. Güntherodt, Springer-Verlag, Berlin, 1993.

40 S. D. Kevan and R. H. Gaylord, *Phys. Rev. B*, 1987, **36**, 5809.
41 N. D. Lang, A. Yacoby and Y. Imry, *Phys. Rev. Lett.*, 1989, **63**, 1499.
42 N. Garcia, J. J. Sáenz and H. De Raedt, *J. Phys.: Condens. Matter*, 1989, **1**, 9931.
43 J. W. Gadzuk, *Phys. Rev. B*, 1993, **47**, 12832.
44 P. Coleman, *Phys. Rev. B*, 1984, **29**, 3035.
45 J. W. Gadzuk, *Phys. Scr.*, 1987, **35**, 171.
46 J. W. Gadzuk, *Annu. Rev. Phys. Chem.*, 1988, **39**, 395.
47 J. W. Gadzuk, *J. Electron. Spectrosc. Relat. Phenom.*, 1999, **98–99**, 321.
48 P. Avouris and B. Persson, *J. Phys. Chem.*, 1984, **88**, 837.
49 *Many-Body Phenomena at Surfaces*, ed. D. Langreth and H. Suhl, Academic Press, Orlando, 1984.
50 Y. Yafet, *Phys. Rev. B*, 1980, **21**, 5023.
51 Z. Crljen and D. Langreth, *Phys. Rev. B*, 1987, **35**, 4224.
52 L. C. Davis, *Phys. Scr.*, 1987, **T17**, 13.
53 J. Faist, F. Capasso, C. Sirtori, K. W. West and L. N. Pfeiffer, *Nature*, 1997, **390**, 589.
54 R. J. Gordon, L. Zhu and T. Seideman, *Acc. Chem. Res.*, 1999, **32**, 1999.
55 *Laser Spectroscopy and Photochemistry on Metal Surfaces*, ed. H.-L. Dai and W. Ho, World Scientific, Singapore, 1995.
56 F. M. Zimmermann and W. Ho, *Sur. Sci. Rep.*, 1995, **22**, 127.
57 H. Ueba, *Prog. Surf. Sci.*, 1997, **55**, 115.
58 H. Petak and S. Ogawa, *Prog. Surf. Sci.*, 1998, **56**, 239.
59 H. Guo, P. Saalfrank and T. Seideman, *Prog. Surf. Sci.*, 1999, **62**, 239.
60 G. R. Darling and S. Holloway, *Rep. Prog. Phys.*, 1995, **58**, 1595.
61 *Proceedings of the International Sympoium on Dynamical Quantum Processes on Solid Surfaces*, ed. A. Okiji, Y. Murata, K. Makoshi and H. Kasai, Elsevier, Amsterdam, 1995.
62 A. Groß, *Surf. Sci. Rep.*, 1998, **32**, 291.
63 G-J. Kroes, *Prog. Surf. Sci.*, 1999, **60**, 1.
64 G. D. Billings, *Dynamics of Molecule Surface Interactions*, Wiley, New York, 2000.
65 *Scanning Tunneling Spectroscopy*, Special Issue of *J. Electron Spectrosc. Relat. Phenom.*, 2000, 109.
66 J. Tully, *Annu. Rev. Phys. Chem.*, 2000, **51**, 153.
67 *Faraday Discuss.*, 1993, **96**.
68 *Faraday Discuss.*, 1998, **110**.

Long-lived adsorbate states on metal surfaces

Jean-Pierre Gauyacq,[a] Andreï G. Borisov,[a] Georges Raşeev[b] and Andrey K. Kazansky[c]

[a] *Laboratoire des Collisions Atomiques et Moléculaires, Unité Mixte de Recherche CNRS-Université Paris Sud UMR 8625, Bât 351, Université Paris-Sud, 91405 Orsay Cedex, France*
[b] *Laboratoire de Photophysique Moléculaire, Unité propre du CNRS, Bât 210, Université Paris-Sud, 91405 Orsay Cedex, France*
[c] *Institute of Physics, St Petersburg University, 198904, St Petersburg, Russia*

Received 3rd April 2000
First published as an Advance Article on the web 22nd August 2000

It has been shown recently that the peculiarities of the band structure of a metal can qualitatively influence the electron tunnelling between an adsorbate and a metal surface, the so-called resonant charge transfer (RCT). The presence of a projected band gap along the normal to the surface in the case of Cu(111) has been shown to lead to a blocking of the RCT in the case of Cs/Cu(111), resulting in the existence of a very long-lived excited state. Such long-lived states are potentially very important for surface reaction mechanisms invoking a transient state as an intermediate. Various systems: Cs, model M$^-$ negative ion of pπ symmetry, CO adsorbed on Cu(111), are investigated in order to determine the conditions for the blocking of the RCT and the existence of long-lived states.

1. Introduction

Transient excited electronic states at surfaces play a very important role in a variety of dynamic processes at surfaces and in surface overlayers. They are very often invoked as intermediate steps in surface reactions, when the excitation or transfer of an electron can trigger the system evolution. The lifetime is a very important feature of a transient state in a reaction mechanism: too short a lifetime can introduce a bottleneck in the reaction.[1] The recent development of time-resolved-two-photon-photo-emission (TR-2PPE) allowed the direct measurement of the energy and lifetime of these transient states.[2] Studies of the alkali metal/Cu systems[3-6] revealed the existence of very long-lived excited electronic states in these systems: up to a few tens of fs in Cs/Cu(111). This is much longer than the lifetimes that have been obtained in theoretical studies on electron tunnelling between an alkali metal adsorbate and a jellium metal[7-9] which are of the order of 0.7 fs. It is thus of paramount importance to understand the origin of the very long-lived states on Cu surfaces and, in general, to analyse the properties of the adsorbate/surface system controlling the lifetime of the transient states.

Recently, the peculiarities of the adsorbate states on Cu surfaces have been interpreted as a consequence of the electronic structure of Cu, in particular of the presence of a projected band gap on the (111) and (100) surfaces.[10] Theoretical studies of the alkali metal/Cu(111)[10] and H$^-$/Cu(111)[11] systems have shown a significant stabilisation of the electronic states, compared to the case of a jellium surface. In contrast, experimental studies on the CO/Cu(111) system,[12,13] did not

DOI: 10.1039/b002630l

show any long-lived states associated with the capture of an electron by the CO molecule on the $2\pi^*$ orbital, although the $2\pi^*$ resonance lies within the Cu(111) band gap. The aim of the present work is to further analyse the origin of the long-lived states in different systems in order to understand the conditions for their existence.

Various decay channels are possible for an electronically excited state in an adsorbate/substrate system. It can decay by one-electron tunnelling into bulk states of the same energy. This is usually referred to as resonant charge transfer (RCT). An excited state can also decay by electron–electron interaction, the excited electron making inelastic collisions with bulk electrons; this is, for example, an important decay channel for the image states at surfaces.[14] The excited state can also relax by transferring energy to the heavy particle motion, *i.e.* by exciting phonons. This is important for the decay of low-energy excited electrons inside bulk metals. When it concerns adsorbate movement (desorption, for example), it can also be viewed as a reaction induced by the excited state.

Usually for excited states located a few eV above the Fermi level, the one-electron transitions are much faster than the many-electron and phonon transitions. Again for the alkali metal jellium system, at chemisorption distances, the one-electron transition rate is in the few fs^{-1} range. These transitions correspond to the electron tunnelling through the potential barrier separating the alkali metal and the metal; it occurs predominantly along the surface normal, where the transparency of the barrier is the largest. However, the electronic band structure of the metal can strongly affect the RCT. Cu(111), for example, presents a projected band gap in the direction normal to the surface (L gap). This forbids electron penetration into the bulk along the normal in the -5.83 to -0.69 eV energy range (with respect to vacuum). In addition, the L-gap leads to the existence of electronic states localised in the surface region and propagating quasi-freely parallel to the surface: the surface state (-5.33 eV for the bottom of the band) and the image states.[15] If a transient state localised on an adsorbate has an energy in this band gap, its decay is only possible by emission of an electron with a certain $k_\parallel$, the electron momentum parallel to the surface (see illustration in Fig. 1). This decay populates either 3D states propagating inside the bulk or the surface state 2D continuum propagating parallel to the surface. The main substrate states involved in the electron tunnelling on a jellium being absent on Cu(111), the RCT process can be efficiently blocked on the Cu(111) surface. In that case, the one-electron decay being much suppressed, the adsorbate state becomes almost a bound state in a one-electron picture and its decay by two-electron interaction has to be considered. This makes such an adsorbate-induced state a localised analogue of the image or surface states. The effect of the two-electron decay has recently been studied for the Cs/Cu(111) and Cu(100) systems using a first principle metal response function and

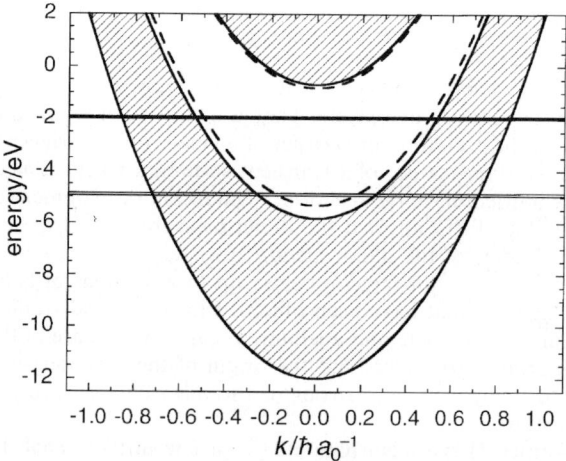

Fig. 1 Electronic band structure for the model Cu(111) surface as a function of $k_\parallel$, the electron momentum parallel to the surface. The hatched area represents the 3D propagating bulk states. The dashed lines represent the first image and the surface states located inside the Cu L-band gap. The horizontal double line is located at the Fermi energy and the horizontal thick line at the energy of the Cs level for the Cs adsorption distance.

an excited state wavefunction obtained by wave-packet propagation.[16] The decay by inelastic electron–electron interaction is found to play a very important role and the total decay rate (one- and two-electron terms) accounts very well for the experimental findings.[3–6] The blocking of the RCT has also a very strong effect on the charge state of reflected particles in the case of collisions on Cu(111) surfaces. The special features of the alkali metal/Cu(111) and H$^-$/Cu(111) systems (time-dependence of the blocking effect, 2D vs. 3D aspects of the RCT) have been experimentally confirmed.[17,18]

The general ideas that can be drawn from the above results are that (i) very long-lived states localised on adsorbates can be observed in the case where the RCT is blocked by a projected band gap and (ii) in these cases, the inelastic e–e interaction can play an important role in the decay. The question then arises whether this is a general feature and whether any excited electron localised in front of a projected band gap will lead to a long-lived state. From experiment,[12,13] the CO/Cu(111) system seems to be a counter example. In an attempt to clarify the situation, we present below a detailed study of the one-electron interaction of a Cs atom and a model M$^-$ negative ion ($p\pi$ symmetry) with a Cu(111) surface, as well as preliminary results on the CO/Cu(111) system. This will provide the basis for a discussion of the conditions required for observing an important blocking of the RCT, *i.e.* a stabilisation of an excited adsorbate state on a metal surface with a projected band gap.

2. Method

The study of the one-electron decay (RCT) of an excited state localised on an adsorbate in front of a Cu(111) surface is performed using the wave-packet propagation (WPP) method[19] or the coupled angular mode (CAM)[20] method. Both methods study the one-electron problem governed by the Hamiltonian H. The WPP method solves the time-dependent Schrödinger equation for the single electron active in the RCT process, whereas the CAM method solves the stationary Schrödinger equation for the same problem. The WPP and CAM approaches are equivalent ways of determining the characteristics of the quasi-stationary states (energy and lifetime) of the system under study. When applied to the same problem, the stationary and non-stationary approaches lead to the same results for the characteristics of the quasi-stationary states. These two methods have been described in detail elsewhere[19,20] and are only briefly presented here.

In the WPP approach, one studies the time evolution of the wavefunction $|\Psi(t)\rangle$ of an electron initially localised around the adsorbate centre. Using a bound state of the free atom $|\Phi_0\rangle$ as the initial state of the propagation

$$|\Psi(t=0)\rangle = |\Phi_0\rangle$$

one obtains by time propagation the wavefunction as a function of time $|\Psi(t)\rangle$, from which one defines the survival amplitude of the electron in the initial state:

$$A(t) = \langle \Phi_0 | \Psi(t) \rangle = \langle \Phi_0 | e^{-iHt} | \Phi_0 \rangle$$

The analysis of the survival amplitude yields the projected density of states of the problem and the energy and the width of the transient states of the system.

In the CAM method, one studies the scattering of an electron in the compound adsorbate–substrate potential in a stationary approach. Because of the cylindrical symmetry of the problem around the z-axis normal to the surface and going through the atom centre, m, the projection of the electron angular momentum on this axis is a good quantum number. The scattering wavefunction of the electron is expanded over spherical harmonics Y_{lm} centred around the adsorbate ($r = (r, \theta, \phi)$ spherical co-ordinates):

$$\Psi(r) = \Psi_m(r) = \frac{1}{r} \sum_l Y_{lm}(\theta, \phi) F_{lm}(r)$$

Bringing this expansion into the stationary Schrödinger equation yields a set of coupled equations for the radial wavefunctions, the solution of which leads to the determination of the scattering S matrix and the time delay matrix Q.[21] The analysis of the energy dependence of Q allow one to evaluate the energy and lifetime of the scattering resonances of the problem *i.e.* of the characteristics of the adsorbate transient states.

The Hamiltonian H governing the evolution of the electron in the WPP and CAM approaches is taken as:

$$H = T + V_S + V_C + \Delta V_S$$

where T is the electron kinetic energy, V_S is the electron–surface interaction potential and V_C the interaction potential between the electron and the adsorbate core. In the case of a positive ion core, the effect of the image of the ionic core (ΔV_S) is also taken into account. Both approaches, WPP and CAM, make the assumption that V_C and V_S can be determined independently. Therefore, both methods are well appropriate for the treatment of collisional charge transfer which is dominated by large projectile–surface distances; they were indeed found to be very successful in the case of free-electron metals (see *e.g.* ref. 9). However, even for the small atom–surface distances typical of chemisorption, in the alkali–free electron metal surface systems, the comparison of CAM and WPP results[9] with fully optimised approaches[7] was very satisfying.

In the present study, the e–Cu(111) surface interaction potential, V_S, is described by a local periodic potential adjusted from first principle calculations.[22] It is only a function of z, the electron–surface distance and takes the periodicity of the bulk (111) planes into account. This potential assumes a free electron motion in the direction parallel to the surface. Since, with this potential, the surface features (energy position of the projected band gap, of the surface and image states) are well represented, it thus contains the main features necessary to describe the RCT blocking, *i.e.* to address the main problem studied here. In order to stress the stabilisation effect of the Cu(111) surface, the results of the study with the model Cu(111) surface are compared with results obtained for a free-electron description of the metal. In that case, the electron–surface interaction, V_S, is described by a local potential taken from the work of Jennings *et al.*[23] on jellium metal surfaces.

The interaction between the electron and the adsorbate core, V_C, is described using pseudo- or model-potential methods designed to represent the case of a free atom or molecule. In the results below, the outer electron in the Cs atom is described by a pseudo-potential. In our earlier study[10] of the Cs/Cu system with the CAM method, we used the l-dependent Bardsley pseudo-potential:[24]

$$V_C(r) = \sum_l |Y_{lm}\rangle U_l(r) \langle Y_{lm}|$$

This is easily implemented in the CAM procedure, but not in the WPP approach which uses a grid of points in cylindrical co-ordinates. We rather used a non-local pseudo-potential of the Kleinman–Bylander form:[25,26]

$$V_C(r) = U_0(r) + \sum_{l=1,2} \frac{|\phi_l^m \Delta U_l\rangle \langle \Delta U_l \phi_l^m|}{\langle \phi_l^m | \Delta U_l | \phi_l^m \rangle}$$

where $\Delta U_l(r) = U_l(r) - U_0(r)$. $U_0(r)$ and $U_l(r)$ are the l-dependent components of the Bardsley potential. ϕ_l^m are the wavefunctions of the 6p and 5d orbitals of the free Cs atom, corresponding to the potentials $U_l(r)$. The $U_l(r)$ potentials have been saturated below $r_{Cut} = 1.0\ a_0$, to avoid the Coulomb singularity in the WPP:

$$U_l(r) = U_l(r_{Cut}), \quad \text{for } r < r_{Cut}$$

We checked that this procedure does not result in any significant shift in the energies of the Cs ground and excited states.

In order to clarify the discussion on the effects generated by the Cu projected band gap, we performed some calculations with a model negative ion M^- of $p\pi$ symmetry and bound by 1.3 eV. In the WPP approach, the model M^- negative ion is described by a pseudo-potential of the form:

$$V_C(r) = -\frac{\alpha}{2r^4} + \beta e^{-0.5r^2}, \quad \text{for } r > r_0$$

$$-V_0 + \gamma r^2 + \beta e^{-0.5r^2}, \quad \text{for } r < r_0$$

where $r_0 = 2\ a_0$ and $\beta = -1.02761$ a.u. V_0 is equal to $1.5\alpha/r_0^4$ and γ to α/r_0^6. The polarisability α is taken equal to that of the CO molecule (13.3 a_0^3). With these parameters, there is a single bound state of pπ symmetry with a binding energy of 1.3 eV and no other bound states in π symmetry. In the CAM method, the model M$^-$ negative ion (pπ symmetry) is described in the effective range approximation[27] (ERT). Namely, one only treats a region outside of the neutral core ($r > r_C$). In this region, the interaction potential is taken as the polarisation potential of the CO molecule and the short-range forces for the pπ symmetry are represented by an energy-independent boundary condition on the radial wavefunction at r_C. The latter is chosen such that the ion binding energy in the pπ symmetry is equal to 1.3 eV.

Finally, for the study of the 2π* resonance of the CO molecule, the low-energy scattering of an electron by a free CO molecule has been described by a two-channel ERT approach,[28] using results from a first principles calculation.

3. Cs/Cu(111) system

The interaction of a Cs atom with a Cu(111) surface has been studied with the WPP method as a function of the atom–surface distance, Z. The results for the energy position and the decay rate of the level correlated at infinity with the Cs ground state (σ symmetry, *i.e.* projection of the electron angular momentum on the axis normal to the surface equal to zero) are presented in Fig. 2(a) and (b) compared with the corresponding results for a free-electron surface. The level characteristics are presented as a function of Z, in order to illustrate the effect of the variation of the strength of the charge transfer interaction from the small distances typical of chemisorption to the large distances probed in collisional charge transfer processes. The present results obtained in the WPP method agree quite well with our earlier CAM results obtained for the small Cs-surface distances.[10] The small differences are attributed to differences in the V_C potential.

In Fig. 2(a), one can see that the Cs level lies inside the Cu(111) projected band gap (-5.83, -0.69 eV, see Fig. 1) in the entire range of Z distances. Nevertheless, the energy positions of the

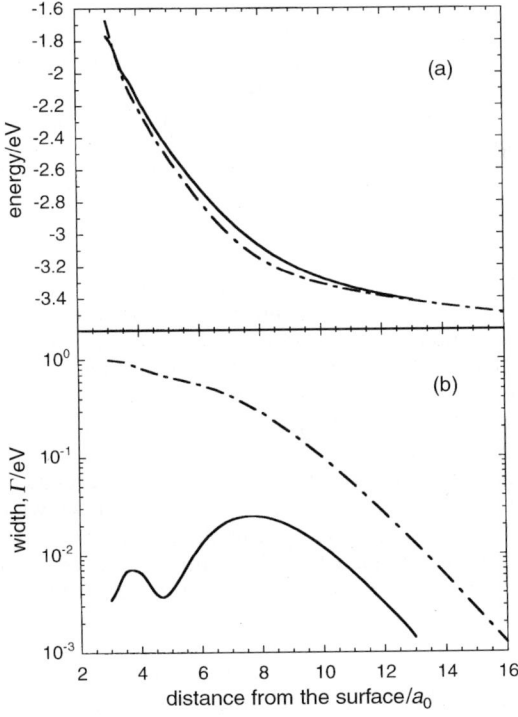

Fig. 2 Energy position (a) and width (b) of the Cs level as a function of the Cs–surface distance, measured from the surface image plane. The level energy is measured with respect to vacuum. (·—·) Cs in front of a free-electron surface; (——) Cs in front of the model Cu(111) surface.

level in the model Cu(111) and in the free-electron surface cases are almost the same, showing that the band gap has very little influence on the level energy in this case. In fact, the upward level shift as a function of Z is roughly given by the image potential at the centre of the Cs atom. The Cs resonance state is above the Fermi level; the neutral atom is then unstable in front of the surface and the computed level width corresponds to the electron transfer rate from the atom to the metal.

The electron transfer rate is dramatically reduced in front of the Cu(111) surface compared to the free-electron surface case. The free-electron result as a function of Z exhibits typical behaviour, *i.e.* an exponential increase when Z decreases and saturation at small Z. For the Cu(111) case, two regimes are clearly visible at large and small Z. At large Z, the atomic level perturbed by the surface resembles very much the unperturbed 6s level of Cs; it is located inside the band gap and is thus much influenced by it. From Fig. 1, one can infer that the Cs state can decay either to the 2D surface state continuum or to the 3D propagating bulk states with a finite $k_\parallel$. The difference in normalisation of the 3D and 2D continua favours decay toward the surface state which is found in the WPP analysis to dominate the state decay at large Z. This result has to be linked with other studies showing the importance of the role played by the surface state and image states 2D continua in various surface processes involving electronic interactions.[29–32]

The blocking effect then results from the balance between two effects: the fact that $k_\parallel$ is not equal to zero decreases the RCT efficiency while the favoured overlap between adsorbate and 2D surface state wavefunctions increases it. The net effect depends on the position of the level in the band gap: the higher the level, the larger the $k_\parallel$ of the transferred electron has to be and the more stable is the level. The Cs 6s level that is high in the gap is stabilised by around one order of magnitude whereas the Li 2s level that is lower in the gap has a larger width at large Z on Cu(111) than on a jellium.[10]

As Z decreases, around 7–9 a_0 from the surface, the Cs level width rapidly decreases. This abrupt change is attributed to a modification of the resonance wavefunction. Indeed, Cs is a highly polarisable atom. The large field close to the metal surface mixes the Cs orbitals, leading to the formation of hybrids. The hybridisation has been discussed in detail for a few systems in front of free-electron metals.[8,33–35] The lowest of these hybrids, studied here, has an electron density significantly shifted away from the ion core towards vacuum; it is thus less coupled to the metal states and more stable against RCT. This feature is illustrated in Fig. 3 which presents the electron wave-packet after a long propagation time, for the Cs-surface distance corresponding to chemisorption ($Z = 3.5\ a_0$ from the image plane[36]). One recognises (i) the region around the atom, which corresponds to the surviving transient state, (ii) the oscillating wave-packet inside bulk Cu which decreases exponentially inside the bulk and which is the direct consequence of the projected band gap and (iii) the electron escaping into Cu at a finite angle from the surface. One can find a discussion of the wave-packet shape in the Li/Cu(111) case in ref. 10. In the present case, the decay of the transient state mainly occurs toward the 3D propagating states of Cu. One can notice that the electronic cloud around Cs is strongly distorted and repelled towards vacuum due to the polarisation (a discussion of similar hybrid formation in high-n ion–surface interaction can be found in ref. 34–37). The polarisation of the electronic cloud has two effects: (i) the cloud is moved away from the surface and this decreases the overlap between the atom and metal states and thus the RCT coupling; (ii) the change in the transient state wavefunction modifies the $k_\parallel$ dependence of the transition rate. The polarisation effect is also present in the case of a free-electron surface and the change of the slope of the corresponding $\Gamma(Z)$ in Fig. 2(b) around 7–9 a_0 is attributed to it. In the case of H ($n = 2$) states interacting with an Al surface, it was shown to lead to a contraction of the dependence of the $k_\parallel$ transition rate around the normal and even to the appearance of a node in the $k_\parallel$ dependence[33] (similar effects have been found for alkali metal atoms). If we assume that the change in the $k_\parallel$ dependence is similar on free-electron metals and on Cu, then we can understand the strong stabilisation of the Cs state. The contraction decreases the level width by enhancing the band gap effect. Furthermore, the node in the $k_\parallel$ dependence prohibits transitions into a certain group of states. For Cs around the chemisorption distance, it results in the almost total blocking of charge transfer into the surface state. Summarising, one then has a quasi-stable Cs state, which mainly decays into the 3D bulk states.

The strong stabilisation at small Z in the model Cu(111) case (around two orders of magnitude) is then attributed to the combined effect of the band gap and the atom polarisation. The Cs atom

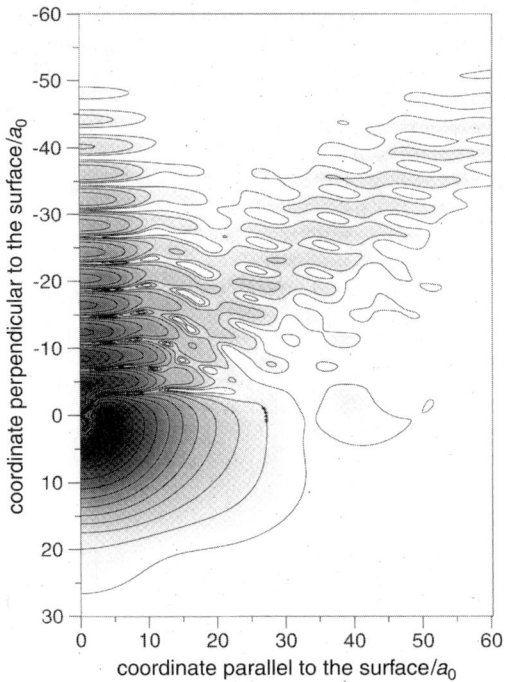

Fig. 3 Wave-packet for the transient state in the Cs/Cu(111) system. It presents the logarithm of the electron wavefunction modulus in cylindrical co-ordinates: normal to the surface (z, positive values in the vacuum) and parallel to the surface (ρ). The dark areas correspond to a large probability of presence of the electron. The contour lines span the -11.5 to -6.0 interval with 0.5 steps. The Cs$^+$ core is located on the $\rho = 0$ axis at $Z = -3.5\ a_0$.

leads to a stronger stabilisation effect on Cu(111) than other alkali metal atoms. For an adsorbate distance surface of 3.0 a_0, the one-electron decay rate on a Cu(111) surface is equal to 82, 99, 12 and 3.5 meV for Li, Na, K and Cs. Compared to the one-electron decay rate on a free-electron metal that is in the 1 eV range, one thus has reductions of one or two orders of magnitude. The differences between the alkali metal atoms can be attributed to differences in the energy position in the gap, the atomic polarisation and the position of the node in the $k_\parallel$ dependence of the transition rate.

At chemisorption distances ($Z \approx 3.5\ a_0$), the Cs level is above the Fermi level and is then an excited state of the Cs/Cu(111) system. Its decay rate by one-electron interaction is around 7 meV and would correspond to a lifetime in the 90 fs range. This being rather long, the effect of multi-electron interaction, namely the inelastic electron–electron interaction inside Cu, has to be taken into account. This leads to a total decay rate[16] around 28 fs, compatible with experiment.[3,4]

4. Model pπ negative ion interacting with a Cu(111) surface

To analyse a situation different from that of a polarisable atom, we investigate the case of a model negative ion M$^-$. The ion corresponds to a single outer electron of π symmetry (angular momentum projection on the normal to the surface equal to ± 1), the same as the 2π^* orbital of CO. The binding energy of the free negative ion (p symmetry) is chosen equal to 1.3 eV, so that the M$^-$ ion energy in front of Cu is similar to that in the Cs case. The results obtained for the M$^-$ ion as functions of the ion–surface distance for the model Cu(111) surface and a free-electron surface are presented in Fig. 4. An ERT description of the ion and the CAM method have been used in these calculations.

The energy of the M$^-$ ion is found to be almost the same for the two surfaces (Fig. 4(a)). The downward energy shift of the level is roughly equal to the image potential at the ion centre. This is

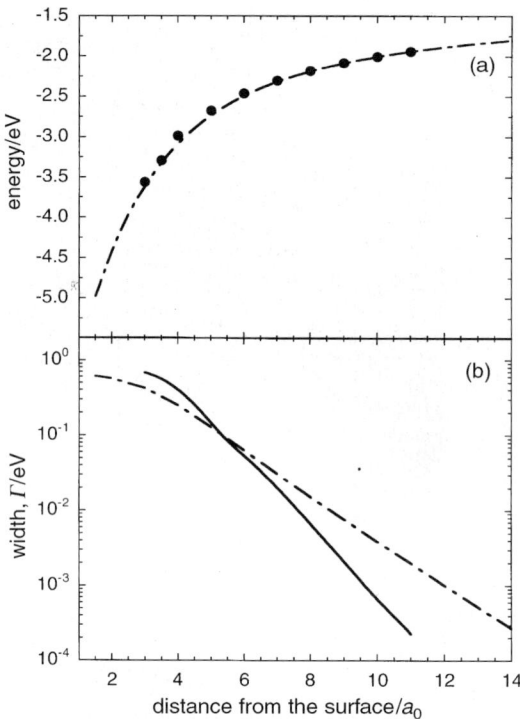

Fig. 4 Energy position (a) and width (b) of the model M⁻ ion level as a function of the ion–surface distance measured from the surface image plane. The level energy is measured with respect to vacuum. Model M⁻ ion in front of a free electron metal surface: (·—·) in a and b. Model M⁻ ion in front of the model Cu(111) surface: (●) in (a) and (———) in (b).

rather different from the case of the H⁻ ion investigated earlier.[11] This difference is directly related to the symmetry of the outer electron of the ion: σ for H⁻ and π for the present M⁻ model ion. In the first case for the Cu(111) surface, because of the 2D nature of the surface state and the σ symmetry of the ion, the ion level was found to lead to the existence of two resonances, the energy position of which is exhibiting an avoided crossing (see discussion in ref. 11). In the alkali metal case discussed above, a second state is also present, however, in an energy range that we did not discuss. The model negative ion M⁻ case is simpler: due to the π symmetry of the electron, the extra state does not exist and in the two cases (Cu(111) and free-electron surface) similar Z-dependences of the ion level energy are obtained.

The decay rate of the M⁻ ion is quite different in the two cases (Fig. 4(b)). The ion level is located inside the band gap and above the Fermi level. In both cases, the ion is unstable in front of the surface and the decay rate corresponds to the RCT rate for electron transfer from the negative ion to the surface. At large Z distances, the situation is comparable to that of Cs and the decay rate is much smaller on Cu(111) than on the free-electron metal, due to the presence of the projected band gap. It should be noted that, due to the π symmetry of the state, decay of the M⁻ ion exactly along the surface normal is suppressed and then the ion has to decay to metal states with a finite $k_{\parallel}$, even in the case of a free-electron surface. Obviously, for a free-electron metal, the decay rate is smaller for a π electron than for a σ electron of comparable energy.[9] On Cu(111), this should decrease the blocking effect of the band gap for π electrons compared to σ electrons.

The difference between the slopes of the $\Gamma(Z)$ functions for the two surfaces deserves discussion. Indeed, the resonance width is due to the interaction between the ion level and the metal states and, therefore, it reflects the overlap between the interaction potential and atomic and metal functions (perturbative view, see *e.g.* ref. 38). In the free-electron case, the metal states are those which correspond to propagation almost along the surface normal. In contrast, in the Cu(111) case, the metal states that are involved are either the surface state or 3D propagating states with a

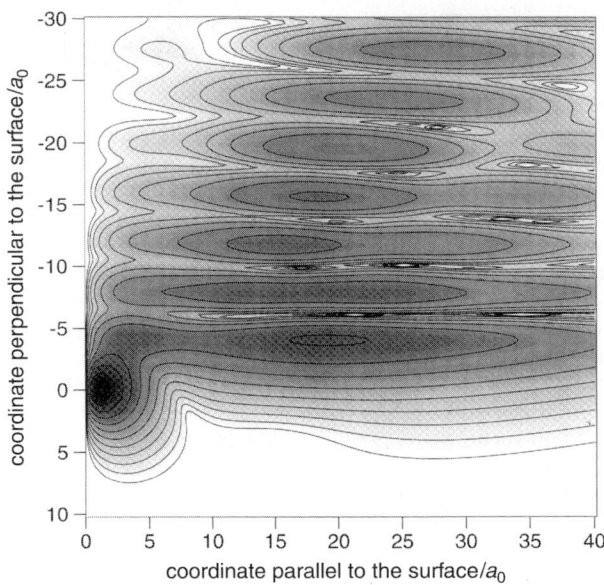

Fig. 5 Wave-packet for the transient state in the model $M^-/Cu(111)$ system. It presents the logarithm of the electron wavefunction modulus in cylindrical co-ordinates: z, normal to the surface (positive values in the vacuum) and ρ, parallel to the surface. The dark areas correspond to a large probability of presence of the electron. The contour lines span the -11 to -5.0 interval with 0.5 steps. The M core is located at the origin of the co-ordinates.

large momentum parallel to the surface. For the same total energy, these states are less extended toward vacuum than the corresponding states for the free-electron case. Bearing in mind the fact that the binding energy of the ion decreases as Z increases, one can explain the steeper behaviour of the $\Gamma(Z)$ function in the Cu(111) case. Because of the opposite variation of the electron binding energy with Z, this difference of slope is weaker in the Cs case. As Z decreases, the M^- decay rate on Cu(111) increases faster and becomes larger than that on the free-electron metal. Fig. 5 presents the wave-packet of the M^- outer electron after a long propagation time for a small ion–surface distance (3.5 a_0 from the image plane). It is similar to that of the Cs case with the electronic cloud around the atom, the oscillating part that penetrates exponentially into bulk Cu and the part that describes the escaping electron. Because of the π symmetry of the problem, the wave-packet has a node along the symmetry axis.

The main decay route of the M^- ion is the decay into the surface state, although the contribution from 3D propagating states can also be seen in Fig. 5. Importantly, the electronic cloud around the atom does not exhibit as strong a distortion as seen in the Cs case. The $p\pi$ shape of the orbital is almost completely conserved, even at this small ion–surface distance, reflecting the very small polarisability of this negative ion. A rather weak distortion is nevertheless visible that corresponds to a displacement of the electronic cloud toward the metal. In the case of a negative ion, the field acting on the outer electron attracts it toward the metal (the image charge of the electron is the leading term) and the negative ion electronic cloud is distorted toward the metal. In addition, no node appears in the $k_\parallel$ dependence of the decay rate (see *e.g.* the case of H^- on Al[39]). Hence, the effect present in the Cs case and responsible for the strong stabilisation at small Z does not exist in the model M^- ion case and even, being opposite, it should lead to an increase of the decay rate by improving the overlap between the ion and metal states.

5. CO/Cu(111) system

In this section, we tentatively discuss the possibility for a stabilisation of the $2\pi^*$ orbital of CO adsorbed on Cu surfaces. For this, we first designed a model description of the low-energy electron scattering by free CO molecules. The low-energy phase shifts have been extracted from a static exchange + cut-off polarisation calculation and used to derive the 2-channel ($p\pi$ and $d\pi$) ERT representation of the CO scattering in the π symmetry (details will be published elsewhere[28]). This ERT description is then included into the CAM procedure. The $2\pi^*$ appears as a resonance in this calculation. In our model treatment of the free molecule at its equilibrium distance, it is located around 1.7 eV with a width of 0.82 eV (see *e.g.* an earlier study in ref. 40). The CO/Cu system is a well-known example of weak chemisorption (see *e.g.* a recent discussion and bibliography in ref. 41). The assumption of an unperturbed CO core is an important approximation of the present study. Nevertheless, we consider that the discussion of the possibility of a stabilisation of the $2\pi^*$ orbital on Cu(111) surfaces can be performed with this model description.

Results obtained with a free-electron surface have shown that the $2\pi^*$ resonance decay rate continuously increases when the CO molecule approaches the metal surface (the CO axis is kept perpendicular to the surface). For a CO–Cu distance typical of chemisorption[42] (the distance between the CO centre of mass and the Cu image plane is around 3 a_0), the decay rate is around 2.3 eV, *i.e.* the life-time for electron transfer from the $2\pi^*$ orbital to the metal is around 0.3 fs. It should be kept in mind that such a short lifetime is difficult to define and hence its value cannot be precise. Preliminary results for the model Cu(111) surface show that the $2\pi^*$ resonance is weakly stabilised by the projected band gap at large CO–surface distance, but around 3.0 a_0 from the surface, the $2\pi^*$ resonance on Cu(111) is slightly less stable than on a free-electron surface (decay rate around 2.5 eV). The present results for the Cu(111) surface overestimate the decay rate of the $2\pi^*$ resonance in CO/Cu(111) (Bartels *et al.*[13] have reported a width of 0.87 eV and a minimum decay rate of 0.13 eV measured by TR-2PPE). This discrepancy in the absolute value is attributed to the approximation of an unperturbed CO core in our calculation.

The comparison between the free-electron and Cu(111) surface effects is then very similar for the CO($2\pi^*$) resonance and the model M^- ion, since both are negative ions of π symmetry. There is however a difference between the two systems: CO($2\pi^*$) is a resonance and M^- is a bound ion at infinite distance from the surface. As a consequence, the qualitative picture of the one-electron decay of the two systems might be different. In the M^- ion, the only decay is *via* the tunnelling of the electron through the potential barrier separating the ion and the surface; this tunnelling is favoured close to the surface normal and is thus very sensitive to the projected band gap. In the CO($2\pi^*$) case, the resonance can decay even without the surface being present, *i.e.* the electron can be emitted in all directions of space according to the angular shape of the resonance; although the latter is favouring the C end of the molecule, *i.e.* the direction of the surface, it should lead to a much weaker sensitivity to the Cu(111) projected band gap

6. Conclusions

We have reported on a theoretical study of the electron transfer between various objects and a Cu(111) surface, with the aim of defining the conditions under which the one-electron transfer between an adsorbate and bulk Cu is blocked. The blocking occurs due to the presence of a projected band gap in Cu(111) that prevents the tunnelling of an electron along the normal to the surface. It is found that, depending on the adsorbate characteristics, the one-electron transition can be modified in different ways by this projected band gap. The cases of transient neutral states and transient negative ions appear to be rather different. Indeed, these two systems are polarised in opposite ways in front of a metal that leads to a qualitatively different effect of the projected band gap. While stabilisation was observed in both systems at large atom–surface distances, at small distances, typical of chemisorption, only highly polarisable neutral transient states are stabilised. One can then expect neutral polarisable adsorbates (atomic or molecular) to lead to stabilised levels. The Cs case appears to be the best example until now with a reduction of the one-electron decay rate by two orders of magnitude compared to the free-electron case.

These effects have been shown on the Cu(111) surface, but should exist on other surfaces with a projected band gap. Some results on the Cu(100) surface[4,10] confirm this statement.

The blocking of the one-electron transitions in excited states at surfaces can lead to various consequences. The band gap effect at large atom–surface distances deeply influences the charge transfer process in the course of atom–surface collisions.[17,18] In adsorbate systems, the existence of very long-lived transient neutral states can help to promote reactions at surfaces. The photon-induced desorption of Cs in Cs/Cu(111) systems, invoked to interpret the TR-2PPE, could be an example of these.

References

1. J. W. Gadzuk and M. Šunjic, in *Aspects of electron-molecule scattering and photo-ionisation*, AIP conference proceedings 204, ed. A. Herzenberg, New York, 1990, p. 118.
2. H. Petek and S. Ogawa, *Prog. Surf. Sci.*, 1997, **56**, 239.
3. M. Bauer, S. Pawlik and M. Aeschliman, *Phys. Rev. B*, 1997, **55**, 10040.
4. M. Bauer, S. Pawlik and M. Aeschliman, *Phys. Rev. B*, 1999, **60**, 5016.
5. S. Ogawa, H. Nagano and H. Petek, *Phys. Rev. Lett.*, 1999, **82**, 1931.
6. S. Ogawa, H. Nagano and H. Petek, *Appl. Phys. B*, 1999, **68**, 611.
7. N. D. Lang and A. R. Williams, *Phys. Rev. B*, 1978, **42**, 616.
8. P. Nordlander and J. C. Tully, *Phys. Rev. B*, 1990, **42**, 5564.
9. A. G. Borisov, D. Teillet-Billy, J. P. Gauyacq, H. Winter and G. Dierkes, *Phys. Rev. B*, 1996, **54**, 17166.
10. A. G. Borisov, A. K. Kazansky and J. P. Gauyacq, *Surf. Sci.*, 1999, **430**, 165.
11. A. G. Borisov, A. K. Kazansky and J. P. Gauyacq, *Phys. Rev. Lett.*, 1998, **80**, 1996; A. G. Borisov, A. K. Kazansky and J. P. Gauyacq, *Phys. Rev. B*, 1999, **59**, 10935.
12. E. Knoesel, T. Hertel, M. Wolf and G. Ertl, *Chem. Phys. Lett.*, 1995, **240**, 409.
13. L. Bartels, G. Meyer, K.-H. Reider, D. Velic, E. Knoesel, A. Hotzel, M. Wolf and G. Ertl, *Phys. Rev. Lett.*, 1998, **80**, 2004.
14. E. V. Chulkov, I. Sarria, V. M. Silkin, J. M. Pitarke and P. M. Echenique, *Phys. Rev. Lett.*, 1998, **80**, 44947.
15. M. C. Desjonquères and D. Spanjaard, *Concepts in Surface Science*, Springer-Verlag Series in Surface Science, 1993, vol. 30.
16. A. G. Borisov, J. P. Gauyacq, A. K. Kazansky, E. V. Chulkov, V. M. Silkin and P. M. Echenique, to be published.
17. L. Guillemot and V. A. Esaulov, *Phys. Rev. Lett.*, 1999, **82**, 4552.
18. T. Hecht, H. Winter, A. G. Borisov, J. P. Gauyacq and A. K. Kazansky, *Phys. Rev. Lett.*, 2000, **84**, 2517.
19. V. A. Ermoshin and A. K. Kazansky, *Phys. Lett. A*, 1996, **218**, 99.
20. D. Teillet-Billy and J. P. Gauyacq, *Surf. Sci.*, 1990, **239**, 343.
21. F. T. Smith, *Phys. Rev.*, 1960, **118**, 349.
22. E. V. Chulkov, V. M. Silkin and P. M. Echenique, *Surf. Sci.*, 1999, **437**, 330.
23. P. J. Jennings, P. O. Jones and and M. Weinert, *Phys. Rev. B*, 1988, **37**, 6113.
24. J. N. Bardsley, *Case Stud. At. Phys.*, 1974, **4**, 299.
25. L. Kleinman and D. M. Bylander, *Phys. Rev. Lett.*, 1982, **48**, 1425.
26. J. R. Chelikowsky, N. Troullier and Y. Saad, *Phys. Rev. Lett.*, 1994, **72**, 1240.
27. J. P. Gauyacq, *Dynamics of negative ions*, World Scientific, Singapore, 1987.
28. J. P. Gauyacq, A. G. Borisov and G. Raseev, to be published.
29. P. J. Rous, *Phys. Rev. Lett.*, 1995, **74**, 1835.
30. A. Hotzel, K. Ishioka, E. Knoesel, M. Wolf and G. Ertl, *Chem. Phys. Lett.*, 1998, **285**, 271.
31. S. Gao and D. C. Langreth, *Surf. Sci.*, 1998, **398**, L314.
32. P. Hyldgaard and M. Persson, *J. Phys.: Condens. Matter*, 2000, **12**, L13.
33. A. G. Borisov, D. Teillet-Billy and J. P. Gauyacq, *Nucl. Instrum. Methods B*, 1993, **78**, 49.
34. P. Nordlander, *Phys. Rev. B*, 1996, **53**, 4125.
35. J. Braun and P. Nordlander, *Surf. Sci.*, 2000, **448**, L193.
36. S. A. Lindgren and L. Wallden, *Phys. Rev. B*, 1983, **28**, 6707.
37. A. G. Borisov, R. Zimny, D. Teillet-Billy and J. P. Gauyacq, *Phys. Rev. A*, 1996, **53**, 2457.
38. J. W. Gadzuk, *Surf. Sci.*, 1967, **6**, 133.
39. A. G. Borisov, D. Teillet-Billy and J. P. Gauyacq, *Surf. Sci.*, 1992, **278**, 99.
40. L. Morgan, *J. Phys. B*, 1991, **24**, 4649.
41. A. Föhlisch, M. Nyberg, P. Bennich, L. Triguero, J. Hasselström, O. Karis, L. G. M. Pettersson and A. Nilsson, *J. Chem. Phys.*, 2000, **112**, 1946.
42. N. Lorente and M. Persson, personal communication.

Effect of the projected band gap on the formation of negative ions in grazing collisions from Cu surfaces

T. Hecht,[a] H. Winter,*[a] A. G. Borisov,[b] J. P. Gauyacq[b] and A. K. Kazansky[c]

[a] *Institut für Physik der Humboldt-Universität zu Berlin, Invalidenstr. 110, D-10115 Berlin, Germany*
[b] *Laboratoire des Collisions Atomiques et Moléculaires (Unité Mixte de Recherche CNRS-Université Paris-Sud UMR8625), Bâtiment 351, Université Paris-Sud, 91405 Orsay Cedex, France*
[c] *Institute of Physics, St. Petersburg University, 198904 St. Petersburg, Russia*

Received 25th April 2000
First published as an Advance Article on the web 13th September 2000

The formation of negative ions (H^-, O^-, S^-, F^-, Cl^-) is studied for grazing scattering of fast ions from Cu(110) and Cu(111) surfaces. In a detailed experimental and theoretical investigation we reveal that the projected L-band gap of the Cu metal affects charge transfer in a specific manner. From the analysis of the negative ion fractions as functions of projectile velocity we conclude that, for the Cu(111) surface the electronic 2D surface state continuum plays an essential role in the projectile–surface electron transfer.

1. Introduction

Electron transfer phenomena between atomic projectiles impinging on the surface of a solid are of considerable interest for fundamental as well as applied research, *e.g.*, many established surface analytical tools are affected by electron loss and capture in a decisive manner. As a consequence, quite a few studies have been devoted over the last decades to this problem as documented in monographs[1–3] and review papers.[4–7] An interesting aspect is the formation of negative ions, bearing in mind possible applications in ion sources and particle detection.[8–10] Already in early studies on negative ion conversion at solid surfaces it was pointed out that, owing to the relatively low electronic binding energies of affinity levels of negative ions, resonant charge transfer (RCT), *i.e.*, energy-conserving one-electron tunneling through the potential barrier between atom/ion and surface, can be considered as a basic electron transfer mechanism.

In their pioneering work, Los and coworkers from FOM Amsterdam provided a basic understanding for the microscopic interaction mechanisms during the formation of negative ions.[5,11–13] Based on combined experimental and theoretical efforts, these authors concluded the importance of the following items concerning formation of negative ions during scattering from solid surfaces:

(1) Resonant charge transfer (RCT) dominates the electron transfer and is affected by: (*a*) The shift of the affinity level of the negative ion owing to image charge potential in front of the surface (enhancement of the effective binding energy of the ion). (*b*) The fast projectile motion parallel to the surface giving rise to kinematic effects that can be taken into account by a frame transformation being nicely described for a free-electron metal (jellium) by the so-called "shifted Fermi-

sphere" concept.[11,12] (c) The slow projectile motion normal to the surface that allows one to describe charge transfer within a rate equation approach. (d) The resulting cycles of capture and loss processes leading to equilibrium charge fractions during the interaction and well defined spatial regimes for final charge state formation at quite some a_0 from the surface ("freezing distance" approximation[13]).

(2) The workfunction typical for clean metals (about 4–5 eV) is clearly larger than affinity energies for most negative ions. Thus, e.g. for H$^-$ ions the negative ion formation can only be triggered by kinematic effects and amounts to only some percent or even less. However, reduction of the workfunction for alkylated metal targets results in substantial negative ion fractions in scattered beams.[5]

(3) The many-body aspects of the RCT associated with the atomic structure of the projectile should be taken into account by inclusion of the statistical factors, different in general for the capture and loss process.[14,15]

On the basis of calculated electron transition rates, application of these concepts leads to a quantitative description of experimental charge fractions for scattered atomic projectiles.[5,6,16] In this respect, we mention the improvement in those calculations by making use of non-perturbative parameter-free methods in order to deduce transition rates for RCT.[16–22] The electronic structure of the target surface is approximated by the free-electron model, i.e., no possible effects of the band structure of the target are taken into account. Al is a prototype for a jellium metal, and for scattering from an Al(111) target quantitative agreement between experimental and theoretical occupations of levels for a number of negative ions[15,16] and alkali atoms[22] has been achieved.

This type of theoretical description has been applied also to cases where the electronic structure of the target surfaces can hardly be approximated by the jellium approach, albeit in certain cases a satisfactory description of experimental data could be reached.[23] Until recently, the influence of the realistic electronic band structure of the target on the RCT process has hardly been explored, except for a few perturbative treatments (see e.g. ref. 24). So it appears obvious that any effort to overcome this deficit would be of relevance for a basic understanding of gas–surface interactions. Recently we started to investigate this intricate problem by implementing a "wave-packet propagation" (WPP) method.[20,25] It turns out that the (111) faces of noble metal crystals are well suited for these kind of studies, since their nearly free sp-electron band has a projected band gap (L gap) in the direction of the surface normal. This gap extends from below the Fermi energy to about vacuum, and electrons of corresponding energies cannot enter the crystal along the surface normal. For the free-electron surface this is, in fact, a preferential direction of the projectile surface electron transfer where the transparency of the potential barrier separating the projectile and the surface is largest. Therefore one would expect a change in the RCT efficiency for the (111) surfaces of noble metals in comparison to a free-electron metal surface. Moreover, the projected band gap leads to the existence of specific electronic states. These states (surface state and image potential states) correspond to the electron localised at the surface and moving parallel to it, which gives rise to 2D continua. The surface and image potential states have been the subject of intense experimental and theoretical research, owing to their importance for a variety of surface phenomena.[26–30] From the analysis based on the WPP calculations it was predicted that the 2D surface state continuum plays an important role in RCT.

The main subject of this paper will be a discussion of the current status of joint theoretical and experimental efforts to investigate the formation of negative ions during grazing scattering from a metal surface. A Cu(111) surface has been chosen as a target, where the L gaps extend from -5.83 to -0.69 eV with respect to vacuum energy and the surface state is located at -5.33 eV at the $\bar{\Gamma}$-point.[31,32] For comparison we also performed experiments with a Cu(110), where the L ($\langle 111 \rangle$-direction) and X ($\langle 100 \rangle$-direction) gaps are clearly tilted from the direction of the surface normal and the effect of the band structure on charge transfer should be reduced. We therefore use this Cu(110) surface as a jellium metal reference.

In recent papers Guillemot et al.[33,34] have presented first experiments on this subject with Ag(111) and Ag(110) surfaces. Pronounced deviations of experimental H$^-$ ion fractions from those calculated in the jellium-approximation were observed for the Ag(111) surface. The authors have concentrated on the angle dependence of H$^-$, O$^-$, and F$^-$ fractions and revealed, for H$^-$, an interaction time effect predicted by theory,[25] whereas, for O$^-$ and F$^-$, no significant effects were found.

In the studies presented here we investigate in detail negative ion formation in the extreme grazing incidence geometry (angle of incidence typically $\Phi_{in} \approx 1°$). Our data are in correspondence with the recent experiments by Guillemot *et al.*, but provide also a number of new aspects. In particular, we will show that the kinematic effects arising from the large velocity component parallel to the surface allow us to reveal the role of the 2D surface state continuum in RCT.

2. Theory of resonant charge transfer

A theoretical description of RCT includes:

(i) The calculation of the parameters (energy E_a and width Γ) of the resonance originated from the active atom (ion) state due to electron tunneling into the surface. The projectile–surface distance, Z, is assumed to be fixed.

(ii) The solution of the rate equation taking into account the frame transformation between the projectile and the metal. The applicability of the rate equation approach was verified in a WPP study.[25]

The calculations of the parameters of negative ion states in front of the free-electron metal surfaces are based, in our study, on the coupled angular modes (CAM) method.[18] This method and its particular application for the negative ion formation in grazing scattering has been described in a number of publications (see *e.g.* refs. 15, 16, 18 and references therein). We do not discuss it here. Rather, on the example of the H$^-$ ion, we give an overview of the theoretical ideas related to RCT between an ion and a metal with a projected band gap.

A simple model of a metal with a projected band gap is provided by a potential $U_{surf}(z)$ acting on the electron and depending only on the electron coordinate along the surface normal, z. For the Cu(111) surface the parameters of this potential were derived from *ab initio* calculations by Chulkov *et al.*[31] This potential exhibits an oscillatory pattern in the bulk and smoothly merges with the image potential outside the metal (see Fig. 1). It supports the correct band structure in the z-direction with the projected band, including image and surface states. The motion of the electron parallel to the surface is considered to be free and, in particular, it generates a 2D surface band.

The H$^-$ ion is treated within the open shell scheme, *i.e.*, it can be described by a simple potential with *s*-symmetry, $U_I(r)$.[20] This potential supports only one bound eigenstate with energy $E_I = -0.75$ eV, in correspondence with the hydrogen affinity.

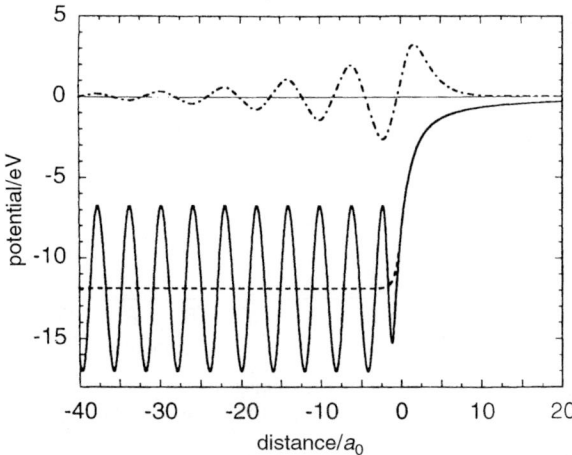

Fig. 1 Model potentials used to represent the electron–Cu(111) (———) and the electron–jellium Cu (— —) interactions as a function of distance between electron and surface ($z = 0$ image plane). The jellium metal potential is defined according to the analytical expression given by Jennings *et al.*[35] (· — ·) represents the wavefunction of the surface state.

The details of the WPP techique can be found elsewhere.[25] Briefly, we solve a non-stationary one-electron Schrodinger equation with the Hamiltonian

$$H = -\frac{1}{2}\frac{\partial^2}{\partial z^2} - \frac{1}{2}\frac{1}{\rho}\frac{\partial}{\partial \rho}\rho\frac{\partial}{\partial \rho} + U_{\text{surf}}(z) + U_{\text{I}}(\sqrt{\rho^2 + (Z-z)^2}) + \frac{m^2}{2\rho^2} \quad (1)$$

In this Hamiltonian the azimuthal symmetry of the problem is explicitly accounted for by cylindrical coordinates with z being along the surface normal. Here, Z is the projectile–surface distance, measured from the image plane; m is a projection of the angular momentum of the electron on the quantization axis z. The time-dependent electron wavefunction $|\Psi(t)\rangle$ has been calculated with implementation of a split-propagation technique[36,37] for a given initial wavefunction $|\Psi(0)\rangle$. For the function $|\Psi(0)\rangle$ we use the wavefunction of a free H$^-$ ion. A straightforward analysis of the autocorrelation function $A(t) \equiv \langle\Psi(0)|\Psi(t)\rangle$ allows one to determine the energy E_a and the width Γ of the quasistationary H$^-$ state coupled with the surface.

The relevant quantities obtained in our calculations for the Cu(111) surface are plotted in Fig. 2 with results obtained for a jellium metal (see Fig. 1). The energy position of the resonance level at large ion–surface distances, Z, follows the image potential curve $E_a(Z) = -1/4Z + E_I$ in both cases. However, close to the surface, the H$^-$/Cu(111) resonance undergoes a pseudo-crossing with

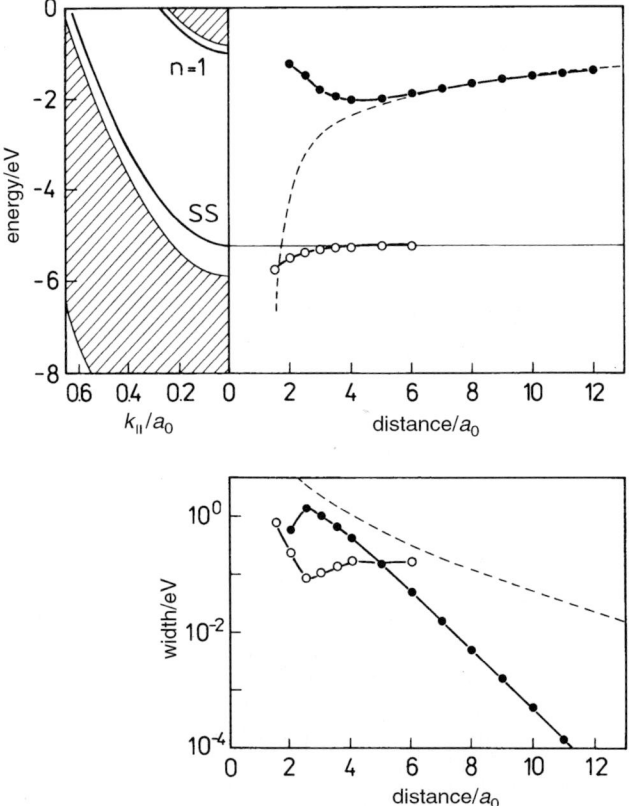

Fig. 2 Upper panel: Energy positions of the H$^-$ resonance states as a function of the distance between the ion and the surface. (———) represents the energy of the resonance obtained within the jellium description of the metal, (●) and (○) are results obtained for the model Cu(111) surface. The structure of electron energies in the model Cu(111) metal as function of the parallel momentum is shown in the left part. SS refers to the surface state, and $n = 1$ refers to the 1st image state. Lower panel: Widths of the H$^-$ level in front of the model Cu(111) surface as a function of the distance between the ion and the surface. Width obtained within the jellium description of the metal (– – –), width of the higher resonance for Cu(111) surface (●), width of the low lying resonance (○).

a state which has been split from the surface band bottom. The origin of this additional state is related to the fundamental property of a 2D system. It is well established that in 2D any weak attractive interaction can support a loosely bound state which is directly related to the non-zero value of the density of the states at the band bottom generic for 2D systems.

The width of the resonance obtained for the model Cu(111) and jellium surface differ by up to several orders of magnitude. Indeed, the most transparent direction for the electron to tunnel from the ion into the metal in the jellium case is the direction along the surface. But, in the case of a Cu(111) surface, tunneling along this direction is blocked by the projected band gap and the electron is compelled to tunnel under some angle to the surface normal. Obviously, the barrier for such propagation is substantially broader and, consequently, the ion resonance decay in front of Cu(111) is slower when compared with the jellium metal. Analysis of the results of our calculations also shows that most (about 90%) of the negative ion decay proceeds into the 2D surface state continuum.

The master equation for the population of the ion state along the trajectory reads

$$\frac{dP^-}{dt} = -\Gamma^l(Z(t))P^-(t) + \Gamma^c(Z(t))(1 - P^-(t)). \tag{2}$$

We assume a trajectory $Z(t) = v_z t$. The quantities $\Gamma^c(Z)$ and $\Gamma^l(Z)$ in eqn. (2) are the electron capture and loss rates, respectively. Close to the surface these rates are large so that the final negative ion fraction is formed in a well defined interval of distances on the outgoing trajectory path ("freezing distance"[13]) and does not depend on initial conditions for P^-. These capture/loss rates are related to the computed resonance parameters in a different manner for the ion state decay into a 3D band and into a 2D band. With use of the "shifted Fermi sphere" approach by van Wunnik et al.[11,12], for the 3D case one obtains:

$$\begin{Bmatrix}\Gamma^c\\ \Gamma^l\end{Bmatrix} = \Gamma(Z)\begin{Bmatrix}g^c\\ g^l\end{Bmatrix}\int_0^\pi |\sigma(\theta, Z)|^2 \sin\theta \, d\theta \int_0^{2\pi} d\varphi \begin{Bmatrix} f[(\mathbf{k}+\mathbf{v}_\parallel)^2/2] \\ 1-f[(\mathbf{k}+\mathbf{v}_\parallel)^2/2] \end{Bmatrix} \tag{3a}$$

where θ is the angle from the surface normal, $g^c = 0.5$ and $g^l = 1$ are capture and loss statistical factors, respectively, and $f(\mathbf{k})$ is the Fermi–Dirac distribution function with the arguments accounting for the Gallilei transformation from the projectile to the metal frame. The momentum $\mathbf{k}$ of the emitted electron is fixed by the resonance condition $k^2/2 + U_0 = E_a(Z)$ where U_0 is the energy of the bottom of the conduction band. The quantity $|\sigma(\theta, Z)|^2$ gives the angular distribution of the transition probability. It reflects the existence of the most transparent direction for electron tunneling, and this function is peaked in a rather narrow θ interval around the surface normal. This leads to $\Gamma^l \gg \Gamma^c$ and the H^- fractions in collision with a jellium metal are on the sub percent level.[16,38]

For the 2D case the corresponding equation reads

$$\begin{Bmatrix}\Gamma^c\\ \Gamma^l\end{Bmatrix} = \Gamma(Z)\begin{Bmatrix}g^c\\ g^l\end{Bmatrix}\int_0^{2\pi} d\varphi \begin{Bmatrix} f[(\mathbf{k}_\parallel+\mathbf{v}_\parallel)^2/2] \\ 1-f[(\mathbf{k}_\parallel+\mathbf{v}_\parallel)^2/2] \end{Bmatrix} \tag{3b}$$

Here $k_\parallel^2/2 + E_s = E_a(Z)$, and E_s is the energy of the surface state (-5.33 eV). There is no angular dependence for the decay rate in the 2D case since the electron is emitted parallel to the surface. In the momentum space the angular integral then directly reflects the fraction of the occupied (capture) or empty (loss) electronic states of the metal which are in resonance with the projectile state. It is not weighted by the angular distribution of the transition probability (see Fig. 3). Assuming the dominant role of the 2D surface state continuum in the H^- formation, which is supported by our WPP calculations, one obtains H^- fractions reaching 6% at maximum for the present case. It should be noted that since the pseudo-crossing of the two states in Fig. 2 is broad, only the upper state is relevant at low normal collision velocities, i.e., for grazing collisions.

The electron capture from the 2D or 3D continuum should manifest itself by the pronounced difference in the dependence of P^- on the parallel velocity. Indeed, the half width of the parallel velocity resonance is given by the Fermi momentum as follows from eqn. (3) and as schematically presented in Fig. 3. The effective energy of the state above the 2D band bottom is about 3 eV which corresponds to an electron momentum along the surface of $0.48\ \hbar\ a_0^{-1}$. Taking into account that the Fermi momentum for the 2D state is $0.17\ \hbar\ a_0^{-1}$, one can conclude that contributions

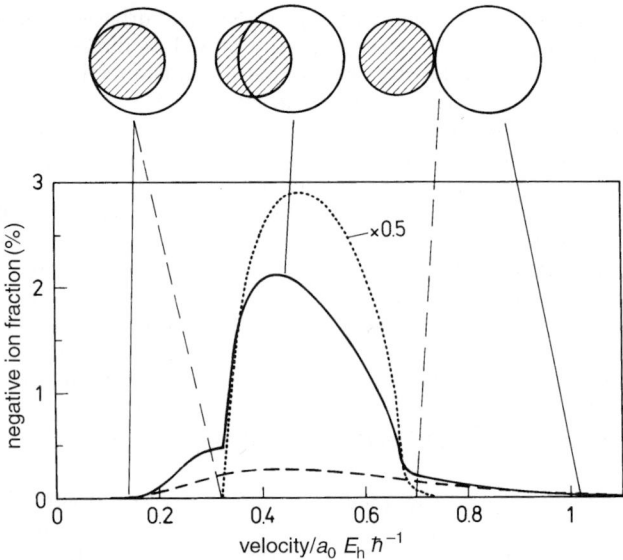

Fig. 3 Upper panel: "Shifted Fermi sphere" model (for details see text). Lower panel: Theoretical results for H$^-$ fractions obtained for different descriptions of the model Cu(111) surface as function of the collision velocity. Jellium model (— —), model Cu(111) surface with contributions from 2D surface band only (· · ·), theory that takes into account also the effect of electron–electron interactions in the bulk and contributions from the 3D band states (———).

from the 2D band to the RCT are confined to an interval of projectile velocities, $0.31 < v/ < 0.65$ $a_0 E_h \hbar^{-1}$. This has to be compared with the velocity range of approximately $0.1 < v/ < 1$ $a_0 E_h^{-1}$ for the 3D case. The calculated theoretical ion yields for the 2D case are plotted in Fig. 3 and compared with those for 3D (jellium).

Going beyond the simple model outlined above, an improved theoretical treatment has been developed that takes into account the contribution from the 3D bulk continuum and electron–electron scattering events in the bulk.[39] Indeed, a reduced electron transfer rate for Cu(111) arises because of the constructive interference of the wavelets reflected owing to the periodic structures of the metal potential. Once in the metal, an electron active in RCT may undergo scattering by metal electrons from the Fermi sea, and will no longer contribute to the interference pattern, but will be lost in the bulk. These many-body interactions are rather different depending upon the process under consideration (capture or loss). The capture process involves the electrons from the Fermi sea. These electrons do not suffer any scattering, since the possible final states are completely filled (we do not address here the electron–phonon scattering process and do not consider electrons in close vicinity to the Fermi level). Therefore many-body interaction should modify electron losses but not electron capture. The effect of the electron–electron interactions is taken into account in a model way. By analogy with low energy electron diffraction (LEED) calculations,[40] the imaginary adsorbing potential was introduced in the bulk with the parameters taken from conventional Fermi-liquid theory and empirical data.[41,42]

The results of the improved theoretical treatment are also presented in Fig. 3. First, owing to the contribution of the 3D bulk continuum the broad background is reproduced (details on the calculations will be given elsewhere[39]). Second, inclusion of the electron–electron scattering effects leads to a substantial increase in the electron loss rate, and to a decrease in calculated H$^-$ fractions. This effect of electron–electron scattering is not important for jellium where, after transfer into the metal, the electron is lost in any case.

In summarizing the theoretical section we recall several important features of the RCT process in front of the surface with projected band gap:

(i) The presence of the projected band gap leads, as a rule, to a decrease in the electron transfer rates. This is of importance not only for collision studies, but also for the understanding of long lived adsorbate induced states at Cu(111) surfaces as observed experimentally.[43,44] For detailed discussion on this subject see the paper by Gauyacq et al. in this volume.

(ii) The broadening of the quasi-stationary state by inelastic electron–electron collisions in the metal appears to be a relevant ingredient of the physical picture.

(iii) The parallel velocity assisted charge transfer can be used as a probe of the dimensionality of the continua involved. In particular, the contribution of the 2D surface state continuum should manifest itself by a narrow structure in the parallel velocity dependence of the negative ion yield. This narrow structure is superimposed on a broader contribution arising from the 3D bulk continuum.

(iv) While the shape of the ion fractions $P^-(v_\parallel)$ is a robust feature reflecting the dimensionality of the continua involved, the absolute negative ion yield, compared to the jellium metal case, depends on parameters of the system. These are the ratio between electron capture and loss rates and relative contribution of the 2D and 3D continua for RCT.

We now present experimental tests on the theoretical predictions. As we have said, we compare the results for a Cu(111) surface with those obtained for a Cu(110) surface. The latter serves here as a free-electron reference against which we test the band structure effects on RCT.

3. Experimental results and discussion

In our experiments we scattered H^+, O^+ (O^{2+}), S^+ (S^{2+}), F^+ (F^{2+}) and Cl^+ (Cl^{2+}) ions from Cu(111) and Cu(110) surfaces at a base pressure in the UHV chamber of about 2×10^{-11} mbar. The ions were produced in an ECR ion source, which is mounted on a high-voltage platform and biased with respect to ground to voltages up to 100 kV. Thus by deceleration and acceleration of the extracted ions a wide range of projectile energies could be used in our measurements, i.e. the (parallel) velocity component, responsible for specific kinematic effects, was tuned over a relatively wide range. The well collimated ion beams are scattered from the target surface under grazing angles of incidence of typically $\Phi_{in} \approx 1°$, and the charge fractions of the scattered beams are analyzed by a pair of electric field plates and a channeltron detector. The entrance aperture of the detector is covered by a thin carbon foil in order to provide an equal response of the detector to projectiles of different charge (stripping and capture in the foil "erases" memory to charge state before penetration). The target surface was carefully prepared by cycles of grazing sputtering with 25 keV Ar^+ ions and subsequent annealing to temperatures up to about 550 °C. Quite a number of cycles were needed in order to obtain an atomically flat surface after a time of preparation of about one month. The final quality was checked by angular distributions of scattered projectiles, which indicate a target surface widely free from defects and mean terrace widths of about 100 nm.[45]

Such a well prepared target surface is an important prerequisite for investigations as presented here, since the final outcome of measurements is rather sensitive to these experimental conditions. The extreme sensitivity to the "quality" of the surface can easily be understood by the large area of the target probed in collisions under grazing incidence. We found that this problem plays an important role in particular for the formation of H^- ions. In Fig. 4 we have plotted the H^- fraction $P^- = n^-/(n^- + n^\circ)$ against the time after preparation of the target (note that for a comparison with theory, the negative ion fractions are normalized to the sum of counts for H^- and H° projectiles). The data show a gradual increase of P^- with time, which we attribute to the adsorption of atoms and molecules from the rest gas in our chamber. This feature allows one to perform measurements only in a time interval of less than 2 h after a preparation cycle. Similar conditions were reported for former experiments with an Al(111) target. In passing we note that P^- amounts to about 0.4% for a target that was kept under UHV overnight and was not treated by a preparation cycle before measuring the data.

In Fig. 5 we show P^- for the scattering of protons from Cu(110) and Cu(111) as a function of projectile velocity. The angle of incidence Φ_{in} was adjusted for these studies in such a manner that the velocity component normal to the surface $v_z = v \cdot \sin \Phi_{in}$ was kept constant: $v_z = 0.02\ a_0 E_h$ $\hbar^{-1}$. The full and open symbols denote data for different azimuthal orientations of the target surface. The solid lines represent calculations for jellium targets, where the workfunctions that enter these calculations are obtained from in situ photoemission studies; $W(Cu(110)) = 4.49(3)$ eV and $W(Cu(111)) = 4.98(3)$ eV. All data sets show the expected kinematic resonance structure for

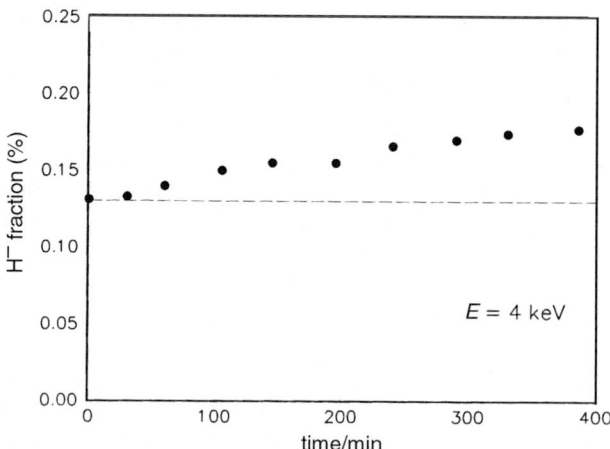

Fig. 4 Dependence of H$^-$ fraction on time after preparation of the target surface for scattering of 4 keV H$^+$ ions from Cu(110) surface with a normal velocity component of 0.02 $a_0 E_h \hbar^{-1}$.

$P^-(v_\parallel)$. However, in obvious contradiction with the jellium-type calculations, a much larger negative ion yield is obtained for the Cu(111), despite its larger workfunction as compared to Cu(110). Furthermore, for Cu(110) the jellium calculations correctly reproduce the $P^-(v_\parallel)$, but they fail in the Cu(111) case. For the latter case one clearly observes the narrow parallel velocity structure superimposed on the broader background, which is in qualitative agreement with the calculations

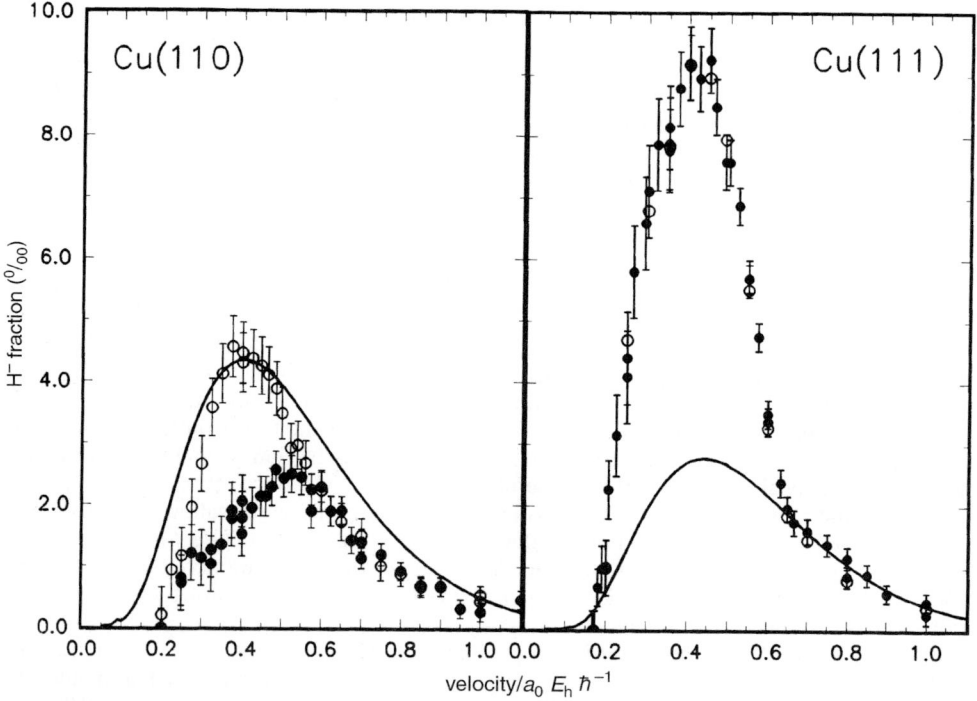

Fig. 5 Left panel: H$^-$ fraction as function of parallel velocity for scattering from Cu(110) surface with a normal velocity component of 0.02 $a_0 E_h \hbar^{-1}$. Calculations within the jellium model, (●) and (○) experimental data for azimuthal settings close to $\langle 001 \rangle$ and $\langle 1\bar{1}0 \rangle$ directions, respectively. Right panel: Same as left panel, but for Cu(111) surface. (———) Calculations within the jellium model, (●) and (○) experimental data for azimuthal settings close to $\langle 1\bar{1}0 \rangle$ and $\langle 121 \rangle$ directions, respectively.

for the model Cu(111) surface. At the same time, the calculated negative ion yield is overestimated by approximately a factor of 2. We attribute this discrepancy to the shortcomings of our modeling of the Cu(111) surface which takes into account only the band structure of the target material in the direction normal to the surface, while the electron motion in the plane parallel to the surface is assumed to be free. To account completely for the three-dimensional band structure of the target material would require a substantial sophistication of the calculations and this has not yet been performed.

In Fig. 6 and 7 we show negative fractions P^- for the formation of O^- and S^- ions, during scattering from Cu(110) and Cu(111) under a fixed incidence angle $\Phi_{in} = 1.3°$. Both data sets are characterized by similar features to those observed for H^- ions. Despite the larger workfunction of Cu(111), the negative ion yields are comparable. For the Cu(111) target one reveals a narrow structure superimposed on a broader background. This we interpret as the contribution of the 2D surface state continuum to the negative ion formation.

One can find qualitative features associated with the 2D surface state continuum for Cu(111) also in the case of the formation of the negative halogen ions F^- and Cl^- (Fig. 8 and 9). Even though the absolute values of P^- are quite well described by a jellium approach for the target surface, the shape of the ion fraction $P^-(v_\parallel)$ is obviously different from the free-electron prediction for the Cu(111) case. It can be considered again as a combination of a narrow 2D and a broad 3D contribution. Note, that the different types of parallel velocity dependence of P^- for Cu(111) and Cu(110) surfaces are related to the different energy position of the affinity level of the negative ion with respect to the Fermi level of the target surfaces at the freezing distance (below or above).[5,6,15,16,46]

As can be seen from the H^-, (O^-, S^-), (F^-, Cl^-) sequence, the relative weight of the contribution from the surface state 2D continuum compared to that from the bulk 3D continuum decreases with increase of the affinity of the projectile. We attribute this to the fact that the energy of the negative ion level is shifted closer to the bottom of the band gap when going from H^- to Cl^- or F^- (see Fig. 10). Then electron transfer to the 3D states can proceed at smaller angles with

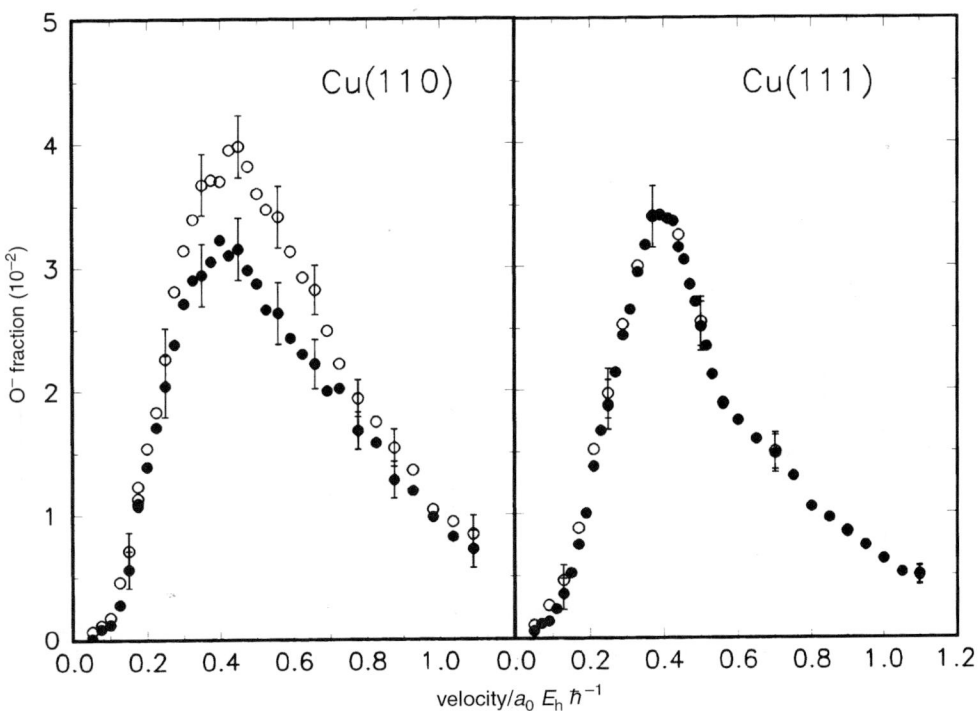

Fig. 6 Same as Fig. 5, but for O^- ions. The angle of incidence was fixed to 1.3°.

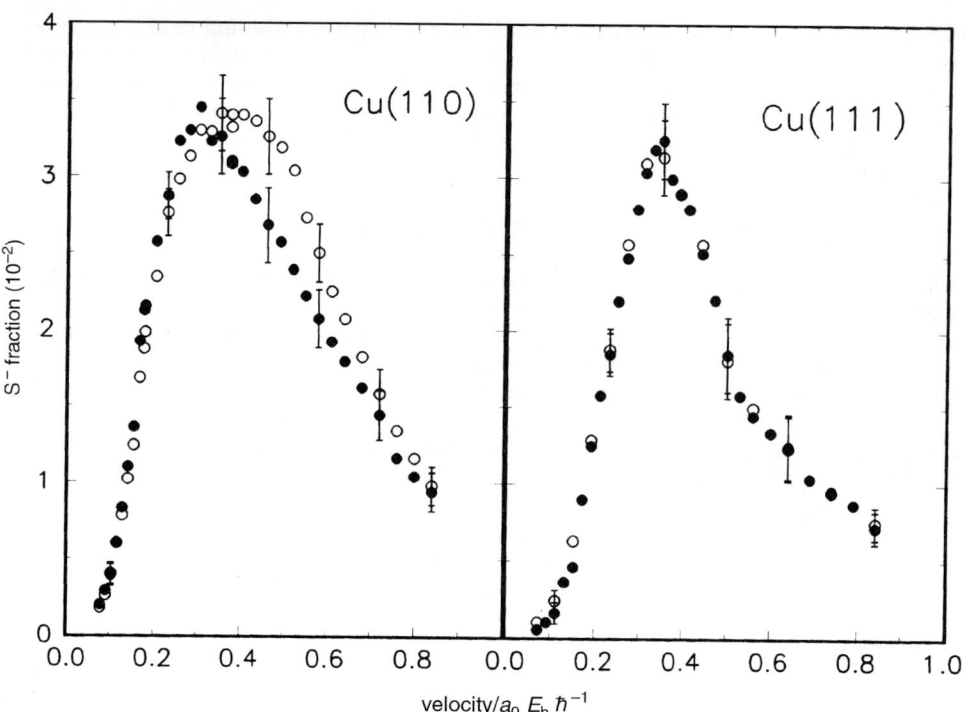

Fig. 7 Same as Fig. 5, but for S$^-$ ions. The angle of incidence was fixed to 1.3°.

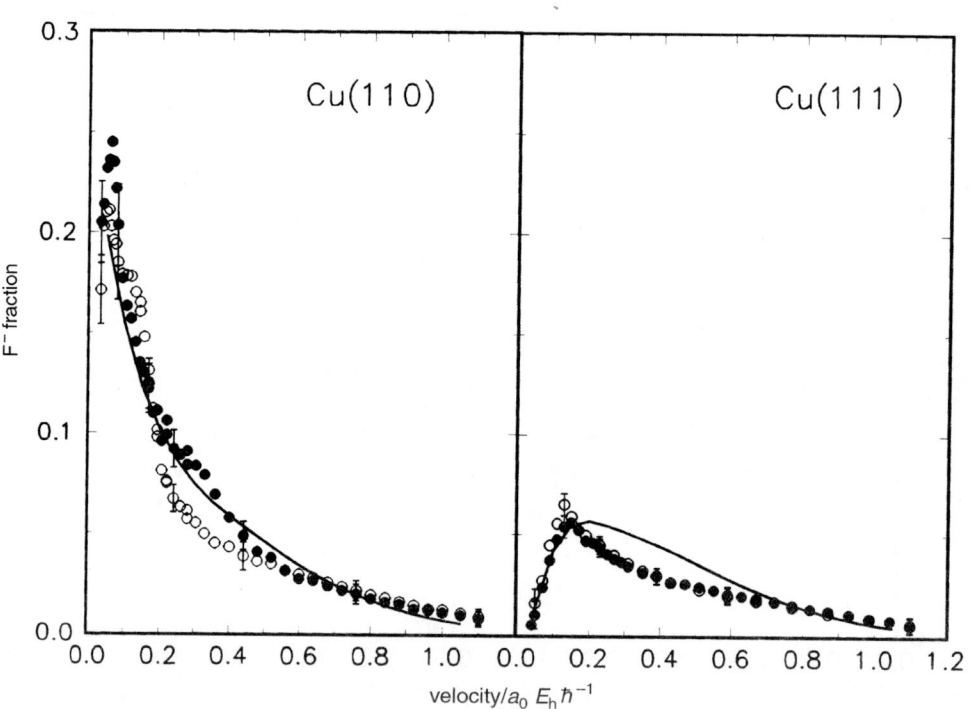

Fig. 8 Same as Fig. 5, but for F$^-$ ions. The angle of incidence was fixed to 1.3°.

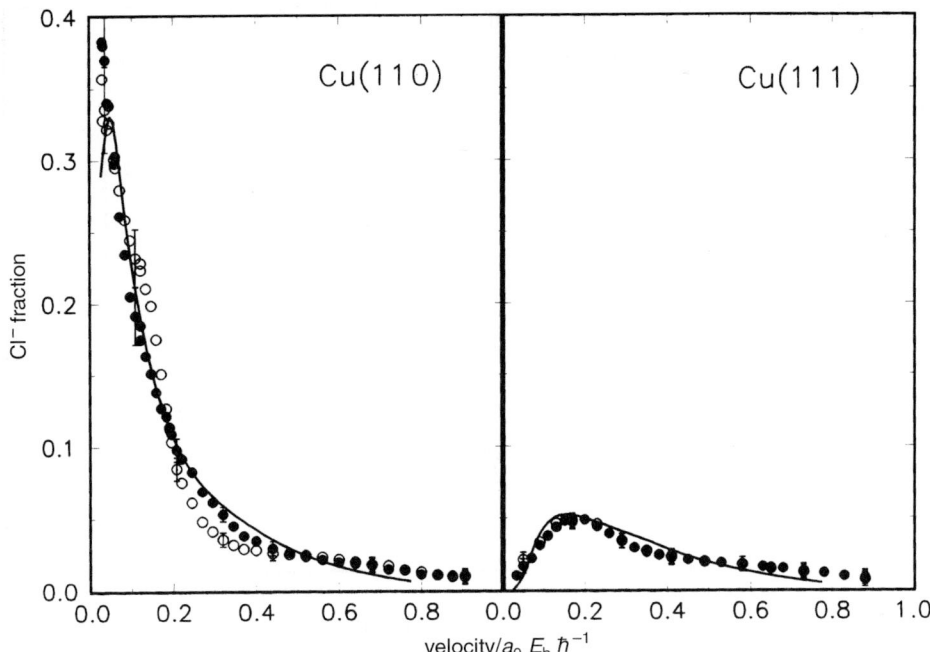

Fig. 9 Same as Fig. 5, but for Cl⁻ ions. The angle of incidence was fixed to 1.3°.

respect to the surface normal, which decreases the blocking effect of the gap. Furthermore, it should be noted that the active electron occupies an s-orbital for H⁻ ions, while p-orbitals of σ and π symmetry are involved in RCT for O⁻, S⁻, F⁻ and Cl⁻ ions. Electrons of π symmetry are *a priori* emitted under a finite angle with respect to the surface normal so that the effect of the band gap for these electrons should be less pronounced.

As a further striking feature, all data for the Cu(110) surface reveal a clear dependence of P^- on the azimuthal settings; this dependence is not found with the Cu(111) target. The azimuthal dependence of P^- observed for the Cu(110) surface was investigated in more detail for H⁻ formation at a projectile energy of 4 keV and $\Phi_{in} = 3°$; *i.e.* $v = 0.4\ a_0 E_h \hbar^{-1}$ and $v_z = 0.02\ a_0 E_h \hbar^{-1}$. The data (full circles) in Fig. 11 show a pronounced oscillation of P^- with the azimuthal angle Θ

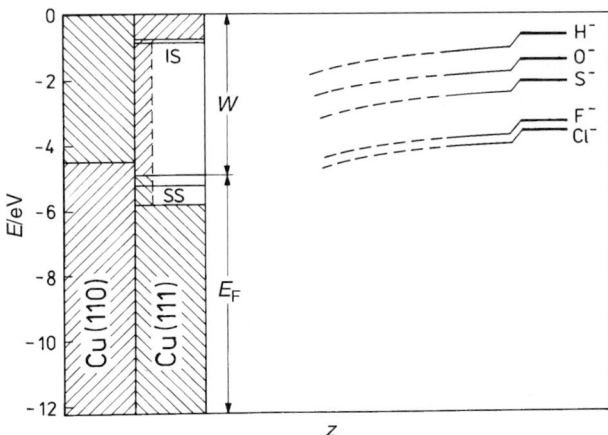

Fig. 10 Sketch of the energies for affinity levels of negative ions considered here and of the electronic band structure of the Cu(111) and Cu(110).

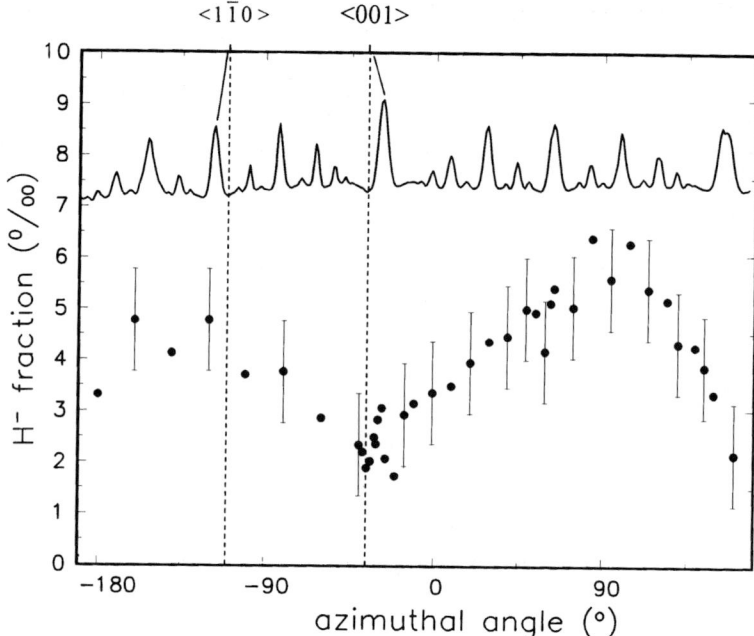

Fig. 11 H⁻ fractions as function of the azimuthal angle for scattering of 4 keV H⁺ ions from a Cu(110) surface under $\Phi_{in} = 3°$ (●) and uncompensated target current (———). The dashed vertical lines represent azimuthal settings for the measurements with Cu(110) presented above.

of period π. For comparison, we have also plotted the (uncompensated) target current (solid line in figure) which shows peak structures for incidence along low index crystal directions. Along those directions the incident protons enter regions of higher electron densities, giving rise to enhanced kinetic electron emission yields.[47] The clearly different azimuthal dependences of P^- and target current indicate that the observed effect for P^- cannot be attributed to trajectories affected by axial surface channeling.

We ascribe this effect to the influence of the Cu(110) electronic structure. Although the projected band gap along the surface normal is absent in this case, the L (⟨111⟩-direction) and X (⟨100⟩-direction) gaps increase the surface reflectivity for the electrons travelling along ⟨111⟩- and ⟨100⟩-directions within a certain energy range. As soon as the translational factors associated with the parallel velocity connect the electron k states in the metal frame with those k_a in the atomic frame ($k = k_a + v_\parallel$ in the free-electron picture), the RCT can be influenced by the orientation of the scattered beam. Therefore, as it follows from our data, even in the case of Cu(110) which approaches the jellium metal, the realistic band structure modifies RCT to a certain extent. It is noteworthy that we do not observe any azimuthal dependences of P^- for Cu(111) where the L gap direction corresponds to the surface normal and the X gap is tilted from the surface normal at a much larger angle than for the Cu(110) so that it cannot be "probed" by transferred electrons.

4. Conclusions

We have presented results from a joint experimental and theoretical study on the negative ion formation in grazing scattering from Cu(111) and Cu(110) surfaces. We find that the absolute values and parallel velocity dependences of the negative ion yields for the Cu(110) surface can be reasonably well described by RCT with an electronic structure of the target approximated by the free-electron model. On the contrary, the RCT in front of the Cu(111) surface is clearly affected by the projected band gap in the direction of the surface normal. This band gap effect is two-fold. First, it modifies the projectile–surface electron transfer rate, as compared to the free-electron case, by blocking the electron loss from the projectile along the surface normal. Second, the existence of the 2D surface state continuum, associated with an electron localized at the surface and moving parallel to it, manifests itself in a narrow structure in the negative ion yield as a function of the velocity component parallel to the surface. This narrow structure is superimposed on a broad one arising from the contribution of the 3D continuum of the bulk states. In this respect it turns out that collision experiments under grazing incidence can serve as a tool for probing the dimensionality of the continua involved in the RCT. As observed for H$^-$ ions, the band gap can lead to a substantial increase in the absolute negative ion yield compared to the free-electron metal. This effect strongly depends on the specific projectile–surface combination. Appearance of the narrow structure in the parallel velocity dependence of P^- should occur whenever the 2D surface state continuum contributes to RCT.

Acknowledgements

The assistance of Dr. A. Mertens, K. Maas, R. A. Noack, A. Bensch, M. Janetzky and A. Laws in the preparation of the experiments and the manuscript is gratefully acknowledged. This work is supported by the Deutsche Forschungsgemeinschaft under contract Wi 1336 and by the Franco-German PROCOPE-programme of DAAD/APAPE.

References

1. M. Kaminsky, *Atomic and Ionic Impact Phenomena on Metal Surfaces*, Springer, Berlin, 1965.
2. Y. H. Ohtsuki, *Charged Beam Interaction with Solids*, Taylor and Francis, London, 1983.
3. *Low Energy Ion–Surface Interactions*, ed. J. W. Rabalais, Wiley, New York, 1994.
4. R. Brako and D. M. Newns, *Rep. Prog. Phys.*, 1989, **52**, 655.
5. J. Los and J. J. C. Geerlings, *Phys. Rep.*, 1990, **190**, 133.
6. J. Burgdörfer, in *Review of Fundamental Processes and Applications of Atoms and Ions*, ed. C. D. Lin, World Scientific, Singapore, 1993, p. 517.
7. H. Niehus, W. Heiland and E. Taglauer, *Surf. Sci. Rep.*, 1993, **17**, 213.
8. C. F. A. van Os, C. Lequijt, A. W. Kleyn and J. Los, *Fusion Technol.*, 1988, **1**, 598.
9. J. Ishikawa, in *Handbook of Ion Sources*, ed. B. Wolf, CRC Press, Boca Raton, FL, 1995, p. 289.
10. P. Wurz, M. R. Aellig, P. Bochsler, A. G. Ghielmetti, E. G. Shelley, S. A. Fuselier, F. Herrero, M. F. Smith and T. S. Stephen, *Opt. Eng.*, 1995, **34**, 2365.
11. J. N. M. van Wunnik and J. Los, *Phys. Scr. T*, 1986, **6**, 27.
12. J. N. M. van Wunnik, R. Brako, K. Makoshi and D. M. Newns, *Surf. Sci.*, 1983, **126**, 618.
13. E. G. Overbosch, B. Rasser, A. D. Tenner and J. Los, *Surf. Sci.*, 1980, **92**, 310.
14. R. Zimny, *Surf. Sci.*, 1990, **233**, 333.
15. J. P. Gauyacq, A. G. Borisov and H. Winter, *Comments At. Mol. Phys.*, 2000, **D2**, 22.
16. A. G. Borisov, D. Teillet-Billy and J. P. Gauyacq, *Phys. Rev. Lett.*, 1992, **68**, 2842.
17. P. Nordlander and J. C. Tully, *Phys. Rev. Lett.*, 1988, **61**, 990.
18. D. Teillet-Billy and J. P. Gauyacq, *Surf. Sci.*, 1990, **239**, 343.
19. P. Kürpick and U. Thumm, *Phys. Rev. A*, 1996, **54**, 1487.
20. V. A. Ermoshin and A. K. Kazansky, *Phys. Lett. A*, 1997, **55**, 466.
21. S. A. Deutscher, X. Yang and J. Burgdörfer, *Nucl. Instrum. Methods Phys. Res. B*, 1995, **100**, 336.
22. A. G. Borisov, D. Teillet-Billy, J. P. Gauyacq, H. Winter and G. Dierkes, *Phys. Rev. B*, 1996, **54**, 17166.
23. N. D. Lang, *Phys. Rev. B*, 1983, **27**, 2019.
24. S. I. Easa and A. Modinos, *Surf. Sci.*, 1987, **183**, 531.
25. (a) A. G. Borisov, A. K. Kazansky and J. P. Gauyacq, *Phys. Rev. B*, 1999, **59**, 10935; (b) *Surf. Sci.*, 1999, **430**, 165.
26. U. Höfer, I. L. Shumay, Ch. Reuß, U. Thomann, W. Wallauer and Th. Fauster, *Science*, 1997, **277**, 1480.
27. E. V. Chulkov, I. Sarría, V. M. Silkin, J. M. Pitarke and P. M. Echenique, *Phys. Rev. Lett.*, 1998, **80**, 4947.
28. H. Petek and S. Ogawa, *Prog. Surf. Sci.*, 1997, **56**, 239.
29. A. Hotzel, K. Ishioka, M. Wolf and G. Ertl, *Chem. Phys. Lett.*, 1998, **285**, 271.
30. Th. Fauster, Ch. Reuß, I. L. Shumay and M. Weinelt, *Chem. Phys.*, 2000, **251**, 111.
31. E. V. Chulkov, V. M. Silkin and P. M. Echenique, *Surf. Sci.*, 1999, **437**, 330.
32. *Landolt Börnstein*, 1994, Vol. III/24b, Springer-Verlag, Heidelberg, p. 120.
33. L. Guillemot and V. A. Esaulov, *Phys. Rev. Lett.*, 1999, **82**, 4552.
34. L. Guillemot, S. Lacombe and V. A. Esaulov, *Nucl. Instrum. Methods Phys. Res. B*, 2000, **164**, 601.
35. P. J. Jennings, P. O. Jones and M. Weinert, *Phys. Rev. B*, 1988, **37**, 3113.
36. M. D. Feit and J. A. Fleck, *J. Chem. Phys.*, 1982, **78**, 301.
37. C. Leforestier, R. H. Bisseling, C. Cerjan, M. D. Feit, R. Friesner, A. Guldberg, A. Hammerich, G. Jolicard, W. Karrlein, H. D. Meyer, N. Lipkin, O. Roncero and R. Kosloff, *J. Comput. Phys.*, 1991, **94**, 59.
38. F. Wyputta, R. Zimny and H. Winter, *Nucl. Instrum. Methods Phys. Res. B*, 1991, **58**, 379.

39 A. G. Borisov, A. K. Kazansky and J. P. Gauyacq, to be published.
40 M. A. van Hove, W. H. Weinberg and C. M. Chan, *Low Energy Electron Diffraction*, Springer-Verlag, Berlin, 1986.
41 T. Hertel, E. Knoesel, M. Wolf and G. Ertl, *Phys. Rev. Lett.*, 1996, **76**, 535.
42 R. Knorren, K. H. Bennermann, R. Burgermeister and M. Aeschlimann, *Phys. Rev. B*, 2000, **61**, 9472.
43 M. Bauer, S. Pawlik and M. Aeshlimann, *Phys. Rev. B*, 1997, **55**, 10040; (b) *Phys. Rev. B*, 1999, **60**, 5016.
44 S. Ogawa, H. Nagano and H. Petek, *Phys. Rev. Lett.*, 1999, **82**, 1931.
45 R. Pfandzelter, *Phys. Rev. B*, 1998, **57**, 15496.
46 H. Winter, *Comments At. Mol. Phys.*, 1991, **26**, 287.
47 H. Winter, G. Dierkes, A. Hegmann, J. Leuker, H. W. Ortjohan and R. Zimny, in *Ionisation of Solids by Heavy Particles*, ed. R. Baragiola, Plenum Press, New York, 1993, p. 253.

Charge-transfer reactions in atom scattering from ionic surfaces: A time-dependent wavepacket approach

G. R. Darling,[a] **Y. Zeiri**[b] **and R. Kosloff**[c]

[a] *Surface Science Research Centre, Department of Chemistry, The University of Liverpool, Liverpool, UK L69 3BX*
[b] *Department of Chemistry, Nuclear Research Center-Negev, P.O. Box 9001, Beer-Sheva 84190, Israel*
[c] *Department of Physical Chemistry and the Fritz Haber Research Center, The Hebrew University, Jerusalem 91904, Israel*

Received 18th May 2000
First published as an Advance Article on the web 11th September 2000

A diabatic description of charge transfer between atoms and ionic surfaces is presented, specifically examining the F/LiF(100) and F/KI(100) systems for which experiment shows ion formation to be very efficient. Potential energy surfaces describing the energetics for these systems have been generated with a semi-empirical scheme. At the site of charge exchange, there is a curve-crossing between the ground state and the state representing charge capture by the projectile. Quantum dynamics calculations with time-dependent wavepacket methods give an initial ion-formation probability of unity for all cases considered. At lowest energies, the ions cannot escape the surface, giving an effective threshold for negative-ion production very close to that observed in experiment. Re-neutralization by charge transfer back to the conduction band of the solid is also examined.

I. Introduction

Charge transfer between an atom and a surface is one of the fundamental processes occurring in gas–surface dynamics. It is best understood on metal surfaces.[1,2] Electron transfer into very low-lying atomic states can be accompanied by ejection of another electron (an Auger mechanism) or emission of light (chemiluminescence), while transfer into atomic affinity levels often occurs by a resonant process. In this paper, we shall focus on resonant transfer between a surface and neutral projectiles. On metal surfaces this is facile because the affinity level of the incoming atom is downshifted by interaction with the electrostatic image charge in the metal surface, as indicated in Fig. 1(a). For low workfunction surfaces, a large fraction of the atoms scatter as negative ions.[2] On surfaces with higher workfunctions (*e.g.* transition metals), if the atoms are directed at grazing incidence (a few degrees from the surface plane) in the hyperthermal energy range ($\sim$ keV), the resonance conditions are assisted by a Doppler upshift of the electrons at the Fermi level.[2]

On insulating ionic surfaces, the situation appears less favourable for electron transfer. The valence electrons are tightly bound to the anion sites, leading to narrow bandwidths, low electron mobilities and high workfunctions. The atom affinity level would have to decrease substantially to become resonant with the valence band of the surface, but the image attraction is small (at least for singly-charged ions). One would expect a very low probability of electron transfer, however the

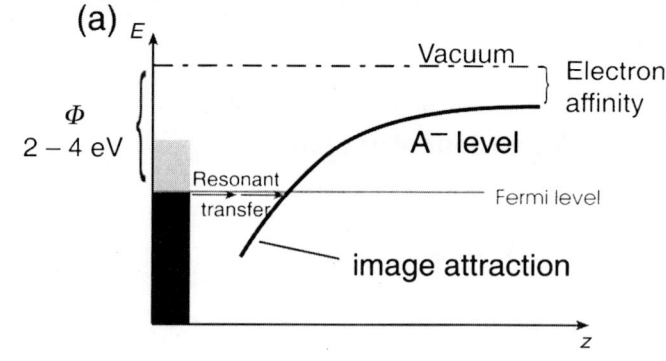

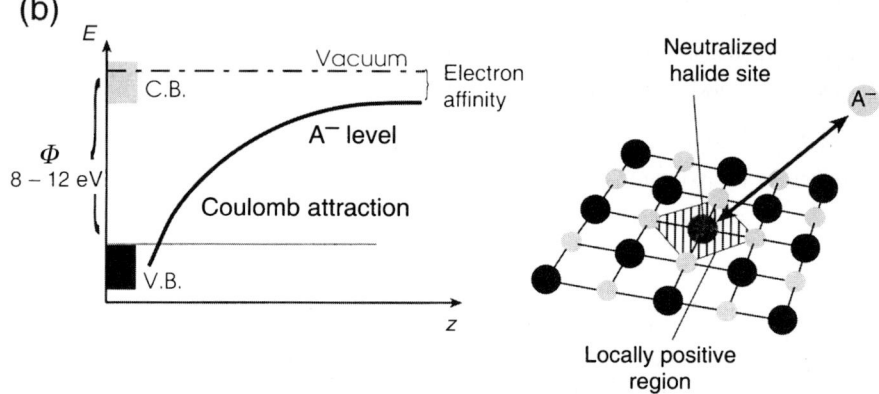

Fig. 1 (a) Schematic of the variation of the atomic affinity level with distance above a metal surface, z. The decrease in energy of the A^- level is due to the interaction with the image charge. Resonant filling can occur when it crosses the Fermi energy. (b) Schematic of the variation of an atomic affinity with distance above an insulating, ionic surface. The work function is very much larger than for a metal, and so the affinity level has further to drop before it can fill resonantly on crossing the top of the valence band. This large drop is effected by a Coulomb attraction between the negative atomic ion and the positively charged region formed on the surface at the site of charge transfer.

opposite has been observed in recent experiments. In grazing incidence scattering experiments with energies up to 100 keV, Winter and coworkers detected very high yields of negative ions for O, and F atoms and positive ions incident on alkali-metal halide surfaces.[3–6] In the scattering of F from KI, for instance, almost complete conversion of neutral atoms into negative ions is observed.[4] Similar results have been seen in scattering of O and F atoms from the MgO(100) surface.[7,8]

A possible explanation of these observations is that the electron is transferred from surface or defect states located in the band-gap of the solid. This would require a much smaller downshift of the affinity level, however, the high yield of negative ions suggests a high density-of-states for these surface/defect states. Complementary studies of the neutralization probability of Na$^+$ ions show that this is not the case. Such states in the band-gap would be expected to give efficient neutralization, yet experiment finds the Na$^+$ scatter as positive ions only.[9]

Simple electrostatic considerations show that the high workfunction of an ionic solid is not, in fact, a barrier to charge transfer from the valence band to a scattering atom. The valence electrons are tightly bound to the anionic sites, therefore the charge transfer from surface to atom is largely a local event involving an electron moving on to the atom from a particular anion site, which then becomes charge neutral. This leaves a charge imbalance in the surface; the cations surrounding the now neutralized anion site create a locally positive region on the surface (Fig. 1(b)). It is the

Coulomb attraction between this positive patch on the surface and the negatively charged scattering atom, that leads to the downshift required to bring the electron affinity and the valence band into resonance,[10,11] as depicted in Fig. 1(b).

A model of these charge-transfer reactions has been developed by Borisov and Sidis[10,11] based on a binary collision formalism with a diabatic description of the energetics.[12] They obtained coupled time-dependent equations for the amplitudes in the two diabatic configurations, generating the time-dependence with classical trajectories, although the motions normal to and parallel to the surface were decoupled and treated separately. For F/LiF, their diabatic potentials representing the F^0 and F^- configurations do not cross. A threshold energy is thus required for negative-ion production, but above this the ions are formed with unit probability. For F/KI there is a curve-crossing and, in consequence, a lower threshold energy. This is in broad agreement with experimental results in the lower part of the energy range considered (below ~ 20 keV), while above this, the negative-ion fraction decreases steadily to zero.

In this paper we consider a similar model of the negative-ion formation. We use diabatic potential energy surfaces (PESs), generated by a semi-empirical procedure, however, we treat the motion of the projectile in a fully quantum fashion allowing full coupling between the normal and parallel motions. We find that in our model, the ion-formation probability is unity at all energies considered, however, we obtain a threshold for the detection due to trapping of the ions at the lowest energies. In the following section, we review the semiempirical scheme used to generate the PESs. Following this, we present some details of the quantum wavepacket calculations that are necessary to deal with the unusually high energies (at least for wavepacket methods) considered in this problem. We then discuss our results and conclusions.

II. The diabatic PESs

A. General considerations

In this work, we employ a diabatic description; the electronic states are not eigenstates of the Hamiltonian for fixed nuclear positions, R, rather we use basis functions which are products of atomic-like orbitals centred on the atom positions. Specifically, for each electronic configuration we use a single Slater determinant. For example, for the ground-state (GS) electronic wavefunction, describing neutral ionic surface and neutral atom, we write

$$\chi_{gs} = \det\left\{\left[\prod_k X_k \bar{X}_k\right] A\right\}. \quad (1)$$

Here, X_k, $\bar{X}_k$ and A are spin eigenfunctions for the valence electrons (we have omitted explicit reference to the core electrons) sited on the halide (X) or projectile (A) sites. For the crystal, this assumes that the valence electrons are fully localized on the anion sites, i.e. the anion and cation sites have integer multiples of charge. There is essentially no band-structure; the valence band is a δ-function. This is a reasonable approximation for the alkali-metal halide surfaces examined in the present paper.[13–15] Similarly, we write for the excited state wavefunction formed when charge is transferred from a halide site to the projectile atom (which we shall call the negative ion state or NIS)

$$\chi_{A^-} = \det\left\{\left[\prod_{k=1}^{N-1} X_k \bar{X}_k\right] X_N A \bar{A}\right\}. \quad (2)$$

In standard fashion, we can employ these electronic orbitals to obtain a matrix equation for the time dependence of the nuclear wavefunction. With a general Hamiltonian of the form

$$\mathcal{H} = \hat{K}_n + \hat{K}_e + v_n + v_e + v_{n-e}, \quad (3)$$

where $\hat{K}_n$ ($\hat{K}_e$) is the kinetic energy operating on the nuclear (electronic) coordinates, v_n (v_e) is the internuclear (interelectronic) Coulomb repulsion and v_{n-e} is the electron–nuclei interaction, the time-dependent Schrödinger equation for the nuclear motion becomes (ignoring the derivatives of electronic wavefunction with respect to nuclear coordinate)

$$i \frac{\partial \psi_v(R)}{\partial t} = (\hat{K}_n + v_n + \langle \chi_v | K_e + v_e + v_{n-e} | \chi_v \rangle)\psi_v + \sum_\mu \langle \chi_v | K_e + v_e + v_{n-e} | \chi_\mu \rangle \psi_\mu, \quad (4)$$

where ψ_v is the nuclear wavefunction corresponding to electronic configuration, v. We can write this more simply as a matrix equation for vector Ψ

$$i\frac{\partial \Psi(R)}{\partial t} = (K + V)\Psi, \tag{5}$$

where K is a diagonal matrix given by $K = \text{diag}(\hat{K}_n)$, and V is a potential energy matrix. Each electronic configuration, represented by one of the X_v, generates a separate PES forming the diagonal elements of V. The off-diagonal elements give the couplings between the different configurations, and induce charge transfer or changes in bonding.

Charge can, of course, be transferred at any halide site, and so there is a NIS PES for every halide site on the surface. Therefore, even when we consider only the GS and NISs, the potential matrix and wavefunction vector are infinite in size, with an NIS matrix element centered on each and every halide site

$$V = \begin{pmatrix} V_{11} & V_{12} & V_{13} & \cdots \\ V_{12} & V_{22} & V_{23} & \\ V_{13} & V_{23} & V_{33} & \\ \vdots & & & \ddots \end{pmatrix}, \tag{6}$$

where V_{11} is the GS PES generated using the wavefunction of eqn. (1). The diagonal elements V_{22}, V_{33}, etc., describe the interaction between a negative ion and the surface with one neutral halogen site, while the off-diagonal elements in the first row and first column, V_{12}, V_{13}, etc., couple these excited states to the electronic GS, i.e., they induce the electron transfer. Fortunately, because of the surface periodicity, the NIS matrix elements are related to one another simply by shifts of whole numbers of lattice vectors, and so need only be computed for one site. The other off-diagonal elements, V_{23}, V_{35}, etc., couple the excitations at different sites; they describe the hopping of electrons (or holes) between the halogens. Although these are non-zero in the Slater orbital basis employed here, they are negligibly small.

B. Computation of the PES matrix elements

To make an explicit computation of the elements of V, we use a semi-empirical valence bond (SEVB) method developed for gas-phase studies of alkali-metal halide PESs and dynamics.[16-18] This has been extended previously to deal with the case of ions bonding to metal halide surfaces.[19] The basis of this approach is to factor the Hamiltonian (excluding the nuclear kinetic energy) into diatomic-like terms in a fashion similar to the diatomics-in-molecules method[20] used in gas-phase dynamics,

$$H = \sum_p H_p + \sum_p \sum_{q<p} V_{pq} \tag{7}$$

where p and q label the atomic sites. H_p is a 'single atom' Hamiltonian given by (using atomic units)

$$H_p = \sum_{i=1}^{n(p)} \left\{ -\frac{1}{2m}\nabla_i^2 - \frac{Z_p}{|r_i - R_p|} \right\} + \sum_{i=1}^{n(p)} \sum_{j<i} \frac{1}{r_{ij}} \tag{8}$$

where i, j label the $n(p)$ electrons on atom p, which has nuclear charge Z_p, m is the mass of the electron and the r_i (R_p) are the electron (nuclear) positions. The first approximation in the SEVB method is to separate this into core and valence electrons, and assume the cores are localized at the nucleus. We can then replace the core terms in eqn. (8) by an energy that is independent of the valence configuration. By choice of energy zero, these core energies can actually be omitted. Additionally, the core electrons act to screen the nuclear charge. In the examples considered in this paper, the effective nuclear charge is reduced to $+1$. Thus eqn. (8) becomes

$$H_p = \sum_{i=1}^{n_v} \left\{ -\frac{1}{2m}\nabla_i^2 - \frac{1}{|r_i - R_p|} + \sum_{j<i} \frac{1}{r_{ij}} \right\} \tag{9}$$

where n_v is the number of valence electrons. In the following we shall always take this to be 0, 1 or 2, corresponding to a singly charged positive ion, a neutral atom and a singly charged negative ion, respectively. With these approximations, the interaction term V_{pq} is given by

$$V_{pq} = -\sum_{i=1}^{n_v(p)} \frac{1}{|r_i - R_q|} - \sum_{j=1}^{n_v(q)} \frac{1}{|r_j - R_p|} + \sum_{i=1}^{n_v(p)} \sum_{j=1}^{n_v(q)} \frac{1}{r_{ij}} + \frac{1}{R_{pq}} + V_{cc}(p, q) \qquad (10)$$

The first two terms in this give the attraction between the core of one atom and the electrons of the other, the second term is the repulsion between electrons on different atoms, and the third term is the nuclear–nuclear repulsion. The final term in eqn. (10) is due to the interaction of the cores with each other and to the core–valence interaction.

Combining eqns. (9) and (10) with the electronic wavefunctions of eqn. (1) and (2) gives us the elements of V in eqn. (6). The GS is given by

$$V_{11} = \langle \chi_{gs} | H | \chi_{gs} \rangle. \qquad (11)$$

Similarly, the NIS elements are given by

$$V_{22} = \langle \chi_{A^-} | H | \chi_{A^-} \rangle. \qquad (12)$$

The partitioning of the electrons into those on atom p and those on atom q is arbitrary. In calculating V_{11} and V_{22} we actually divide the electrons differently: for V_{22} we 'move' one electron from a halide site onto A. We do not, however, explicitly compute any of the integrals in eqns. (11) and (12), rather at this point we introduce the empirical part of the SEVB method, evaluating the integrals by the point-charge approximations.[21] For example,

$$\langle A | H_A | A \rangle = -E_i(A), \qquad (13)$$

$$\langle \bar{X}_i X_i | H_{X_i^-} | \bar{X}_i X_i \rangle = -E_i(X) - E_{ea}(X) \qquad (14)$$

and

$$\langle \bar{X}_i X_i \bar{A} A | H_{X_i^- A^-} | \bar{X}_i X_i \bar{A} A \rangle = V_{cc}(X_i, A) + \frac{1}{R_{AX_i}}, \qquad (15)$$

where E_i and E_{ea} are the ionization potential and electron affinity, respectively. We neglect all three-centre integrals and higher.

The only remaining unknown in this scheme is the 'core–core repulsion' term, V_{cc}. In the form we have partitioned the Hamiltonian, this is a purely diatomic property, we have a different V_{cc} for each pair of atom types. It is an assumption of the SEVB method that these core–core terms are transferrable to all environments in which the atom-pair finds itself. As in previous work,[16] we repeat the algebra above, deriving a PES matrix for the ground and excited states of the diatomic alone. The eigenvalues of this are then fit to the known binding energy curve, which after some algebraic rearrangement yields V_{cc}. Further details of this may be found in ref. 16 and 22. By construction, the SEVB method gives the exact binding energy curve for the diatomic molecule, V_{cc} is the correction which makes this so. Although we call it the core–core repulsion, it is not purely repulsive, but rather has a shallow well some 10 s of eV deep. To have a convenient functional form for later work, we fit V_{cc} to a sum of Gaussians.

Combining all of the above, and by choosing the energy zero to be the energy of the neutral atom and neutral surface at infinite separation, the GS potential for the atom–surface interaction is

$$V_{11} = \sum_p \{V_{cc}(A, X_p) + V_{cc}(A, M_p)\}, \qquad (16)$$

where M_p are the cations (metal ions). This is predominantly repulsive, but has a very shallow chemisorption well. We add to this a component accounting for the dipole induced on A by the surface charges, which does not arise in the SEVB scheme. The form of this is discussed below.

The induced-dipole terms are more important for the NIS, where we must account for the polarization of the A^- by the surface and for the polarization of the surface by the A^-. The polarizations of the crystal atoms also change when one halide site is neutralized, giving

$$V_{22} = E_i(X) - E_{ea}(A) + E_{mad} + \Delta V_{xpol} + \sum_p \{V_{cc}(A, X_p) + V_{cc}(A, M_p)\}$$

$$+ \sum_p \left\{\frac{1}{R_{AX_p}} - \frac{1}{R_{AM_p}}\right\} - \frac{1}{R_{AX_N}} + V_{dip}(A^-)$$

$$- \sum_p \left\{\frac{\alpha(X^-)}{R_{AX_p}^4} + \frac{\alpha(M^+)}{R_{AX_p}^4}\right\} + \frac{1}{2R_{AX_N}^4}\{\alpha(X^-) - \alpha(X)\}, \quad (17)$$

where E_{mad} is the Madelung energy of the crystal, ΔV_{xpol} the change in the polarization energy of the crystal and the αs are the polarizabilities of the respective ions and atoms. The change in crystal polarization is not easy to obtain, however the potential is constructed semi-empirically, so it makes perfect sense to replace the first four terms by the known asymptotic value of the potential, i.e. by $\Phi - E_{ea}(A)$, the energy required to move an electron from the surface to infinity (the work function Φ) and then to deposit it onto A ($-E_{ea}(A)$). The eighth term is the energy of the dipole induced on the A^- ion by the crystal pointcharges, which has basically the same form as the correction for the GS. The final two terms describe the polarization of the crystal by the A^- ion. Clearly these are treated as diatomic-like again, i.e. we have neglected any screening of the induced dipoles by the surrounding crystal. This term will therefore be an overestimate of the true induced-dipole energies.

Finally we write

$$V_{22} = \Phi - E_{ea}(A) + V_{11} + V_{pc} + V_{dip}(A^-) + V_{pol}, \quad (18)$$

where V_{pol} represents all the ion induced crystal polarization terms.

$$V_{pc} = \sum_p \left\{\frac{1}{R_{AX_p}} - \frac{1}{R_{AM_p}}\right\} - \frac{1}{R_{AX_N}} \quad (19)$$

is the interaction of A^- with all the point charges in the surface. The sum is taken over all surface sites (so that we may express this as a Fourier series, as discussed below), and so we must explicitly account for the neutral atom at the site labelled X_N. The $1/R$ Coulomb attraction into the locally positive, active halide site is especially clear in eqn. (19).

For consistency with the determination of V_{cc} we have also included in the diagonal matrix elements the terms proportional to $S_{AX}^2 = |\langle X|A\rangle|^2$. These are, however, small compared to the other terms, so we omit discussion of them here.

The greatest advantage of the SEVB method is that we can obtain the off-diagonal matrix elements within the same scheme and with the same approximations used for the diagonal matrix elements. The coupling between V_{11} and V_{22} is

$$V_{12} = \tfrac{1}{2} S_{AX_N}\left\{V_{11} + V_{22} - V_{pc} + E_{mad} - \tfrac{1}{2}(\rho_A + \rho_X) + \frac{1}{R_{AX_N}}\right\}, \quad (20)$$

where ρ_A and ρ_X are the screening constants of the outer electrons of A and X.[23] Analytic approximations for the overlap integral $S_{AX_N} = \langle X|A\rangle$ have been given by Mulliken et al.[24]

C. Fourier expressions for lattice summations

To evaluate the summations over lattice positions, we employ Fourier expansions, checked by explicit summation over finite clusters. In the following, we assume the crystal has a square NaCl type lattice. As detailed by Steele[25] we split the summations into those over atoms in identical positions, i.e. we sum anion and cation sites separately, and odd and even planes separately.

$$\sum_p V_{cc}(A, X_p) = \sum_p \sum_n B_n \exp(-\alpha_n R_{AX_p}^2) = F_{cc}^{even} + F_{cc}^{odd} \quad (21)$$

where

$$F_{cc}^{even} = \sum_{g_1 g_2} \sum_n \sum_k \frac{\pi B_n}{a_s \alpha_n} e^{-\alpha_n z_k^2} e^{-g^2/4\alpha_n} C_1(g_1, g_2, x, y) \quad (22)$$

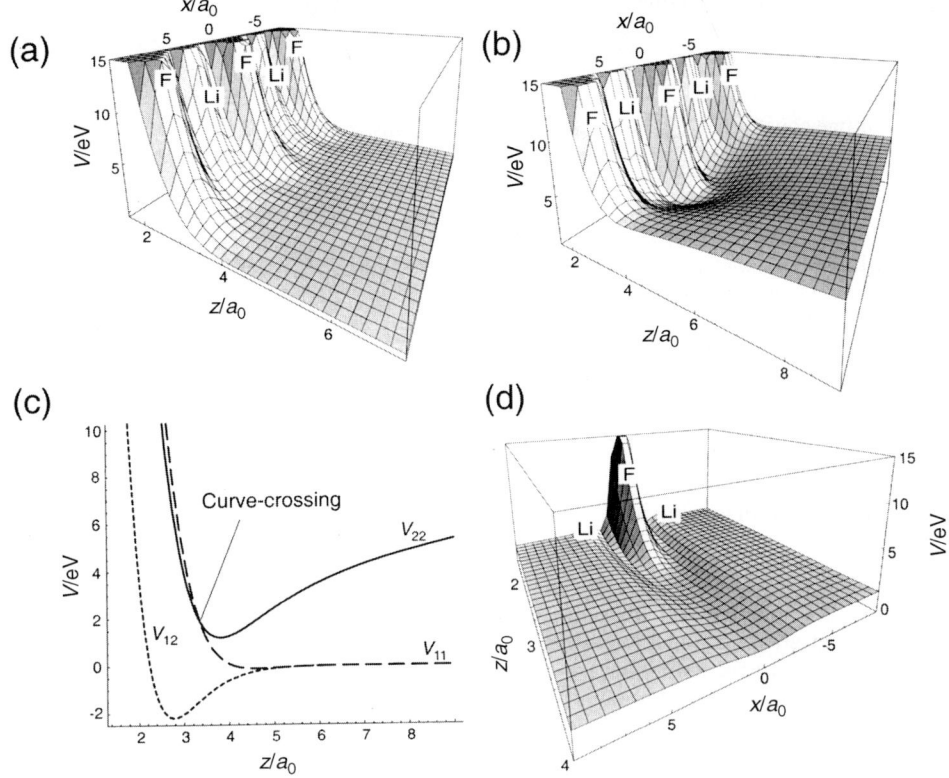

Fig. 2 The SEVB PESs for the F/LiF(100) system. (a) The ground-state PES representing the interaction of neutral surface with neutral atom. The cut shown runs along the tops of the surface atoms in the ⟨010⟩ direction. The spacing between alkali-metal and halide sites is 3.81 a_0. (b) The negative-ion state PES representing the interaction of an F$^-$ ion with an LiF surface with one neutralized F site (located at the origin). The coordinates are the same as for panel (a). (c) The PESs as a function of distance above the site of charge transfer. (d) The off-diagonal coupling, i.e. V_{12}, between the ground and NIS states.

where g_1 and g_2 are reciprocal lattice vectors, $g = |g_1 + g_2|$, k sums over even planes only, a_s is the area of the unit cell, x and y are coordinates in the surface plane and z_k is the vertical distance to the kth plane, and

$$C_1(g_1, g_2, x, y) = \cos(g_1 x)\cos(g_2 y)\{1 + (-1)^{G_1+G_2}\}2^{2-\delta_{0,G_1}-\delta_{0,G_2}} \tag{23}$$

with $g_1 = 2\pi G_1/a_x$ where a_x is the lattice constant. For the odd planes, we obtain an identical expression, but with C_1 replaced by

$$C_2(g_1, g_2, x, y) = \cos(g_1 x)\cos(g_2 y)\{(-1)^{G_1} + (-1)^{G_2}\}2^{2-\delta_{0,G_1}-\delta_{0,G_2}}, \tag{24}$$

and for the summations over M-sites we simply exchange C_1 and C_2. Similarly we can get

$$V_{pc} = \frac{32\pi}{a_s} {\sum_{g_1 g_2}}' \frac{e^{-gz} \cos g_1 x \cos g_2 y}{g(1 + \exp(-g a_x/2))}, \tag{25}$$

where the prime indicates summation over odd reciprocal lattice vectors only. The polarization term, V_{pol} also splits into separate summations for M and X atoms differing only by the symmetry factors C_1 and C_2. For the X sites we get

$$V_{pol}(X) = \frac{-\pi\alpha(X^-)}{2a_s} \sum_{g_1 g_2} \sum_k \left\{ \frac{g}{z_{2k}} K_1(g z_{2k}) C_1 + \frac{g}{z_{2k+1}} K_1(g z_{2k+1}) C_2 \right\}, \tag{26}$$

where K_1 is a modified Bessel function of the second kind. Finally,[25]

$$V_{\text{dip}}(A) = -\frac{1}{2}\alpha(A)\frac{256\pi^2}{a_s^2}\left\{\frac{\exp(-4\pi z\sqrt{2}/a_x)}{(1+\exp-\pi\sqrt{2})^2}\right\}\left\{2+\cos\frac{4\pi x}{a_x}+\cos\frac{4\pi y}{a_x}\right\}, \qquad (27)$$

where the lattice constant, a_x, is the same in both directions, x and y.

The end result of this lengthy, but largely straightforward, manipulation, is shown in Fig. 2 for the F/LiF(100) system. The GS PES, panel (a), is predominantly repulsive, more so at the halide than at the alkali-metal sites, and is perfectly periodic. The NIS PES in contrast, is not periodic because of the $1/R$ attraction from the active site, located at the origin of the coordinates. Moving away from this site, however, the halide and alkali-metal sites are in turn repulsive and attractive to the A$^-$ ion. Asymptotically, V_{22} lies $\Phi - E_{ea}(A)$ above V_{11}. The Coulomb attraction serves to decrease this energy gap. For F/LiF(100), we actually get a curve-crossing between the PESs at $z \sim 3.35\ a_0$, as can be seen in panel (c). This is in contrast to Borisov and Sidis[11] who found no curve-crossing for this system. The influence of the curve-crossing on the subsequent ion formation probability remains to be fully ascertained. As can be seen from panel (d), the off-diagonal matrix element is only significant at the active site.

III. Wavepacket propagation

To follow the quantum motion of the projectile atom, we have used time-dependent wavepacket methods with fast Fourier transforms (FFTs) to switch between real and momentum space to apply, in turn, the potential and kinetic energy operators. The use of FFTs imposes restrictions on the finite element grids on which the nuclear wavefunction and operators are represented. Firstly, the grid is periodic. In the z-direction, we must remove the wavepacket from the end of the grid, which we do by grid-cutting.[26] Parallel to the surface, we must ensure that the negative ion cannot return and re-interact with the active site, and become a neutral again, so the grid must be sufficiently long that when the ion returns, it does so at large z where V_{12} is negligible. Secondly, however, the lengths of the momentum and real space grids are linked since the maximum momentum, $K_{\max} = N\pi/L$, where L is the length of the real-space grid, and N the number of grid points. At the grazing incidence conditions of the experiment, the momentum parallel to the surface is very high. To accommodate large L with large $K_{\max}$ we employ the shift-theorem of Fourier transforms to centre the momentum at a high value, K_0:[27] before the FFT to momentum space, we multiply the wavefunction by the shift factor $e^{-iK_0 x}$ and after determining the kinetic energy and FFTing back to real space, we shift back with the factor $e^{+iK_0 x}$. We also reduce the momentum-space grid by employing a projectile mass of 1 u rather than 19 u. As shown below, this makes little difference to the results.

As noted above, to an incoming particle, all surface sites appear equivalent; ion formation can occur at any. The GS PES and wavefunction are therefore periodic. The NIS PES is not periodic because a charge defect is left at the active halide site, this site thus appears different from the remainder of the surface (Fig. 2(b)). A separate NIS PES is required for each site, therefore a separate NIS wavefunction is also required for every halide site. As for the PES, however, the NIS wavefunctions are identical up to a shift of a lattice vector parallel to the surface. Consider a halide site, labelled 2, centred at $x = 0$; the equation-of-motion for the NIS state, ψ_2, is

$$i\frac{\partial \psi_2(x,y,z,t)}{\partial t} = (\hat{K}_n + V_{22}(x,y,z))\psi_2(x,y,z,t) + V_{12}(x,y,z)\psi_1(x,y,z,t). \qquad (28)$$

Similarly, if ψ_3 represents the excitation at a halide site one lattice constant away, it obeys the same equation-of-motion with the label 3 replacing 2. But, $\psi_3(x,y,z,t) = \psi_2(x - a_x, y, z, t)$, $V_{33}(x, y, z) = V_{22}(x - a_x, y, z)$ and $V_{13}(x, y, z) = V_{12}(x - a_x, y, z)$. As the GS wavefunction, ψ_1, is always periodic, so the equation-of-motion for ψ_3 becomes

$$i\frac{\partial \psi_2(x - a_x, y, z, t)}{\partial t} = (\hat{K}_n + V_{22}(x - a_x, y, z))\psi_2(x - a_x, y, z, t)$$

$$+ V_{12}(x - a_x, y, z)\psi_1(x - a_x, y, z, t), \qquad (29)$$

i.e., the same as eqn. (28) but with a coordinate shift. The equation-of-motion for the GS

$$i\frac{\partial \psi_1(x, y, z, t)}{\partial t} = (\hat{K}_n + V_{11}(x, y, z))\psi_1(x, y, z, t) + V_{12}(x, y, z)\psi_2(x, y, z, t)$$
$$+ V_{13}(x, y, z)\psi_3(x, y, z, t) + V_{14}(x, y, z)\psi_4(x, y, z, t) + \cdots \quad (30)$$

becomes

$$i\frac{\partial \psi_1(x, y, z, t)}{\partial t} = (\hat{K}_n + V_{11}(x, y, z))\psi_1(x, y, z, t) + V_{12}(x, y, z)\psi_2(x, y, z, t)$$
$$+ V_{12}(x - a_x, y, z)\psi_2(x - a_x, y, z, t)$$
$$+ V_{12}(x - 2a_x, y, z)\psi_2(x - 2a_x, y, z, t) + \cdots \quad (31)$$

Clearly then, ψ_2 and V_{22} can perform the role of all of the excited states if we simply add the off-diagonal contribution (shifted appropriately in x) to ψ_1 many times over.

The explicit time-dependence of the wavefunction is solved using the Chebychev method[28]

$$\Psi(t + \Delta t) = e^{-iH\Delta t}\Psi(t) = \sum_n a_n(t)T_n(\tilde{H})\Psi(t). \quad (32)$$

The Chebychev polynomials, T_n, calculated recursively, are functions of the Hamiltonian, renormalized to have eigenfunctions in the range $[-1, 1]$. Care must be taken when renormalizing H because although only two NISs are stored, there are effectively very many (>50) and so H has a much larger spectral range than might be expected. In common with all other Chebychev implementations, we limit the eigenvalues of H by imposing a cut-off on the potential (values of 50–100 eV seem adequate), and also by saturating the parallel momentum contribution to the kinetic energy operator when it is >50 eV from $K_0^2/2M$.[29]

A general approximation in models of grazing incidence scattering from surfaces is to limit the dimensionality in that the projectile can only move in a plane perpendicular to the surface, parallel to the incident beam direction, *i.e.* the calculations are performed on a two-dimensional slice corresponding to the PESs shown in Fig. 2. The initial wavefunction is taken to be the product of a plane-wave state in x (running in the $\langle 010 \rangle$ direction) with a Gaussian-weighted plane-wave in z. Grid lengths of 240 × 240 points are used at the lower energies, increasing to 480 × 240 at higher energies.

IV. Results and discussion

A. The threshold region

Fig. 3 shows a snapshot of the wavepacket on the two surfaces part way through a scattering event, for an initial translational energy of 4 keV at an incidence angle of 1° relative to the surface plane (the "normal energy", $\varepsilon_i \cos^2 \theta = 1.22$ eV in this case). The ground-state wavefunction, shown in Fig. 3(a), is periodic in x (actually it appears structureless at these incidence conditions). On the GS, it never quite reaches the crossing point (which becomes a line, or seam, in two dimensions) as it has insufficient normal energy (the crossing is at an energy of 1.98 eV at the active site). Instead, it can be seen to transfer efficiently onto the NIS PES, as in Fig. 3(b), which shows the NIS for the halide site at $x = 0$. The transfer appears to be very strongly localised on the active site, the wavefunction streaming rapidly away from a region just before the highest point of the crossing seam, *i.e.* from the classical turning point on the GS PES. Transforming the wavefunctions and PESs to an adiabatic representation gives a clearer picture of the local nature of the charge exchange. We can see from the lower two frames of Fig. 3 that the wavepacket jumps diabatically from the lower adiabatic state (Fig. 3(c)) to the upper adiabatic state (Fig. 3(d)) in the turning region (smeared out in quantum mechanics because of the tunneling into the repulsive wall). (Note that the adiabatic transformation is strictly only valid in the neighbourhood of $x = 0$ because we have not transformed with the full PES and wavefunction, having NIS elements at every halide site, but only with the NIS state at the origin.)

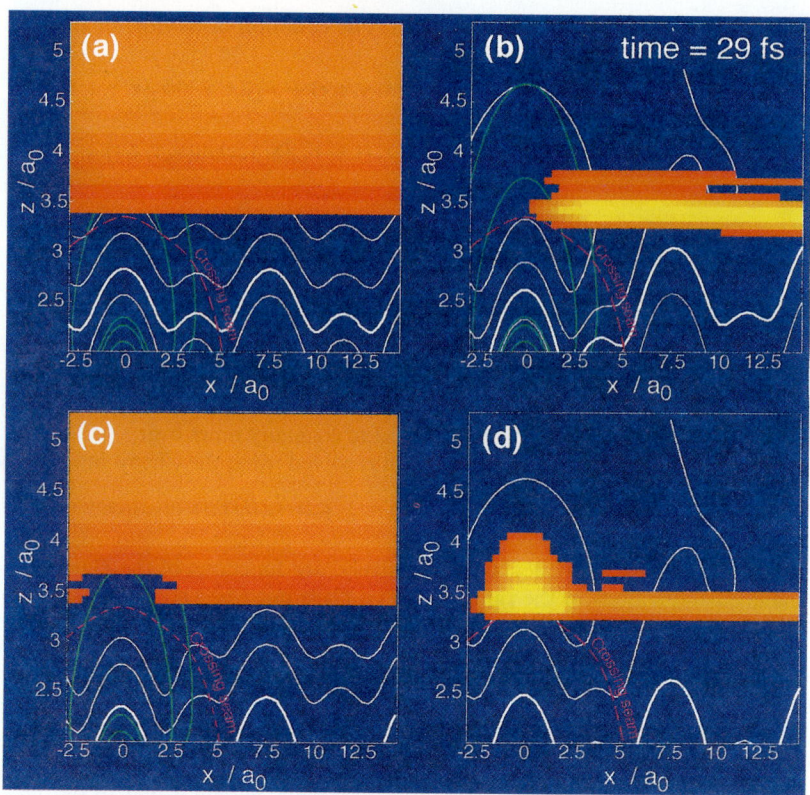

Fig. 3 (a) The wavefunction on the ground-state PES incident at 4 keV at an angle of 1° to the surface plane, with normal energy of 1.22 eV. The white lines show a contour plot of the PES, the green lines are a contour plot of the V_{12} matrix element, and the purple line shows the locus of points where V_{11} and V_{22} cross. The wavepacket intensity decreases from white, though yellow to red, with the very lowest values removed for clarity. (b) The wavefunction on the NIS located on the halide site at the origin. The white lines now show a contour plot of V_{22}. The transfer between states clearly occurs on the active site, just before the curve crossing. (c) and (d) The wavepacket from (a) and (b) transformed into an adiabatic representation, with corresponding adiabatic PES shown as white contour lines. The charge transfer occurs diabatically at the active site, as evidenced by the sudden jump from (c) to (d) at $x = 0\ a_0, z \sim 3.5\ a_0$.

Following the wavepacket in time, we find that the intensity on the GS gradually disappears, transferring onto the NISs. In other words, the incoming atoms convert, almost entirely, to negative ions, as can can be seen in Fig. 4, which shows the integrated intensities on the GS and NISs (summed over all halide sites) for the same incidence conditions as Fig. 3. Quite clearly, the intensity on the NIS increases smoothly at the expense of that on the GS, *i.e.*, *the probability of forming a negative ion in the initial collision, P_i, is approximately 1.* We have found this to be so across the entire energy range considered here (1 eV–15 keV) for both the F/LiF(100) and F/KI(100) systems. The intensity does not return to the GS at the active site because the newly formed negative ions move very rapidly away from there in the *x*-direction, passing out the side of the region of strong coupling.

To follow the evolution of the NIS wavefunction with minimum disturbance from periodic return to the active site, we stop the full propagation once the transfer to the NIS state is complete, and then propagate the wavefunction on this state alone, with the origin displaced far from the active site. Long propagation times are then possible, as can be seen in Fig. 4. The intensity on the NIS state is stable up until ~100 fs when there is a slight decline. This is due to the grid-cutting in the *z*-direction which removes wavepacket intensity before the end of the grid. The remainder of the wavepacket fails to reach the end of the grid. It has failed to acquire the full 8.58

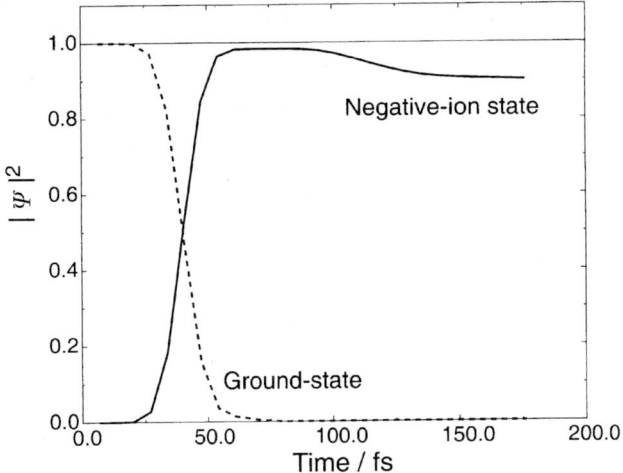

Fig. 4 The total normalization of the wavepackets on the ground and negative-ion states *vs.* simulation time for an initial energy of 4 keV incident 1° from the surface plane. After ∼25 fs, the incoming neutral atoms gradually convert, almost entirely, to negative ions. After ∼100 fs, some ions have escaped from the surface (the wavepacket is cut from the edge of the grid causing the decrease in normalization), but almost 90% remain trapped.

eV of normal energy required to escape the surface and remains trapped in the weak attraction of the polarization and point-charge contributions to the PES. In other words, for these incidence conditions, only ∼10% of the incident atoms escape the surface as negative ions *in the first bounce*, although there is initially 100% conversion to negative ions.

Clearly, the initial ion-formation probability, P_i, is a poor approximation to the experimental results for the F/LiF(100) system, lacking as it does any threshold. However, we do obtain a threshold if we count only those negative ions escaping the surface in the first bounce. This effective negative ion fraction, P_N, is shown in Fig. 5, top panel, for the F/LiF(100) system. The threshold we obtain is very close to that in experiment, and shifts to lower energy as the angle of incidence is increased because there is then more normal energy to aid escape from the surface. It can also be seen from Fig. 5, top panel, that for this system, averaging the results over several impact planes, *i.e.* averaging over the *y*-coordinate not explicitly treated in the dynamics, has little effect on the results.

The normal energy corresponding to the threshold region is far below that required by the NIS asymptotically; even at 14 keV, the normal energy at 1° incidence angle is only 4.36 eV, compared to the 8.58 eV separation of V_{11} and V_{22} in the F/LiF system. The extra energy comes from the very high parallel motion, in quite straightforward fashion. As can be seen in Fig. 3, the wavepacket on the upper state leaves the active site travelling predominantly in the *x*-direction. When it reaches $x = 7.62\ a_0$, it encounters the next halide site along from the active site, and experiences a strong Coulomb repulsion of like charges. In other words, it moves uphill away from the active site, using up parallel momentum to do so. Some of the potential energy gained in doing this (if we assume the negative ion moves horizontally away from the curve-crossing it encounters a potential of ∼6 eV at the next halide site along) is channelled into normal motion, thereby reducing the normal energy escape threshold. The ultimate fate of the trapped ions is difficult to determine in the present model. We assume that the energy is much too high for permanent trapping and that on return to the surface, the ions re-neutralize (which costs more energy) and return to the gas phase after the second bounce. This remains to be fully tested.

The escape probability of the negative ion can also be enhanced by reducing the energy required asymptotically for the NIS. This can be done by changing to a surface with a lower workfunction, such as the KI(100) surface. As shown in the lower panel of Fig. 5, the threshold in P_N for F/KI(100) occurs at much lower incident energies, in agreement with experiment.[4] Once again, this threshold arises solely from the trapping behaviour, P_i is always close to unity. In the present model, the shift in threshold is due solely to the smaller workfunction—the PESs for

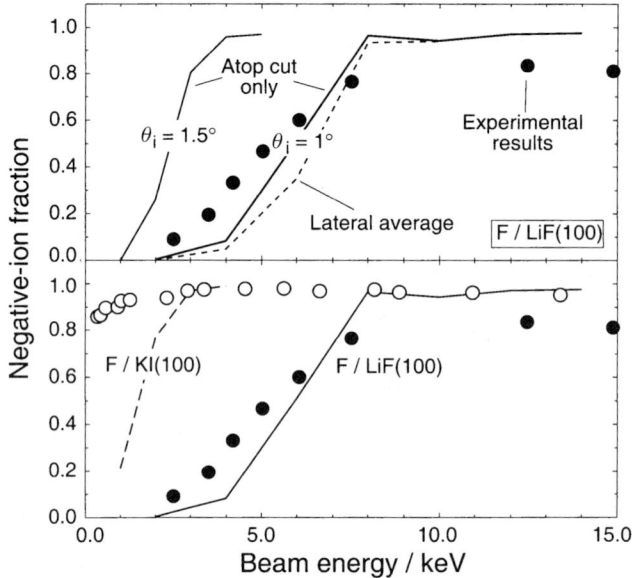

Fig. 5 Top panel: Negative-ion fraction *vs.* incident energy obtained by counting only the promptly scattered ions (i.e. excluding the trapped fraction) for the F/LiF(100) system. In all cases, the initial ion formation probability is high. The atop cut runs across the surface atoms in the ⟨010⟩ direction (*cf.* Fig. 2). Motion is restricted to a plane. Averaging over several such planes in the ⟨010⟩ direction clearly makes little difference to the results. The threshold energy decreases with initial angle because the increase in normal energy makes escape more likely. Experimental results are shown as filled circles. Bottom panel: Negative-ion fraction *vs.* incident energy for F scattering from LiF(100) and KI(100) at an incident angle of 1°. Trapped ions are not counted, and results are for the atop slice only. Experimental results are indicated by filled/open circles for LiF/KI.

F/LiF and F/KI do not differ greatly, except that for the latter system, there is no translational threshold to reach the curve-crossing point. In general, from this trapping model, we predict that the lowest energy *thresholds* will occur for atoms with high electron affinities incident on surfaces with low workfunctions, as this yields the smallest asymptotic separation between ground and excited states.

B. Re-neutralization at the surface

The final yield of negative ions will also be determined by the re-neutralization (RN) probability. There are two possible mechanisms for this: charge transfer into the conduction band of the solid, and electron emission into the gas phase.[30] To explore the influence of RN on the threshold behaviour, we have included the former process, charge transfer from the projectile back to the surface, although the latter can also be expected to be important, especially for LiF, which has a negative electron affinity. We incorporate RN into the present model by adding another diabatic state to represent electron transfer into the conduction band of the solid. Again, we assume that this is a purely local event, occurring this time into an atomic-like orbital on one of the alkali-metal sites. The extra diagonal and off-diagonal PES matrix elements are generated using the electron wavefunction

$$\chi_{RN} = \det\left\{\left[\prod_{k=1}^{N-1} X_k \bar{X}_k\right] X_N M_L \bar{A}\right\}. \tag{33}$$

The diagonal PES for RN is basically identical to the GS PES, but shifted up by the band-gap energy, E_{gap} of the crystal

$$V_{RN} = E_{gap} + V_{11} - \left(\frac{1}{R_{X_N M_P}} - \frac{2}{a_x}\right), \tag{34}$$

where the last term accounts for the reduction in the Madelung energy cost when the ion-formation and re-neutralization sites are near one another.

The NIS wavefunction is not periodic, so the alkali-metal sites are not all equivalent in this process. To approximate the effect of interacting with many alkali-metal sites, we compute the RN for each in a separate calculation, assuming it is the only RN site and use recursion to compute the net survival probability, S, of the negative-ion

$$S = \lim_{n \to \infty} [1 - P_{RN}(n)]S(n), \qquad (35)$$

where $P_{RN}(n)$ is the probability of RN at site n, and $S(n)$ is the probability that the negative ion has survived re-neutralization at all sites before n. The implicit assumption here is that the fraction not re-neutralizing is unaffected by V_{RN}.

Since another charge defect is created in the surface, RN requires additional energy to be taken from the projectile motion, *i.e.* the PES lies above the NIS PES at large z. There is no crossing between the NIS and RN PESs, so transfer is by a near-resonant Demkov-type process.[31] At the lowest energies, there is simply not enough normal energy, RN does not occur and the threshold for ion detection is the same as Fig. 5. At the highest energies we have considered, as many as 20 alkali-metal sites had to be included to get convergence in the P_N. The net results are shown in Fig. 6. For F/LiF(100), the results appear reasonable however P_N is decreasing much too rapidly at high energy. Partly this is due to the use of an H atom mass rather than an F atom mass. This is entirely consistent with RN being a Demkov process. We can write the diabatic transfer probability as

$$P = \tfrac{1}{2}\mathrm{sech}^2\left(\frac{\pi \Delta E}{2\gamma v}\right), \qquad (36)$$

where v is the velocity of the projectile. Increasing the mass simply decreases v, which in turn leads to smaller probabilities for curve-hopping. This mass dependence is not enough to correct the P_N completely. We can see that for F/KI(100), the discrepancy is very large and occurs at much lower energies, because the PESs are much closer in energy.

What is clear from this example of a possible RN path is that we must consider more than three PESs to describe the net effect of the atom–surface reaction. Especially for the F/KI system, with its lower workfunction and band-gap, we can expect the ion to be formed and re-neutralized many times over at higher energies. Each time this occurs, there will be some energy cost to be paid from

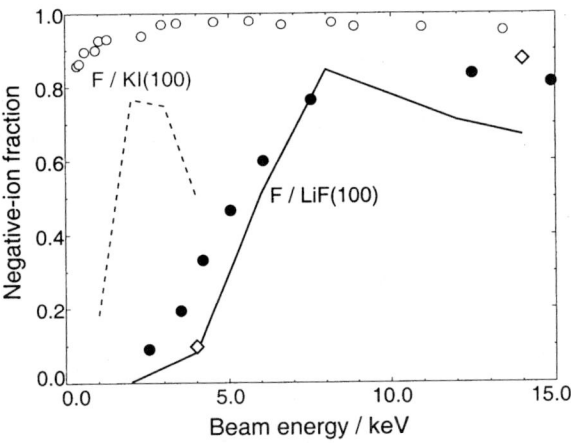

Fig. 6 Net negative-ion fraction *vs.* incident beam energy when re-neutralization by charge transfer into the conduction band is included. The lines represent theoretical results using a mass of 1 u, filled (open) circles represent experimental results for LiF(100) (KI(100)) and the open diamonds represent theoretical results for a projectile mass of 4 u.

the translational motion. At the very highest energies, it is impossible to say whether the atom will end-up as a negative or a neutral. Most likely then the decline of P_N observed in experimental results between ~20 and 100 keV is due to a decrease in probability of the initial ion formation stage. This remains to be tested by further calculation.

V. Conclusions

We have presented a diabatic description of the charge exchange reaction occuring between atoms and alkali-metal surfaces, deriving PESs with a flexible and consistent semi-empirical scheme and solving the atom motion with exact wavepacket dynamics. For the F/LiF(100) system, we find a curve-crossing between the PESs for (neutral atom, neutral surface) and (negative ion, positive surface) configurations. There is efficient charge transfer, in fact the initial ion-formation probability is unity. At lowest energies, these ions fail to escape from the surface promptly. If we count only the fraction which do escape on the first bounce, we obtain an ion detection probability behaving very similarly to that of experiment. Increasing the angle of incidence moves the threshold to lower energy, and the threshold energy is lower for a surface with a lower workfunction, as we have demonstrated for the F/KI(100) system, in agreement with experiment. In future work, we shall examine the importance of the curve-crossing to the dynamics, and try to establish the fate of the trapped negative ions.

Acknowledgements

This work was supported in part by a British Council—Israel Ministry of Science Fellowship for GRD.

References

1 R. Brako and D. M. Newns, *Rep. Prog. Phys.*, 1989, **52**, 655.
2 J. Los and J. J. C. Geerlings, *Phys. Rep.*, 1990, **190**, 133.
3 C. Auth, A. G. Borisov and H. Winter, *Phys. Rev. Lett.*, 1995, **75**, 2292.
4 H. Winter, A. Mertens, C. Auth and A. G. Borisov, *Phys. Rev. A*, 1996, **54**, 2486.
5 H. Winter, C. Auth and A. G. Borisov, *Nucl. Instrum. Methods Phys. Res. B*, 1996, **115**, 133.
6 C. Auth, A. Mertens, H. Winter, A. G. Borisov and V. Sidis, *Phys. Rev. A*, 1998, **57**, 351.
7 S. Ustaze, R. Verucchi, S. Lacombe, L. Guillemot and V. A. Esaulov, *Phys. Rev. Lett.*, 1997, **79**, 3526.
8 S. Ustaze, L. Guillemot, R. Verucchi, S. Lacombe and V. A. Esaulov, *Nucl. Instrum. Methods. Phys. Res. B*, 1998, **135**, 319.
9 A. Mertens, C. Auth, H. Winter and A. G. Borisov, *Phys. Rev. A*, 1997, **55**, R846.
10 A. G. Borisov, V. Sidis and H. Winter, *Phys. Rev. Lett.*, 1996, **77**, 1893.
11 A. G. Borisov and V. Sidis, *Phys. Rev. B*, 1997, **56**, 10628.
12 V. Sidis, *Adv. Chem. Phys.*, 1992, **82**, 135.
13 J. M. Adams, S. Evans and J. M. Thomas, *J. Phys. C*, 1973, **6**, L382.
14 A. B. Kunz, *Phys. Rev. B*, 1982, **26**, 2056.
15 G. K. Wertheim, J. E. Rowe, D. N. E. Buchanan and P. H. Citrin, *Phys. Rev. B*, 1995, **51**, 13675.
16 Y. Zeiri and M. Shapiro, *Chem. Phys.*, 1978, **31**, 217.
17 Y. Zeiri, M. Shapiro and R. Tenne, *Chem. Phys. Lett.*, 1983, **99**, 11.
18 M. Shapiro and Y. Zeiri, *J. Chem. Phys.*, 1986, **85**, 6449.
19 Y. Zeiri, R. Tenne and M. Shapiro, *J. Chem. Phys.*, 1984, **80**, 5283.
20 F. O. Ellison, *J. Am. Chem. Soc.*, 1963, **85**, 3540.
21 H. A. Pohl, R. Rein and K. Appel, *J. Chem. Phys.*, 1964, **41**, 3385.
22 G. R. Darling, R. Kosloff and Y. Zeiri, in preparation.
23 R. Pariser, *J. Chem. Phys.*, 1953, **21**, 568.
24 R. S. Mulliken, C. A. Rieke, D. Orloff and H. Orloff, *J. Chem. Phys.*, 1949, **17**, 1248.
25 W. A. Steele, *The Interaction of Gases with Solid Surfaces*, Pergamon, New York, 1974, ch. 2.
26 R. Heather and H. Metiu, *J. Chem. Phys.*, 1994, **86**, 5009.
27 G. J. Kroes and R. C. Mowrey, *J. Chem. Phys.*, 1994, **101**, 805.
28 H. Tal-Ezer and R. Kosloff, *J. Chem. Phys.*, 1984, **81**, 3967.
29 R. Kosloff, in *Numerical Grid Methods and Their Application to the Schrödinger Equation*, ed. C. Cerjan, Kluwer, Amsterdam, 1997, p. 175.
30 J.-P. Gauyacq, personal communication.
31 Yu. N. Demkov, *Sov. Phys. JETP*, 1964, **18**, 138.

General Discussion

Prof. Kleyn opened the discussion of Dr Gauyacq's and Prof. Winter's papers: Experiments and theory are reported concerning lifetimes or level width of excited states at surfaces. The theory behind the explanation for two technically very different experiments, ion scattering and pump–probe, is the same. The experiments so far have been done for different systems. Are there results for the same system by the two methods?

Prof. Winter responded: At present there are no data available for comparable systems. However, it is planned to investigate *e.g.* Cs on Cu(111) and obtain from image charge effects on the angular distributions the electronic transition rates as a function of distance from the surface plane. For the adsorption position these rates can be directly compared with the lifetimes obtained from fs-pump–probe studies.

Dr Gauyacq added: The theory behind the two different experiments (adsorbate lifetime measurement by TR-2PPE and charge state of reflected particles in grazing angle scattering) is the same. However, one can stress that the two experiments are looking at two different aspects of the process. First, by TR-2PPE, one looks at the adsorbate at the equilibrium distance close to the surface, whereas the charge state of grazingly scattered particles is mainly sensitive to what happens at large projectile–surface distances. Secondly, the TR-2PPE experiment directly probes the excited state lifetime; in contrast, grazing angle scattering probes the balance between capture and loss processes for the projectile during the collision and due to the sensitivity of these processes to the dimensionality of the electron wavefunction, it probes the relative role of the 2D surface state and 3D bulk states. Nevertheless, although the two aspects probed by the two experiments are different, they are both induced by the same feature, the projected band gap. The situation will be different with the new scattering experiment mentioned by Prof. Winter which directly looks at the excited state lifetime.

Prof. Persson asked: Why does your approximation of an unperturbed core in your calculations influence the resonance width of the $2\pi^*$ of CO on Cu?

Dr Gauyacq replied: In our model treatment of the CO/Cu system, we use an e^-–CO interaction extracted from an *ab initio* study of electron scattering by a free CO molecule. Assuming that this interaction is the same in the adsorbed system neglects the CO transformation due to its weak chemisorption on the Cu surface. This approximation can quantitatively influence the results on the energy and lifetime of the $CO(2\pi^*)$ resonance on Cu. However, the main aim of this model study was to investigate whether the $CO(2\pi^*)$ resonance located in front of the Cu band gap would lead to a long-lived state such as Cs (see paper). The absence of a long-lived state and the corresponding interpretation are clear enough and this qualitative result should not be influenced by the approximation made in describing CO. As for the quantitative aspects, we can nevertheless notice that the width we found for CO/Cu(111) is very close to the one you and Lorente found,[1] the energies being different.

1 N. Lorente and M. Persson, *Faraday Discuss.*, 2000, **117**, 277.

Prof. Gadzuk commented: In past days of tunnelling spectroscopy, when some feature in an energy distribution or current–voltage curve was thought to be due to a surface state or surface resonance, it was submitted to the "crud test": adsorb anything on the surface and if the feature disappeared, then it was declared to have been due to a surface state. The impurity perturbed the surface state eigenvalue, pushing it up or down towards one or the other of the gap edges. However, as the surface state energy approached a gap edge, the surface state wavefunction became more delocalized into the bulk which, by normalization, necessarily reduced the amount of charge per surface state in the exponential tail sticking out into vacuum which was able to participate in the surface sensitive spectroscopic process; hence the disappearance of its spectroscopic signature.[1] Presumably a similar sensitivity to surface conditions should be observed in your charge transfer scattering processes that involve substrate surface states, for the same reason: the perturbed and diminished exponential tail does not overlap with the atomic states of the external projectile as well. Have you seen any effects of this sort and do you have any comments about this possibility?

1   J. W. Gadzuk, *J. Vac. Sci. Technol.*, 1972, **9**, 591.

Dr Gauyacq answered: We have not looked at this point in detail; however, among other effects, it should have an influence on the coverage dependence of the excited state lifetime. A similar effect could also play a role for the $e^- - e^-$ interaction terms: when the excited state energy is changing in the band gap, the excited state wavepacket penetration into the bulk changes which should be reflected in the decay rate by inelastic $e^- - e^-$ collisions in the bulk. As for the Berlin charge transfer scattering studies, they were performed with clean Cu surfaces, so the surface conditions were always the same. Scattering studies on metal surfaces partially covered with adsorbates exhibit strong variations with the surface conditions (see *e.g.* a review in ref. 1); however, the problem is somewhat different from the one addressed in the papers presented here, since it corresponds to studying the effect of a projectile approaching an adsorbate.

1   J.-P. Gauyacq and A. G. Borisov, *J. Phys. Condens. Matter*, 1998, **10**, 6585.

Dr Shluger said: We are discussing the atom scattering from surfaces in terms of one atom interacting with an infinite surface. However, there are many atoms interacting with the surface at the same time. How important is the interaction between these atoms both in the initial and final state, *i.e.* after trapping of an electron at excitation? What are the limitations of the "one atom" model?

Dr Gauyacq responded: Our study has been performed for a single adsorbate in front of the surface. This is perfectly valid in a collision context, but changes of the alkali coverage could influence the results on adsorbate induced excited states. Two aspects could limit the validity of this approach. First, in the case of low coverage, the adsorbates are far away one from the other and one can assume that the various adsorbates only interact *via* long range electrostatic interactions. As a rule of thumb, this should correspond to the region where the surface work function varies roughly linearly with the adsorbate coverage. We performed a model study of this problem,[1] in which we analysed how the long range dipolar potential of the alkali adsorbates is perturbing the energy and lifetime of the excited state. The energy varies with the alkali coverage in excellent agreement with the results of Bauer *et al.*[2] for surface work function changes below 1.5 eV. As for the level width, we found a rather limited variation for the resonant charge transfer rate in the low coverage region, in particular the blocking of the charge transfer by the projected band gap was always present. The equivalent study of the coverage dependence of the $e^- - e^-$ interaction has not been performed. Secondly, for large alkali coverage, lateral interactions become more important and the layer depolarises. In this range, the modelling we have been using, with a single positive alkali ion core is not appropriate anymore. This second aspect is the real limitation of our present approach.

1   A. G. Borisov, A. K. Kazansky and J.-P. Gauyacq, *Surf. Sci.*, 1999, **430**, 165.
2   M. Bauer, S. Pawlik and M. Aeschlimann, *Phys. Rev. B*, 1997, **55**, 10040.

Prof. Aeschlimann commented: Increasing the Cs coverage on the Cu(111) surface will drastically affect the electron dynamics of the probed system. I assume that after the photo-induced excitation of a substrate electron into the excited Cs state, the Cs will become a transient neutral state at very low coverage, and a negatively charged ion at higher coverage. In other words, we may move from a GMR scenario to an Antoniewiez scenario.

My question is whether, in the wavepacket propagation calculation of Gauyacq et al.,[1] this drastic change of the potential energy surface as a function of Cs coverage can be included or not?

1   A. G. Borisov, A. K. Kazansky and J.-P. Gauyacq, *Surf. Sci.*, 1999, **430**, 165.

Dr Gauyacq replied: This question is linked to the previous one by Dr Shluger. At high coverage, the structure of the Cs adsorbate layer is very different from that at low coverage, so that the photo-excitation processes should be different: excitation in the Cs adsorbate or excitation of a substrate electron into a Cs excited state would then be possible. On the theoretical side, the wavepacket propagation approach is a one electron approach that relies on a potential to describe the electron time evolution, the choice of the potential is a direct consequence of the structure of the system. The case of high coverage then requires a modelling different from the one used in the present low coverage study: provided this modelling is available, it can then be included in the wavepacket propagation study.

Dr Dujardin asked: Concerning the blocking of charge transfer on metal surfaces, I am wondering whether people have investigated the role of temperature? The reason why I am asking this, is that on real band gap materials such as semiconductors when blocking of charge transfer also exists, strong temperature effects have been observed.

Dr Gauyacq answered: A very clear temperature dependence of the excited state characteristics has been observed in the case of Cs/Cu(111) by Ogawa et al.[1] The situation should nevertheless be different from semiconductors, since Cu is a conductor and only exhibits projected band gaps. In Cs/Cu(111), the transient state lifetime is found by Ogawa et al.[1] to increase when the temperature is decreased. Their interpretation invokes the possible existence of different adsorbate sites on the surface that would be populated differently at different temperatures and changes of the relative Cs–Cu distance with the temperature.

1   S. Ogawa, H. Nagano and H. Petek, *Phys. Rev. Lett.*, 1999, **82**, 1931.

Prof. Wolf commented: Let me add that very pronounced temperature effects arise for image potential states on Cu(111) due to the changes of the energetic position of the band gap as a function of temperature. On this surface the ($n = 1$) image state is located close to the upper edge of the gap. With increasing temperature the band gap shrinks, the ($n = 1$) state shifts closer to the band edge and hence the penetration of the image state wavefunction increases. The increased coupling to electron–hole pair excitations in the Cu bulk leads to a decrease of the inelastic lifetime.

Dr Gauyacq replied: This is another possibility for a temperature effect in the case of Cs/Cu(111), the variation of the substrate band structure with temperature should also influence the adsorbate induced state characteristics.

Dr Saalfrank said: I would like to ask Dr Gauyacq a general question. In his theoretical treatment of RCT and electron dynamics in general, a single "active" electron is considered. In this approach the antisymmetry of the many electron wavefunction is not accounted for, *i.e.* the Pauli principle for indistinguishable particles is not taken care of. The question is: do exchange interactions in any way affect the electron dynamics, *e.g.* the computed negative ion resonance lifetime?

Dr Gauyacq responded: The approach we are using relies on a model (or pseudo) potential description of the various interactions, allowing one to reduce the problem to a single electron interacting with the rest of the system *via* an effective Hamiltonian which takes the exchange interaction into account in an effective way. One can stress that such an approach that considers one electron and freezes the rest of the system has a projection character; it allows for the study of excited states as well as for an easy study of the charge transfer process, since, for example, the charge state of an adsorbate is well defined. The situation is different if the problem is not really a one electron problem. This happens, for example, for atoms (ions) with a few equivalent electrons interacting with a surface or with charge transfer processes in the case of degenerate states. Modifications of the one electron picture approximation can handle these for the projectile (adsorbate) side, but not for the substrate.

Dr Shluger asked: Is there any direct theoretical (*e.g.* from band structure calculations) or experimental evidence that the projected band gap at the surface and in the bulk exists in the presence of projectile species on the surface? In other words, how do the projectile species at real concentrations affect the electronic structure of the Cu substrate?

Dr Gauyacq replied: The bulk band structure should not be modified by the presence of adsorbates; in particular, the projected band gap, *i.e.* the fact that electrons cannot propagate in certain directions at certain energies in the bulk is linked to the crystal periodicity and should always exist. However, the electronic structure of Cu in the surface region is locally modified by the presence of an adsorbate; this is what our calculation is looking at (see my earlier answer to Dr Shluger for the case of a finite coverage). The local surface properties, for example, the image state or surface state characteristics, are modified by the presence of adsorbates (see *e.g.* all the experimental studies of image states on rare gas covered Cu surfaces), but not the bulk properties, such as the band gap.

Prof. Aeschlimann commented: The electronic band structure as shown in Fig. 1 of your paper is based on Block states. My question is whether we can really use this "static" band structure picture in the case of an electron transfer which happens within a few fs. I assume that after the electron transfer from the electronically excited Cs atom into the substrate, the electron will first build up a wavepacket superposition of Block states, which will at first be strongly localized in the surface region. For this wavepacket, k is not a good quantum number anymore and hence, the "static" band structure picture may be meaningless.

Dr Gauyacq answered: Yes, in principle, the "static" band structure can only be defined on an infinite time scale and one could observe some deviations on short time scales. We observed such deviations when we looked at the decay of H^- ions in front of Cu(111) surfaces.[1] Two decay regimes were observed; at short time, the H^- ion decays as in front of a free electron metal, whereas at large times, it decays with a much smaller decay rate due to the characteristics of the Cu(111) surface. This late decay rate is the one we defined as the decay rate in front of Cu(111). The change of decay regime corresponds to the point raised by your question. In the initial stage of the decay (the initial state is an unperturbed free H^- ion), the electron wavepacket does not overlap with the oscillating potential inside bulk Cu. Tunnelling then occurs from the ion to the metal as on a free electron metal surface. It is only after the wavepacket has spread inside the metal and has been partially reflected by the successive (111) planes in the bulk, that it fully experiences the Cu band structure and in particular the projected band gap. Then after an intermediate period of readjustment of the wavepacket to accommodate the band gap, the ion decay occurs with the slow rate. For the H^-/Cu system, the time required for the band gap effect to switch on is of the order of 0.5 fs. Thus, it is possible to observe different decays on different time scales, however, the short time decay should depend on the way the initial state is prepared, whereas the large time decay is governed by the "static" lifetime of the excited state.

1   A. G. Borisov, A. K. Kazansky and J.-P. Gauyacq, *Phys. Rev. B*, 1999, **59**, 10935.

Prof. Matsumoto said: When we think about the bonding at surfaces, it is important to specify the adsorption site, since the absorbate–substrate interactions strongly depend on where an atom or a molecule is adsorbed. In fact, some adsorbate–substrate systems clearly show site dependence in photochemical activities. This can be due to the difference in the excited-state lifetime among the adsorption sites. Thus, the site dependence of the lifetime is very crucial. Could you tell how the lifetime is affected by the adsorption site in your calculations?

Dr Gauyacq responded: In our calculations, the Cu surface is represented by a 1D potential, *i.e.* it includes the crystal periodicity along the surface normal and assumes free electron motion parallel to the surface. Such an approach does not separate different adsorption sites and thus cannot predict how the excited state lifetime is site dependent. However, one can think that the blocking of the resonant charge transfer due to the projected band gap should be present and lead to long lived states in any adsorption site.

Dr Lee commented to Prof. Winter: I noticed that there is a reasonable azimuthal dependence on H^- fraction as a function of velocity (Fig. 5 of your paper) for scattering from the Cu(110) surface, but not from the Cu(111) surface. My question is whether this different azimuthal dependence from each surface is related to the existence of a band gap or something else? Are there any theoretical explanations for this?

Prof. Winter replied: From what we have explored so far, it is likely that the observed azimuthal dependence is related to the band gap. For the (110)-face, the projected band gap is somewhat tilted with respect to the surface normal. To clear up things from theory, the present model with a constant surface potential parallel to the surface has to be expanded to a full 3D description of the metal which is quite a challenging task.

Dr Shluger said: In this session we were comparing two metal surfaces and one ionic surface. Surfaces of ionic crystals, such as LiF, are known to be rough after cleaving and contain many steps. Surfaces of metals are probably easier to prepare. Could you please comment on the role of surface preparation and surface roughness in grazing collision experiments?

Prof. Winter responded: Surface roughness is an essential item for collisions of fast ions under grazing angles, because the projectiles probe a considerable projection of the surface plane. Indeed for LiF cleaving is hardly sufficient to fulfil these requirements. In our case we performed, after cleaving under air, mechanical polishing and prepared the target in UHV by a large number of cycles of grazing angle sputtering with noble gas ions and subsequent annealing. After a typical period of about one month an atomically flat and clean surface is obtained, as can be easily checked by recording very defined angular distributions for scattered projectiles.

Dr Kolasinski opened the discussion of Prof. Gadzuk's paper: (1) Could you discuss the meaning of corrugation in STM images. Does your theory mean that wherever there are two or more electron tunnelling paths, we cannot interpret images in terms of topography.
(2) What is imaged at the other focus in a corral—enhanced electron density or simply enhanced electron tunnelling probability.

Prof. Gadzuk responded: Corrugation, as it appears in STM images, is a measure of the position (in the surface-normal direction) of the classical turning points of the tunnelling electron entering or leaving the surface, exponentially averaged (due to the exponential tunnelling probability) over energies of the participating electron states, and subject to well understood resolution limitations in the transverse direction due to the trade off between $k_\parallel$ needed for transverse resolution and the decrease in tunnelling current due to smaller k_z-dominated tunnelling probabilities which follows when k_z is used up as $k_\parallel$.
Concerning the question of multiple-tunnelling paths and topography, in my talk I presented a simple model for generic STM-tip-position-dependent resonance tunnelling involving atoms adsorbed on (jellium) surfaces that were translationally invariant in the transverse direction. The problem was set up, in the spirit of Fano's theory for discrete state mixing with a continuum, in terms of interfering direct and resonance tunnelling paths. Because the clean surface was translationally invariant in the surface plane (hence no topography!), the direct tip-to-surface tunnelling amplitude was invariant with transverse position, and as a result, my argument could proceed as if there was only a single value for the magnitude of the direct tunnelling path amplitude (which can be obtained in terms of Feynman path integrals over many interfering adjacent paths, but basically given by the action along the single "classical path", suitably reinterpreted for classically forbidden tunnelling). If corrugation hence topography was part of the mix, then not only tunnelling involving adsorbate-related paths, but also direct tip-to-surface tunnelling would depend

upon the transverse position of the tip, showing a periodic variation in the direct tunnelling amplitude as the tip scans across the periodically arranged unit cells. For any given tip position one can still formulate the resonance tunnelling in terms of a direct and an adsorbate-involved amplitude, but now the magnitudes of both amplitudes as well as their phases are transverse-position-dependent.

Finally, Manoharan *et al.*[1] have observed so-called spectroscopic mirages in Kondo systems of magnetic Co atoms adsorbed on non-magnetic Cu(111) surfaces: when the Co atoms are arranged in an elliptic shaped "corral", one atom is placed at a foci of the ellipse, and the STM tip is placed over the other foci. The fact that similar (but much reduced in strength) tunnelling characteristics are seen for the tip directly over the vacant foci as when over the adsorbate-occupied-foci has raised questions as to what is being imaged when at the clean foci? In simple terms what is probably occurring is that the tunnelling electrons arriving at (or departing from) the clean foci, while moving in the plane of the surface, are constructively scattered and refocused by the elliptical corral atoms back to the covered foci, where they then attempt to resonantly tunnel. All this happens coherently. When translated into your terminology, I guess this is equivalent to saying that enhanced tunnelling probability "is imaged at the other focus of the corral".

1 H. C. Manoharan, C. P. Lutz and D. M. Eigler, *Nature*, 2000, **403**, 512.

Mr Al-Halabi commented: In Fano formalism, the excited state of an atomic system can be written as a sum of a discrete eigenstate and continuous eigenstates. What is the importance of the relative phase between the discrete and the continuous state?

Prof. Gadzuk replied: The relative phase between the discrete and the continuum state is in fact the crucial factor in determining the asymmetry of the Fano resonance profile. For instance, the amplitude for interfering resonance processes involving a continuum and a quasi-discrete resonance can be written as:

$$A_l = a_0 \left(1 + \frac{a(\Gamma/2)\exp(i\delta_l)}{(\varepsilon - i\Gamma/2)} \right),$$

where δ_l is the relative phase between them, taken to be slowly varying with energy, and a is a measure of the inherent strength of the resonance compared to continuum processes. The resonance denominator can be expressed as $(\varepsilon - i\Gamma/2) = R \times \exp(i\delta_r(\varepsilon))$ with $R = (\varepsilon^2 + (\Gamma/2)^2)^{1/2}$ and $\delta_r(\varepsilon) = \tan^{-1}(\Gamma/2\varepsilon)$. Thus $|A_l|^2/|a_0|^2 = 1 + a^2(\Gamma/2)^2/R^2 + (a\Gamma/R)\cos(\delta_l - \delta_r(\varepsilon)) \equiv Y(\varepsilon)$ which, in terms of the scaled energy $\varepsilon' \equiv 2\varepsilon/\Gamma$, can be brought to the form:

$$Y(\varepsilon) = \frac{(\varepsilon' + a \cos \delta_l)^2 + f(\delta_l)}{(\varepsilon'^2 + 1)},$$

where $f(\delta_l)$ is independent of energy, in the spirit of the appropriate Fano-like expression. Note that the asymmetrical contribution to the lineshape depends explicitly on $\cos \delta_l$, a function of the relative phase between the discrete and continuum processes. For further elaboration there is a most insightful treatment of this issue in Chapter XVIII of ref. 1 on resonance phenomena which I highly recommend.

1 A. Böhm, *Quantum Mechanics*, Springer-Verlag, New York, 1979, ch. XVIII.

Prof. Holloway asked: What progress is being made for first principles calculations of excited states at surfaces?

Prof. Gadzuk answered: To the best of my knowledge, there is very little progress to report on first principles calculations of excited states at surfaces. Avouris *et al.*[1] obtained potential energy curves for various charge (hence excited) states of fluorine on aluminum, with a focus towards DIET processes. More recently Klüner *et al.*[2] have considered ground and excited states of NO/NiO(100) (an insulating oxide rather than the conducting metal substrate which many here would like to see treated) with an eye towards further understanding of laser-excited, hot-electron-induced desorption. For the most part, the limited results obtained in both studies are compatible with the requirements and expected behavior implied by the negative ion resonance models for excitation and desorption that we have discussed over the past decade. Other than these two studies, I draw a blank but would be delighted to hear that more work has been completed.

1   Ph. Avouris, R. Kawai, N. D. Lang and D. M. Newns, *J. Chem. Phys.*, 1988, **89**, 2388.
2   T. Klüner, H.-J. Freund, J. Freitag and V. Staemmler, *J. Chem. Phys.*, 1996, **104**, 10030.

Prof. Palmer addressed Prof. Gadzuk, Dr Gauyacq and Prof. Winter: This is a leading question for all three speakers. We have seen how the coupling between molecular resonances and the band structure of the substrate—in particular, the surface projected band gap on the (111) surfaces of the noble metals—can lead to a significant enhancement of the lifetime of the molecular excited state. Might it therefore be possible to design multilayer substrate structures (similar to X-ray mirrors) in which the band structure is "engineered" in order to control the interaction with the molecular states? Maybe we could select the negative ion resonance lifetime by this method, and thus control the dynamics?

Prof. Gadzuk replied: Certainly the electron band structure in the surface region of the substrate determines the elastic reflectivity of the surface, hence the propensity for an electron wavepacket initially localized in an adsorbate excited state resonance or in an (excited) image–potential surface state to elastically delocalize into the bulk. The timescale controlled by these mirror-like aspects of the surface can be thought of as a T_2^* pure dephasing time. In addition, any construction (such as an insulator film) which produces a gap in the local surface electronic excitation spectrum will also hamper decay of the excited state back to its ground state, this being characterized by a T_1 energy relaxation time, due to the blocking off of the electron–hole pair heat bath, also known as the metallic substrate. I have previously proposed control in hot electron femtochemistry at surfaces by exactly the method you raise, varying the negative ion resonance lifetime (see refs. 1 and 2). Extensive studies have been carried out within the group of Charles Harris at UC Berkeley, in which noble gas and alkane adsorbed multilayers and quantum well structures, in the spirit of your question, have been used to modify excited electronic state lifetimes, albeit image potential surface states rather than adsorbate resonances. Finally, your old colleague Phil Rous has considered the effect of such "designer surfaces" (his term) on negative ion resonance lifetimes. See ref. 3 and references therein for further elaboration.

1   J. W. Gadzuk, *Femtosecond Chemistry*, ed. J. Manz and L. Wöste, VCH, Weinheim, 1995, vol. 2, p. 603.
2   J. W. Gadzuk, *Surf. Sci.*, 1995, **342**, 345.
3   P. J. Rous, *Phys. Rev. Lett.*, 1999, **83**, 5086.

Prof. Aeschlimann added: Being able to use a metallic multilayered structure with electron dynamics which may favour a specific chemical reaction on surfaces has been a goal for many years. From the experimental point of view, it is possible today to produce high quality metallic multilayer structure. However, theoretically the electron dynamics of a multilayer structure are not understood and therefore a prediction of the dynamic behaviour in this structure is not possible.

We studied the electron dynamics of a few monolayers of Ag and Co on a Cu(100) surface by means of TR-2PPE. We found a huge, unexpected difference in the electron dynamics between the two investigated model systems, which so far can not be theoretically explained. Therefore, using multilayer structure would end in a "trial-and-error" experiment with no chance of predicting what the resulting electron dynamics would be.

As an experimentalist I hope that the progress in wavepacket propagation calculations will allow, in the near future, the calculation of the behaviour of an optically excited electron in a simple metallic double layer structure.

Prof. Winter opened the discussion on Dr Darling's paper: Your results are in conflict with earlier interpretation of the data, where the threshold in the ion-fractions as a function of projectile velocity can be reproduced by transitions between parallel potential curves (*e.g.* the Demkov approach) and no additional trapping of F^- ions is needed. Do you have an idea why your transition probabilities for electron transfer (per site) are so much higher? Where do the F^- ions (in particular the electron) finally end up?

Dr Darling responded: The disagreement in the behaviour in the threshold region is mainly about the initial ion formation probability, which we find to be always very high, whereas the earlier work of Borisov et al.[1-7] has a threshold due to the reduced probability of transition between the curves at low velocity. I think everyone would agree that if the transitions can occur at low energies there must be some trapping.

We have considered a number of reasons why there might be a discrepancy. The most obvious is that the potential energy surfaces are different for F/LiF(100); ours has a curve crossing at ca. 2 eV, whereas that of Borisov et al. does not. Since the energy deficit, ΔE, appears in the Demkov[8] formula for the diabatic transfer probability,

$$P = \frac{1}{2} \operatorname{sech}^2\left(\frac{\pi \Delta E}{2\gamma v}\right), \qquad (1)$$

where v is the velocity of the projectile, it seems sensible to increase the energy deficit in our case. We have tried this, pushing the crossing point up to 10 eV. For an initial energy of 2 keV at 1° incidence (normal energy of ~ 0.7 eV) we still find an initial ion formation probability of ~ 1. We have also halved the coupling, V_{12}, between the states for this adjusted PES with the crossing at 10 eV. For energies between 2 and 6 keV at 1° incidence, we find no threshold, but interestingly P_i is reduced to 0.6. Finally, for technical reasons associated with the wavepacket methods used, most calculations have been performed using a hydrogen atom mass for the projectile rather than the correct mass of F. For the same energy, eqn. (1) shows that the higher velocity of an H atom mass should lead to higher transition probability. Using the PESs from our paper, we have performed calculations at energies of 2 and 8 keV with an F mass: the ion formation probability in both cases is still 1. We have yet to consider all the changes together, so we cannot conclusively rule out the possibility that the threshold behaviour is entirely explicable in a 'standard' picture. It may also be that the coupling is still too high.

Another reason for the discrepancy could be in the different treatments of the dynamics; we have used a fully quantum method, earlier workers have employed the semi-classical approaches common in the gas-phase. It is known that semi-classical methods can fail in systems with strong coupling.[9] While it would be surprising if this were true at the comparatively high energies considered here, we cannot yet rule it out. We are currently engaged in further work in this direction.

1 C. Auth, A. G. Borisov and H. Winter, *Phys. Rev. Lett.*, 1995, **75**, 2292.
2 C. Auth, A. Mertens, H. Winter, A. G. Borisov and V. Sidis, *Phys. Rev. A*, 1998, **57**, 351.
3 A. G. Borisov, V. Sidis and H. Winter, *Phys. Rev. Lett.*, 1996, **77**, 1893.
4 A. G. Borisov and V. Sidis, *Phys. Rev. B*, 1997, **56**, 10628.
5 A. Mertens, C. Auth, H. Winter and A. G. Borisov, *Phys. Rev. A*, 1997, **55**, R846.
6 H. Winter, A. Mertens, C. Auth and A. G. Borisov, *Phys. Rev. A*, 1996, **54**, 2486.
7 H. Winter, C. Auth and A. G. Borisov, *Nucl. Instrum. Methods Phys. Res. B*, 1996, **115**, 133.
8 Y. N. Demkov, *Sov. Phys. JETP*, 1964, **18**, 138.
9 Y. Zeiri, G. Katz, R. Kosloff, M. S. Topaler, D. G. Truhlar and J. C. Polanyi, *Chem. Phys. Lett.*, 1999, **300**, 523.

Dr Gauyacq said: In your calculations and those of Borisov et al.[1] different transition probabilities are defined and computed: either transition probabilities for the entire collision, probing a large number of F^- sites on the surface or transition probabilities for the interaction between the projectile and a single F^- site on the surface—could this influence the discussion about the existence of an energy threshold in the change transfer process as a function of the collision energy? On a similar point, did you find an energy threshold for the charge transfer process when the collision energy is lowered or is the transition probability always equal to one?

1 A. G. Borisov, V. Sidis and H. Winter, *Phys. Rev. Lett.*, 1996, **77**, 1893.

Dr Darling replied: The way we have approached the problem it is not really possible to calculate a transition probability for a single site. If we tried, then in these two-dimensional calculations, the active site would be surrounded by many inactive ones. Some of the wavepacket will simply not make a transition because it hits the surface far from the active site. How much of it does this depends on exactly how long we make the x-coordinate, in the extreme, when the x-coordinate is very, very long, the effective ion fraction will be zero because most of the wavepacket will never see the active site.

In this particular geometry and for the cases considered in our paper, we have not looked at energies below 1 keV. It would seem reasonable that a threshold must exist at very low energies, but we have not verified this.

Prof. Groß commented: I have two somehow related questions:
(1) In your calculations the corrugation is apparently very important since it allows the parallel kinetic energy to be used to surmount the ionisation threshold. How much parallel kinetic energy is actually converted into perpendicular kinetic energy by diffraction?
(2) Your explanation of the role of the corrugation for the ionisation process is expressed in terms of classical physics. How important are quantum effects in the motion of the atoms at such high kinetic energies?

Dr Darling answered: It is not clear how much energy exchange comes from diffraction. Asymptotically far from the surface several electron volts are required to produce a negative-ion by transferring charge from the surface. However, if we consider only parallel motion starting near the curve-crossing at the active site, the projectile must initially move uphill as it approaches the next halide site where it experiences a repulsion from the coulomb forces. If it continues to move parallel to the surface, then very far from the active site it only requires another ~ 1 eV to escape. The 'extra energy' has come from parallel motion, but it is not really diffraction, simply moving up hill. As you pointed out, this is a very classical picture. We have not yet, however, implemented any semi-classical methods. The difficulty with them is that if the trapping is important, then it is essential to have conservation of energy (as you imposed in your paper explicitly), but also the correct balance between the normal and parallel momenta. The quantum mechanics simply does this if one has programmed it correctly, in semi-classical approaches we felt this might have to be imposed in some way. We are trying out some semi-classical schemes at present to try and determine the discrepancy in the threshold behaviour between our calculations and the earlier ones of Borisov et al.,[1-7] so we shall be able to give you a clearer answer in the future.

1 C. Auth, A. G. Borisov and H. Winter, *Phys. Rev. Lett.*, 1995, **75**, 2292.
2 C. Auth, A. Mertens, H. Winter, A. G. Borisov and V. Sidis, *Phys. Rev. A*, 1998, **57**, 351.
3 A. G. Borisov, V. Sidis and H. Winter, *Phys. Rev. Lett.*, 1996, **77**, 1893.
4 A. G. Borisov and V. Sidis, *Phys. Rev. B*, 1997, **56**, 10628.
5 A. Mertens, C. Auth, H. Winter and A. G. Borisov, *Phys. Rev. A*, 1997, **55**, R846.
6 H. Winter, A. Mertens, C. Auth and A. G. Borisov, *Phys. Rev. A*, 1996, **54**, 2486.
7 H. Winter, C. Auth and A. G. Borisov, *Nucl. Instrum. Methods Phys. Res. B*, 1996, **115**, 133.

Prof. Kleyn asked: (1) Is it correct to say that a quantum description is necessary, because the states are very strongly mixed due to the high value of V_{12}, as shown in Fig. 2d of your paper? Therefore, it is not like a Landau–Zener transition where (except in the crossing region) a single diabatic potential is followed?
(2) Is the high value of V_{12} realistic?

Dr Darling responded: The coupling is strong, and although it appears from the figures in our paper that a diabatic and adiabatic representations can be used to explain the details of the dynamics at the active site, when we attempted this at higher energy (75 keV) this turned out not to be so. The coupling is so strong that both representations are unclear. In any case the curve crossing itself is not so important, and the process is more by the Demkov mechanism,[1] as originally noted by Borisov et al.[2] We have tried reducing the coupling (by 50%), but this did not result in a threshold (even when the crossing was moved to higher energy) but it did reduce the saturation value of P_i to ~ 0.6. This was just a one-off test and we have not yet fully explored the possibilities of this.

1 Y. N. Demkov, *Sov. Phys. JETP*, 1964, **18**, 138.

2   C. Auth, A. G. Borisov and H. Winter, *Phys. Rev. Lett.*, 1995, **75**, 2292.

Dr Shluger commented: Self-trapping, *i.e.* localisation of holes on one or two lattice sites in alkali halides, is known to take more than several hundred femtoseconds. How are hole delocalisation and mobility likely to affect your predictions?

Dr Darling replied: We have assumed that in our case, the hole is always localised at the site where it formed, this is implicit in the way we generate the PESs. We have considered the possibility that the hole could be 'dragged along' by the projectile, however it is extremely difficult to implement such a process in the wavepacket formalism we have employed, and we would have to be sure that it was going to be significant *before* we attempted to do this.

Dr Gauyacq commented: Dr Darling's approach to the negative ion formation on LiF surfaces assumes that the charge transfer process occurs between the projectile and localised states on the surface anion sites and it does not involve electronic states delocalised on the LiF crystal. Recently we performed a theoretical study of this problem using a tight binding approach[1] for F^- formation on LiF surfaces. It showed that the electron capture process is not influenced by the hole movement in the crystal. The interaction between the various particles is lifting the degeneracy between the states in the valance band: when the incident projectile approaches a given anion site on the surface, the hole state on this site splits from the band, leading to the localised character of the charge transfer. These results justify the use of a localised description in this system (F^- formation on LiF).

1   A. G. Borisov, J.-P. Gauyacq, V. Sidis and A. K. Kazansky, *Phys. Rev. B*, in the press.

Direct and indirect DIET and DIMET from semiconductor and metal surfaces: What can we learn from 'toy models'?

Peter Saalfrank,[*a] Gisela Boendgen,[a] Cécile Corriol[a] and Tohru Nakajima[b]

[a] *University College London, Chemistry Department and Centre for Theoretical and Computational Chemistry (CTCC), 20 Gordon Street, London, UK WC1H OAJ*
[b] *The University of Tokyo, Department of Chemical Engineering, Graduate School of Engineering, Tokyo 113-8656, Japan*

Received 6th April 2000
First published as an Advance Article on the web 11th September 2000

Desorption induced by electronic transitions (DIET) and its variant DIMET (M = 'Multiple'), are among the simplest possible "reactions" of ad-species involving ultra-short lived electronically excited states at surfaces. The non-adiabatic bond-cleavage can be enforced, for example, with laser irradiation or with electrons or holes emitted from the tip of a scanning tunnelling microscope (STM). The transient creation of excited intermediates can proceed directly (localised to the adsorbate–substrate complex), or indirectly (*i.e.*, through the substrate). To understand the basic processes, simple one-mode two-state "toy models" such as the Menzel–Gomer–Redhead (MGR) or the Antoniewicz scenarios have proven very useful in the past. We adopt and extend MGR- and Antoniewicz-type models together with numerically exact open-system density matrix theory to address a few actual problems/experiments in DI(M)ET: (1) Direct, laser-induced desorption of H(D) from Si(100) surfaces which has been realised in the continuous-wave DIET regime only recently [T. Vondrak and X.-Y. Zhu, *Phys. Rev. Lett.*, 1999, **82**, 1967], is studied and compared to so-far hypothetical femtosecond laser desorption. The possibility of controlling the reaction by shaping the laser pulses is addressed. (2) For the same system, temperature effects are studied for electron- or hole-stimulated desorption with an STM [T. C. Shen, C. Wang, G. C. Abeln, T. R. Tucker, J. W. Lyding, Ph. Avouris and R. E. Walkup, *Science*, 1995, **268**, 1590; C. Thirstrup, M. Sakurai, T. Nakayama and K. Stokbro, *Surf. Sci.*, 1999, **424**, L329]. A modified version of Gadzuk's "sudden transition and averaging" approach is adopted which accounts for temperature dependent excited state lifetimes. (3) For photodesorption of NO from Pt(111), based on quantum dynamical simulations possible experimental tests involving static electric fields are suggested to address the relevance of the recently challenged [F. M. Zimmermann, *Surf. Sci.*, 1997. **390**, 174], "negative ion resonance" model of the Antoniewicz type.

1 Introduction

Lasers[1] and STM[2] are powerful tools for modern physics and physical chemistry in general, and for surface science in particular. These tools serve to image or spectroscopically detect ad-species, but also to actively *manipulate* them.[3] The manipulation of adsorbates can be initiated by photons, electrons, holes, or ions. The possible outcomes are simple vibrational excitation of ad-species, the rotation of molecules, the lateral motion of adsorbates on a surface, their dissociation, or the desorption of atoms or molecules from the substrate.[1,3] Desorption is clearly the best-studied "reaction" of them all: It is still comparatively simple but at the same time already of practical relevance for the nanostructuring of materials or the passivation of microelectronic devices.[4]

Laser- and STM-induced desorption *via* short-lived electronically excited states is what we are concerned with also in this paper. The initial excitation can be *direct*, or *indirect*, that is substrate-mediated.[5] In surface *photo*chemistry, the direct mechanism prevails typically for insulator and semiconductor surfaces and is due to the coupling of the radiation field to the electronic transition dipole moment of the adsorbate. In contrast when ultraviolet (UV) or visible photons are directed toward a metal substrate one normally creates "hot electrons" in the solid first, which then drive the dynamics of the adsorbate–substrate complex indirectly.[5]

Direct and indirect excitation mechanisms can cause very different outcomes; great differences emerge also when working either in the single- or the multiple-excitation limits (on the timescale of the vibrational relaxation of the adsorbate), which leads to DIET and DIMET, respectively.[6] In the realm of photochemistry continuous wave (cw) or nanosecond (ns) lasers lead to singular excitations and DIET, while intense laser pulses in the picosecond (ps) and femtosecond (fs) regime cause DIMET. With an STM, DIET and DIMET can be realised as well, *e.g.*, by tip-emitted electrons or holes above a threshold energy (DIET), or below threshold (DIMET). The DIET and DIMET yields, product branching ratios, and product energy distributions can be very different, hence opening the possibility to control surface chemistry by the laser or STM parameters.[7] Also, DIET and DIMET product properties (yields, translational and internal energy content) scale differently, sometimes non-linearly with laser fluence (photochemistry) or current (STM), respectively.[7]

Traditionally, desorption resulting from electronically non-adiabatic transitions was described with simple models, including only a few, typically two, electronic states, and a single mode, the desorption coordinate, *i.e.*, usually the distance of the centre-of-mass of the desorbing species from the substrate. In their original variants, these henceforth called "toy models" treat the motion of the adsorbate classically. The most prominent two-state one-mode models are the MGR[8] and the Antoniewicz models,[9] respectively. In the former, the adsorbate moves initially *outward* after excitation, *e.g.*, when the excited state potential is repulsive, before rapid quenching (energy relaxation) by the coupling of the adsorbate to the electronic or phonon degrees of freedom of the substrate takes place which reduces the escape probability. In the latter, the excited state is bound and the excited state equilibrium bond length shorter than in the ground state, hence the excited adsorbate moves initially *toward* the surface before relaxation to the ground state and desorption occurs.

A system for which MGR-type potentials are assumed to be at work is Si(100)-H(2 × 1); here, either with STM-electrons,[10] or directly with photons,[11] a Si–H $\sigma \rightarrow \sigma^*$ transition may be enforced which eventually leads to desorption of hydrogen. Avouris and co-workers have constructed, based on multi-configuration self-consistent field (MCSCF) cluster calculations,[10] potential energy curves for the ground and excited states which are shown in Fig. 1(a). The bound excited potentials typical for the Antoniewicz mechanism should be realised for systems which readily form positive or, more frequently, negative ion resonances that are stabilised by image charge forces. Image-charge stabilised negative ion resonances are often encountered when the substrate is a metal and the adsorbate has a positive electron affinity. A system frequently quoted as a prototype example for a negative-ion-resonance mediated indirect photodesorption process, is NO/Pt(111).[12] Model potentials have been constructed for this system[12,13] which are shown for the one-mode variant, in Fig. 1(c). Finally in Fig. 1(b) an MGR-type model for the desorption of H from Si(100) *via* a hole resonance which is not image-charge stabilised is shown, the significance of which will be outlined below.

Since the early days the MGR and Antoniewicz models have been improved in many ways:[14] (1) Quantum effects for the nuclear motion have been included; (2) multi-state (rather than two-state) models were developed; (3) the excitation/quenching was treated more thoroughly; (4) multi-dimensional models including internal adsorbate modes were proposed. All of these improvements are necessary to explain specific observations for specific systems, but altogether it is fair to say that the "toy models" were remarkably successful in explaining a wealth of phenomena. Among these are the often small desorption yields, the non-thermal energy distribution of

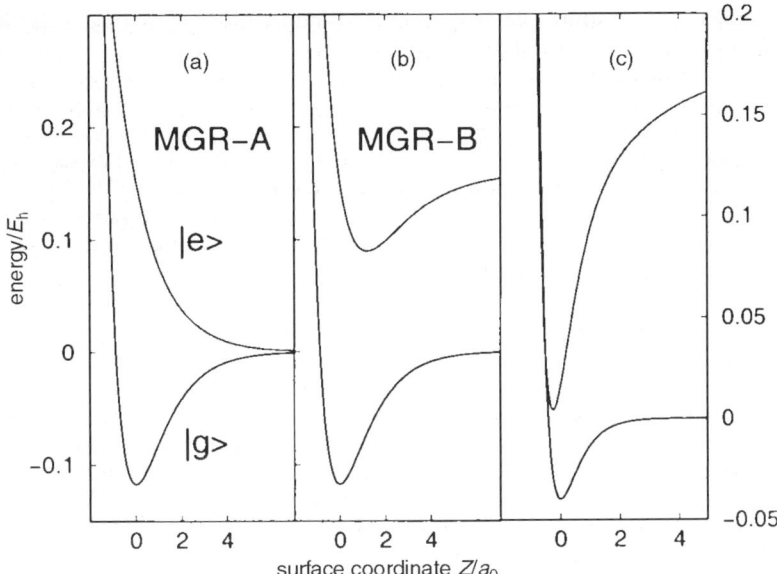

Fig. 1 The three 1D "toy models" used in this work. In (a), the MGR model for Si(100)-H(D)(2 × 1) is shown, when the excited state is of the σ → σ* type (MGR-A). (b) shows the MGR potential curves for Si(100)-H(D)(2 × 1) with the excited state being of the "hole resonance type" (MGR-B). Fig. (c) gives the Antoniewicz-type potentials for NO/Pt(111), when the excited state is assumed to be an image charge stabilised negative ion resonance. Note the different energy scale on Fig. 1(c).

the desorbates, "threshold" and "resonance" behaviour for excitation wave lengths (in photon-stimulated desorption, PSD) or electron energies (in electron-stimulated desorption, ESD), and subtle isotope and temperature effects.

This success of the MGR and Antoniewicz "toy models" will be exploited in the following to make predictions which go beyond the ordinary DIET and DIMET scenarios. Specifically, we will consider the following open questions/problems:

1. The desorption of H and D atoms from a Si(100)-H(D)(2 × 1) surface was recently achieved by a direct ns laser experiment.[11] This process will be modelled by MGR-type potentials in connection with an open-system density matrix description. The excited state is assumed to be the repulsive σ → σ* resonance as suggested by Avouris et al.[10] (see Fig. 1(a)). We further propose to use fs laser pulses rather than ns pulses to control the reactivity of the system.

2. (a) For the same system, STM-induced DIET and DIMET had been realised already earlier by Avouris and coworkers, who worked at positive sample bias (i.e., electrons where travelling from the tip to the sample).[10] The occurrence of DIET[10] (accessible with electrons with energies above the excitation threshold of about 7.3 eV), and DIMET[15] (below threshold), was established. One of the results in ref. 10 was that in the surface temperature range between 11 and 300 K, no temperature effect on the desorption characteristics (yields, isotope effects) in the DIET regime could be found. In contrast a clear negative temperature effect was measured in the DIMET regime, i.e., the desorption yield decreases with increasing substrate temperature.[15] Avouris et al. developed also the σ → σ* MGR-type model mentioned above which lead to the potentials shown in Fig. 1(a).[10] (b) More recently Stokbro et al. STM-desorbed H from Si(100)-H(2 × 1), however, this time at negative sample bias. In this case holes travel from the tip to the sample.[16] A distinct suppression of the desorption rate at larger sample temperatures was found, even in the DIET regime.[17] Stokbro et al. assume that in their case the excited state relevant for STM-desorption is a positive-ion, or hole resonance. As long as this state is not stabilised by image charges, which is reasonable to assume for Si, the excited potential is expected to be still bound but the bond length should be larger than in the ground state. From this argument one may construct an approximate excited state potential (see below), which is shown in Fig. 1(b) and which is still of the MGR-type in the sense that an initially excited adsorbate will move outward.

We will investigate temperature effects during STM-induced DIET of H from Si(100)-H(2 × 1) using these MGR-type models (a) and (b) below.

3. Finally, UV-laser induced desorption of NO from Pt(111) will be considered. This process has been studied theoretically in the past.[12–14,18–29] Despite these many investigations the exact nature of the dominant electronically excited state is still under dispute—is it the above mentioned image-charge stabilised negative ion resonance, or not? We will suggest, based on dynamical calculations, an experiment to resolve this "negative ion resonance dilemma."[25,28,29]

In the next section, general aspects of the models and theoretical simulation will be outlined. The three following sections are devoted to the three specific questions/problems listed above. A final Section 6 summarises and concludes this work.

2 Models and quantum dynamical methods: Generic aspects

For the desorption of H and D from Si(100) and NO from Pt(111) we employ here one-dimensional two-state models with $|g\rangle$ denoting the ground state, and $|e\rangle$ denoting an excited state. $|e\rangle$ is either of the repulsive MGR-type (problems 1 and 2(a), see Fig. 1(a)), the non-stabilised hole resonance (problem 2(b), see Fig. 1(b)), or the image-charge stabilised negative ion resonance á la Antoniewicz (problem 3, see Fig. 1(c)). As in previous work[14] we solve an open-system Liouville-von-Neumann (LyN) equation of the form[30]

$$\frac{\partial}{\partial t}\hat{\rho} = -\frac{i}{\hbar}[\hat{H}, \hat{\rho}] + \frac{\partial}{\partial t}\hat{\rho}_{\text{env}}, \quad (1)$$

which, when solved subject to some initial condition $\hat{\rho}(0)$ for the reduced density operator $\hat{\rho}(t)$, gives any expectation values of interest from the trace relation

$$\langle \hat{A} \rangle(t) = \text{tr}\{\hat{A}\hat{\rho}(t)\}. \quad (2)$$

The first term on the right of eqn. (1) stands for the dissipation-free *Hamiltonian* evolution of the system density operator, with the system Hamiltonian $\hat{H}$ given by:

$$\hat{H} = \hat{H}_g|g\rangle\langle g| + \hat{H}_e|e\rangle\langle e| + V_{ge}|g\rangle\langle e| + V_{eg}|e\rangle\langle g|. \quad (3)$$

The diagonal blocks of $\hat{H}$ are

$$\hat{H}_i = -\frac{\hbar^2}{2m}\frac{d^2}{dZ^2} + V_i(Z), \quad i = e, g, \quad (4)$$

with m being the adsorbate mass and $V_i(Z)$ the ground and excited state potentials depending on the adsorbate–surface distance, Z. The off-diagonal terms induce *direct* transitions between the states. Here we include only transitions stimulated by an external photon field, giving in semi-classical dipole approximation for the adsorbate-field coupling

$$V_{eg} = V_{ge} = -\mu_{eg}(Z) \cdot F(t), \quad (5)$$

where $\mu_{eg}(Z)$ is the component of the transition dipole moment parallel to the adsorbate–surface bond, and $F(t)$ the external electric field which is assumed to be polarised along the same direction. In practice all operators occurring in eqn. (1) are represented on an equidistant, spatial grid in Z.[13,20,22,31] In this case the diagonal elements of the density matrix are probability densities in configuration space.[31] A similar block structure as for the Hamiltonian holds for the density operator $\hat{\rho}$ itself, and the traces over the $\hat{\rho}_e$ and $\hat{\rho}_g$ blocks give the excited and ground electronic state populations N_e and N_g, respectively.

For the dissipative term in eqn. (1) we employ a Markovian Lindblad form[32]

$$\frac{\partial}{\partial t}\hat{\rho}_{\text{env}} = \sum_k^K (\hat{C}_k \hat{\rho} \hat{C}_k^\dagger - \tfrac{1}{2}\hat{C}_k^\dagger \hat{C}_k \hat{\rho} - \tfrac{1}{2}\hat{\rho}\hat{C}_k^\dagger \hat{C}_k), \quad (6)$$

which guarantees by construction, for a probabilistic interpretability of the density matrix at all times. At most $K = 2$ "dissipative channels" and hence also Lindblad operators $\hat{C}_k$ are considered. The first one, $\hat{C}_1$, accounts for the ultrafast quenching of the excited state $|e\rangle$ toward the ground state $|g\rangle$, with a lifetime $\tau = 1/\Gamma_{ge}$[13,20]

$$\hat{C}_1 = \sqrt{\Gamma_{ge}}|g\rangle\langle e|. \qquad (7)$$

Generalisation to coordinate-dependent quenching is straightforward.[13,22] The second Lindblad operator, $\hat{C}_2$, models the substrate-mediated indirect excitation of the adsorbate in DIMET,

$$\hat{C}_2 = \sqrt{\Gamma_{eg}(t)}|e\rangle\langle g|, \qquad (8)$$

where the excitation rate

$$\Gamma_{eg}(t) = \Gamma_{ge} \cdot \exp\{-\Delta V/k_B T_{el}(t)\} \qquad (9)$$

is explicitly time-dependent through the time-dependence of the substrate electronic temperature $T_{el}(t)$. The latter results from the response of the substrate electrons to the ultrafast laser pulse. The "upward" rate Γ_{eg} is related to the "downward" rate Γ_{ge} through the principle of detailed balance. (ΔV in eqn. (9) is the energy difference between V_e and V_g, and k_B Boltzmann's constant.)

The initial density operator $\hat{\rho}(0)$ is different for the different problems/experiments to be considered. For DIET, irrespective of whether induced by a low-intensity ns laser pulse in direct photodesorption (see problem 1), or by STM-emitted electrons or holes (problem 2(a) and 2(b)), or by hot electrons in a ns-laser driven indirect photodesorption process (problem 3), the initial, impulsive excitation is treated as a Franck–Condon transition of the ground state density to the excited state:

$$\hat{\rho}(0) = |e\rangle\langle e| \otimes \sum_{v_g} p_{v_g}(T_s)|v_g\rangle\langle v_g|. \qquad (10)$$

Here, $|v_g\rangle$ is the v_gth vibrational wavefunction in the ground electronic state, and $p_{v_g}(T_s)$ its Boltzmann weight at substrate (phonon) temperature T_s. Further, in the case of DIET, only the quenching process described by the Lindblad operator $\hat{C}_1$ is included.

To model DIMET or the direct fs laser scenarios the excitation process is explicitly included in the formalism, either by the direct coupling term V_{eg} in eqn. (5) for direct photodesorption (problem 1 when fs lasers are used), or by the upward Lindblad operator $\hat{C}_2$ in eqn. (8). In these cases,

$$\hat{\rho}(0) = |g\rangle\langle g| \otimes \sum_{v_g} p_{v_g}(T_s)|v_g\rangle\langle v_g|. \qquad (11)$$

The LvN equation (1) was solved numerically either by a direct density matrix propagation scheme involving fast Fourier transform (FFT) techniques and Newton polynomial interpolation of the Liouvillian time-propagation superoperator,[13,20,31] or, in selected cases (DIET with coordinate-independent quenching), indirectly by a "sudden transition and averaging" (STA) algorithm as introduced by Gadzuk[12] (see below). The latter can be shown to be equivalent to the former for those cases considered here.[21,25]

3 Direct photoinduced DIET and DIMET of H and D from Si(100)-H(D)(2 × 1)

3.1 Nanosecond vs. femtosecond laser desorption

Recent experiments of Vondrak and Zhu[11] have demonstrated that the H–Si bond at a Si(100)-H(2 × 1) surface can be dissociated directly by photons of wavelength 157 nm (corresponding to $\hbar\omega = 7.9$ eV). That the coupling of the photon field to the transition dipole is in fact direct, was verified by changing the laser polarisation. In ref. 11 ns lasers were used leading to DIET. A large isotope effect

$$I_{des} = \frac{P_{des}(H)}{P_{des}(D)} \qquad (12)$$

was found of about 10 ± 3. An even larger isotope effect of about 50 had been reported earlier by Avouris et al. for their STM-DIET experiments. In both ref. 10 and 11 also quantum dynamical simulations for zero surface temperature were carried out, with an MGR-type model and potentials based on ab initio cluster calculations.[10] These calculations explained the large isotope effect by assuming that the excited state lifetime $\tau = 1/\Gamma_{ge}$ (see eqn. (17)) is ultrashort, $\tau < 1$ fs. In both papers the theoretical modelling was essentially the same, since the excitation leading to DIET was always treated as an initial Franck–Condon transition, *i.e.*, no specifics of the laser or the STM entered the formalism.

In our own work we also investigated DIET of H and D from Si(100)-(2 × 1).[33] In addition to the earlier work by Avouris *et al.*[10] and the later one by Zhu *et al.*,[11] we demonstrated that coordinate-dependent quenching, $\tau = \tau(Z)$ with $\lim_{z \to \infty} \tau(Z) = \infty$, leads to increased desorption probabilities P_{des} in contrast to an Antoniewicz scenario where the opposite is true.[14] Further, for coordinate-independent quenching an approximate semiclassical expression for the isotope effect was derived,

$$I_{des} = e^{-\gamma/\tau}, \qquad (13)$$

with γ being related to the excited state potential parameters and the masses of H and D respectively. The analytic expression (13) demonstrates the exponential increase of I_{des} with decreasing τ; hence for very short τ tiny differences in τ (or γ) will cause huge differences in I_{des}. For example, according to ref. 33 I_{des} is about 12 at $\tau = 0.75$ fs, but already 52 at $\tau = 0.45$ fs. Thus, even small differences of the experimental conditions may lead to large differences in I_{des}, which may explain why Avouris *et al.* and Zhu *et al.* find quantitatively different isotope ratios in their STM- and photo-induced DIET experiments.

In the following, we propose to directly photodesorb H or D from Si(100) by using ultrashort, high-intensity fs lasers rather than the low-intensity ns lasers of ref. 11. By variation of the laser parameters we may then be able to control the desorption process, *i.e.*, the desorption yields and isotope effect I_{des}, for example. To this end we adopt the direct coupling model with a coupling $V_{eg}(t)$ as in eqn. (5), with the electric field chosen specifically as

$$F(t) = F_0 s(t) \cos(\omega t). \qquad (14)$$

Here, $s(t)$ is a shape function, for which we choose two half Gaussians possibly separated by a plateau region,

$$s(t) = \begin{cases} e^{-(t-t_0)^2/(2\sigma^2)}, & t \leq t_0 \\ 1 & t_0 \leq t \leq t_0 + t_p, \\ e^{-(t-t_0-t_p)^2/(2\sigma^2)}, & t \geq t_0 + t_p. \end{cases} \qquad (15)$$

and ω is the laser carrier frequency. Finally, the Condon approximation is made, *i.e.*, we neglect the coordinate dependence of the transition dipole moment μ_{ge} in eqn. (5) such that the product $\mu_{ge} F_0$ can be interpreted as a coordinate-independent coupling amplitude, A_0. Hence, in the present study we have five laser parameters, A_0, σ, t_0, t_p and ω, four of which are varied independently to optimise a desired outcome. (The Gaussian peak time parameter, t_0, is chosen such that for a given width parameter, σ, the field is practically zero at $t = 0$.) We do not use freely optimised pulses from optimal control theory,[34] mostly because optimal control theory is anything but straightforward and numerically efficient for the dissipative problem at hand.

Dissipation enters again through the Lindblad quenching operator $\hat{C}_1$, whilst the indirect-excitation operator $\hat{C}_2$ is neglected under the assumption that for H/Si photodesorption only the direct mechanism plays a role. The initial condition was given by eqn. (11), with $T_s = 0$ K, *i.e.*, we start in the ground vibrational state of the ground electronic state initially.

The LvN equation (1) was solved by direct propagation using a Newton polynomial integrator and grid methods combined with FFT.[35] The time-resolved desorption probability was calculated from

$$P_{des}(t) = \text{tr}\{h(Z - Z_d)\hat{\rho}(t)\}, \qquad (16)$$

where $h(Z - Z_d)$ is a step function which is zero for $Z < Z_d$, and 1 for $Z \geq Z_d$ ($Z_d = 8.38 \, a_0$ in the following.) The observable desorption probability P_{des} itself is given by $P_{des} = \lim_{t \to \infty} P_{des}(t)$. For the potentials, Avouris' *ab initio* data were used,[10] giving for the ground state a Morse potential and for the excited state a repulsive exponential:

$$V_g(Z) = D[1 - e^{-\alpha Z}]^2 - D \qquad (17)$$

$$V_e(Z) = A e^{-\beta Z}, \tag{18}$$

with parameters,[33] $D = 0.117\,24\ E_h = 3.2$ eV, $\alpha = 0.8123\ a_0^{-1}$, $A = 0.153\,26\ E_h = 4.2$ eV, $\beta = 0.693\ a_0^{-1}$.

With a coordinate-independent excited state lifetime of $\tau = 0.75$ fs we obtain, for the sudden-transition DIET model a desorption probability for H of $P_{des} = 2.4 \times 10^{-3}$; for D, $P_{des} = 2.0 \times 10^{-4}$, giving $L_{des} = 12$. Hence, $\tau = 0.75$ fs reproduces the isotope effect found in the DIET experiments of ref. 11 and therefore this lifetime will be used in the following unless stated otherwise. Fig. 2(b) shows $P_{des}(t)$ for H and $\tau = 0.75$ fs on a semi-logarithmic scale, demonstrating that the desorption which occurs almost exclusively in the ground state, is complete after about 100 fs.

Next, simple Gaussian fs pulses with $\sigma = 400\ \hbar/E_h$ were used to model a fs laser desorption experiment. The laser pulse with $A_0 = 0.0145\ E_h$ and a width of ≈ 25 fs shown in Fig. 2(a), has been optimised for the dissipation-free case, i.e., for $\tau = \infty$ ($\Gamma_{ge} = 0$). Without dissipation this pulse gives, for H, an excited state population $N_e = \text{tr}\{\hat{\rho}_e(t)\}$ larger than 0.99 (see Fig. 2(c)) and, consequently, also a desorption probability of ≈ 1 (Fig. 2(b)). For the pulse shown, the "best" carrier frequency ω is $0.262\ E_h/\hbar$ (7.1 eV/$\hbar$), which is almost the classical resonance frequency ω_0 at the Franck–Condon point $Z = 0$, $\hbar\omega_0 = V_e(0) - V_g(0) = 0.269\ E_h$ (7.3 eV). When this same pulse is used but dissipation "switched on", then the excited state population never exceeds $N_e \approx 0.1$ and also the final desorption probability P_{des} is clearly well below 1 ($P_{des} = 7.5 \times 10^{-3}$, see Fig. 2(b)). Still, the desorption probability for the fs laser is larger than the desorption probability in the (ns laser or STM above threshold) DIET cases, which was $P_{des} = 2.4 \times 10^{-3}$. This enhancement of reactivity appears to be even more evident when considering that the desorption probability in DIET as calculated here is per excitation event and therefore an upper limit, whereas for the fs laser the finite excitation probability is explicitly accounted for.

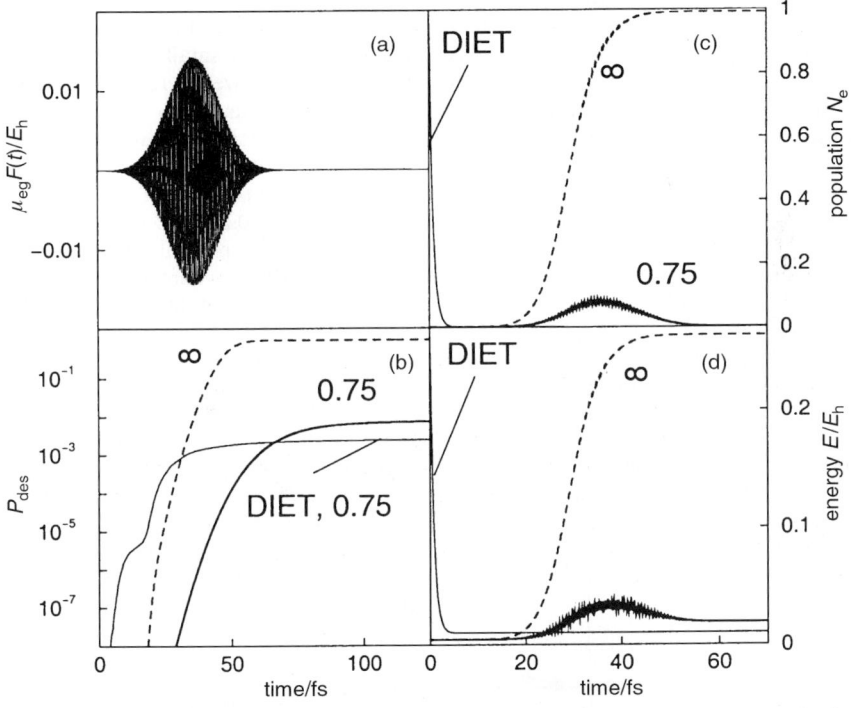

Fig. 2 Direct photodesorption of H from Si(100)-H(2 × 1). In (a), a Gaussian laser pulse is shown with parameters $A_0 = 0.0145\ E_h$, $\sigma = 400\ \hbar/E_h$ (9.7 fs), $t_0 = 1500\ \hbar/E_h$ (36.3 fs), $\hbar\omega = 0.262\ E_h$ (7.1 eV). (b) gives the desorption probability P_{des} as a function of time, for DIET (impulsive excitation limit) and $\tau = 0.75$ fs, and for desorption with the pulse shown in (a) when $\tau = 0.75$ fs (thick solid line) and $\tau = \infty$ (thick dashed line), respectively. (c) gives the corresponding excited state populations $N_e(t)$ and (d) the total system energies $E(t)$.

3.2 Toward optimal femtosecond laser pulses

As a next step some of the laser pulse parameters were partly optimised for the dissipative case in order to enhance the desorption yield for H/Si(100). Unfortunately these calculations are quite costly, hence a complete and systematic variation of the full parameter space cannot be presented here. Nevertheless, basic trends are already obvious.

First, we vary the field amplitude F_0 in eqn. (14), which means that under the Condon approximation we effectively vary $A_0 = F_0 \mu_{ge}$, keeping all other field parameters fixed. In Fig. 3 it is shown that P_{des} increases non-linearly with the laser-molecule coupling amplitude A_0. Closer inspection shows that P_{des} grows approximately quadratically with A_0, and hence quasi-linearly with the laser intensity and laser fluence, indicative for a "DIET-like" scaling.

To simply increase the laser intensity is, however, not always a practical way to go because, apart from possible technical problems to reach high intensities, side-reactions may take place such as thermal heating and ionisation, which are simply not considered here. Alternatively one may resort to Gaussian pulses with a plateau region, i.e. $t_p \neq 0$ in eqn. (15). Fig. 4 demonstrates what happens when such a plateau pulse operates. For the example shown, a plateau region with $t_p = 400 \; \hbar/E_h$ (97 fs) was introduced, and $A_0 = 0.0193 \; E_h$. Without a plateau the corresponding simple Gaussian pulse gives $P_{des} = 0.012$ for H (see Fig. 3). With the plateau region, however, the desorption probability for hydrogen reaches 0.32, and that of deuterium 0.077 (Fig. 4(b)). This is an increase of P_{des} for H relative to the DIET case by a factor of about 160—again without taking the finite excitation probability for the latter into account. From Fig. 4(c) which gives the population of the excited state both for H and D, we note that the cw-like plateau pulse enforces damped Rabi oscillations, in fact typical for a dissipative 2-level system when driven by a cw-field. The plateau pulse is more successful than the simple Gaussian because the field continuously feeds energy into the system thus "beating" dissipation. In fact from the system energy curves in Fig. 4(d) we see that the system can gain much more energy than by a simple Gaussian pulse (Fig. 2(d)), and this leads to enhanced desorption yields.

For the plateau pulse shown in Fig. 4(a) not only the yield increases but also the isotope effect I_{des} goes down, because generally large yields are associated with small isotope ratios. In the DIET case we had $I_{des} = 12$, while with the plateau pulse of Fig. 4 we get $I_{des} \approx 4$, demonstrating that also isotope ratios can be influenced by using fs rather than ns lasers.

The question arises as to whether it is possible to really control the isotope effect, and in particular to achieve $I_{des} < 1$, i.e., preferred deuterium desorption. For that purpose we have experimented with linearly *chirped* plateau pulses, for which the laser carrier frequency is a function of time, $\omega(t) = \omega_0 + \kappa t$. In particular negative chirps with $\kappa < 0$ are expected to be useful. For them the laser frequency ω decreases with time and hence adapts optimally to the reduced resonance frequency of a wave packet moving away from the surface where the energy difference $V_e(Z) - V_g(Z)$ becomes progressively smaller. The optimal chirp will be different for H and for D

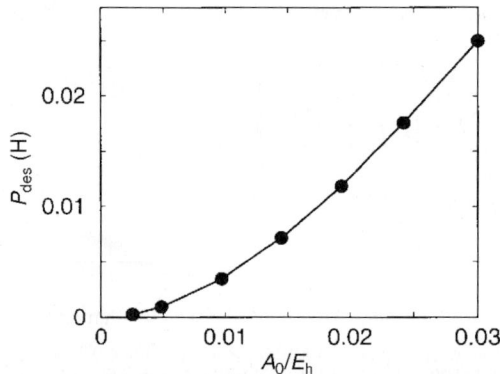

Fig. 3 Direct photodesorption of H from Si(100)-H(2 × 1) with Gaussian fs laser pulses. Shown is the non-linear scaling of the desorption yield with the molecule-field coupling amplitude, A_0. The other Gaussian parameters were $\sigma = 400 \; \hbar/E_h$, $t_0 = 1500 \; \hbar/E_h$, $\hbar\omega = 0.262 \; E_h$.

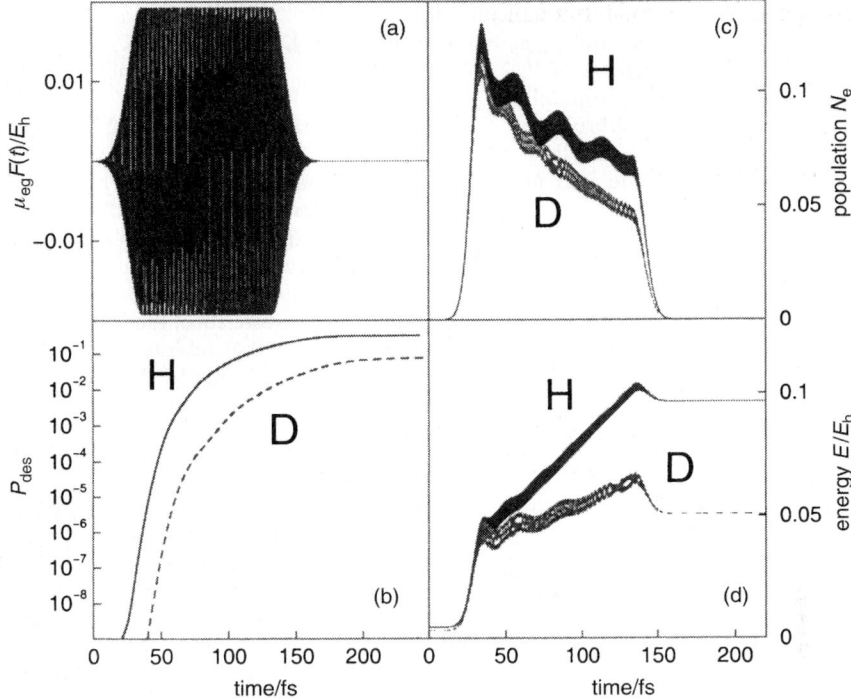

Fig. 4 Direct photodesorption of H and D from Si(100)-H(D)(2 × 1) with a "plateau" pulse. (a) shows the pulse with parameters $A_0 = 0.0193\ E_h$, $\sigma = 400\ \hbar/E_h$, $t_0 = 1500\ \hbar/E_h$, $\hbar\omega = 0.262\ E_h$ and $t_p = 400\ \hbar/E_h$. In (b), the resulting desorption probabilities for H and D are given. (c) and (d) show the corresponding excited state populations $N_e(t)$ and system energies $E(t)$, respectively.

because the latter moves more slowly. So far, however, our numerical experiments with chirped pulses were of only limited success.

Nevertheless, reaction control at surfaces using ultrashort lasers seems possible, even when the dissipation is "ultrafast". Clearly a more systematic investigation of the parameter space is required at this stage together with a more detailed treatment of the process (coordinate-dependent quenching, vibrational relaxation, multi-dimensionality), to arrive at conclusions directly relevant for immediate quantitative experimental verification.

4 STM-DIET of H and D from Si(100)-H(D)(2 × 1): Temperature effects

As mentioned earlier, Avouris and coworkers had prior to Zhu et al.'s PSD experiments used an STM, operated at positive sample bias, to desorb H and D from Si(100)-(2 × 1). In the DIET limit with high-energy electrons above the threshold for $\sigma \rightarrow \sigma^*$ excitation, no temperature effect in the yields was observed between $T_s = 11$ K and 300 K in contrast to the DIMET ("below threshold") regime. This finding is a surprise since, according to the MGR model, surface heating causes vibrational excitation of the adsorbate; after electronic excitation those lobes of the vibrationally excited wave packet which are already farther away from the surface have therefore a better chance to desorb and consequently P_{des} should increase with increasing T_s. Such a temperature effect which for slightly different reasons[14] is expected to hold also in the Antoniewicz case, was predicted years ago.[36] Since then the positive correlation of desorption yields with substrate temperature has been found for a number of systems, both by experiment[37,38] and theory.[13,14,20] In ref. 33, temperature effects for DIET of H and D from Si(100) were theoretically investigated within the impulsive Franck–Condon excitation/MGR $\sigma \rightarrow \sigma^*$ model. It was found that indeed up to $T_s = 300$ K, the desorption probability for both H and D is rather unaffected by temperature. This is due to the large energy of the H–Si and D–Si vibrations ($\hbar\omega_v \approx 0.24$ and 0.17 eV, respectively), which causes the Boltzmann ratio $p_1/p_0 = e^{-(\hbar\omega_v/k_B T_s)}$ to be very small: $p_1/p_0 = 0.93 \times 10^{-4}$ for H at $T_s = 300$ K, and $p_1/p_0 = 1.39 \times 10^{-4}$ for D at $T_s = 300$ K. Only above this temperature, the expected increase of P_{des} sets in, first for D, then for H. Hence, the desorption probabilities increase and the isotope effect I_{des} decreases with increasing T_s.[33]

Recently, Thirstrup and Stokbro et al.[17] STM-desorbed H from Si(100)-H(2 × 1) at negative sample bias when holes are emitted from the STM tip rather than electrons. They report a strong *suppression* of the desorbed rate after heating the surface from 300 to 610 K. This negative temperature effect is most pronounced in the DIET ("above threshold") limit and therefore opposite to what we predicted in ref. 33 for STM-DIET at positive sample bias. Stokbro et al. assume that in their case (i) the desorption proceeds *via* a "hole resonance",[16] and that (ii) the excited state lifetime, τ, decreases with increasing substrate temperature, thus reducing the desorption rate at higher T_s. This mechanism is different from the negative temperature effect found by Avouris et al. in the positive sample bias DIMET regime, where faster vibrational relaxation at higher T_s was made responsible for a similar effect.[15] According to Stokbro et al.'s model the decrease of τ with T_s is due to the enhanced coupling of the 5σ hole resonance to both the electrons and the phonons of the Si substrate. In ref. 17, the excited state lifetime is parametrised as

$$\tau^{-1} = \Gamma_{ge} = a + b(e^{c/T_s} - 1)^{-1}, \tag{19}$$

with $a = (0.67 \text{ fs})^{-1}$, $b = (0.91 \text{ fs})^{-1}$ and $c = 402$ K. This parametrisation gives $\tau(0 \text{ K}) = 0.67$ fs, $\tau(300 \text{ K}) = 0.53$ fs and $\tau(600 \text{ K}) = 0.38$ fs, respectively. These lifetimes are of the same order of magnitude as in Avouris' model, but are now temperature dependent. The higher T_s, the smaller τ and the smaller the desorption probability per excitation event.

We have modelled the temperature dependence of the STM-DIET process at negative sample bias using a hole resonance model. This may have implications also for the temperature dependence at positive sample bias which proceeds *via* the σ → σ* resonance. To do so, the proposed temperature dependence of the excited state lifetime is included.[17] In a hole resonance model, the ground state potential $V_g(Z)$ should still be the same as for the σ → σ* model of Fig. 1(a). In contrast the excited state is parametrised as a displaced and shifted Morse potential:

$$V_e(Z) = D_+[1 - e^{-\alpha_+(Z - Z_+)}]^2 + K. \tag{20}$$

For the parameters, we take $D_+ = 0.0694 \ E_h = 1.9$ eV, $\alpha_+ = 0.557 \ a_0^{-1}$, $Z_+ = 1.2 \ a_0$, and $K = 0.207 \ E_h = 5.6$ eV. This parametrisation is based on an idea used by Jennison et al. who constructed a hole resonance state for the dissociation of Si–H bonds in the bulk,[39] exploiting an analogy with the formation of H_2^+ from H_2 by "hole attachment". For the H_2 molecule the equilibrium bond length increases by a factor of 1.418, the binding energy decreases by a factor of 1.69, and the vibrational frequency by a factor of 1.895 when going from H_2 to H_2^+. Assuming (i) that the same scaling applies to the Si–H bond when ionised, (ii) using the harmonic approximation to relate the Morse exponents to vibrational frequencies and (iii) determining K such that the Franck–Condon energy difference $V_e(0) - V_g(0)$ matches the experimental resonance energy of $0.257 \ E_h$ (7.0 eV),[16] we end up with the potential parameters listed above and with the potential shown in Fig. 1(b). This modified MGR model is henceforth denoted by "MGR-B", in contrast to the σ → σ* one which is denoted as "MGR-A" (Fig. 1(a)).

With these potentials and under the further assumption of an initial Franck–Condon transition (DIET limit), the LvN equation (1) can be solved for finite substrate temperature T_s and with only the "quenching operator" $\hat{C}_1$ included. ($\hat{C}_2$ is assumed to be zero since the electronic temperature is small as compared to $V_e(0) - V_g(0)$.) The solution of the LvN equation with the DIET initial condition (10) can be done either by direct density matrix propagation, or by Gadzuk's efficient STA algorithm.[12] The latter has to be generalised to account for the two temperature dependences encountered here, namely the one imposed by the initial condition (10) and that of the excited state lifetime anticipated by eqn. (19). In the ordinary STA model, different "residence times" τ_R and initial vibrational quantum numbers v_g are selected to generate, by standard wave-packet propagation, time-dependent nuclear wavefunctions of the type

$$|\psi(t; \tau_R, v_g)\rangle = e^{-i\hat{H}_g(t - \tau_R)/\hbar} e^{-i\hat{H}_e \tau_R/\hbar} |v_g\rangle. \tag{21}$$

These reflect (1) the sudden transition of the initial wave function $|v_g\rangle$ at $t = 0$, (2) propagation in the excited state for a time τ_R, (3) sudden transition of the wave packet back to the ground state, and (4) propagation to a final time, t. From these "quantum trajectories", expectation values

$$A(t; \tau_R, v_g) = \langle \psi(t; \tau_R, v_g) | \hat{A} | \psi(t; \tau_R, v_g) \rangle \tag{22}$$

can be calculated. In a second step an incoherent averaging over the $A(t; \tau_R, v_g)$ is performed to model a continuous residence time distribution. At finite T_s we have to weight each v_g with the corresponding Boltzmann factor. Additionally, to account for the possible T_s-dependence of the excited state lifetime, $\tau = \tau(T_s)$, we calculate observables from

$$\langle \hat{A} \rangle(t) = \sum_{v_g} p_{v_g}(T_s) \frac{\int_0^\infty e^{-\tau_R/\tau(T_s)} A(t; \tau_R, v_g) \, d\tau_R}{\int_0^\infty e^{-\tau_R/\tau(T_s)} \, d\tau_R}. \tag{23}$$

Note that once the $A(t; \tau_R, v_g)$ are known (for instance $P_{des}(t; \tau_R, v_g)$), the treatment of various finite T_s is trivial—only the Boltzmann weights and the lifetimes $\tau(T_s)$ have to be recalculated. This is much easier than performing a separate density matrix propagation for every temperature, which is demanded for the direct solution of the LvN equation (1). The STA wave-packet approach outlined above is equivalent to the direct solution of (1), however, only for DIET and when the decay rates are coordinate-independent.[21,25]

This open-system density matrix approach[40] was adopted to rationalise, with the MGR model of Fig. 1(b) (MGB-B), the hole resonance experiments reported in ref. 17 at various temperatures T_s. First, however, in Fig. 5(a) the desorption probabilities P_{des} for $T_s = 0$ K are shown as a function of excited state lifetime τ, for H and D. Both the desorption *via* a "hole resonance" (MGR-B) and a $\sigma \to \sigma^*$ resonance (MGR-A) is considered. (For the latter, see also ref. 33).

We observe that, for all cases, the desorption probability increases dramatically with increasing τ. For realistic lifetimes ($\tau < 1$ fs), the MGR-B model predicts yields at a given τ that are somewhat larger than for the $\sigma \to \sigma^*$ model, largely because the slope of the excited state potential at the Franck–Condon point $V'_e(0) = (dV_e/dZ)|_0$, is larger for the former (Fig. 1(b)) than for the latter (Fig. 1(a)).

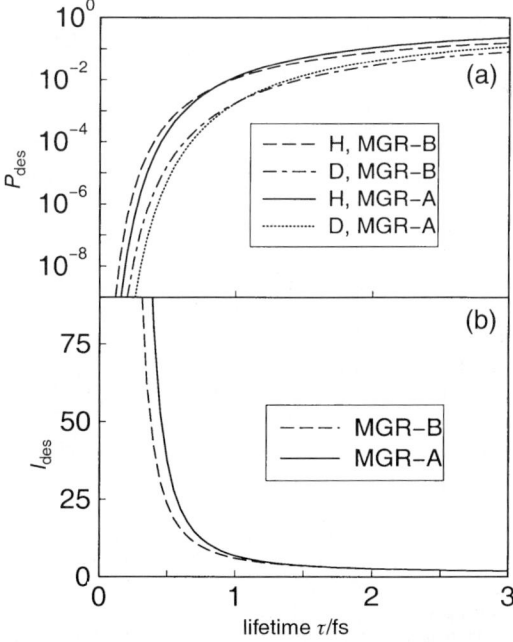

Fig. 5 Temperature effects during the DIET of H and D from Si(100)-H(D)(2 × 1). In (a) the (averaged) desorption probabilities for H and D are shown as a function of the excited lifetime, τ, for the $\sigma \to \sigma^*$ model of Fig. 1(a) (MGR-A), and the hole resonance model of Fig. 1(b) (MGR-B). (b) gives the corresponding isotope effects I_{des} for MGR-A and MGR-B.

The desorption yields of Fig. 5(a) translate into the isotope effects I_{des} shown in Fig. 5(b). The isotope effect decays exponentially with τ as predicted by eqn. (13). For a given lifetime, I_{des} is smaller for the MGR-B hole model than for the MGR-A model. This follows again from the simple rule that small yields are associated with large isotope effects and *vice versa*. At a lifetime of $\tau = 0.75$ fs (which was used in Section 3), we find $I_{des} = 12$ for the MGR-A model (see above), and $I_{des} = 10$ in the MGR-B hole model. Of course there is no strong reason why the hole resonance state should have the same lifetimes as the $\sigma \to \sigma^*$ resonance state. In fact, there is even some uncertainty for the lifetime of the $\sigma \to \sigma^*$ state itself considering that in Avouris *et al.*'s STM-experiments $I_{des} \approx 50$ (giving $\tau \approx 0.45$ fs according to Fig. 5(b)), while $I_{des} = 10 \pm 3$ in Zhu *et al.*'s laser desorption experiments suggesting $\tau \approx 0.75$ fs. In the STM-DIET experiments of Stokbro *et al.*, according to eqn. (19), $\tau(T_s = 0) = 0.67$ fs (see above), giving $I_{des} = 12$ for MGR-B. (The isotope effect was not reported in ref. 16 and 17.) Despite these differences in detail, in any of the examples above the isotope ratio is large and the excited state lifetime shorter than 1 fs.

In Fig. 6 finite temperature effects are considered. For the MGR-B hole model we have knowledge from ref. 17 of the dependence of the excited state lifetime on T_s. With the parametrisation (19) for $\tau(T_s)$, a lifetime of the excited state as a function of temperature is obtained as in Fig. 6(a) (dotted line). With this information the desorption probability P_{des} for hydrogen can be calculated as a function of T_s (Fig. 6(b)). Note that in fact the experimentally observed[17] temperature suppression of the desorption yield occurs as a consequence of the decreasing lifetime. Quantitatively, the H desorption probability decreases by a factor of about 10 when heating from $T_s = 0$ K to $T_s = 450$ K. In the experiments reported in ref. 17, the desorption rate became smaller by a factor of about 200 (for a voltage of -7 V and a tunnelling current of 2 nA), when heating from $T_s = 300$ to $T_s = 450$ K. Hence the proposed simple MGR(-B) open-system density matrix model accounts qualitatively for the experimental observations. The theoretical model further predicts that the isotope effect I_{des} should considerably increase with temperature, simply because the yields become smaller thus favouring larger I_{des} (see Fig. 6(c)).

In ref. 33 an opposite temperature effect was predicted for the STM-DIET experiments of Avouris and coworkers (MGR-A), *i.e.*, the isotope effect should decrease above $T_s \approx 300$ K

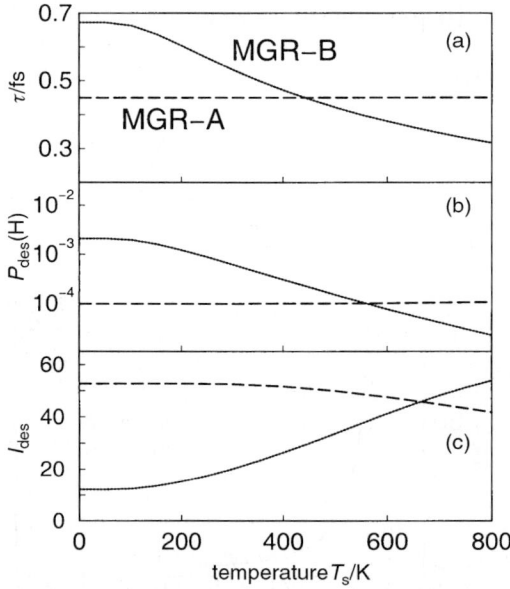

Fig. 6 Temperature effects during the DIET of H and D from Si(100)-H(D)(2 × 1). In (a) the excited state lifetime is shown as a function of surface temperature, for the "MGR-B" hole model using the parametrisation of ref. 17. For MGR-A, a T_s-independent model (horizontal line at $\tau = 0.45$ fs) is used. (b) gives the desorption probabilities for H for the two cases of (a), when only the "Boltzmann effect" is included (MGR-A), or both the Boltzmann and the "lifetime effect" (MGR-B) are accounted for. (c) gives the corresponding isotope ratios I_{des}.

because P_{des} starts to increase. However, this calculation did not take account of possible temperature dependences of the σ → σ* resonance lifetime—only the Boltzmann averaging was done. At higher T_s higher vibrational states of Si–H and Si–D are appreciably populated, hence P_{des} increases with T_s and I_{des} decreases. Fig. 6(a) to (c) illustrate these trends for STM-DIET (model MGR-A, dashed), when only the Boltzmann effect is considered and otherwise the excited state lifetime is assumed to be τ = 0.45 fs, irrespective of T_s.

The question arises as to what might happen once possible additional, but opposite effects due to a T_s-dependent lifetime are taken into account. Unfortunately, we can only speculate on this point because the variation of τ on T_s is not known for the σ → σ* resonance. Test calculations with "reasonable" T_s-dependences, however, reveal that the second, yield-diminishing temperature effect usually dominates over the Boltzmann effect. As a consequence the isotope ratio would increase rather than decrease upon surface heating. Again, conclusive evidence on the T_s-dependence of τ is lacking for MGR-A, hence it is only safe to say that the isotope effect for H/D desorption in the (STM-)DIET limit should undergo measurable changes for $T_s > 300$ K. If I_{des} decreases with increasing T_s, that would indicate that even at temperatures above 300 K the lifetime of the σ → σ* resonance is fairly independent of T_s; if I_{des} goes up, then one may suspect a strong decrease of τ with T_s similar to the hole resonance case of ref. 17. Altogether, the temperature dependence of the desorption dynamics appears to be a potentially rich source for the microscopic unravelling of non-adiabatic processes at surfaces.

5 Indirect photoinduced DIET and DIMET of NO from Pt(111)

5.1 General aspects

We will now use an Antoniewicz "toy model" to make predictions that may help to clarify an actual dispute on the relevance of negative ion resonance states for photodesorption from metal surfaces, for the specific example of NO/Pt(111). The standard model of photodesorption of NO from Pt(111)[12] assumes a ground state potential V_g for the chemisorption of neutral NO on Pt, and a short-lived negative ion resonance formed by temporary attachment of a photoexcited substrate electron to NO to give NO⁻. In the 1D model to be used below, the potential of the neutral ground state is of Morse form,[17] with parameters,[12,13] $D = 0.0397$ $E_h = 1.08$ eV and exponent $\alpha = 1.708$ a_0^{-1}. For the excited metal-to-ligand-charge-transfer state the parametrisation is[12,13]

$$V_e(Z) = V_g(Z) - \frac{\delta^2}{4(Z - Z_{im})} + \varepsilon. \qquad (24)$$

Here, $\varepsilon = 0.1838$ $E_h = 5$ eV is the approximate ionisation potential of Pt(111) minus the electron affinity of NO. The negative ion resonance is *image charge stabilised* by the term $-(\delta^2/4(Z - Z_{im}))$. δ is the charge transferred from the metal to NO, and $Z_{im} = 1.757$ a_0 is the location of the image plane.[13]

With the 1D model major experimental facts for DIET,[41] namely a small desorption probability per excitation event (of $P_{des} \approx 10^{-4}$), and a non-thermal translational energy distribution of the desorbing NO molecules can be explained when it is assumed that the excited state lifetime is ultra-short, $\tau = 1/\Gamma_{ge} < 10$ fs.[12,13] Also, the non-linear scaling of the desorption probability in DIMET with laser fluence,[7] as well as other DIMET hallmarks such as the quasi-linear scaling of the translational energy of the desorbates with laser fluence,[7] emerge from the modelling.[13]

Unfortunately in a 2D model when the N–O vibration is also included the assumption of a full one-electron charge transfer leads to a computed vibrational energy content of NO which is by orders of magnitude larger than measured.[22,25,27,29] The too-pronounced vibrational excitation results from too big a N–O bond-lengthening in the excited state after a full one-electron charge transfer into the antibonding 2π* level of NO. To resolve this "negative ion resonance dilemma",[25,28,29] we have suggested that coordinate-dependent quenching in the excited state can lead to a reduction of the computed vibrational overpopulation.[22] An even more efficient "cooling mechanism" arises from relaxing the full one-electron charge transfer assumption; in fact as argued on the basis of tight-binding Green's function calculations in ref. 24, a full charge transfer should occur only *asymptotically*, i.e., $\delta = \delta(Z)$ and $\lim_{z\to\infty}\delta(Z) = -1|e|$. Closer to the surface the charge transfer is incomplete, $|\delta(Z)| < 1|e|$. Still, the negative ion resonance model for the photo-desorption of NO from Pt(111) and similar non-adiabatic surface reactions and systems,[1] has been challenged by several groups.[28,29]

We wish to suggest here an experiment which may help to resolve this puzzle. The suggestion is to carry out the indirect laser desorption experiment in the presence of a strong, slowly varying or even static electric field. A static electric field strong enough to cause a measurable effect can be provided by bringing an STM close to the adsorbate, for example. The external field is assumed to be homogeneous and oriented perpendicular to the surface. The field will lead, by dipole coupling, in general to the "distortion" of both the ground and the excited state potentials. In semiclassical dipole approximation we have

$$V_i(Z) \to \tilde{V}_i(Z) = V_i(Z) - \mu_i(Z)F, \qquad (25)$$

where μ_i is the z-component of the Z-dependent electric dipole moment in state $i =$ g, e. These dipole moments can be calculated by quantum chemistry methods, but are simply approximated here as follows. The ground state dipole moment is neglected, $\mu_g = 0$, because in the neutral ground state the charge transfer between adsorbate and substrate is small. In contrast in the excited state, if of the negative ion resonance type, the dipole moment is large and strongly Z-dependent. In a classical point charge approximation we get, for two charges (the negative charge on NO plus the image charge in the metal), a dipole moment of

$$\mu_e(Z) = 2(Z - Z_{im})\delta. \qquad (26)$$

In the following a full charge transfer will be assumed for simplicity, i.e., $\delta = -1|e|$, hence μ_e is defined here as *negative* when an electron is transferred from the metal to NO. As argued above, the full charge transfer assumption introduces an error for small Z, but the conclusions to be drawn below are sufficiently robust against this approximation. We further neglect the polarisability of the molecule–surface bond—this leads merely to higher-order field terms which may be introduced on desire.

With a static external field F, we may "tune" the excited state potential V_e provided it is of the negative ion resonance type; this tuning will have a strong influence on the desorption dynamics. If the excited state most relevant for desorption is not a charge-transfer state, then μ_e should be small and the influence of the field be a minor one. If the excited state is an image-charge stabilised positive ion resonance (which is not to be expected), then the influence of the field will be large too, but the dependence of the desorption dynamics on the *polarity* of the field will be opposite to the negative ion case. If more than one excited states are important then one may tune in and out of the negative ion resonance state to enhance or diminish its relevance for the entire process. Hence we expect to learn a lot about the nature of the relevant excited state(s) by a combined field-laser experiment. This is important because it is difficult to get reliable excited states for an adsorbate/metal system from *ab initio* calculations.

If the excited state is the above-mentioned negative ion resonance, then a *positive field* $F > 0$ for which the field vector points from + at the surface to − in the vacuum, will cause a *destabilisation* of the excited state potential at $Z > Z_{im}$, relative to the ground state because then the interaction term $-\mu_e F$ in eqn. (25) is positive. When the field is generated by an STM, this corresponds to a positive sample bias. Conversely, for negative field (or negative sample bias), $\mu_e(Z)$ and F are parallel and the effective potential $\tilde{V}_e$ in eqn. (25) is *stabilised* relative to V_g. This is illustrated in Fig. 7 which shows the effective potential $\tilde{V}_e$ of eqn. (25) for various field strengths $F \in [-0.04, +0.04] E_h (ea_0)^{-1}$. We note that $F = 0.0194 E_h (ea_0)^{-1}$ corresponds to 1 V Å^{-1}. Field strengths of the order of 1 V Å^{-1} can be produced with an STM operating above a semiconductor surface; for metals, they are typically somewhat lower.[3]

Already simply from Fig. 7 the following trends can be predicted for indirect photodesorption of NO from Pt in the presence of a static field, as long as the most relevant excited state is the assumed negative ion resonance:

1. The *threshold* energy to excite the adsorbate, i.e., $\approx \Delta V = \tilde{V}_e(Z_0) - V_g(Z_0)$ becomes larger for $F > 0$. In this case the negative ion resonance is destabilised and hence, even for an indirect mechanism, higher-energy photons are needed to enforce the $|g\rangle \to |e\rangle$ excitation. In contrast for $F < 0$, the upper state becomes more stable and lower energy photons are sufficient. Therefore, this first effect determines the *excitation cross-section*.

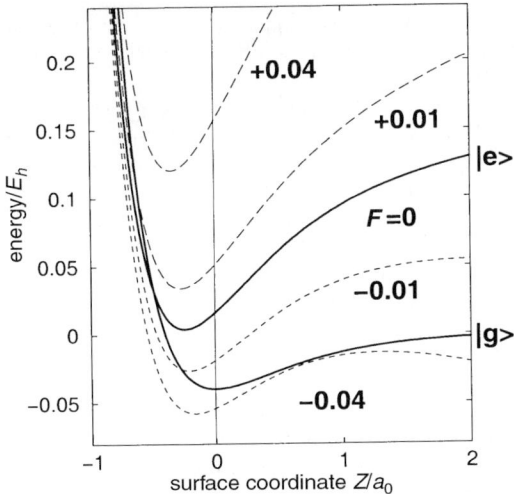

Fig. 7 Indirect photodesorption of NO from Pt(111): Effective potentials V_{eff} (eqn. (25)) for various field strengths F (attached numbers, in units $E_h (ea_0)^{-1}$). The field-free ground and excited state potentials are also shown.

2. According to the "standard model",[12] the desorption probability per excitation event depends, also in the Antoniewicz case, critically on the gradient of the excited state at the Franck–Condon point, $V'_e(0)$. The more positive $V'_e(0)$, the more the excited state wave packet is accelerated toward the substrate and hence the higher P_{des}. The gradient $\tilde{V}'_e(0)$ in the "field-on" case is given here analytically from eqn. (25) and (26), as

$$\tilde{V}'_e(0) = V'_0(0) - 2\delta F, \qquad (27)$$

where $V'_e(0)$ is the gradient in the field-free case. Hence for the negative ion ($\delta = -1|e|$), the gradient becomes more positive with increasingly positive field F (Fig. 7). This second field effect, therefore, influences the *desorption probability per excitation* event, i.e., the post-excitation dynamics.

3. A third and perhaps minor effect arises for strongly negative fields. In this case the excited state may become even more stable than the former ground state. One would then expect, even without the presence of a laser, spontaneous emission from $|g\rangle$ to $|e\rangle$, i.e., the spontaneous creation of NO$^-$ ions. These would desorb, however, only when the barrier to desorption is low or absent (i.e., for fields $F < -0.04 E_h (ea_0)^{-1}$), otherwise they remain trapped. Also, this process would proceed on the "long" timescales typical for spontaneous emission.

5.2 DIET dynamics

The dynamical consequences of the proposed experiment are investigated now in more detail. We first concentrate on item (2) of above, i.e., on the post-excitation dynamics, by studying an indirect, photoinduced DIET process in which the excitation is treated as an initial Franck–Condon transition. To this end the LvN equation (1) is solved by the STA wave-packet algorithm,[42] with the initial condition (10) (for $T_s = 0$ K), and with the "field-on" potentials shown in Fig. 7. In the calculations a coordinate-independent excited state lifetime of $\tau = 2$ fs was assumed, which was determined in ref. 13 to roughly reproduce experimental data for the field-free case.

In Fig. 8(a) the desorption probability per excitation event is given for various external field strengths F. We consider both ^{14}N^{16}O (see also ref. 13), and its isotopically substituted relative, ^{15}N^{18}O. We find that, as expected, P_{des} is larger for positive fields than without a field, and smaller for negative fields. This is due to the progressively larger gradients $\tilde{V}'_e(0)$ for increasingly positive F. In a realistic field range of $F \in [-0.01, +0.01] E_h (ea_0)^{-1}$, P_{des} increases for ^{14}N^{16}O, by a factor of about 200 from 3.4×10^{-6} at $F = -0.01 E_h(ea_0)^{-1}$, to 6.8×10^{-4} at $F = +0.01$

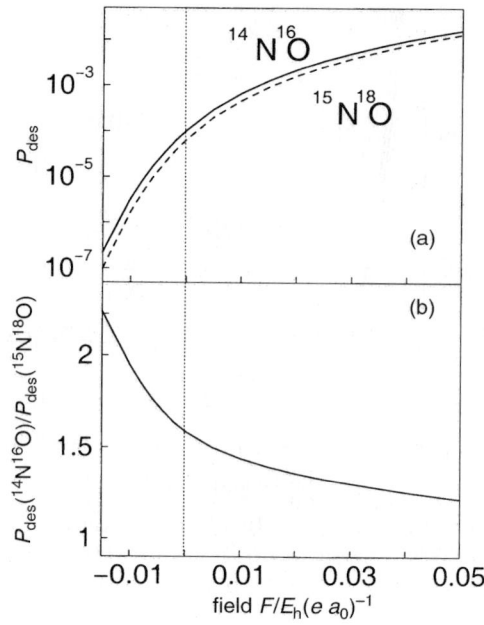

Fig. 8 Indirect photodesorption of NO from Pt(111): DIET model. In (a) the desorption probabilities are shown as a function of the field strengths, F, for two isotopomers of NO. In (b), the corresponding isotope effect in the yields is shown.

$E_h(ea_0)^{-1}$. We note further that the desorption probability for the heavier isotopomer is consistently smaller than for $^{14}N^{16}O$—for reasons similar to those mentioned for H and D desorption from Si(100). Finally, the isotope effect in the yields, $I_{des} = P_{des}(^{14}N^{16}O)/P_{des}(^{15}N^{18}O)$ turns out to be strongly field-dependent, too. I_{des} gradually decreases with increasingly positive F (Fig. 9(b), later), again simply because large yields cause small isotope effects and *vice versa*. For negative fields, isotope effects as large as 2 can be expected which is remarkable when considering the small mass mismatch.

5.3 DIMET dynamics

In carrying out an actual experiment and comparing it to the theoretical predictions made by Fig. 8, a word of caution is in order. Namely, the field will not only influence the post-excitation dynamics, but even more so the excitation process itself (see point (1) in section (5.1)). Therefore, to directly compare theory and experiment for DIET the excitation cross-section has to be known.

This problem does not arise in DIMET, because here we actually take the excitation process explicitly into account in the theoretical modelling. We have solved the LvN equation (1) for the indirectly induced DIMET case, using the direct Newtonian density matrix propagation algorithm,[43] and the initial condition (11). The indirect excitation process was modelled by the "upward" Lindblad operator $\hat{C}_2$ of eqn. (8). The upward rate $\Gamma_{eg}(t)$ was determined from eqn. (9); for the electronic temperature $T_{el}(t)$ occurring in the latter equation, a model form[13] based on the well-known two-temperature model[44] was used. We chose for $T_{el}(t)$ the functional form given as eqn. (2.18) in ref. 13. Two specific choices for the parameter T_m were made, namely $T_m = 10\,000$ K, and $T_m = 6000$ K. All other parameters are the same as in Table I of ref. 13. With these choices $T_{el}(t)$ peaks after about 100 fs with temperature T_{el}^{max} of about 7650 K (for the larger T_m), and 4600 K (for the lower T_m), respectively (see Table III of ref. 13). The excited state lifetime used below was again $\tau = 2$ fs.

Fig. 9 gives the analogous information to Fig. 8(a), however, for the DIMET case, and for $^{14}N^{16}O$ only. We note that for $T_m = 10\,000$ K and $F = 0$, $P_{des} = 1.02 \times 10^{-3}$.[45] This value is about a factor of 10 larger than for DIET (Fig. 8), indicating once more that DIMET usually favours larger yields.

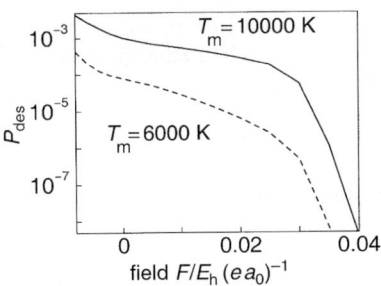

Fig. 9 Indirect photodesorption of NO from Pt(111). The same as Fig. 8(a), however, for the DIMET case and the ^{14}N^{16}O isotopomer only. The solid curve refers to a high laser fluence ($T_m = 10\,000$ K, see text), the dashed one to a lower one ($T_m = 6000$ K).

When finite fields are considered we find that, in contrast to Fig. 8, the desorption probability *decreases* with increasingly positive field strength F. This means that the suppression of the excitation cross-section (effect (1) of Section 5.1), outperforms the simultaneous increase of the desorption probability per excitation event reported in Fig. 8. The decrease of the excitation cross section in our DIMET model is due to the exponential decrease of the excitation rate $\Gamma_{eg}(t)$ with the energy gap $\Delta V = \tilde{V}_e(0) - V_g(0)$ in eqn. (9). Quantitatively this difference depends linearly on F for the present "toy model", according to ($\delta = -1|e|$):

$$\Delta V = \frac{e^2}{4Z_{im}} + \varepsilon + Z_{im}|e|F. \tag{28}$$

For $F < 0$, on the other hand, the desorption probability is larger than in the field-free case because the enhanced excitation probability beats the reduction effect due to the smaller gradient in the excitation state.

As seen from Fig. 9, if the positive fields are only large enough the desorption can almost be suppressed. Hence, there appears to be threshold behaviour in the DIMET case: For fields larger than a critical field strength F_c, desorption comes to a halt. For the specific example of $T_m = 10\,000$ K, $F_c \approx 0.035\, E_h\, (ea_0)^{-1}$, according to Fig. 9.

More "accessible" field strengths for actual observation of the predicted threshold effect are expected when the maximum electronic peak temperature T_{el}^{max} is reduced by using a (femtosecond) laser with a smaller fluence. Then the exponential factor $e^{-\Delta V/k_B T_{el}(t)}$ in eqn. (9) is exponentially small already for small ΔV, i.e., for less positive F. This expectation is borne out by the DIMET calculation with the smaller T_m, $T_m = 6000$ K. From Fig. 9 we note that, in this case, already without a field the desorption probability is substantially lower than for $T_m = 10\,000$ K, in accordance with ref. 13. When the static field is switched on, we find that in fact a shift of the threshold field strength toward less positive F occurs. We believe that the threshold field strength F_c can be measured more easily than the variation of P_{des} with F in the plateau-like middle region of Fig. 9.

In passing we note that the isotope effects I_{des} in the DIMET case are found to vary only weakly with F, except at field strengths $F > F_c$ where the exponentially small desorption yields can cause large isotope ratios.

6 Summary and conclusions

Three "toy models" of the MGR (Fig. 1(a) and 1(b)), and of the Antoniewicz type (Fig. 1(c)), have been used to make predictions/simulations for non-adiabatic desorption systems of actual interest. While quantum effects and a proper treatment of electronic relaxation are accounted for, the simplicity of the original MGR and Antoniewicz models is largely retained. Despite these simplifications a number of predictions can be made which may be verified or disproved by experiments.

(1) In particular we suggest to use femtosecond rather than nanosecond lasers to directly photodesorb H and D from Si(100) surfaces. By shaping the ultrashort pulses, the possibility to control

the photodesorption process, i.e., yields and isotope ratios, emerges. This was not obvious from the outset, because the quenching of the excited state proceeds within less than one fs, which is considerably shorter than the pulses employed.

(2) For the same system temperature effects during DIET have been investigated. We find that two major sources contribute to a possibly strong surface temperature dependence of the dynamics—a "Boltzmann" and an electronic "lifetime" effect. Both oppose each other (the first one favouring, the latter suppressing desorption), and it is not *a priori* clear which of these effects will dominate for a specific system. Additional complexity is expected in the DIMET case, where as a third source the exponential decrease of the vibrational lifetime of the adsorbate-surface bond with increasing temperature can become dominant.[15] In general, the T_s-dependence of the desorption process appears to be a potentially rich source to learn about the vibrational and electronic relaxation and the desorption mechanism itself.

(3) Finally, for NO/Pt(111) an experiment was proposed which may help to resolve the ongoing dispute on the relevance of image charge stabilised negative ion resonances for non-adiabatic processes at metallic surfaces. Several effects can be predicted when a laser-induced desorption experiment is carried out in the presence of static or slowly varying electric fields. Due to their ionic nature, charge transfer states are very sensitive to external fields and this sensitivity may be exploited as a probe to analyse the nature of photophysically relevant excited states.

Perhaps the greatest advantage of the "toy models" is their simplicity, which allows one to clearly separate between various, often opposing trends during non-adiabatic surface reactions. However, it is also true that nature is not always simple and hence effects neglected here may be important for quantitative predictions. Multi-dimensionality, vibrational relaxation, the participation of more than one excited state, and a more detailed treatment of the excitation process during the STM-driven and substrate-mediated reactions are issues which have to be addressed in the future. Perhaps the biggest challenge is the calculation of reliable excited state potentials. Various groups are already working along these lines. Due to the complexity of non-adiabatic dynamics at solid surfaces, however, we do expect that "toy models" such as those advertised above will continue to be most useful in surface science—both as a starting point for a more sophisticated treatment, and as a rich source for physical insight by itself.

We thank the Deutsche Forschungsgemeinschaft (Sonderforschungsbereich 450, project C1: "Theory of Control of Chemical Reactions by Ultrashort Laser Pulses") for support.

References

1. *Laser Spectroscopy and Photochemistry on Metal Surfaces*, ed. H.-L. Dai and W. Ho, World Scientific, Singapore, 1995.
2. G. Binning, H. Rohrer, Ch. Gerber and E. Weibel, *Phys. Rev. Lett.*, 1982, **49**, 57.
3. W. Ho, *Acc. Chem. Res.*, 1997, **31**, 567.
4. B. N. J. Persson and Ph. Avouris, *Surf. Sci.*, 1997, **390**, 45.
5. F. Weik, A. de Meijere and E. Hasselbrink, *J. Chem. Phys.*, 1993, **99**, 682.
6. J. A. Prybyla, T. F. Heinz, J. A. Misewich, M. M. T. Loy and J. H. Glowina, *Phys. Rev. Lett.*, 1990, **64**, 1537.
7. W. Ho, *Surf. Sci.*, 1996, **363**, 166.
8. (a) D. Menzel and R. Gomer, *J. Chem. Phys.*, 1964, **41**, 3311; (b) P. A. Redhead, *Can. J. Phys.*, 1964, **42**, 886.
9. P. R. Antoniewicz, *Phys. Rev. B*, 1980, **21**, 3811.
10. (a) T. C. Shen, C. Wang, G. C. Abeln, J. R. Tucker, J. W. Lyding, Ph. Avouris and R. E. Walkup, *Science*, 1995, **268**, 1590; (b) Ph. Avouris, R. E. Walkup, A. R. Rossi, T. C. Shen, G. C. Abeln, J. R. Tucker and J. W. Lyding, *Chem. Phys. Lett.*, 1996, **257**, 148; (c) Ph. Avouris, R. E. Walkup, A. R. Rossi, H. C. Akpati, P. Nordlander, T. C. Shen, G. C. Abeln and J. W. Lyding, *Surf. Sci.*, 1996, **363**, 368; (d) T. C. Shen and Ph. Avouris, *Surf. Sci.*, 1997, **390**, 35.
11. T. Vondrak and X.-Y. Zhu, *Phys. Rev. Lett.*, 1999, **82**, 1967.
12. J. W. Gadzuk, L. J. Richter, S. A. Buntin, D. S. King and R. R. Cavanagh, *Surf. Sci.*, 1990, **235**, 317.
13. P. Saalfrank and R. Kosloff, *J. Chem. Phys.*, 1996, **105**, 2441.
14. H. Guo, P. Saalfrank and T. Seideman, *Prog. Surf. Sci.*, 1999, **62**, 239, and references therein.
15. E. T. Foley, A. F. Kam, J. W. Lyding and Ph. Avouris, *Phys. Rev. Lett.*, 1998, **80**, 1336.
16. K. Stokbro, C. Thirstrup, M. Sakurai, B. Yu-Kuang Fu, F. Perez-Murano and F. Grey, *Phys. Rev. Lett.*, 1998, **80**, 2618.
17. C. Thirstrup, M. Sakurai, T. Nakayama and K. Stokbro, *Surf. Sci.*, 1999, **424**, L329.

18 N. Chakrabarti, V. Balasubramanian, N. Sathyamurthy and J. W. Gadzuk, *Chem. Phys. Lett.*, 1995, **242**, 490.
19 S. M. Harris, S. Holloway and G. R. Darling, *J. Chem. Phys.*, 1995, **102**, 8235.
20 P. Saalfrank, R. Baer and R. Kosloff, *Chem. Phys. Lett.*, 1994, **230**, 463.
21 P. Saalfrank, *Chem. Phys.*, 1996, **211**, 265.
22 K. Finger and P. Saalfrank, *Chem. Phys. Lett.*, 1997, **268**, 291.
23 P. Saalfrank, *Surf. Sci.*, 1997, **390**, 1.
24 T. Klamroth and P. Saalfrank, *Surf. Sci.*, 1998, **410**, 21.
25 P. Saalfrank, G. Boendgen, K. Finger and L. Pesce, *Chem. Phys.*, 2000, **251**, 51.
26 O. Kühn and V. May, *Chem. Phys.*, 1996, **208**, 117.
27 H. Guo, *J. Chem. Phys.*, 1997, **106**, 1967.
28 H. Guo and G. Ma, *J. Chem. Phys.*, 1999, **111**, 8595.
29 F. M. Zimmermann, *Surf. Sci.*, 1997, **390**, 174.
30 K. Blum, *Density Matrix Theory and Applications*, Plenum Press, New York, 1981.
31 M. Berman, R. Kosloff and H. Tal-Ezer, *J. Phys. A: Math. Gen.*, 1992, **25**, 1283.
32 G. Lindblad, *Commun. Math. Phys.*, 1976, **48**, 119.
33 G. Boendgen and P. Saalfrank, *J. Phys. Chem. B*, 1998, **102**, 8029.
34 (*a*) S. Shi, A. Woody and H. Rabitz, *J. Chem. Phys.*, 1988, **88**, 6870; (*b*) Y. J. Yan, R. E. Gillian, R. M. Whitnell, K. R. Wilson and S. Mukamel, *J. Phys. Chem.*, 1993, **97**, 2320; (*c*) A. Bartana, R. Kosloff and D. J. Tannor, *J. Chem. Phys.*, 1993, **99**, 196.
35 An equidistant grid with 512 points extending from $Z = -2.87\ a_0$ to $Z = 47.77\ a_0$ was used. A Newton polynomial integrator of degree 16, and a timestep of $4\hbar/E_h$ was employed. Test calculations with shorter timesteps showed that, with the high-frequency fields considered here the $P_{des}(t)$ are accurate to within 5%, and systematic trends are reliable.
36 D. Menzel, *Surf. Sci.*, 1969, **14**, 340.
37 Q.-S. Xin and X.-Y. Zhu, *Chem. Phys. Lett.*, 1997, **265**, 259.
38 S. Thiel, T. Klüner and H.-J. Freund, *Chem. Phys.*, 1998, **236**, 263.
39 D. R. Jennison, J. P. Sullivan, P. A. Schultz, M. P. Sears and E. B. Stechel, *Surf. Sci.*, 1997, **390**, 112.
40 For the calculations using the STA algorithm, a grid of 1024 points (twice as large as for the direct density matrix propagation) was used, as well as a split-operator propagator for the time-integration of the Schrödinger equation. The numerical details are the same as those in ref. 33. With the enlarged grids the desorption probabilities can be calculated very accurately, thus allowing for the treatment of lifetimes of shorter than 0.7 fs. In contrast, the (more general) treatment according to ref. 35 was only sufficiently accurate for $\tau > 0.7$ fs or so.
41 S. A. Buntin, L. J. Richter, R. R. Cavanagh and D. S. King, *Phys. Rev. Lett.*, 1988, **61**, 1321.
42 For the STA-DIET wave packet calculations an equidistant grid consisting of 1024 points was used, with a grid spacing $\Delta Z = 0.025\ a_0$. The integral over the different residence times τ_R in eqn. (23) was discretised on a τ_R grid, in the interval $\tau_R \in [0.5, 25]$ fs. The total propagation time was 600 fs. Those parts of the wave packet which had reached $Z_{des} = 4\ a_0$ after this time were counted as desorbed.
43 For the DIMET calculations the same numerical procedures and parameters as in ref. 13 were used, except that the total propagation time was extended from 500 to 1000 fs. Also, for $F < 0$ shorter timesteps than those reported in ref. 13 had to be chosen to ensure numerical stability.
44 S. I. Anisimov, B. L. Kapeliovich and T. L. Perel'man, *Sov. Phys. JETP*, 1974, **39**, 375.
45 The value of $P_{des} = 1.02 \times 10^{-3}$ found here for $F = 0$ is slightly larger than the one of 0.94×10^{-3} of ref. 13. The present value is more accurate, mostly due to the longer propagation time.

Photoinduced charge-transfer reaction at surfaces
Part I. $(HCl)_m \cdot \cdot Na_n/LiF(001) + h\nu$ (640 nm) → $(HCl)_{m-1}Cl^-Na_n^+/LiF(001) + H(g)$

Javier B. Giorgi, Tae Geol Lee, Fedor Y. Naumkin, John C. Polanyi,* Sergei A. Raspopov and Jiaxi Wang

Department of Chemistry, University of Toronto, Toronto, Ontario, Canada M5S 3H6

Received 17th May 2000
First published as an Advance Article on the web 18th August 2000

A sub-monolayer of atomic sodium, Na_n, was deposited on LiF(001) at 50 K and characterized by temperature-programmed desorption, X-ray photoelectron spectroscopy, and titration with HCl. The Na_n was dosed with HCl to form $(HCl)_m \cdot \cdot Na_n/LiF(001)$, which was then irradiated by 640 nm laser-radiation to induce a charge-transfer (CT) reaction. Reaction-product atomic H(g) was observed leaving the surface, by two-color Rydberg-atom time-of-flight (TOF) spectroscopy. These H-atoms gave evidence of arising from the photoinduced harpooning reaction between the sodium clusters, Na_n, on the substrate, and $(HCl)_m$ adsorbed on the Na_n. The translational energy distribution, its vibrational structure, and the angular distribution of H(g) gave information regarding the harpooning event. Translationally and vibrationally excited HCl(g) was shown, by resonance-enhanced multiphoton ionization (REMPI), to be formed as an alternate product; by way of $(HCl)_m \cdot \cdot Na_n/LiF(001) + 602$ nm → $(HCl)_{m-1}Na_n/LiF(001) + HCl(g)(v \geq 0)$.

I. Introduction

The study of simple CT reactions in gases can be traced back to the 1930s when Michael Polanyi and co-workers proposed that reactions involving transitions from covalent to ionic potential-energy surfaces provided an instructive example of a well-defined transition state.[1,2] The jump of an electron from an atom of low ionization potential such as an alkali metal to a molecule of high electron affinity, with the subsequent reeling-in of the anionic collision-partner by the cation came to be known as "harpooning". Such reactions in gases have been studied extensively. The novelty of the present work is that the alkali metal electron-donor (Na_n) is present as small clusters at an inert surface (LiF(001)), and the electron acceptor, $(HCl)_m$, is also present in the adsorbed state.

The reaction of Na + HCl → NaCl + H is endoergic. It has been shown to be efficiently promoted in the gas phase by vibrational excitation of the HCl bond, under attack.[3,4] This was interpreted in terms of the dynamics to be expected for the crossing of the type of 'late' energy-barrier characteristic of endoergic reactions.[5,6] The dynamics of the harpooning reaction have also been explored by experiment,[7,8] and theory[9] for the case of Na in its lowest Na*(^{2}P) electronically-excited state.

In recent years harpooning reactions have been examined by direct excitation into the reactive transition-state region, using an approach introduced by Soep[10,11] and Wittig[12,13] and applied to the case of alkali metal atom reactions in this laboratory. An alkali metal beam is crossed with a halide RX to form van der Waals complexes Na··XR which are then excited by tunable visible

DOI: 10.1039/b003973j

laser-radiation to the transition state, $[Na^{*} \cdots XR]^{\ddagger}$, broadly defined to include states intermediate between the reagent, $Na^{*} + XR$ and the products, $Na^{+}X^{-} + R$. In a series of papers this approach, which falls under the heading of 'Transition State Spectroscopy',[14] has been applied to the cases of $Na_n \cdots (ClCH_3)_m$,[15,16] $Na \cdots$fluorobenzene,[16] $Na \cdots (FCH_3)_n$,[17–19] $Na \cdots BrCH_3$,[19] $Na \cdots FH$.[19,21]

The path to products for this general category of reaction is indicated in eqn. (1):

$$Na \cdots XR \xrightarrow{h\nu} [Na^{*} \cdots XR]^{\ddagger} \longrightarrow [Na^{+} \cdots XR^{-}]^{\ddagger} \longrightarrow Na^{+}X^{-} + R \quad (1)$$

where $\ddagger$ is used to symbolize transition states. The potential-energy surfaces, PES, for this process have been computed by high-level *ab initio* calculations for the case that $XR = FH$.[20–23] These same PES govern the alternate inelastic pathway leading to $Na + FH(v \geqslant 0)$ as products.

Until now, the nearest parallel to these studies in the realm of surface science has been the investigation of photoinduced CT from metals to physisorbed adsorbates.[24–32] This has included alkali metal surfaces.[28,33] In order to approach the process of eqn. (1) still more closely in the adsorbed state, we have undertaken the present study (which we plan as the first of a series) of photoinduced CT from small sodium clusters, Na_n, to co-adsorbed hydrogen halides $(HCl)_m$ and, in later work, $(HF)_m$. The Na_n is adsorbed on a LiF(001) crystal cooled to 50 K to render the Na immobile (see Section III.A, below). Since Na_n is held in the adsorbed state, the $Na_n \cdots (ClH)_m$ complex is fixed in the laboratory frame of reference, rendering the observed product angular distributions informative.

Photoinduced CT reaction between Na_n and $(HCl)_m$ was induced by using 640 nm radiation (1.94 eV), well below the work function of sodium metal (2.75 eV[34]) but in the range of wavelengths used for electronic-excitation in the comparable gas-phase experiments. In a subsequent paper we will describe the effect of varying the wavelength of the incident radiation over a wide range. The results presented here are principally for the reactive pathway in which the product of eqn. (1) is $Na_n^{+}Cl^{-}(HCl)_{m-1}(ad) + H(g)$. For simplicity the discussion will for the most part be couched in terms of the case $m, n = 1$, *i.e.*, $Na \cdots ClH(ad)$ as reagent. However, evidence will be presented to suggest the importance of $(HCl)_2$, and the possible involvement of Na_2. The evidence for $(HCl)_2$ comes from observed vibrational fine-structure in the translational-energy of $H(g)$ from the reactive process. This structure gives a measure of the vibrational energy spacing in the $Na_n^{+}Cl^{-}(HCl)_{m-1}(ad)$ product. This vibrational spacing is found to be intermediate between that of NaCl and HCl, suggestive of $Na_n^{+}Cl^{-}(HCl)_{m-1}(ad)$. The structure and motion of this product is discussed on the basis of DIM (diatomics-in-molecules) computation, supported by *ab initio* calculations. Experimental evidence is also presented of the involvement of a second, non-reactive, pathway leading to $Na_n(ad) + HCl(v = 0, 1)$ as product (or, more generally, $Na_n(HCl)_{m-1}(ad) + HCl(v = 0, 1)$); detection of $HCl(v = 0, 1)$ was by REMPI–TOF, as described in the following section.

II. Experimental

The experiments were carried out in ultra-high-vacuum (UHV), with a base pressure better than 2×10^{-10} Torr. The LiF(001) surface was cleaved in air from a LiF single crystal (Harshaw Chemical Co.) and then annealed in UHV for 1 h at 700 K. During experiments, the crystal was cooled to 40–50 K by a closed-cycle He refrigerator (CSW-204SLB-6.5, APD Cryogenics Inc.). Sodium was dosed from a commercial SAES-getter mounted on a holder with a small defining aperture between it and the LiF(001) crystal. The Na-getter was degassed by passing 4.50 A through it for 5 min prior to each dose. Thereafter the Na itself was dosed by passing 6.50 A through the getter for 25 or 50 s to provide the appropriate dosage. The quantity of Na dosed was not linear with time, moreover it depended on the precise position of the hot wire, consequently it was necessary to calibrate the doser by various measures of the Na dosed (see below and Section III.A). The HCl gas was obtained from Matheson with stated purity of 99.0%, and was further purified by freeze–pump–thaw cycles.

For characterization of the Na adlayer, three types of experiments were performed: temperature-programmed desorption (TPD), X-ray photoelectron spectroscopy (XPS) and titration with known amounts of HCl. A doubly-differentially-pumped quadrupole mass spectrometer

(QMS) was used as detector in the TPD experiments. The QMS was set for mass = 23, with a temperature ramp of 2 K s^{-1}. The XPS spectra of the surface for increasing amounts of Na were obtained using Mg-Kα (1253.6 eV) radiation, with the electron analyzer at 8° off normal. In titration experiments, a fixed amount of Na was used and the H-atom yield from photoinduced reaction with HCl was measured for increasing amounts of HCl. The end-point of titration (see Section III.A) was signaled by termination of the increase in H(g) with increased HCl dose.

The reactive pathway was studied by H-atom Rydberg TOF spectroscopy, recently applied to surface experiments in UHV.[35,36] Briefly stated, upon excitation of the Na$_n$·(ClH)$_m$(ad) complex at 640 nm, the outgoing H-atom reaction-product was tagged by a two-color excitation (121.6 and approx. 365 nm) to a high-n Rydberg state, H**. The translationally-hot H-atoms traveled as neutral H** in UHV to a microchannel plate (MCP) detector where they were field-ionized and counted. The H-atom TOF obtained in this manner has high sensitivity and energy-resolution (better than 50 meV). The angular distribution of the ejected H-atoms was measured by rotation of the MCP detector around the crystal. Typically 200 laser pulses (640 nm, $\sim$0.8 mJ pulse^{-1}) were used for each data point, prior to depositing a fresh Na and HCl surface. This corresponded to an estimated $\sim$50% conversion of the adsorbate to products. To improve the signal-to-noise ratio in the evaluation of the observed vibrational spectra, this procedure was repeated 10 times and the 10 spectra were averaged for a total of 2000 laser pulses.

The non-reactive pathway was studied by monitoring the translationally-excited HCl leaving the surface in $v = 0$ and $v = 1$ vibrational states. This was done using mass-resolved REMPI to measure the TOF of the HCl for the specified vibrational and rotational states.[37,38] In the example shown in Section III.D, the HCl($v = 0$ and $v = 1$) TOF was obtained by irradiation of the Na$_n$·(ClH)$_m$(ad) complex with 602 nm, using (2 + 1) REMPI through the F($^1\Delta_2$) intermediate state of HCl.

III. Results and discussion

A The adlayer

Earlier CT reaction studies from alkali metal-atoms to HX were performed using several layers of metal. The CT electrons were made to travel through a layer of xenon to the HX target as an obstacle to halogenation of the metal.[28,33] In the present work we have dosed a *sub-monolayer* of Na atoms or clusters, Na$_n$, onto the surface and physisorbed the halide *directly onto* the sodium islands. The objective is to obtain information on the photoinduced CT reaction between Na$_n$(ad) and HCl(ad) (or (HCl)$_m$(ad)), extending to the adsorbed state the studies of harpooning in gaseous complexes, Na·$\cdot$XH(g) or Na·$\cdot$(XH)$_m$(g).

Träger and co-workers have shown that sodium atoms adsorbed on LiF(001) at 100 K aggregate to form large clusters.[39–41] Optical spectroscopy indicated that Na clusters of diameter $\simeq$10 to 90 nm were formed.[41,42] In the present work, a surface temperature of 40–50 K was used to reduce Na-atom mobility on the surface. The rate of Na diffusion was calculated on the basis of Na residence time ($\tau_{res} = \tau_o \exp(E_a/kT)$, with $E_a = 0.66$ eV) on the surface to be 10^{45} less than under the conditions used by Träger.[43] Since the Na coverage was low (see below), this slow diffusion of Na atoms should result in the presence of individual Na(ad) atoms or Na$_n$(ad) having low n.

TPD spectra for the system Na$_n$/LiF(001) are shown in Fig. 1. For the lowest observable Na coverage, the maximum in the TPD occurs at $\sim$400 K. Träger et al.[44] obtained TPD peaks that were invariably approx. 100 K lower in temperature ($\sim$300 K) than those observed here, indicative of binding energies to their LiF(001) surface in the range of $E_d = 0.5$–0.8 eV in contrast to $E_d \geqslant 1$ eV observed here. The stronger binding to the surface observed by us is probably related to the smaller clusters, or the individual Na(ad), obtained when the LiF surface was maintained at the low temperature used in our experiments. As already noted, it is expected that at 40–50 K the Na(ad) or low-n Na$_n$(ad) does not aggregate into the large clusters reported by Träger et al. Surprisingly, increase of coverage (either in the low-n range of these experiments, or the high-n range of Träger's work) is associated with an increase in the peak temperature of the TPD—compare peaks a and b in Fig. 1, and see refs. 39–41. This indicates that increase in coverage has another effect than simply increase in n, which decreases the peak in the TPD.

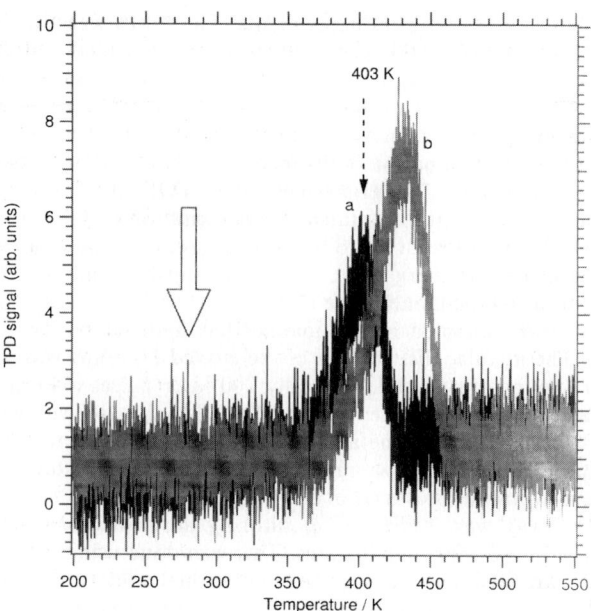

Fig. 1 TPD of Na from LiF(001). The initial surface temperature was 50 K. A ramp of 2 K s^{-1} was used. The peak for the smallest observable Na coverage was at 403 K, $E_d \sim 1$ eV. Peaks a and b are for doses of 7.00 A for 40 s and 7.00 A for 120 s respectively (with the usual 4.50 A for 5 min prior degassing time). The hollow arrow indicates the peak temperature for desorption observed by Träger et al.[44]

Fig. 1 shows two (50 K deposition) TPD spectra for Na coverages higher (for example, Na coverage of 7.0 A for 40 s is approx. 8 × higher than that of 6.5 A for 25 s) than those used in the photoinduced CT experiments reported here, since we were not able to detect smaller Na coverages by TPD. The contrast between the 280 K peak TPD in Träger's work (large clusters) as against the present work for which desorption peaks at ⩾400 K is clearly evident.

Characterization of the amount of adsorbed Na was undertaken in the first place by XPS. The change in energy and intensity (peak area) of the individual XPS peaks can be used to determine

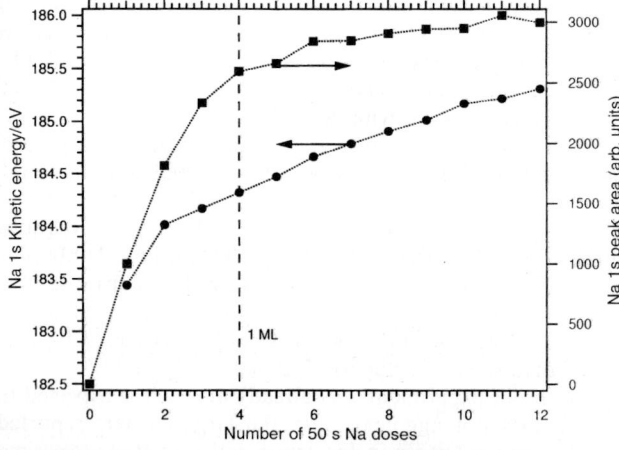

Fig. 2 Changes in the XPS peaks as a function of Na coverage. (●) Na 1s kinetic energy (left axis); (■) Na 1s peak area (right axis). Both curves are plotted as a function of the number of 50 s Na doses (bottom axis). Peaks shift as Na coverage is increased, until a plateau is reached and no more shifts are observed; a monolayer (1 ML, indicated by a vertical broken line) is assigned to this coverage.

the amount of Na deposited on the LiF(001) surface[45] (see Fig. 2 which has dosage along the abscissa and energy plus intensity along the ordinate scales at the left and the right, respectively). Both the Na 1s and the Na $KL_{23}L_{23}$-Auger peaks (the latter is not shown) shifted as the Na coverage was increased, reaching a plateau after ~4 consecutive 50 s doses of Na. Interpreting the plateau region to indicate an environment in which the Na has only similar Na atoms as neighbors, we can define the dosage at which the plateau is reached as being one monolayer (ML).[45] Similarly, the peak intensity shows a plateau after ~4 consecutive doses of Na for 50 s (90% of the maximum intensity). On this basis we estimate that one dose of Na for 50 s is equivalent to 0.25 ML. Under the conditions of our photoinduced CT reaction, namely 25 s Na-dose, the XPS intensity indicates a coverage of 0.05 ML.

Secondly, the sodium coverage was independently estimated by a titration procedure using the photoinduced reaction yielding atomic H(g). The method consisted of, first, dosing a reproducible amount of Na (for example: 6.50 A for 25 s, following a standard 4.50 A for 5 min of degassing of the getter). Next, an increasing amount of HCl was dosed. Following each HCl dose CT reaction was induced by illumination at 640 nm. The amount of HCl required to 'saturate' the H-atom signal (*i.e.* reach a maximum H-atom yield) provided a measure of the amount of Na on the surface (see Fig. 3). In earlier work using TPD, we have shown that 1 ML of HCl is equivalent to ~2 L (L = Langmuir) measured in the same UHV system.[35] Since the H-atom yield for a 25 s Na-dose saturates at ~0.1 L of HCl, a 1:1 relationship between Na and HCl implies a Na coverage of 0.05 ML. A 50 s Na-dose saturates at 0.5 L of HCl implying a coverage of 0.25 ML. Both these coverages are in excellent agreement with those determined from the XPS experiments.†

These very small Na coverages (*e.g.*, 0.05 ML), together with the high desorption energy seen in the TPD spectra and the extremely small rate calculated for the diffusion for Na atoms on the LiF(001) surface at 50 K (see discussion of the Träger group's findings, above) indicate that under our conditions we can deposit Na either as separate atoms or as small clusters. The Na coverages of 0.05 and 0.25 ML were those used for the photoinduced CT experiments described below.

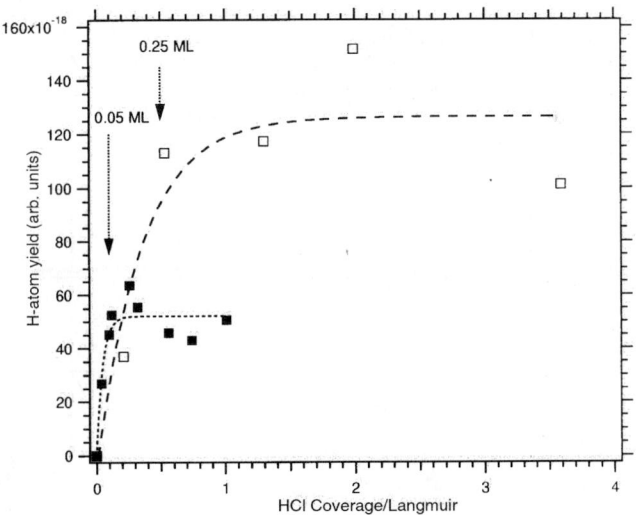

Fig. 3 Titration of Na coverage. (■) Sodium dose of 6.50 A for 25 s, (□) 6.50 A for 50 s. Broken lines are intended as guides to the eye (not fitted). Saturation of the H-atom yield occurs at 0.05 and 0.25 ML of HCl for the two Na doses; these coverages are indicated by arrows in the graph.

---

† The density of HCl on the LiF(001) surface has been estimated as 0.37 molecules per surface ion-pair[35] (using an ion-gauge sensitivity of $S_{HCl}/S_{N_2} = 1.5$,[46] and the area per adsorption site on LiF(001) = 16.08 Å^2 per two ion-pairs).[47,48] This suggests the following absolute converges of Na: for 25 s dosage, 0.0185 Na atoms per surface ion-pair, and for 50 s dosage, 0.0925 Na atoms per surface ion-pair.

B The reactive channel

The reactive channel was monitored by H-atom Rydberg TOF spectroscopy (HRTOF).[49,50] The $(HCl)_m \cdot \cdot Na_n/LiF(001)$ was excited close to the Na D-line, since Na is the main chromophore in the adsorbate. Excitation of the Na$\cdot\cdot$FH(g) complex in gas-phase studies[20,21] showed a maximum absorption at 640 nm, so this same wavelength was used in the present work to excite the $Na_n \cdot \cdot (ClH)_m(ad)$. The LiF(001) surface is transparent at this wavelength, and does not contribute to the CT reaction.

The HRTOF method, used extensively in the gas phase, has been shown recently to yield accurate TOF spectra for H-atoms leaving a surface upon reaction, provided that H-atoms of all speeds are within the width of the HRTOF excitation laser beams.[35,36] The measured flux of H atoms can be transformed into translational energy distributions, $P(E'_T)$, by multiplication by a factor of t^3. This, however, introduces substantial uncertainty at large t, i.e. at low energies.[51]

The observed translational-energy distributions, $P(E'_T)$, for H(g) coming from the photoinduced CT reaction in $(HCl)_m \cdot \cdot Na_n/LiF(001)$ is recorded in Fig. 4 for two observation angles, 0° and 35° to the crystal-surface normal. The translational energies observed extend over a range of approx. $E'_T = 0$–1.5 eV, with a most-probable E'_T of approx. 0.5 eV.

As is evident in Fig. 4, reproducible structure was found superimposed on the $P(E'_T)$ in the region ≥ 0.5 eV, where the error-bars are small. This structure is suggestive of vibrational excitation, both as regards the magnitude of the energy-spacing (0.1–0.25 eV) and its gradation. This gradation consists of a systematic narrowing of the spacing between the peaks toward lower E'_T, i.e., in the region where the energy available for vibrational-excitation is greatest. This is consistent with vibrational anharmonicity.

The H-atom kinetic energy, E'_T, arising from the photoinduced reaction $Na_n \cdot \cdot (ClH)_m(ad) \to Na_n{}^+Cl^-(HCl)_{m-1}(ad) + H(g)$, depends on the CT photon energy, $h\nu$, the

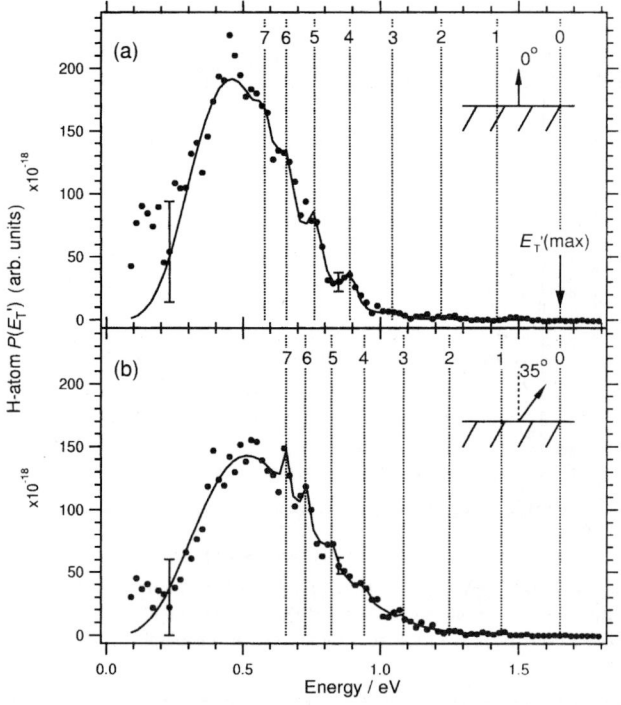

Fig. 4 H-atom translational energy distributions measured at 0° and 35° off-normal ((a) and (b) respectively) for 0.05 ML Na coverage. Broken lines in each case indicate the vibrational peak progression as estimated by a fitted curve consisting of one Boltzman function, and eight Gaussian functions superimposed on the Boltzman function (see text).

endothermicity of the reaction, ΔH, and the energy, $E_{v,J}(\text{ad})$, remaining behind in the reaction-product at the surface, $\text{Na}_n{}^+\text{Cl}^-(\text{HCl})_{m-1}(\text{ad})$. Eqn. (2) gives the H-atom translational energy available for the case where the complex contains only one HCl molecule ($m = 1$):

$$E'_T(H) = \left(\frac{\text{mass}(\text{Na}_n\text{Cl})}{\text{mass}(\text{Na}_n\text{ClH})}\right)(h\nu - \Delta H - E_{v,J}(\text{ad})) \qquad (2)$$

The maximum available energy of the departing H atoms, $E'_T(\text{max})$, can be estimated from eqn. (2) for $E_{v,J} = 0$ using the known endothermicity of the gas-phase reaction $\text{Na}(g) + \text{ClH}(g) \rightarrow \text{NaCl}(g) + \text{H}(g)$ of 0.18 eV.[3,4] This endothermicity would be only slightly modified by the prior formation of the $\text{Na}_n\cdots\text{ClH}(g)$ complex. (For comparison, $\text{Na}\cdots\text{FH}$ has been found to be bound by less than 0.076 eV.[20]) The endothermicity will also be subject to some modification due to the heat of adsorption of the reagent, $\text{Na}_n\cdots\text{ClH}(\text{ad})$, and the product, NaCl(ad). For the simplest complex reagent, $\text{Na}_n\cdots\text{ClH}(\text{ad})$, taking $\Delta H = 0.18$ eV, the value of $E'_T(\text{max})$, using 640 nm CT radiation, is 1.67 eV. This is much higher than the observed translational energy of much of the photoproduced H(g) (Fig. 4), leaving sufficient energy for other modes of product excitation.

The vibrational structure superimposed on the observed translational energy distributions was fitted using a vibrational Morse progression starting at the maximum available energy for the H atoms, $E'_T(\text{max})$, and a broad Boltzman function with a most-probable energy at ~ 0.5 eV. Each peak in the vibrational progression was fitted using a Gaussian function. The peak position for the higher energy peaks was extrapolated from the anharmonicity observed in the lower-energy vibrational peaks. The value of $E'_T(\text{max})$ was taken to be 1.67 eV. The results of the fit yielded estimates of the vibrational constants for each detection angle as follows: for $\theta = 0°$, $\omega_e = 0.25(1)$ and $\omega_e\chi_e = 0.012(3)$ eV; and for $\theta = 35°$, $\omega_e = 0.23(5)$ and $\omega_e\chi_e = 0.011(6)$ eV.

The observed vibrational spacing was much larger than the vibrational spacing reported for gaseous NaCl ($\Delta E_v(1 \leftarrow 0) = 0.045$ eV[52,53]). The fundamental transition was not observed in our vibrational progression, but using the Morse fitting $\Delta E_v(1 \leftarrow 0)$ is 0.22(6) eV for 0° (Fig. 4(a)), and 0.21(2) eV for 35° off-normal detection (Fig. 4(b)), approximately 5× that for NaCl(g). The $\Delta E_v(1 \leftarrow 0)$ for gaseous HCl is 0.35(8) eV.[54] Therefore, the observed spacings in the H-atom $P(E'_T)$ seem to be too large to be due to residual $\text{Na}_n\text{Cl}(\text{ad})$ (see the following section for a comparison of NaCl vibration with Na_2Cl), and too small to be due to scattering of H off adjacent HCl.

A possible source of the observed spacing is the $\text{Na}_n{}^+\text{Cl}^-$–HCl product of CT in a $\text{Na}_n\cdots(\text{ClH})_2(\text{ad})$ reagent. For this HCl-dimer reagent, the value of the H-atom maximum available energy is higher, so that the observed progression ($v = 0$ to 7 in Fig. 4) extends to approx. two vibrational levels above the values of v assigned in Fig. 4. *Ab initio* calculations have been made of this product (Section III.C). Briefly, the Na^+Cl^-–HCl product is calculated to be bound by ≈ 0.5 eV, with a vibrational spacing $\Delta E_v(1 \leftarrow 0) \approx 0.23$ eV for the asymmetric (H–Cl) stretch, which is comparable to the observed vibrational spacing (re-calculated, as noted above, for the larger value of $E'_T(\text{max})$ with this product).

The angular distribution of the H-atom yield summed over E'_T (Fig. 5) shows, for the lower Na coverage of 0.05 ML, a bimodal distribution with maxima at 0° and 35° off normal. This suggests a bimodal distribution of the Cl–H alignment within the $\text{Na}_n\cdots(\text{ClH})_2$ (ad) reagent, with respect to the surface. As is evident in Fig. 4 for 0.05 ML, the vibrational spacing is narrower at the off-normal angle of 35°. This, in turn, indicates that the reaction product left behind at the surface for the low coverage of 0.05 ML of Na is different for 0° and 35° ejection of H(g). We tentatively suggest alternate reagent configurations in the following section.

When the Na coverage is increased from 0.05 to 0.25 ML, the peak progression in $P(E'_T)$ becomes indistinct, consistent with a larger variety of clusters undergoing reaction. At this higher Na coverage, the bimodal angular distribution of H atoms also disappears, being replaced by a $\cos^n\theta$ distribution with a single maximum at $\theta = 0°$ (Fig. 5).

C Theory

In this section we present *ab initio* configuration-interaction (CI) calculations carried out with extensive aug-cc-pVTZ basis sets[55] using the Molpro package[56] with the objective of identifying the CT reaction through the vibrational structure of the product left behind at the surface. The simplest $\text{Na}\cdots\text{ClH}(\text{ad})$ complex would yield, following CT, the product NaCl(ad). As already

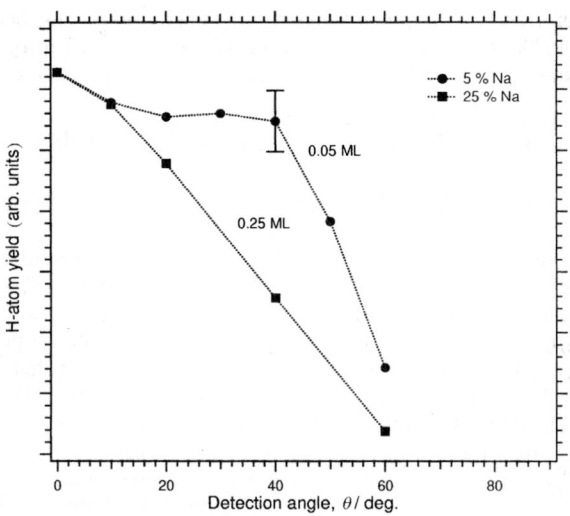

Fig. 5 Angular distributions of the H-atom yield for two Na coverages, 0.05 and 0.25 ML. Sample error-bar is shown.

noted, this has a vibrational frequency $\omega_e \approx 0.045$ eV in the gas, far less than the observed value of 0.25 eV. To estimate the influence of the electric field generated by the LiF crystal surface ions on this gas-phase vibrational frequency, multireference internally contracted CI (MRCI) calculations were carried out for this molecule in isolation, and above a 2D grid of 11 × 11 alternating positive and negative charges. The results obtained for several distances from the surface showed only minor variations of the potential shape and hence of ω_e.

To check, in addition, the effect of chemical binding to the surface, we have considered a hypothetical F^-–Na^+ Cl^- system. Since NaCl is a typical ionic molecule with a strong charge separation, the interaction in such a triatomic system can be approximated by a direct sum of the ground state NaF and NaCl potentials plus the Coulomb repulsion between the Cl^- and F^- ions. The vibrational levels of $(FNaCl)^-$ were evaluated using the Triatom package.[57] The vibrational spacing was again too small to account for the experimental observations. These results assume that the F^- ion in $(FNaCl)^-$ is completely isolated from other surface ions of LiF(001). If we were to include other ions, thereby hindering the motion of the F^- ion in the complex, the effective increase in F^- mass would reduce the vibrational spacing in $(FNaCl)^-$ and would have the effect of further increasing the discrepancy with the experimental finding.

To consider the consequences of possible clustering of sodium, DIM[58] and MRCI calculations were performed to obtain the PESs for Na_2Cl, which could be present as a reaction product of an $Na_2 \cdot\cdot ClH(ad)$ complex. The results showed a relatively small change of the PES shape in the favoured symmetric T-shaped (C_{2v}) geometry as compared to an isolated NaCl molecule. The direction of the effect is to decrease ω_e, since the reduced mass is larger, once again increasing the discrepancy with experiment, as compared with NaCl as product.

Next we considered an indirect collisional scenario in which the ejected H atom struck some plentiful co-adsorbate, losing a part of its energy before leaving the crystal surface. For this we need a sufficiently strongly bound co-adsorbate to survive the transfer of ≥ 0.5 eV of vibrational excitation. This eliminates collision partners which are weakly bound, such as $Na\cdot\cdot ClH(ad)$, $Na_2\cdot\cdot ClH(ad)$, etc. Though the large vibrational spacing suggests a hydrogen-containing collision-partner, HCl (which is plentiful) has too high an ω_e value (≈ 0.36 eV). This could, however, be reduced in a sufficiently strongly-bound complex in which the effective mass of H might be greater. At the low temperatures characteristic of the present experiments, HCl is likely to dimerise forming $(HCl)_2$. This complex is, however, too weakly bound ($D_e < 0.1$ eV) to be a candidate to receive ≥ 0.5 eV of vibration.

We have looked, therefore, for a more strongly bound complex of HCl. The Na^+Cl^- product of CT has a large dipole moment capable of providing a strong attraction for HCl. Accordingly we computed the properties of the $Na^+Cl^-\cdots HCl$ complex which may be left behind at the surface as reaction product following photoinduced CT in $Na\cdot(ClH)_2(ad)$.

We report calculations for the $Na^+Cl^-\cdots HCl$ complex at the CISD (single-reference CI) level. The interaction potential between the NaCl and HCl molecules fixed at their equilibrium internuclear separations is shown in Fig. 6(a) for the collinear geometry corresponding to the (classical) minimum-energy orientation of two interacting dipoles. The binding between Na^+Cl^- and HCl even in this less than optimal configuration is substantial; $D_e \approx 0.3$ eV at $r_e(Cl^-\cdots H) = 2.28$ Å. The actual equilibrium configuration is found to be planar (Fig. 7) and have an Na^+Cl^-H angle $\approx 68°$ and Cl^-HCl angle $\approx 150°$, with the HCl and Na^+Cl^- slightly stretched (by ≈ 0.05 Å). For this configuration $D_e \approx 0.5$ eV, as indicated in Fig. 6(a). The minimum-energy configuration shown in Fig. 7 can be compared to the rhombic geometry of $(NaCl)_2$ and the near-L-shaped[59] geometry of $(HCl)_2$.

As expected, the ω_e value for the relative motion of the two rigid molecules, NaCl and HCl, in the collinear geometry is too low to match the experimental value. The vibrational mode associated with the relative motion of the H atom is the one that is of interest. To estimate its frequency we moved H along the HCl axis while keeping all other atoms fixed in their equilibrium positions. This approximately corresponds to the asymmetric stretch normal mode. The associated potential is shown in Fig. 6(b).

Evaluation of the vibrational levels for the effective 1D potential of the asymmetric stretch gives rise to significantly smaller vibrational spacing than for HCl, as can be seen in Fig. 6(b). The ω_e value is about 25% lower than for HCl(g), and the $\omega_e\chi_e$ value about four times larger than for HCl(g). The computed values are $\omega_e \approx 0.28$ eV and $\omega_e\chi_e \approx 0.025$ eV, in satisfactory agreement

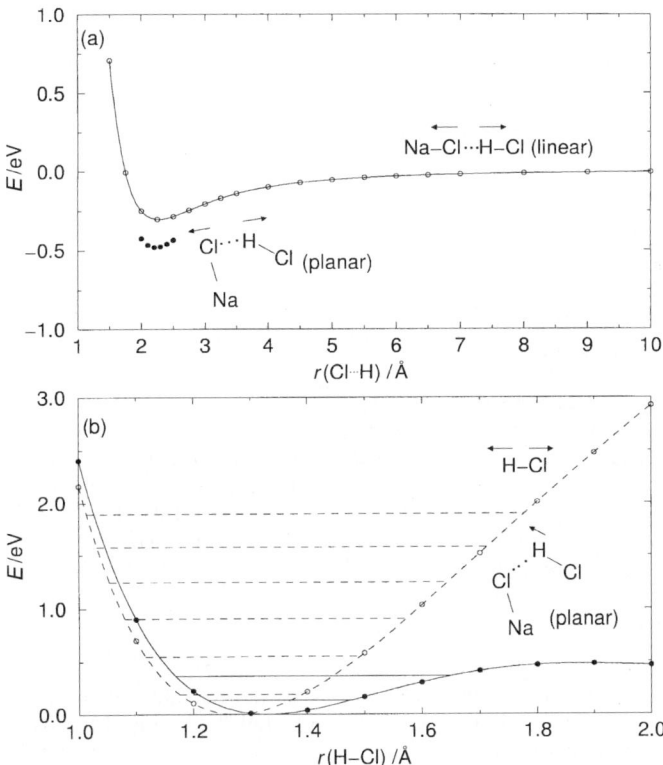

Fig. 6 (a) The *ab initio* ground state interaction potential between NaCl and HCl fixed at their equilibrium interatomic separations for the collinear geometry (○) and for the equilibrium planar configuration (●). (b) The interaction potential and associated vibrational levels for HCl(g) (– –) and for the H motion (an asymmetric stretch) within the complex (———), with other atoms fixed in their equilibrium positions.

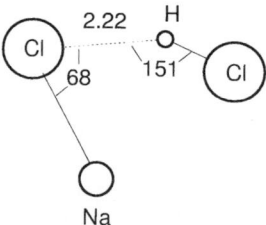

Fig. 7 Equilibrium configuration of the ground state NaCl··HCl complex, with r_e(NaCl) = 2.46 Å and r_e(HCl) = 1.33 Å.

with the experimental values of $\omega_e \approx 0.25$ eV and $\omega_e \chi_e \approx 0.012$ eV. (The calculated large anharmonicity may require higher order terms.)

The equilibrium configuration of the excited state NaClH complex associated with the reaction is predicted to have a Na–Cl–H angle of about 90°. Combined with the nearly-L-shaped equilibrium configuration of (HCl)$_2$ (with the Cl–H–Cl and H–Cl–H angles close to 180° and 90°, respectively), we arrive at the following schematic picture of the CT photoreaction:

$$\text{Na}\cdot\cdot\text{Cl}\cdots\text{HCl}\overset{\text{H}}{\phantom{\text{Cl}}} + h\nu \longrightarrow \text{Na}^+\text{Cl}^-\cdots\text{HCl}\overset{\text{H}}{\phantom{\text{Cl}}} \quad (3)$$

The reaction transforms the Na and its nearest Cl atom into a strongly bound Na$^+$Cl$^-$ molecule, while ejecting the central H atom which adjoins Cl$^-$. An alternative adsorbate configuration would involve two Na atoms in binding the (HCl)$_2$ to the surface:

$$\begin{array}{c}\text{ClH}\cdots\text{ClH}\\ \cdot \quad \cdot \\ \cdot \quad \cdot \\ \text{Na} \quad \text{Na}\end{array} + h\nu \rightarrow \begin{array}{c}\text{ClH}\\ \cdot \quad \cdot \\ \cdot \quad \text{Cl}^-\\ \text{Na} \quad \text{Na}^+\end{array} + \text{H} \quad (4)$$

The Na–Na spacing of ≈ 3 Å in Na$_2$(g) matches the separation between fluorine negative ions in the underlying LiF(001) crystal, and makes the involvement of Na$_2$ plausible. The schematic of eqn. (4) shows photoinduced CT ejecting a terminal H-atom that is able to escape at high translational energy without collision (compare eqn. (3)).

The differing adsorbate structures may perhaps be associated with the different components of the angular distributions for fast H(g) observed at low coverage, one normal to the surface (eqn. (3)) and one at an angle to the normal (eqn. (4)). Photoinduced CT to release the central H atom (though not shown in eqn. (4)) will also occur. However, since this H is 'caged' between two Cl atoms its translational energy will be broadened and lessened; it could contribute to the apparently unstructured low-energy portion of the H atom energy-distribution in Fig. 4.

D The non-reactive channel

The non-reactive channel has been observed by (2 + 1) REMPI, of the rotationally and vibrationally excited HCl leaving the surface. The TOF spectra of HCl($v = 0$ and $v = 1$) were measured using the F($^1\Delta_2$)–X$^1\Sigma^+$(0,0) and (1,1) bands for REMPI.[60–62] Fig. 8(a) gives the translational energy distribution of HCl in the ground vibrational level ($v = 0$). By observing a higher rotational state ($J = 5$) the background signal from gas-phase HCl already present in the chamber, was largely eliminated.

For discussion we consider the simplest adsorbate complex, Na$_n$··ClH(ad) as the reagent in the CT process. The available translational energy for the case in which the reagent contains only one

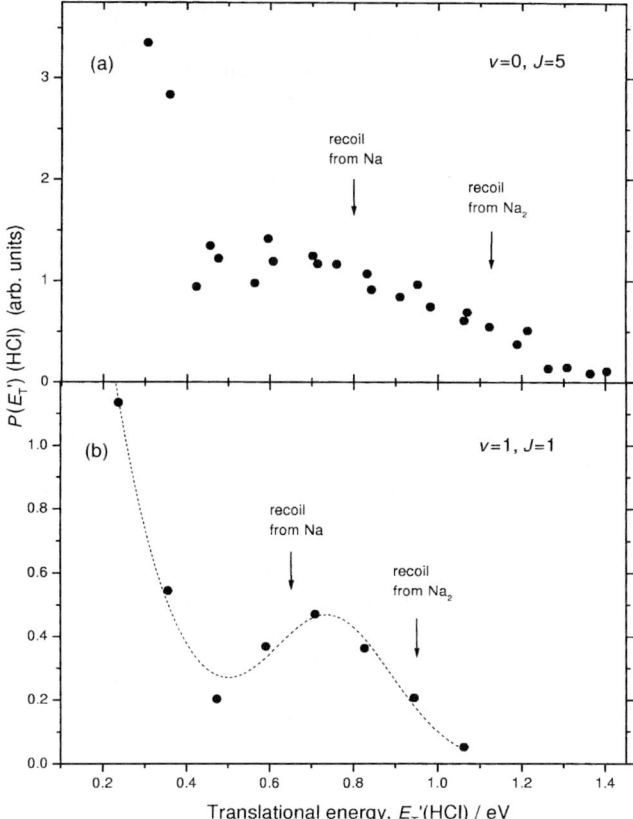

Fig. 8 Translational energy-distribution of HCl, produced by the non-reactive CT pathway; (a) HCl($v = 0$; $J = 5$); (b) HCl($v = 1$; $J = 1$). The photolysis wavelength was 602 nm, and the Na coverage 0.12 ML. Arrows show maximum available energies, calculated by eqn. (5) for recoil from Na and Na$_2$ as indicated.

HCl molecule can be obtained from eqn. (5) assuming impulsive recoil of HCl from Na$_n$:

$$E'_T(\text{HCl}) = \left(\frac{\text{mass(Na}_n)}{\text{mass(Na}_n\cdot\cdot\text{ClH})}\right)(hv - E_b(\text{Na}_n\cdot\cdot\text{ClH(ad)}) - E_{v,J}(\text{HCl}) - E_{v,J}(\text{Na}_n)) \quad (5)$$

where E_b is the physisorption binding energy of HCl to Na$_n$.

In Fig. 8(a), translationally hot HCl($v = 0$) can be observed up to $\sim$1.2 eV of translational energy. The expected maximum translational energy if the HCl originated from Na$\cdot\cdot$ClH would be 0.78 eV (assuming a negligible binding energy in the complex, E_b) which is indicated by an arrow in Fig. 8(a). If HCl recoiled from Na$_2$ the corresponding maximum translational energy would be 1.13 eV, indicated by the high-energy arrow in Fig. 8(a). Recoil from the heavier Na$_2$ may explain the higher energy contribution to E'_T(HCl), by simple momentum conservation. This gains some credence from the observation of desorbing Na$_2$ (46 u) and Na$_2$Cl (81.4 u) by photoionisation, in the photolysis experiments.

The low-energy-channel contribution to $P(E'_T)$(HCl) in Fig. 8 (average E'_T about 0.2 eV) is likely to be due to HCl molecules produced by the non-reactive pathway that lose energy in collisions with adjacent adsorbate (a pathway that we have called 'indirect' in other work[36]).

In addition to HCl($v = 0$), vibrationally excited HCl($v = 1$) was observed with smaller yield (approx. 25–35% compared to HCl($v = 0$)). The translational energy distribution and the calculated available energies (eqn. (5)) are shown in Fig. 8(b). As for HCl($v = 0$), a high-energy channel is observed which can be presumed to be due to direct escape of HCl($v = 1$) from the surface. The high energy HCl($v = 1$) may also arise from a $Na_2 \cdot \cdot ClH$ complex, since its maximum translational energy is well in excess of the limit of 0.65 eV calculated for $Na \cdot \cdot ClH(ad)$.

Some contribution to the observed HCl may be due to the excitation of larger clusters, $Na_n \cdot \cdot (ClH)_m(ad)$. Since the $HCl \cdot \cdot HCl$ binding is weak (0.05 eV in gas phase), the HCl molecules ejected from such clusters are expected to have similar energies to those arising from $Na_n \cdot \cdot ClH(ad)$. The tail in $P(E'_T)$ extending up to 0.2 eV beyond the limit set by eqn. (5) is understandable if the effective mass of Na_2 is increased by the near-by presence of the surface.

IV. Conclusion

The CT reaction between small sodium clusters, Na_n, and hydrogen halide, $(HCl)_m$, has been studied in the adsorbed state using a cold 40–50 K LiF(001) surface. The excitation of the adsorbed complex, $Na_n \cdot \cdot (ClH)_m(ad)$, was by 640 nm radiation. The product, H(g), from photoinduced CT reaction to give sodium halide was monitored by HRTOF spectroscopy. The H-atom translational energy distributions exhibited a vibrational progression that could reflect the excitation in the residual $Na_n \cdot \cdot Cl(ClH)_{m-1}(ad)$ at the surface. An analysis of the observed H-atom $P(E'_T)$ was performed assuming the simplest reagent ($n = 1$ and $m = 1$), to obtain values for ω_e and $\chi_e \omega_e$. The vibrational spacing was too large to be due to NaCl and too small to be due to HCl. With the aid of *ab initio* theory, the $Na \cdot \cdot (ClH)_2$ complex was identified as a possible source of the observed vibrational structure since its product, $Na^+Cl^-HCl(ad)$, was calculated to be significantly bound ($D_e \approx 0.5$ eV) and to have a $\omega_e \sim 0.28$ eV for the normal mode associated with the H–Cl stretch, close to the observed value of ~ 0.25 eV. The non-reactive channel of the photoinduced CT, yielding $Na + HCl(E'_T, v, J)$, was characterized by REMPI of $HCl(v = 0)$ and $HCl(v = 1)$. Translationally hot HCl was observed up to approximately the limit set by momentum conservation in the recoil of HCl from Na and Na_2.

Acknowledgements

The authors are indebted to the Natural Sciences and Engineering Research Council of Canada (NSERC), to the Canadian Institute for Photonics Innovation (CIPI) a National Center of Excellence, to Photonics Research Ontario (PRO) an Ontario Center of Excellence, to the Ontario Research and Development Challenge Fund (ORDCF), and to the Gennum Corporation of Burlington, Ontario, for their support of this work. We would also like to thank Professor Agust Kvaran for valuable advice in the implementation of the REMPI detection, and Professor Yehuda Zeiri for helpful discussions of these results.

References

1. R. A. J. Ogg and M. Polanyi, *Trans. Faraday Soc.*, 1935, **31**, 604.
2. M. G. Evans and M. Polanyi, *Trans. Faraday Soc.*, 1938, **34**, 11.
3. B. A. Blackwell, J. C. Polanyi and J. J. Sloan, *Chem. Phys.*, 1978, **30**, 299.
4. F. E. Bartoszek, B. A. Blackwell, J. C. Polanyi and J. J. Sloan, *J. Chem. Phys.*, 1981, **74**, 3400.
5. J. C. Polanyi, *Science*, 1987, **236**, 680.
6. M. Shapiro and Y. Zeiri, *J. Chem. Phys.*, 1979, **70**, 5264.
7. M. F. Vernon, H. Schmidt, P. S. Weiss, M. H. Covinsky and Y. T. Lee, *J. Chem. Phys.*, 1986, **84**, 5580.
8. J. M. Mestdagh, B. A. Balko, M. H. Covinsky, P. S. Weiss, M. F. Vernon, H. Schmidt and Y. T. Lee, *Faraday Discuss. Chem. Soc.*, 1987, **84**, 145.
9. K. Yamashita and K. Morokuma, *Chem. Phys. Lett.*, 1990, **169**, 263.
10. C. Jouvet and B. Soep, *Chem. Phys. Lett.*, 1983, **96**, 426.
11. C. Jouvet, M. Boiveneau, M. C. Duval and B. Soep, *J. Phys. Chem.*, 1987, **91**, 5146.
12. S. Buelow, G. Radhakrishnan, J. Catanzarite and C. Wittig, *J. Chem. Phys.*, 1985, **83**, 444.
13. C. Wittig, S. Sharpe and R. A. Beaudet, *Acc. Chem. Res.*, 1988, **21**, 341.
14. J. C. Polanyi and A. H. Zewail, *Acc. Chem. Res.*, 1995, **28**, 119.
15. K. Liu, J. C. Polanyi and S. Yang, *J. Chem. Phys.*, 1992, **96**, 8628.
16. K. Liu, J. C. Polanyi and S. Yang, *J. Chem. Phys.*, 1993, **98**, 5431.
17. J. C. Polanyi, J.-X. Wang and S. H. Yang, *Isr. J. Chem.*, 1994, **34**, 55.
18. J. C. Polanyi and J.-X. Wang, *J. Phys. Chem.*, 1995, **99**, 13691.
19. X. Y. Chang, R. Ehlich, A. J. Hudson, J. C. Polanyi and J.-X. Wang, *J. Chem. Phys.*, 1997, **106**, 3988.
20. X. Y. Chang, R. Ehlich, A. J. Hudson, P. Piecuch and J. C. Polanyi, *Faraday Discuss.*, 1997, **108**, 411.

21 M. S. Topaler, D. G. Truhlar, X. Y. Chang, P. Piecuch and J. C. Polanyi, *J. Chem. Phys.*, 1998, **108**, 5378.
22 M. S. Topaler, D. G. Truhlar, X. Y. Chang, P. Piecuch and J. C. Polanyi, *J. Chem. Phys.*, 1998, **108**, 5349.
23 Y. Zeiri, G. Katz, R. Kosloff, M. S. Topaler, D. G. Truhlar and J. C. Polanyi, *Chem. Phys. Lett.*, 1999, **300**, 523.
24 W. Ho, in *Surface Photochemistry*, ed. H.-L. Dai and W. Ho, World Scientific, Singapore, 1995.
25 X. L. Zhou and J. M. White, in *Photodissociation and Photoreaction of Molecules Attached to Metal Surfaces*, ed. H.-L. Dai and W. Ho, World Scientific, Singapore, 1995.
26 I. Harrison, in *Adsorbate Photochemical Dynamics on Pt(111)*, ed. H.-L Dai and W. Ho, World Scientific, Singapore, 1995.
27 E. P. Marsh, T. L. Gilton, W. Meier, M. R. Schneider and J. P. Cowin, *Phys. Rev. Lett.*, 1988, **61**, 2725.
28 S. J. Dixon-Warren, J. C. Polanyi, C. D. Stanners and G. Q. Xu, *J. Phys. Chem.*, 1990, **94**, 5664.
29 S. J. Dixon-Warren, E. T. Jensen and J. C. Polanyi, *Phys. Rev. Lett.*, 1991, **67**, 2395.
30 S. J. Dixon-Warren, E. T. Jensen and J. C. Polanyi, *Desorption Induced by Electronic Transitions, DIET V*, ed. A. R. Burns, E. B. Stechel and D. R. Jennison, Springer-Verlag, Berlin, 1993.
31 S. J. Dixon-Warren, D. V. Heyd, E. T. Jensen and J. C. Polanyi, *J. Chem. Phys.*, 1993, **98**, 5954.
32 S. J. Dixon-Warren, E. T. Jensen and J. C. Polanyi, *J. Chem. Phys.*, 1993, **98**, 5938.
33 S. J. Dixon-Warren, E. T. Jensen, J. C. Polanyi, G.-Q. Xu, S. H. Yang and H. C. Zeng, *Faraday Discuss. Chem. Soc.*, 1991, **91**, 451.
34 *CRC Handbook of Chemistry and Physics*, ed. D. R. Lide, CRC Press, Boca Raton, FL, 1996–1997, pp. 12–122.
35 J. B. Giorgi, R. Kühnemuth, J. C. Polanyi and J.-X. Wang, *J. Chem. Phys.*, 1997, **106**, 3129.
36 J. B. Giorgi, R. Kühnemuth and J. C. Polanyi, *J. Chem. Phys.*, 1999, **110**, 598.
37 R. C. Jackson, J. C. Polanyi and P. Sjovall, *J. Chem. Phys.*, 1995, **102**, 6308.
38 J. B. Giorgi, F. Y. Naumkin, J. C. Polanyi, S. A. Raspopov and N. S.-K. Sze, *J. Chem. Phys.*, 2000, **112**, 9569.
39 W. Hoheisel, U. Schulte, M. Vollmer, R. Weidenauer and F. Träger, *Appl. Surf. Sci.*, 1989, **36**, 664.
40 M. Vollmer, R. Weidenauer, W. Hoheisel, U. Schulte and F. Träger, *Phys. Rev. B*, 1989, **40**, 12509.
41 W. Hoheisel, U. Schulte, M. Vollmer and F. Träger, *Appl. Phys. A*, 1990, **51**, 271.
42 T. Gotz, W. Hoheisel, M. Vollmer and F. Träger, *Z. Phys. D*, 1995, **33**, 133.
43 M. Vollmer and F. Träger, *Z. Phys. D*, 1986, **3**, 291.
44 M. Vollmer and F. Träger, *Surf. Sci.*, 1987, **187**, 445.
45 X. Shi, D. Tang, D. Heskett, K.-D. Tsuei, H. Ishida and Y. Morikawa, *Surf. Sci.*, 1993, **290**, 69.
46 R. L. Summers, *NASA technical note*, 1969, TN D-5285.
47 *Gmelins Handbuch der Anorganischen Chemie, Lithium*, 8. Auflage, Ergänzungsband, Verlag Chemie, Weinheim, 1960.
48 *Landolt-Börnstein, Zahlenwerte und Funktionen aus Physik, Chemie, Astronomie*, Springer-Verlag, Berlin, 1971 Band II, 1. Teil, S. 479.
49 L. Schnieder, W. Meier, K. H. Welge, M. N. R. Ashford and C. M. Western, *J. Chem. Phys.*, 1990, **92**, 7027.
50 L. Schnieder, K. Seekamp-Rahn, F. Liedeker, H. Steuwe and K. H. Welge, *Faraday Discuss. Chem. Soc.*, 1991, **91**, 259.
51 M. Nooney, J. M. Price, R. M. Martin and A. M. Wodtke, *Chem. Phys. Lett.*, 1995, **245**, 377.
52 P. Brumer and M. Karplus, *J. Chem. Phys.*, 1973, **58**, 3903.
53 R. S. Ram, M. Dulick, B. Guo, K. Q. Zhang and P. F. Bernath, *J. Mol. Spectrosc.*, 1997, **183**, 360.
54 K. P. Huber and G. Herzberg, *Constants of Diatomic Molecules*, Van Nostrand Reinhold, New York, 1979.
55 T. H. Dunning, *J. Chem. Phys.*, 1989, **90**, 1007.
56 Molpro is a package of *ab initio* programs written by H.-J. Werner and P. J. Knowles, with contributions from J. Almlof, R. D. Amos, A. Berning, M. J. O. Deegan, F. Eckert, S. T. Elbert, C. Hampel, R. Lindh, W. Meyer, A. Nicklass, K. Peterson, E.-A. Reinsch, R. M. Pitzer, A. J. Stone, P. R. Taylor, M. E. Mura, P. Pulay, M. Schuetz, H. Stoll, T. Thorsteinsson and D. L. Cooper.
57 J. Tennyson, S. Miller and C. R. LeSueur, *Comput. Phys. Commun.*, 1993, **75**, 339.
58 F. O. Ellison, *J. Am. Chem. Soc.*, 1963, **85**, 3540.
59 M. J. Elrod and R. J.Saykally, *J. Chem. Phys.*, 1995, **103**, 933.
60 D. S. Green, G. A. Bickel and S. C. Wallace, *J. Mol. Spectrosc.*, 1991, **150**, 303.
61 W. R. Simpson, T. P. Rakitzis, S. A. Kandel, A. J. Orr-Ewing and R. N. Zare, *J. Chem. Phys.*, 1995, **103**, 7313.
62 A. Kvaran, H. Wang and A. Logadottir, *J. Chem. Phys.*, 2000, **112**, 10811.

Semiclassical treatment of reactions at surfaces with electronic transitions

Christian Bach and Axel Groß*

Physik-Department T30, Technische Universität München, D-85747 Garching, Germany

Received 3rd April 2000
First published as an Advance Article on the web 29th September 2000

The semiclassical treatment of reactions at surfaces with electronic transitions based on the fewest-switches algorithm is compared with full quantum mechanical results. As a model system the ionization probability in I_2 scattering from a diamond surface is chosen. In the calculations we treat the molecular distance from the surface and one surface oscillator coordinate explicitly. Furthermore, we also consider molecular rotation in the semiclassical calculations. The semiclassical results agree with the quantum results although some discrepancies remain, as far as the phase coherence is concerned. We identify energy transfer to molecular and surface degrees of freedom as a possible mechanism that could explain the experimental dependence of the ionization probability on the incident kinetic energy of the molecule.

I. Introduction

There has been tremendous progress in the theoretical treatment of reactions at surfaces in recent years. In particular, for hydrogen dissociation on metal surfaces, high-dimensional dynamical calculations have been performed on potential energy surfaces which were derived from density-functional calculations.[1–4] These calculations allow quantitative comparison with experiment. However, they rely on one basic approximation, namely the Born–Oppenheimer approximation, *i.e.* in these simulations it is assumed that the electrons follow the motion of the nuclei adiabatically.

Dynamical calculations in the Born–Oppenheimer approximation exclude the treatment of processes such as surface photochemistry or charge transfer at surfaces. In these fields theory is far behind experiment since the theoretical treatment of reactions at surfaces with electronic transitions still represents a great challenge. Any reasonable theoretical description faces three major problems: (i) the potential energy surface of the ground *and* the excited states has to be determined; (ii) matrix elements between the different electronic states or the lifetimes of the excited states, respectively, have to be known; (iii) a simulation of the reaction dynamics including the electronic transitions has to be performed.

Quantum chemical algorithms still represent the main method to address the first two problems (see, *e.g.*, ref. 5), but methods based upon density-functional theory are also starting to be used to address this issue.[6,7] However, in this contribution we will focus on the third problem, reaction dynamics with electronic transitions at surfaces. On the one hand, these processes should be treated quantum mechanically because of the light mass of the electrons. On the other hand, to get a full and accurate understanding of dynamical processes on surfaces, a high-dimensional treatment including all relevant degrees of freedom is necessary. Low-dimensional studies might not only be quantitatively wrong, they can even prevent the detection of the correct qualitative mechanisms (see, *e.g.*, ref. 2). Unfortunately, despite all the progress in quantum dynamical methods in recent years a high-dimensional quantum treatment of reactions of interest with electronic transitions is not yet possible. These systems often involve oxygen or heavier atoms. Now the quantum effects in the motion of the nuclei are often negligible as long as hydrogen is not concerned, hence semiclassical schemes might be the method of choice to tackle these high-dimensional problems. In semiclassical methods the motion of the nuclei should be treated by classical methods, while the electronic transitions have to be described quantum mechanically. Still in such a scheme the feedback between classical and quantum mechanical degrees of freedom has to be treated self-consistently.

There have been quite a number of different methods proposed for the semiclassical treatment of reactions with electronic transitions which are not necessarily able to reproduce the correct quantum results (see, *e.g.*, ref. 8). Here we present a semiclassical study of charge transfer processes in the scattering of molecules at surfaces using the fewest-switches algorithm proposed by Tully.[9] This algorithm, which has already been tested for a number of model potentials,[10–12] impresses by its elegance and conceptual simplicity.

The goal of this investigation is twofold. On the one hand we want to check the performance and accuracy of this semiclassical treatment of reactions with electronic transitions. In particular, we are focusing on a situation with more than one nuclear degree of freedom that is treated classically and compare it with the exact quantum mechanical solution, *i.e.*, we are going beyond one-dimensional two-surface problems that is usually done.[9–15] On the other hand, we are also interested in gaining some qualitative insight into a particular reaction. Hence we have chosen the ionization probability in I_2 scattering from a diamond surface as our model system. This ionization probability has been measured as a function of the incoming kinetic energy of the scattered molecule by Danon and Amirav.[16]

For a low-dimensional treatment of the scattering process we compare semiclassical results with full quantum results. The comparison shows that the semiclassical method is capable of adequately reproducing the quantum results. Quantitative differences are identified, as far as quantum interference phenomena are concerned. In higher-dimensional applications, however, these differences will disappear. This becomes evident in the semiclassical calculations in which the rotational motion of the molecule is also taken into account.

In contrast to the quantum method, the semiclassical treatment can easily be extended to take into account surface degrees of freedom in the simulation. This makes a rather realistic description of the charge transfer in molecule–surface scattering possible. The method is, however, not limited to scattering processes. It represents a versatile tool for the description of reaction dynamics with electronic transitions. For example, we are currently planning to apply the semiclassical method to the description of electron-stimulated or photon-stimulated desorption from surfaces.

This paper is organized as follows. In the next section we briefly introduce the semiclassical and quantum methods that we have used. Then we describe the model system that we have chosen. In the main part of the paper the results of the semiclassical and quantum calculations are compared and discussed.

II. Method

In this section we briefly summarize the most important aspects of the theoretical methods that are essential for the following discussion.

The total Hamiltonian is written as

$$H = T_R + H_0(r, R), \qquad (1)$$

where R and r refer to the coordinates of the nuclei and the electrons, respectively. The wave function $\psi(r, R, t)$ is expanded in terms of some electronic basis functions

$$\psi(r, R, t) = \sum_j c_j(t) \phi_j(r, R). \qquad (2)$$

The matrix elements with respect to the electronic Hamiltonian H_0 are given by

$$V_{ij}(R) = \langle \phi_i(r, R) | H_0(r, R) | \phi_j(r, R) \rangle. \qquad (3)$$

$V_{ii}(R)$ describes the potential energy surface of the system in the electronic state ϕ_i. In surface hopping methods the system evolves for a particular period of time on one specific potential energy surface and the classical particles follow a trajectory $R(t)$ that is determined by the integration of the classical equation of motion

$$M \frac{d^2}{dt^2} R = -\nabla_R \langle \phi_i | H_0(r, R) | \phi_i \rangle \qquad (4)$$

Along the trajectory $R(t)$ sudden hops between different potential energy surfaces occur according to some instruction that has to be specified. In the diabatic representation the non-diagonal matrix elements $V_{ij}(R)$ lead to these transitions to other potential energy surfaces. In the adiabatic picture it is the change of the eigenfunctions that cause these transitions. This change is described by the nonadiabatic coupling vector d_{ij}:

$$d_{ij} = \langle \phi_i(r, R) | \nabla_R | \phi_j(r, R) \rangle \qquad (5)$$

The electronic coefficients c_j are determined according to the time-dependent Schrödinger equation for the electronic Hamiltonian $H_0(r, R)$ which is now time dependent through the trajectory $R(t)$. This time-dependent Schrödinger equation can be written as[9]

$$i\hbar \dot{c}_k = \sum_j c_j (V_{kj} - i\hbar \dot{R} \cdot d_{kj}). \qquad (6)$$

In this equation the basis functions ϕ_i can be any mixture of diabatic and adiabatic states. The probability of finding the system at time t in the electronic level k is then given by

$$P_k(t) = |c_k(t)|^2. \qquad (7)$$

In Tully's fewest-switches algorithm,[9] electronic transitions between different levels can occur at any point along the trajectories $R(t)$. The transition probability is constructed in such a way that the number of state switches is minimized, under the constraint that in an ensemble of trajectories the average population of each level is given by the square modulus of the expansion coefficients c_k (eqn. (7)).

Since transitions can occur at any point along the trajectories, the potential energies of the electronic states can well be different at the moment of the state switch, i.e. $V_{kk}(R(t)) \neq V_{ll}(R(t))$. In order to conserve the total energy of the system, the velocities of the classical degrees of freedom have to be readjusted. It is not *a priori* clear how this adjustment has to be done. In the fewest switches algorithm as suggested by Tully the velocities are changed in the direction of the non-adiabatic coupling vector d_{ij}. This choice has been proposed by Herman[17] and later been reconfirmed by Coker and Xiao.[18]

Per total energy, 1000 to 2000 trajectories are calculated to determine the transition probabilities. This means that all semiclassical probabilities have a statistical uncertainty of ± 0.02 to ± 0.03. The quantum mechanical calculations of the ionization probability have been performed by solving the time-independent Schrödinger equation within a coupled-channel scheme.[19] The convergence of the quantum dynamical results with respect to the basis set has been carefully tested. In the calculations including the surface oscillator (see below) up to 250 oscillator channels per electronic level had to be taken into account to reach convergence.

III. Model system

As already stated in the Introduction, the goal of this investigation is twofold: to test the semiclassical treatment in comparison to full quantum calculations and, rather than just study some theoretical model system, to try to learn something about a real system. We have chosen the ionization probability in I_2 scattering from surfaces as our model system. This ionization probability has been measured as a function of the incoming kinetic energy of the scattered molecule by Danon and Amirav.[16] The results are plotted in Fig. 1. They show a threshold at ~ 3.0 eV and an absolute I_2^- yield of $\sim 1\%$ at a kinetic energy of 10 eV.

To our knowledge, there are no theoretical data on the interaction potential of I_2 with a diamond surface. Hence we are left with inventing potential energy surfaces based on empirical

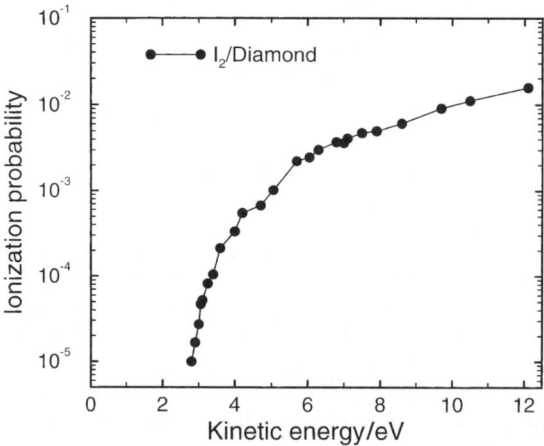

Fig. 1 Experimental results of the ionization probability in the scattering of I_2/diamond as a function of the incident kinetic energy of the molecule (after ref. 16).

data, experience and intuition. In our simulation we model the molecular center of mass distance from the surface z and one surface oscillator coordinate x. Without the surface oscillator the one-dimensional two-surface potential can be written as

$$V_{1D}(z) = \begin{pmatrix} V_{11}(z) & V_{12}(z) \\ V_{12}(z) & V_{22}(z) \end{pmatrix} \qquad (8)$$

Our chosen model potential is plotted in Fig. 2. It corresponds to a typical one-dimensional curve-crossing problem. The diabatic potential energy surfaces V_{ii} are parametrized as Morse potentials

$$V_{ii}(z) = D_{ii}(e^{-2\alpha_{ii}(z-z_{ii})} - 2e^{-\alpha_{ii}(z-z_{ii})}) + S_{ii} \qquad (9)$$

while the diabatic coupling between the surfaces V_{12} has an exponential form

$$V_{12}(z) = D_{12}e^{-\alpha_{12}z}. \qquad (10)$$

The parameters we have chosen are listed in Table 1. We assume that in the experiment the diamond surface has been hydrogen-covered.[16] Due to this passivation we have chosen only a

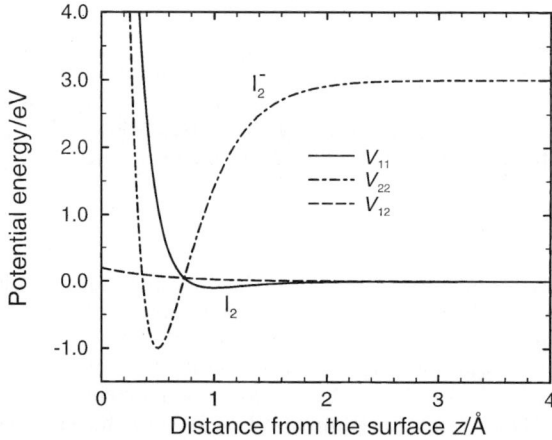

Fig. 2 One-dimensional model potential for the ionization of I_2 in the scattering from diamond surfaces. V_{11} corresponds to the potential energy surface for the neutral molecule (——) and V_{22} to the potential for the charged molecule (—·—). V_{12} is the coupling between the two potentials (— —).

Table 1 Parameters of the model potential for I_2/diamond scattering according to eqn. (9)

Potential	D_{ij}/eV	$\alpha_{ij}/\text{Å}^{-1}$	$z_{ii}/\text{Å}$	S_{ii}/eV
V_{11}	0.1	3.0	1.0	0.0
V_{22}	4.0	3.0	0.5	3.0
V_{12}	0.2	2.0	—	—

small physisorption well of 0.1 eV for neutral I_2. As for the I_2^-/diamond well-depth, we have to rely on speculation. Since diamond has a large band gap of 5.4 eV, image charge effects should be small. However, during the charge transfer process there will be a locally charged complex at the surface which could cause an attractive potential of the order of 1 eV. The strength of the coupling V_{12} had to be guessed. The shift of $S_{22} = 3.0$ eV between the two Morse potentials for $z \to \infty$ was chosen to reproduce the threshold of the I_2 ionization at a kinetic energy of ~ 3.0 eV.

The surface oscillator is coupled to the molecular motion via $V_{1D}(z) \to V_{1D}(z - x)$. The whole 2D potential is given by

$$V_{2D}(z, x) = \begin{pmatrix} V_{11}(z-x) + V_{osc}(x) & V_{12}(z-x) \\ V_{12}(z-x) & V_{22}(z-x) + V_{osc}(x) \end{pmatrix} \quad (11)$$

with

$$V_{osc}(x) = \frac{m_{osc}}{2}\omega^2 x^2 \quad (12)$$

For the surface oscillator we have chosen $\hbar\omega = 50$ meV and $m_{osc} = 180$ u. These parameters are not very realistic for a diamond surface. For example, m_{osc} corresponds to 15 times the mass of a carbon atom. Usually one selects one to two times the mass of a surface atom for the surface oscillator model.[20–23] We had tried more realistic parameters like for example used in ref. 22. However, with these parameters the surface oscillator did not really participate dynamically in the scattering process. It mainly recoiled adiabatically upon impingement of the iodine molecules with very little energy transfer. This is caused by the stiffness of the diamond surface. We have selected parameters for the surface oscillator that result in a larger energy transfer from the impinging molecule to the surface; this oscillator should, rather, be regarded as a general mode for energy transfer during the scattering process.

We have, furthermore, coupled the molecular rotation to the scattering process via

$$V_{1D}(z) \to \tfrac{1}{2}\left[V_{1D}\left(z - \frac{r}{2}\cos\theta\right) + V_{1D}\left(z + \frac{r}{2}\cos\theta\right)\right], \quad (13)$$

where $r = 2.66$ Å is the bondlength of I_2 and θ is the angle of the molecular axis with the surface normal. Note that in this model I_2 is treated as a rigid rotor, i.e. no molecular vibrations are included. We would like to emphasize at this point that we are trying to reproduce qualitative trends of the experiment. We are looking for a qualitative explanation that is *consistent with the experiment*. This does not necessarily exclude that the true explanation might be quite different. However, by such an approach the number of possible mechanisms underlying an experimental finding can be narrowed down.

IV. Results and discussion

The quantum and semiclassical results using the one-dimensional two-surface potential of eqn. (8) are plotted in Fig. 3. The semiclassical results have been obtained in the adiabatic representation. First, the oscillatory structure of the quantum and semiclassical results is evident. These are typical Stückelberg oscillations due to the fact that the molecule can be ionized on the way either to or from the surface, and these two paths interfere. It is evident that the quantum and semiclassical results show the same amplitude of the Stückelberg oscillations, the phase, however, does not

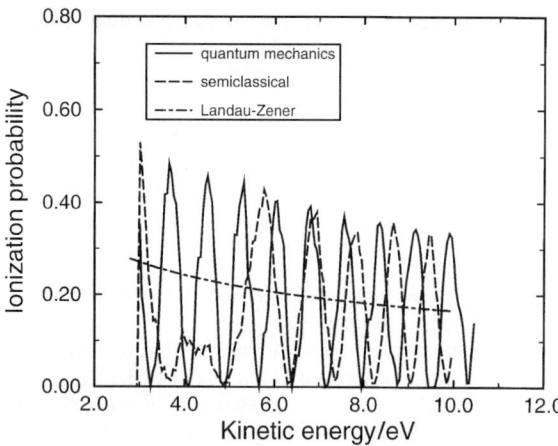

Fig. 3 Theoretical results of the ionization probability of I_2/diamond as a function of the incident kinetic energy of the molecule using the one-dimensional two-surface potential of eqn. (8). (———) Quantum mechanical result; (— —) semiclassical result; (— · —): Landau–Zener approximation.

agree. In addition, in the semiclassical calculation one peak of the Stückelberg oscillations at approximately 4 eV is missing. The fact that the phases do not agree is not too surprising considering that the semiclassical approximation breaks down at the classical turning points where the de Broglie wavelength of the molecule becomes infinite.

In addition, we have included results according to the Landau–Zener approximation.[24] In this approximation the transition between two *adiabatic* states which we denote by a and b is given by

$$w_{ab} = \exp\left(-\frac{2\pi V_{12}^2}{\hbar v \left|\frac{dV_{11}}{dz} - \frac{dV_{22}}{dz}\right|}\right), \quad (14)$$

where the values of the coupling V_{12}, of the derivatives of the potentials and of the velocity v are all taken at the curve-crossing between the *diabatic* states. Now the molecule passes the location of the curve-crossing twice before it is scattered back into the gas phase, hence the total probability of the ionization in the Landau–Zener approximation is given by

$$P_{12} = 2w_{ab}(1 - w_{ab}). \quad (15)$$

Note that in the gas phase the diabatic states which we have denoted by 1 and 2 and the corresponding adiabatic states are the same. In the Landau–Zener approximation the Stückelberg oscillations are absent since no phase information is included. But the Landau–Zener results correspond rather accurately to the averaged quantum and semiclassical results. This tells us that indeed surface hopping occurs in both methods rather close to the surface crossing point. This is also confirmed by an analysis of the semiclassical trajectories. The Landau–Zener probability w_{ab} alone rises with increasing kinetic energies. However, P_{12} has its maximum for $w_{ab} = 0.5$ (see eqn. (15)). Hence the decreasing ionization probability is due to the fact that w_{ab} is larger than 0.5.

The experimental results, on the other hand, showed an increase with increasing energy. From eqn. (14) and (15) we see directly that the coupling V_{12} between the *diabatic* curves actually has to become larger so that w_{ab} becomes smaller than 0.5 in order to reproduce the rising behavior of the experimental ionization probability. This means that we would get closer to the adiabatic limit. In fact, with $D_{12} = 1.5$ eV we can almost exactly reproduce the experimental curve. But first, this coupling seems to be unrealistically large, however, as there are no reliable calculations for the coupling such a value cannot be excluded. Even more importantly, very simple low-dimensional model calculations should not try to reproduce experimental data exactly in order to leave enough room for the influence of all the other neglected degrees of freedom. In fact, we believe that the electronic coupling is not responsible for this rising behavior but energy transfer to other degrees of freedom, as we will discuss below.

There are no Stückelberg oscillations apparent in the experimental results. Usually one argues that quantum mechanical interference effects wash out in realistic high-dimensional situation (although this is not always true[3]). And indeed, if the surface oscillator is included in the quantum mechanical calculations these oscillations disappear almost entirely as is shown in Fig. 4. These results correspond to the surface oscillator initially in its ground state. It is remarkable how close the quantum mechanical results follow the simple Landau–Zener expression at higher kinetic energies (note that in the Landau–Zener results the surface oscillator is not considered).

Now the quantum mechanical results actually show an initial increase at low kinetic energies. This is simple due to energy transfer to the oscillator. Due to this energy transfer the number of molecules that have enough energy to become ionized is reduced. In fact we believe that this is the qualitative explanation for the rising behavior of the ionization probability in the experiment. Regarding the facts that I_2 has very soft vibrational modes with $\hbar\omega = 20$ meV, that the dissociation energy of I_2 is only 1.5 eV which results in a large I_2 dissociation probability[25] and also that rotations can be excited very efficiently in the scattering (see ref. 26 and below), we see that a lot of energy is transferred to other degrees of freedom during the scattering event. This limits the number of molecules that have enough energy to be ionized, in particular close to the threshold energy.

Turning to the semiclassical results, we see that the Stückelberg oscillations are not washed out at all, their amplitude is almost the same as in the rigid surface case. It seems to be a paradox that in the semiclassical calculations more phase coherence is retained than in the quantum calculations, however, this has already also been observed in one-dimensional two-surface calculations.[9] Now in these calculations the surface oscillator has been treated fully classically, i.e., the surface oscillator was initially at rest. In the quantum calculations, on the other hand, the surface oscillator has zero-point motion according to a zero-point energy of 25 meV. This actually corresponds to a temperature of 300 K in a classical picture. It is possible that this uncertainty in the position and momentum of the oscillator contributes to the suppression of the Stückelberg oscillations. Hence we have performed semiclassical calculations in which the surface initially had a vibrational energy of 25 meV and in which the initial phase of the oscillator was sampled randomly. Indeed, the Stückelberg oscillations are reduced significantly if the zero-point energy is taken into account in the semiclassical calculations, as is demonstrated in Fig. 5. But still these oscillations are much stronger than in the full quantum calculations. However, the mean value of the semiclassical results closely follows the quantum results.

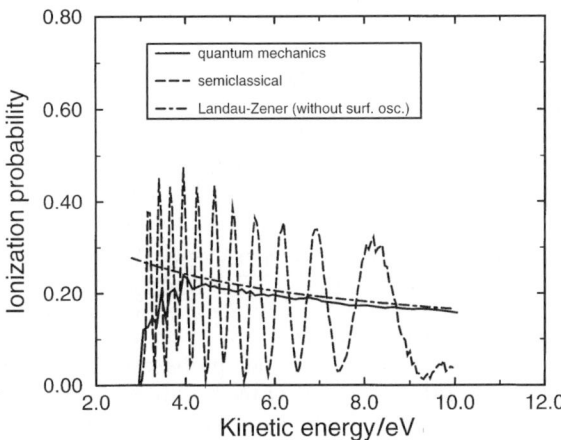

Fig. 4 Theoretical results of the ionization probability of I_2/diamond as a function of the incident kinetic energy of the molecule including the surface oscillator according to eqn. (11). The notation is the same as in Fig. 3. No zero-point energy corrections for the surface oscillator have been taken into account. The results in the Landau–Zener approximation in which no surface oscillator is considered are also plotted as a guide to the eye.

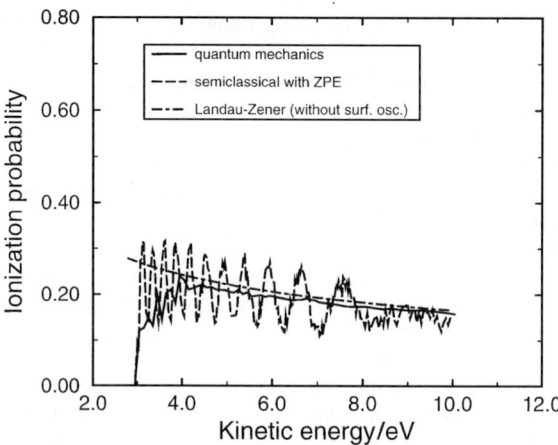

Fig. 5 Theoretical results of the ionization probability of I_2/diamond as a function of the incident kinetic energy of the molecule including the surface oscillator according to eqn. (11). The notation is the same as in Fig. 1. In the semiclassical calculations a zero-point energy of 25 meV of the surface oscillator was taken into account.

There may be several reasons why the Stückelberg oscillations in the semiclassical calculations are not suppressed in the same way as in the quantum calculations. First, in the scattering at the surface oscillator a superposition of oscillator states is excited in the quantum dynamics which causes a distribution in the kinetic energy of the scattered molecule. In the semiclassical calculations, for a fixed kinetic energy a fixed amount of kinetic energy is transferred to the surface oscillator, if the surface oscillator is initially at rest, so that no distribution of kinetic energies results which could suppress the Stückelberg oscillations. If the surface oscillator is initially already vibrating with the zero-point energy, a certain distribution in the energy transfer is the consequence, but apparently this distribution is not wide enough to fully suppress the Stückelberg oscillations, as Fig. 5 demonstrates.

Secondly, the coupling $V_{12}(z - x)$ also leads to transitions between different surface oscillator states in the quantum dynamics. One can show[27] that

$$\langle m | V_{12}(z - x) | n \rangle = \langle m | D_{12} e^{-\alpha_{12}(z-x)} | n \rangle$$

$$= D_{12}^*(z) \tilde{\alpha}^{|m-n|} \sum_{k=0}^{\infty} \frac{\tilde{\alpha}^{2k}}{k!(k+|m-n|)!} \frac{(\max(m,n)+k)!}{(m!n!)^{1/2}}, \quad (16)$$

with

$$D_{12}^*(z) = D_{12} e^{-\alpha_{12} z} e^{\tilde{\alpha}^2/2} \quad (17)$$

and

$$\tilde{\alpha} = \alpha_{12} \left(\frac{\hbar}{2 m_{osc} \omega} \right)^{1/2}, \quad (18)$$

where $|m\rangle$ and $|n\rangle$ are harmonic oscillator states.

Eqn. (16) demonstrates that there is a non-vanishing probability that the transition between the two electronic states is accompanied by a change in the surface oscillator states at *any* point along the trajectories. This leads to a change in the kinetic energy of the molecule in the quantum dynamics even if $V_{11}(R(t)) = V_{22}(R(t))$ at the point of the transition between the two electronic states in the quantum dynamics. On the other hand, in the semiclassical calculations the kinetic energy of the molecule will not be altered in such a situation. These oscillator transitions will contribute to the loss of phase coherence in the quantum dynamics, while it is absent in the semiclassical calculations.

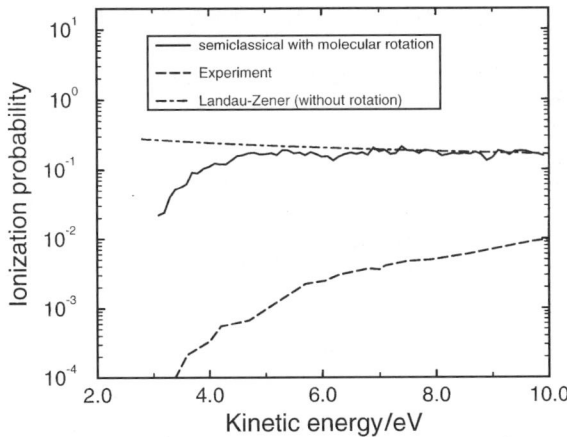

Fig. 6 Theoretical results of the ionization probability of I_2/diamond as a function of the incident kinetic energy of the molecule taking the molecular rotation into account according to eqn. (13) (——). In addition, the experimental results from ref. 16 (— —) and the Landau–Zener results without rotations (—·—) are plotted as a guide to the eye.

However, instead of trying to change the semiclassical algorithm in order to suppress the artificial oscillations we are led by the notion that, in a high-dimensional application which we have in mind, such oscillations will be washed out. Indeed, if we consider the molecular rotation in the semiclassical calculations according to eqn. (13) instead of the surface oscillator, there are no longer any Stückelberg oscillations evident, as Fig. 6 demonstrates. First, in the semiclassical calculation the initial orientation of the molecule is sampled randomly, which increases the stochastic nature of the scattering event; and secondly, while the energy transfer to the surface oscillator is less than 1 eV at all kinetic energies for the chosen parameters, up to more than 3 eV is transferred to the molecular rotation in the scattering. Both effects cause a suppression of the Stückelberg oscillations. Note that we have not performed quantum dynamical calculations taking into account the molecular rotation. Due to the small rotational energy quantum of I_2 more than 1000 channels have to be considered in the quantum calculations per electronic levels which makes these calculations computationally much too expensive. This is actually precisely the reason why we are applying the semiclassical treatment to the electron transfer problem.

In Fig. 6 we have also included the experimental results. Due to the low dimensionality of the calculations we have not tried to reach quantitative agreement with experiment. However, it is evident that due to the energy transfer to the molecular rotation the ionization probability is suppresssed by more than one order of magnitude close to the ionization threshold compared to the Landau–Zener probability without rotations. The qualitative trend in the calculated ionization probability is rather similar to experiment. The inclusion of the molecular vibrations and also the molecular dissociation channel will further suppress the ionization probability. This confirms our qualitative explanation of the experiment that it is not the electronic coupling *per se* that causes the strong increase in the ionization probability with rising kinetic energy. Instead, the large energy transfer to other degrees of freedom during the scattering event leads to a large suppresion of the ionization probability, in particular for energies close to the ionization threshold.

V. Conclusions

We have calculated the ionization probability in the scattering of molecules from surfaces by a quantum mechanical and a semiclassical treatment. In our low-dimensional description the molecular center of mass distance from the surface and either a surface oscillator coordinate or the molecular orientation have been considered. The parameters of the model potential have been chosen to resemble the system I_2/diamond.

The semiclassical results agree with the quantum results although some discrepancies remain as far as the phase coherence is concerned. These discrepancies, however, will most probably be absent in high-dimensional applications, as our semiclassical calculations in which the molecular rotation is taken into account demonstrate. Thus the semiclassical method opens the way to the description of electronically non-adiabatic processes in realistic high-dimensional simulations.

The results of these particular low-dimensional simulations suggest that the strong increase in the ionization probability in the scattering of I_2/diamond with increasing kinetic energy observed in the experiment is not caused by the electronic coupling *per se*. Due to the efficient energy transfer to other degrees of freedom in the scattering process the number of molecules that have enough kinetic energy to become ionized is strongly reduced, in particular for energies close to the ionization threshold. At higher kinetic energies the fraction of molecules retaining sufficient kinetic energy in the scattering process becomes larger, leading to the increase in the ionization probability.

Acknowledgements

We thank E. Kolodney for providing us with the unpublished data of the I_2 dissociation probability in the I_2/diamond scattering.

References

1. G. R. Darling and S. Holloway, *Rep. Prog. Phys.*, 1995, **58**, 1595.
2. A. Groß, *Surf. Sci. Rep.*, 1999, **32**, 291.
3. A. Groß, S. Wilke and M. Scheffler, *Phys. Rev. Lett.*, 1995, **75**, 2718.
4. G. J. Kroes, E. J. Baerends and and R. C. Mowrey, *Phys. Rev. Lett.*, 1997, **78**, 3583.
5. T. Klüner, H.-J. Freund, V. Staemmler and R. Kosloff, *Phys. Rev. Lett.*, 1998, **80**, 5208.
6. E. Runge and E. K. U. Gross, *Phys. Rev. Lett.*, 1984, **52**, 997.
7. X. Gonze and M. Scheffler, *Phys. Rev. Lett.*, 1999, **82**, 4416.
8. M. S. Topaler, T. C. Allison, D. W. Schwenke and D. G. Truhlar, *J. Chem. Phys.*, 1998, **109**, 3321.
9. J. C. Tully, *J. Chem. Phys.*, 1990, **93**, 1061.
10. S. Hammes-Schiffer and J. C. Tully, *J. Chem. Phys.*, 1994, **101**, 4657.
11. D. Kohen, F. H. Stillinger and J. C. Tully, *J. Chem. Phys.*, 1998, **109**, 4713.
12. J.-Y. Fang and S. Hammes-Schiffer, *J. Chem. Phys.*, 1999, **110**, 11166.
13. J. C. Tully, *Faraday Discuss.*, 1998, **110**, 407.
14. O. V. Prezhdo and P. J. Rossky, *J. Chem. Phys.*, 1997, **107**, 825.
15. M. F. Herman, *J. Chem. Phys.*, 1982, **76**, 2949.
16. A. Danon and A. Amirav, *Phys. Rev. Lett.*, 1988, **61**, 2961.
17. M. F. Herman, *J. Chem. Phys.*, 1984, **81**, 754.
18. D. F. Coker and L. Xiao, *J. Chem. Phys.*, 1995, **102**, 496.
19. W. Brenig, T. Brunner, A. Groß and R. Russ, *Z. Phys. B*, 1993, **93**, 91.
20. M. Hand and J. Harris, *J. Chem. Phys.*, 1990, **92**, 7610.
21. A. Groß and W. Brenig, *Chem. Phys.*, 1993, **177**, 497.
22. A. Groß and W. Brenig, *Surf. Sci.*, 1994, **302**, 403.
23. M. Dohle and P. Saalfrank, *Surf. Sci.*, 1997, **373**, 95.
24. L. D. Landau and E. M. Lifshitz, *Quantum Mechanics, Non-relativistic Theory*, Pergamon Press, London, 1958.
25. E. Kolodney, personal communication.
26. R. B. Gerber and R. Elber, *Chem. Phys. Lett.*, 1993, **102**, 466.
27. A. Groß, PhD Thesis, Technical University Munich, 1993.

Quantum dynamics of the dissociation of H_2 on Cu(100): Dependence of the site-reactivity on initial rovibrational state

Drew A. McCormack,[a] Geert-Jan Kroes,[a] Roar A. Olsen,[b] Jeroen A. Groeneveld,[b] Joost N. P. van Stralen,[b] Evert Jan Baerends[b] and Richard C. Mowrey[c]

[a] *Leiden Institute of Chemistry, Gorlaeus Laboratories, Leiden University, P.O. Box 9502, 2300 RA Leiden, The Netherlands*
[b] *Theoretical Chemistry, Free University, De Boelelaan 1083, 1081 HV Amsterdam, The Netherlands*
[c] *Theoretical Chemistry Section, Code 6189, Naval Research Laboratory, Washington, DC 20375-5342*

Received 29th March 2000
First published as an Advance Article on the web 18th August 2000

We perform six-dimensional (6D) quantum wavepacket calculations for H_2 dissociatively adsorbing on Cu(100) from a variety of rovibrational initial states. The calculations are performed on a new potential energy surface (PES), the construction of which is also detailed. Reaction probabilities are in good agreement with experimental findings. Using a new flux analysis method, we calculate the reaction probability density as a function of surface site and collision energy, for a variety of initial states. This approach is used to study the effects of rotation and vibration on reaction at specific surface sites. The results are explained in terms of characteristics of the PES and intrinsically dynamic effects. An important observation is that, even at low collision energies, reaction does not necessarily proceed predominantly in the region of the minimum potential barrier, but can occur almost exclusively at a site with a higher barrier. This suggests that experimental control of initial conditions could be used to selectively induce reaction at particular surface sites. Our predictions for site-reactivity could be tested using contemporary experimental methods: The calculations predict that, for reacting molecules, there will be a dependence of the quadrupole alignment of j on the incident vibrational state, v. This is a direct result of PES topography in the vicinity of the preferred reaction sites of $v = 0$ and $v = 1$ molecules. Invoking detailed balance, evidence for this difference in preferred reaction site of $v = 0$ and 1 molecules could be obtained through associative desorption experiments.

I. Introduction

Surface reactions are exceedingly complex. Even a seemingly simple diatom–surface reaction, such as the benchmark $H_2 + Cu$ system, includes six molecular degrees of freedom, not to mention countless surface degrees of freedom; understanding the intricacies of such a reaction is difficult at best. In spite of this, considerable progress has been made by studying reduced-dimensional models of these reactions;[1–22] many of the important effects observed in the full reaction are also exhibited in the lower-dimensional model systems. This approach has demonstrated how steric effects,[8,10] corrugation of the potential,[12,13] steering of molecules,[14] and 'lateness' of barrier[6,7,9] influence reaction and back-scattering.

DOI: 10.1039/b002507k

The usefulness of the knowledge garnered from reduced-dimensional studies cannot be overstated (see, for instance, ref. 23), but they have been less successful for making quantitatively accurate predictions. With the recent advent of the first quantum calculations to treat all molecular degrees of freedom,[24–38] a new phase of theoretical research has begun. The first full-dimensional calculations utilized time-independent methods to study the barrierless H_2 + Pd(111) reaction;[24–31] later, time-dependent wavepacket methods were used to study activated systems such as H_2 on Cu(100)[32–36] and Cu(111).[37,38]

Calculations can now do more than simply help interpret experiment: they can be used to make new predictions, and show experimentalists 'where to look' in the atomic game of hide-and-seek. To date, full-dimensional calculations have largely been used either to confirm that observations seen in reduced-dimensional studies persist with the inclusion of more degrees of freedom, or to make quantitative comparisons with experiment in order to test the accuracy of the theory used [*e.g.* the accuracy of the PES]. But as they become more tractable, the possibility exists to perform ever more detailed analyses, and this is now beginning to contribute to improved understanding.

Here we introduce a flux-based analysis technique to study the reactivity of individual surface sites in the dissociation of H_2 on Cu(100). Earlier studies of site-specific reaction are relatively few and far between. Four-dimensional (4D) 'fixed-site' quantum calculations have been performed for H_2 reaction on Cu(100)[20,21] and Cu(111),[10,17] and we have recently presented results of 6D classical trajectory calculations that address the issue of site-specific reaction directly.[39] No study involving full-dimensional quantum calculations has yet been undertaken.

Results presented here are resolved with respect to the incident rovibrational state of H_2 and the collision energy, allowing a detailed picture of the reaction to be constructed. We show how the findings can be interpreted with reference to effects observed in earlier reduced-dimensional calculations. Certain initial conditions are found to lead to a strong preference for reaction at particular surface sites, not necessarily the site of lowest reaction barrier, which could be exploited experimentally to probe the surface site-selectively. We also report and explain an interesting finding for the rotational alignment of reacting molecules; namely, that it depends on the incident vibrational state of H_2. We show how this dependence, which results from qualities of the PES at certain reaction sites, could be used as an experimental signature for our predictions for site-reactivity.

In Section II the wavepacket method is briefly discussed, along with a detailed description of the flux analysis method employed and the methodology used to calculate the PES. The PES is completely new, so details of the electronic structure calculations and fitting procedure are given. Section III presents results of the dynamics calculations, along with some details of the new PES. In Section IV findings are discussed and related to experiment, and Section V concludes.

II. Method

A. PES

To describe the H_2/Cu(100) PES we have calculated 14 two-dimensional (2D) PESs employing a two-layer slab with H_2 on one of the high-symmetry sites above the slab within a 2 × 2 surface unit cell. The two geometric parameters that are varied are Z and r (Fig. 1(a)). Each 2D PES is based on between 55 and 75 calculated points and fitted using bicubic splines. As we will show below, the 14 2D PESs are then used to obtain the full 6D H_2/Cu(100) PES through a symmetry-adapted basis set expansion method.

The electronic structure calculations presented in this study were performed using BAND.[40–42] In the program the Kohn–Sham equations[43,44] are solved self-consistently for a periodic system, in our case a semi-infinite slab with translational symmetry in two directions. A flexible basis set of numerical atomic orbitals (NAOs) obtained from numerical Herman–Skillman type calculations,[45] Slater-type orbitals (STOs), or a combination of both, are used in the expansion of the one-electron states. The frozen core approximation can be used for the core electrons of the heavier atoms, avoiding the use of pseudopotentials. An accurate Gauss-type numerical integration scheme[42] is used to calculate the matrix elements of the Hamiltonian, and the *k*-space integration can be done accurately using the quadratic tetrahedron method.[46]

The exchange-correlation energy in the local density approximation (LDA) is calculated using the Vosko–Wilk–Nusair formulae.[47] The generalized gradient approximation (GGA) we use is the

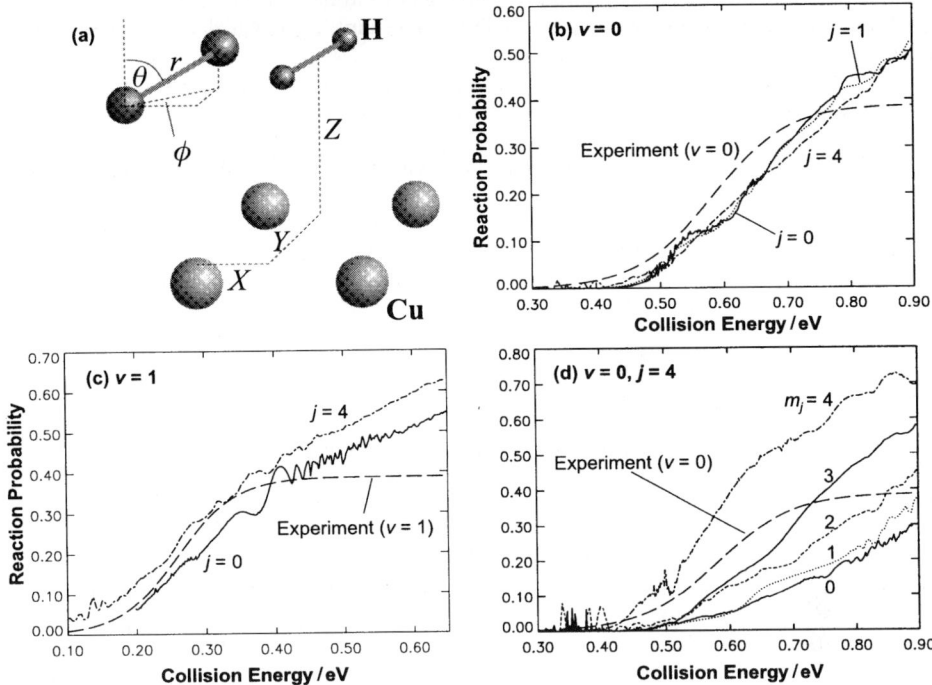

Fig. 1 The coordinate system and collision-energy-dependent reaction probabilities. (a) The 6D coordinate system used in the calculations, shown with respect to a single surface unit cell. (b) Computed reaction probabilities for various degeneracy-averaged $(v = 0, j)$ states as a function of collision energy, compared with the $v = 0$ reaction probability resulting from an experimental analysis.[63] The experimental results are not rotationally resolved. (c) Computed reaction probabilities for various degeneracy-averaged $(v = 1, j)$ states, shown with rotationally-unresolved experimental results for $v = 1$. (d) Reaction probabilities resolved with respect to m_j for $(v = 0, j = 4)$. Also shown are the experimental results for $v = 0$.

combination of the Becke correction[48] for the exchange energy with the Perdew correction[49] for the correlation energy (BP). The gradient corrections are calculated from the self-consistent LDA density, which has been shown to be an excellent approximation to the binding energies calculated from the self-consistent nonlocal densities.[50]

The basis set used here is the same as in earlier calculations.[51,52] This basis set gives binding energies accurate to about 0.10 eV when compared to calculations using a basis set giving results very close to the basis set limit. The ACCINT parameter (a parameter that is a general real space integration parameter[41,42]) has been set to 4.0. This results in a real space integration accuracy for the binding energies of about 0.01 eV when compared to more accurate calculations. (In the previous calculations ACCINT was set to 3.5, giving an accuracy of about 0.03 eV in the binding energies with respect to the real space integration.) The KSPACE parameter governs the number of integration points in the surface Brillouin zone (SBZ) and has been set to 5. This gives 25 integration points in the irreducible wedge of the SBZ, corresponding to a total of 41 points in the whole SBZ, giving an accuracy in the k-space integration of about 0.02 eV in the binding energies when compared to more accurate calculations. (In the previous calculations KSPACE was set to 3, resulting in 9 integration points in the irreducible wedge of the SBZ, giving an accuracy of about 0.10 eV in the binding energies with respect to the k-space integration.) Increasing the number of layers to 3 changes the binding energies by about 0.13 eV. Seeing these (basis set, real space integration, k-space integration, and number of layers) as independent errors gives an overall accuracy of about 0.17 eV. But for all test geometries the changes resulting from increasing the quality of the basis set and increasing the number of layers are of opposite sign. We therefore believe that the binding energies we find are converged to about 0.1 eV of the GGA limit for the H_2/Cu(100) system. The experimental Cu bulk lattice constant, $a_{lat} = 4.824\ a_0$,[53] has been used for the slab.

The full 6D PES is written as a linear combination of symmetry-adapted basis functions (see Fig. 1(a) for coordinate definitions)

$$V_{6D} = \sum_{jm_jp=00,\ 20,\ 40,\ 44e}\ \sum_{nm=00,\ 10,\ 11} V_{jm_jpnm}(r, Z)Y_{jm_jp}(\theta, \phi)H_{nm}(X, Y)$$

$$+ \sum_{jm_jp=22e,\ 42e} V_{jm_jpB_110}(r, Z)Y_{jm_jp}(\theta, \phi)H_{B_110}(X, Y) \quad (1)$$

with

$$H_{00}(X, Y) = \sqrt{\frac{1}{A}},$$

$$H_{10}(X, Y) = \sqrt{\frac{1}{A}}\,(\cos GX + \cos GY),$$

$$H_{11}(X, Y) = 2\sqrt{\frac{1}{A}}\,(\cos GX \cos GY),$$

$$H_{B_110}(X, Y) = \sqrt{\frac{1}{A}}\,(\cos GX - \cos GY),$$

$$G = 2\pi/a_{lat} \quad (2)$$

where A is the area of the surface unit cell, and

$$Y_{22e}(\theta, \phi) = \sqrt{\frac{1}{2}}\,[Y_{22}(\theta, \phi) + Y_{2-2}(\theta, \phi)],$$

$$Y_{42e}(\theta, \phi) = \sqrt{\frac{1}{2}}\,[Y_{42}(\theta, \phi) + Y_{4-2}(\theta, \phi)],$$

$$Y_{44e}(\theta, \phi) = \sqrt{\frac{1}{2}}\,[Y_{44}(\theta, \phi) + Y_{4-4}(\theta, \phi)], \quad (3)$$

with Y_{jm_j} being spherical harmonics. The parity p can take the values 'e' for even and 'o' for odd, and is for reflection in the YZ-plane. The expansion coefficients V in eqn. (1) are obtained in a two-step procedure. First, three 4D PESs in (r, Z, θ, ϕ) are obtained above the two-fold bridge site,

$$V_{4D}^{bridge} = V_{00b}(r, Z)Y_{00}(\theta, \phi) + V_{20b}(r, Z)Y_{20}(\theta, \phi) + V_{22eb}(r, Z)Y_{22e}(\theta, \phi)$$

$$+ V_{40b}(r, Z)Y_{40}(\theta, \phi) + V_{42eb}(r, Z)Y_{42e}(\theta, \phi) + V_{44eb}(r, Z)Y_{44e}(\theta, \phi), \quad (4)$$

the four-fold top site,

$$V_{4D}^{top} = V_{00t}(r, Z)Y_{00}(\theta, \phi) + V_{20t}(r, Z)Y_{20}(\theta, \phi)$$

$$+ V_{40t}(r, Z)Y_{40}(\theta, \phi) + V_{44et}(r, Z)Y_{44e}(\theta, \phi), \quad (5)$$

and similarly for the four-fold hollow site. We have calculated six 2D PESs above the bridge site $[(\theta, \phi) = (90,0), (90,45), (90,90), (122.58,0), (122.58,90)$, and $(154.98,45)$, all angles in degrees] and four 2D PESs above the top and hollow sites $[(\theta, \phi) = (90,0), (90,45), (122.58,22.5)$, and $(154.98,22.5)]$. By inverting the corresponding set of linear equations, the expansion coefficients in eqns. (4) and (5) are determined, as are those for the hollow site. The coefficients pertaining to Y_{00} in eqn. (1) are then given by

$$V_{0000}(r, Z) = \sqrt{\frac{A}{4}} [V_{00t}(r, Z) + V_{00h}(r, Z) + 2V_{00b}(r, Z)],$$

$$V_{0010}(r, Z) = \sqrt{\frac{A}{4}} [V_{00t}(r, Z) - V_{00h}(r, Z)],$$

$$V_{0011}(r, Z) = \sqrt{\frac{A}{8}} [V_{00t}(r, Z) + V_{00h}(r, Z) - 2V_{00b}(r, Z)], \quad (6)$$

with a similar expression for the expansion coefficients related to Y_{20}, Y_{40} and Y_{44e}. The coefficients related to Y_{22e} and Y_{44e} are given by

$$V_{22eB_110}(r, Z) = \sqrt{\frac{A}{2}} V_{22eb}(r, Z),$$

$$V_{42eB_110}(r, Z) = \sqrt{\frac{A}{2}} V_{42eb}(r, Z). \quad (7)$$

For further detail see ref. 52 where the method has been extensively described.

B. Dynamical method

The Hamiltonian used to describe the motion of H_2, in terms of the coordinates shown in Fig. 1(a), is given by

$$\hat{H} = -\frac{1}{2M}\left(\frac{\partial^2}{\partial Z^2} + \frac{\partial^2}{\partial X^2} + \frac{\partial^2}{\partial Y^2}\right) - \frac{1}{2\mu}\frac{\partial^2}{\partial r^2} + \hat{H}_{\text{rot}} + V_{6D}(Z, X, Y, r, \theta, \phi), \quad (8)$$

where M is the mass of H_2; μ, its reduced mass; and $\hat{H}_{\text{rot}}$, the rotational Hamiltonian of the molecule. We adopt the convention that $\hbar = 1$. In deriving eqn. (8), the usual approach of multiplying the wavefunction by r has been adopted, with an appropriate redefinition of the inner product on r.[54] The potential energy, V_{6D}, is the PES described in Section II.A.

In these calculations we only treat the case of normal incidence. This allows for a dramatic reduction in computational cost through adaptation of the basis used to represent the wavefunction to the symmetry of the surface (C_{4v}).[32,36,55] Wavepacket propagations are performed for one or two symmetries, depending on the initial non-symmetry-adapted (NSA) state for which results are sought, and the results combined.[36,55] A close-coupling representation is used for the wavefunction, with a grid in Z and r, and basis functions in X, Y, θ and ϕ:

$$\Psi_{\Gamma^\alpha}(t) = \sum_{jm_j\Gamma_d^\beta nm} f_{\Gamma^\alpha jm_j\Gamma_d^\beta nm}(Z, r; t) g_{\Gamma^\alpha jm_j\Gamma_d^\beta nm}(X, Y, \theta, \phi). \quad (9)$$

Quantum numbers used here are v for vibration, j and m_j for rotation, and n and m for diffraction (i.e. translation parallel to the surface). The wavefunction symmetry is Γ^α, the αth partner of symmetry species Γ in the C_{4v} point group. The symmetry-adapted (SA) rotation–diffraction functions, $g_{\Gamma^\alpha jm_j\Gamma_d^\beta nm}(X, Y, \theta, \phi)$, also have overall Γ^α symmetry, with the diffraction component having Γ_d^β symmetry. The rotation function symmetry is uniquely determined by the Γ^α and Γ_d^β labels.[55] Detailed theory for setting up the SA bases has been given elsewhere, for $m_j = 0$[55] initial states, other m_j-even states,[34] and m_j-odd states.[36]

The initial wavepackets are chosen to be real, and take the form of superimposed Gaussians in Z, one incident upon the surface, and the other moving away.[32] This choice facilitates the use of real algebra throughout the expensive part of the calculations. The absorbing boundary condition (ABC) evolution operator,[56]

$$\exp(-i\hat{H}t) = \sum_{n=0} \hat{\Omega}_n = \sum_{n=0} (2-\delta_{n0})\exp(-i\bar{H}t)(-i)^n J_n(\Delta Ht)\hat{Q}_n, \quad (10)$$

is used to propagate the wavepackets.[32] In eqn. (10), $\bar{H}$ and ΔH are estimates of the midpoint and halfwidth of the spectrum of $\hat{H}$; the $\hat{Q}_n$ are modified Chebyshev polynomials; and the J_n are Bessel functions. The ABC operator incorporates an optical potential, which absorbs the wavepacket at the edge of the grid.[56]

From the SA wavepacket propagations, we use the scattering amplitude formalism[57,58] to determine components of the S-matrix for SA asymptotic states, S^{SA}, for scattering of molecules back to the gas phase. A simple transformation to NSA states gives elements of the actual S-matrix, S, which can be used to compute, amongst other things, the reaction probability.

C. Analysis method for site reactivity

In addition to the analysis method used during the wavepacket propagation to establish the reaction probability and state-to-state probabilities for back-scattering, we here introduce another analysis technique for calculating the reaction probability density as a function of surface site. This new method is flux-based, and is adapted from a widely-used approach for calculating reaction probabilities in which the net probability flux across a cut in the product channel is calculated as a function of energy.[59–62] Here we perform a similar calculation, but rather than integrating over all degrees of freedom to determine the total reaction probability, we exclude the integration over the diffraction coordinates, X and Y, thereby extracting the reaction probability density as a function of surface site, which can serve as a measure of site reactivity.

At any point in time, the flux across a cut at a single value, r_{cut}, of the bondlength coordinate, r, is given by

$$P(Z, X, Y, \theta, \phi; t) = \frac{1}{\mu} \text{Im}\left(\Psi^*(t) \frac{\partial \Psi(t)}{\partial r}\right)\bigg|_{r=r_{cut}}. \quad (11)$$

To get the flux as a function of energy, it is necessary to calculate the stationary scattering states from the time-dependence of the wavefunction. The total-energy-resolved flux through the cut then becomes[62]

$$P(Z, X, Y, \theta, \phi; E) = \frac{2\pi M}{|k_Z|\mu} \text{Im}\left(\Phi^*(E) \frac{\partial \Phi(E)}{\partial r}\right)\bigg|_{r=r_{cut}}, \quad (12)$$

where k_Z is the momentum conjugate to Z and corresponding to total energy E. The stationary scattering states are defined by

$$\Phi(E) = \frac{|k_Z|}{2\pi M b(-k_Z)} \int_0^\infty dt\, e^{iEt} \Psi(t), \quad (13)$$

with the derivatives given by

$$\frac{\partial \Phi(E)}{\partial r} = \frac{|k_Z|}{2\pi M b(-k_Z)} \int_0^\infty dt\, e^{iEt} \frac{\partial \Psi(t)}{\partial r}. \quad (14)$$

$b(k_Z)$ is the momentum space representation of the initial wavefunction in the scattering coordinate, and is thus related to the incident wavepacket in Z by

$$\Psi(Z; t=0) = \frac{1}{\sqrt{2\pi}} \int dk_Z\, b(k_Z) e^{ik_Z Z}. \quad (15)$$

Integrating eqn. (12) over all remaining coordinates gives the reaction probability:

$$P_R(E) = \frac{2\pi M}{|k_Z|\mu} \int dZ\, dX\, dY\, d\theta\, d\phi\, \text{Im}\left(\Phi^*(E) \frac{\partial \Phi(E)}{\partial r}\right)\bigg|_{r=r_{cut}}. \quad (16)$$

However, in this case we do not wish to calculate the reaction probability, but the reaction probability density. We therefore integrate eqn. (12) over all remaining coordinates *except* X and Y, giving

$$P_D(X, Y; E) = \frac{2\pi M}{|k_Z|\mu} \int dZ\, d\theta\, d\phi\, \text{Im}\left(\Phi^*(E) \frac{\partial \Phi(E)}{\partial r}\right)\bigg|_{r=r_{cut}}. \quad (17)$$

We now discuss how this approach is incorporated into the symmetry-adapted close-coupling wavepacket method employed here. The scattering states and derivatives given by eqn. (13) and (14) are first accumulated during the propagation at $r = r_{cut}$ for a handful of discrete energies. The derivatives are calculated using a Fourier transform in r. The ABC method utilized to propagate the wavepacket is a single time step method, so 'accumulating' involves summing the contributions of each of the terms in the series expansion [eqn. (10)]. Thus,

$$\Phi(E) = \frac{|k_z|}{2\pi M b(-k_z)} \sum_{n=0} \int_0^{T_{max}} dt\, e^{iEt} \hat{\Omega}_n \Psi(t=0), \tag{18}$$

with an analogous expression for the scattering state derivatives. The time T_{max} is the total propagation time. When a new term in the expansion has been calculated, the integration over time is performed numerically, and the contribution added.

Having accumulated the scattering states and their derivatives, we can use them to evaluate eqn. (17). But first we need to consider whether coherence between SA wavefunctions in calculations involving multiple propagations, one for each symmetry, needs to be taken into account in the evaluation of $P_D(X, Y; E)$. Take a NSA wavepacket that can be represented as a linear combination of two orthogonal SA wavefunctions,

$$\Psi_{NSA}(t) = c_1 \Psi_{\Gamma_1}(t) + c_2 \Psi_{\Gamma_2}(t). \tag{19}$$

The label Γ_1 (Γ_2) is for the symmetry species and partner type of the first (second) SA wavefunction. The energy-resolved flux for this wavepacket, in terms of the scattering states, is given by

$$P(Z, X, Y, \theta, \phi, E) = \frac{2\pi M}{|k_z|\mu} \operatorname{Im}\left(\Phi_{NSA}^*(E) \frac{\partial \Phi_{NSA}(E)}{\partial r}\right)\bigg|_{r=r_{cut}}$$

$$= \frac{2\pi M}{|k_z|\mu} \operatorname{Im}\left(c_1^* c_1 \Phi_{\Gamma_1}^*(E) \frac{\partial \Phi_{\Gamma_1}(E)}{\partial r} + c_1^* c_2 \Phi_{\Gamma_1}^*(E) \frac{\partial \Phi_{\Gamma_2}(E)}{\partial r}\right.$$

$$\left. + c_2^* c_1 \Phi_{\Gamma_2}^*(E) \frac{\partial \Phi_{\Gamma_1}(E)}{\partial r} + c_2^* c_2 \Phi_{\Gamma_2}^*(E) \frac{\partial \Phi_{\Gamma_2}(E)}{\partial r}\right)\bigg|_{r=r_{cut}} \tag{20}$$

The cross-terms reduce to zero when eqn. (20) is integrated over all remaining coordinates in calculating the total reaction probability [eqn. (16)], and the total flux for the NSA wavefunction is then simply an appropriately weighted sum of the fluxes from the respective SA wavefunctions. However, in calculating the reaction probability density as a function of X and Y, the cross-terms cannot necessarily be neglected: Although the net flux integrated over X and Y may be zero, this does not exclude the possibility that there are nonzero contributions at specific points in the (X, Y)-plane, which could cause a redistribution of $P_D(X, Y; E)$.

Here, we are saved the inconvenience of calculating the cross-terms in eqn. (20) by a property of the PES used in the calculations. Only rotation and diffraction basis functions of symmetry A_1 and B_1 have been used in the fitting of PES IV [eqn. (1)], which restricts the basis-function symmetries required to represent the wavefunctions.[36,55] In particular, the SA rotation–diffraction basis function $g_{\Gamma^\alpha j m_j \Gamma_d^\beta nm}(X, Y, \theta, \phi)$ can always be written as the product of a single SA rotational basis function, $R_{\Gamma^\alpha j m_j \Gamma_d^\beta}(\theta, \phi)$, and a single SA diffractional basis function, $D_{\Gamma_d^\beta nm}(X, Y)$:

$$g_{\Gamma^\alpha j m_j \Gamma_d^\beta nm}(X, Y, \theta, \phi) = D_{\Gamma_d^\beta nm}(X, Y) R_{\Gamma^\alpha j m_j \Gamma_d^\beta}(\theta, \phi). \tag{21}$$

Under these conditions, the cross-terms in eqn. (20) take the form

$$\Phi_{\Gamma_1}^*(E) \frac{\partial \Phi_{\Gamma_2}(E)}{\partial r} = \sum_{jm_j\Gamma_{d,1}nm j'm_j'\Gamma_{d,2}n'm'} \left(\chi_{\Gamma_1 jm_j\Gamma_{d,1}nm}^*(r, Z; E) \frac{\partial \chi_{\Gamma_2 j'm_j'\Gamma_{d,2}n'm'}(r, Z; E)}{\partial r}\right.$$

$$\left. \times D_{\Gamma_{d,1}nm}^* D_{\Gamma_{d,2}n'm'} R_{\Gamma_1 jm_j\Gamma_{d,1}}^* R_{\Gamma_2 j'm_j'\Gamma_{d,2}}\right) \tag{22}$$

where $\Gamma_{d,1}$ ($\Gamma_{d,2}$) is the symmetry of a diffraction function for the first (second) SA wavefunction. The energy-dependent functions, $\{\chi_{\Gamma_\kappa jm_j\Gamma_{d,\kappa}nm}, \kappa = 1, 2\}$, are related to the time-dependent functions in eqn. (9) by

$$\chi_{\Gamma_\kappa jm_j\Gamma_{d,\kappa}nm}(r, Z; E) = \frac{|k_Z|}{2\pi Mb(-k_Z)} \int_0^{T_{max}} dt e^{iEt} f_{\Gamma_\kappa jm_j\Gamma_{d,\kappa}nm}(Z, r; t). \qquad (23)$$

A consequence of the PES symmetry alluded to above is that the rotational basis functions used to represent the first SA wavefunction [$\Psi_{\Gamma_1}(t)$ in eqn. (19)] are always of a different symmetry to those used to represent the other SA wavefunction [$\Psi_{\Gamma_2}(t)$ in eqn. (19)] in the same calculation.[36,55] This means that integrating eqn. (22) over θ and ϕ, as required by eqn. (17), reduces the cross-terms in eqn. (20) to zero, because

$$\langle R_{\Gamma_1 jm_j\Gamma_{d,1}} | R_{\Gamma_2 j'm'_j\Gamma_{d,2}} \rangle = 0. \qquad (24)$$

With the PES used, the reaction probability density is therefore given by

$$P_D(X, Y; E) = \frac{2\pi M}{|k_Z|\mu} \text{Im}\left(c_1^* c_1 \sum_{jm_j\Gamma_{d,1}nmn'm'} C_{\Gamma_1 jm_j\Gamma_{d,1}nmn'm'} D_{\Gamma_{d,1}nm}^* D_{\Gamma_{d,1}n'm'} \right.$$
$$\left. + c_2^* c_2 \sum_{jm_j\Gamma_{d,2}nmn'm'} C_{\Gamma_2 jm_j\Gamma_{d,2}nmn'm'} D_{\Gamma_{d,2}nm}^* D_{\Gamma_{d,2}n'm'} \right) \qquad (25)$$

where

$$C_{\Gamma_1 jm_j\Gamma_{d,1}nmn'm'} = \int dZ \chi_{\Gamma_1 jm_j\Gamma_{d,1}nm}^*(r = r_{cut}, Z; E) \frac{\partial \chi_{\Gamma_1 jm_j\Gamma_{d,1}n'm'}(r = r_{cut}, Z; E)}{\partial r} \qquad (26)$$

with an analogous expression for $C_{\Gamma_2 jm_j\Gamma_{d,2}nmn'm'}$. Calculating the reaction probability density thus involves calculating the (X, Y)-resolved flux from each SA wavefunction propagation,

$$P_D^{\Gamma_i}(X, Y; E) = \frac{2\pi M}{|k_Z|\mu} \text{Im} \sum_{jm_j\Gamma_{d,i}nmn'm'} C_{\Gamma_i jm_j\Gamma_{d,i}nmn'm'} D_{\Gamma_{d,i}nm}^* D_{\Gamma_{d,i}n'm'}, \qquad (27)$$

and performing a weighted average.

D. Details of dynamics calculations

We have performed calculations here for the following incident states of H_2: $(v = 0, j = 0)$, $(v = 0, j = 1, m_j = 0,1)$, $(v = 0, j = 4, m_j = 0\cdots 4)$, $(v = 0, j = 11, m_j = 0,11)$, $(v = 1, j = 0)$, $(v = 1, j = 4, m_j = 0\cdots 4)$. Numerical details of the calculations are similar to those used in earlier studies,[32] with a few variations.

For each $v = 0$ incident state, a single wavepacket propagation has been performed for each symmetry involved. In other studies, we have tended to break the energy range into two separate domains, propagating longer for the low-energy range, and using a larger basis for the high-energy range.[32,34,36] Here we have not adopted this practice, but instead treat the whole energy range in each propagation. For even-j incident states, the rotational basis set was constructed such that $j \leq 28$, and the diffraction basis such that $|n| + |m| \leq 11$; for odd-j states, $j \leq 29$. Tests indicate that under these conditions results are fully converged with respect to basis size.

Calculations for $v = 1$ states have also been performed with one propagation yielding results for the entire energy range of interest. An exception is the $(v = 1, j = 4, m_j)$ states, for which extra propagations were performed in order to extend the range to lower energies. The same basis specifications were used for $v = 1$ incident states as $v = 0$.

Generally, wavepackets were propagated for 60 000 au of time, but the $(v = 0, j = 4, m_j = 1, 2, 3)$ incident states and the $(v = 1, j = 4, m_j)$ states were only propagated 20 000 au.

The value of r_{cut} used in the calculation of the reaction probability density was 2.77 a_0. This value was chosen because it places the cut in the product channel at each site, but still in the vicinity of the barrier, thus avoiding redistribution which may occur after the barrier has been crossed. Tests with smaller basis sets indicated that site-reactivity results were not very sensitive to the position of r_{cut} in the product channel.

III. Results

A. PES IV

Fig. 2 shows various 2D cuts through PES IV, as well as several plots for the potential *vs.* reaction coordinate for different high-symmetry surface sites. The use of bicubic splines in the fitting procedure (Section IIA) makes the PES smoother than was previously the case:[52] there are less artifacts in the 2D contour plots. The most significant difference between PES IV and earlier PESs, however, is that PES IV is anisotropic in ϕ at the top and hollow sites. Previous versions were based on the assumption that the potential was azimuthally isotropic at these four-fold sites (the expansion used included up to 2nd order spherical harmonics only). As it turns out, this assumption was quite reasonable: Fig. 2(f) and (h) show that the potential does not depend strongly on ϕ at these sites in front of and close to the barrier.

Table 1 shows that some barriers, namely for the top-to-bridge and hollow-to-bridge geometries, have decreased in height. Since this occurs in geometries for which the PES is not interpolated in θ or ϕ, in either PES III or PES IV, some explanation is required. The discrepancy is due mostly to improvements in the DFT/GGA calculations used, which, as discussed in Section IIIA, were performed here with more accurate numerical integrations than in the previous calculations.

B. Total reaction probabilities

Fig. 1 gives the reaction probability as a function of collision energy, E_c, for various initial rovibrational states of H_2. Results, degeneracy-averaged with respect to m_j, are compared with a fit to experimental data.[63] The experimental fit is based mostly on molecular beam data up to around 0.5 eV collision energy,[64] with the resulting fit tested for consistency against data taken from an associative desorption experiment.[65] The fit is resolved with respect to v, but not j; at the highest collision energy (0.5 eV), the average rotational energy of the molecules in the beam corresponded to the rotational energy of $j = 4$ H_2.[64]

The agreement between the experimental fit and degeneracy-averaged results calculated here for various $v = 0$ (Fig. 1(b)) and $v = 1$ (Fig. 1(c)) states is better than seen for our earlier PESs.[32–34,36,39] The agreement for $v = 1$ is particularly good. The calculated $v = 0$ curves appear translated toward higher energies by around 0.05 eV relative to experiment.

Table 1 Comparison between barrier heights for PES III[52] and the current PES IV. The barrier heights, given in eV, are taken relative to the minimum of the gas-phase H_2 potential. The values given are for H_2 fixed at one of the three high-symmetry sites, and in the specific orientation indicated. Angles are in degrees. The results for PES III which are not based on DFT calculations for the corresponding geometry, but instead are interpolated, are marked with an asterisk

Surface site	θ	ϕ	Barrier PES III	Barrier PES IV
Bridge	90	0	0.48	0.50
	90	45	1.32*	0.84
	90	90	1.37	1.19
Top	90	0	0.70	0.63
	90	45	0.70*	0.73
Hollow	90	0	0.64	0.58
	90	45	0.64*	0.65

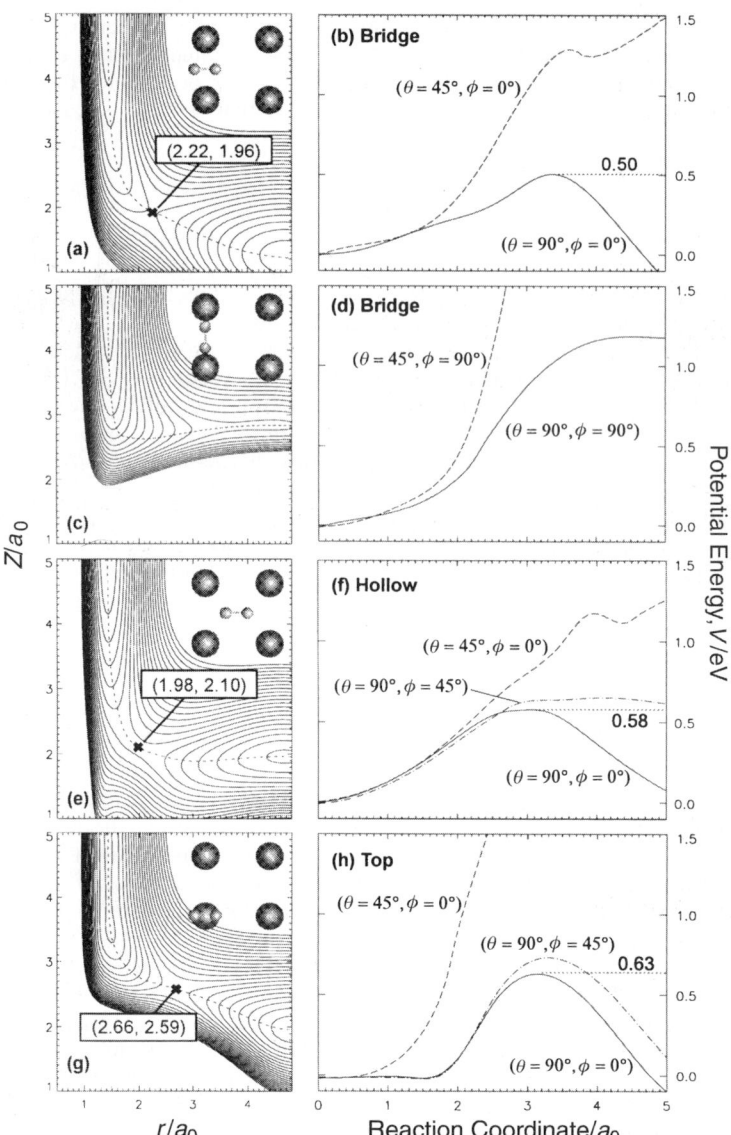

Fig. 2 Aspects of PES IV. (a), (c), (e) and (g). Contour plots of 2D cuts through PES IV for the bridge-to-hollow, bridge-to-top, hollow-to-bridge and top-to-bridge configurations, respectively. The insets show the geometry of the molecule relative to the surface unit cell. In all cases, the molecule is parallel to the surface. The contour lines are for potential energies 0.1 eV apart. The reaction path is shown in each plot as a dotted line, with the barrier position (r, Z) indicated. (b), (d), (f) and (h). The dependence of potential energy on the reaction path coordinate, which is here taken to be zero at $Z = 5.0\ a_0$ and increasing moving toward the surface. The respective plots are for the geometries given in the corresponding plots on the left. Each panel includes plots for two values of θ, and panels (f) and (h) also include plots for two values of ϕ. In panels (b), (f) and (h), the barrier heights (in eV) are indicated for the bridge-to-hollow, hollow-to-bridge and top-to-bridge geometries with $\theta = 90°$.

Calculated results for $v = 0$ do not show a strong dependence on j (Fig. 1(b)), and there is no clear trend for the j-states shown. Increasing j enhances reaction for molecules incident in $v = 1$ at all collision energies shown (Fig. 1(c)).

Results for the $(v = 0, j = 4, m_j)$ states (Fig. 1(d)) show the strong preference for reaction of 'helicoptering' molecules (i.e., $|m_j| = j$) that is characteristic of these reactions.[10,34,36,38,66]

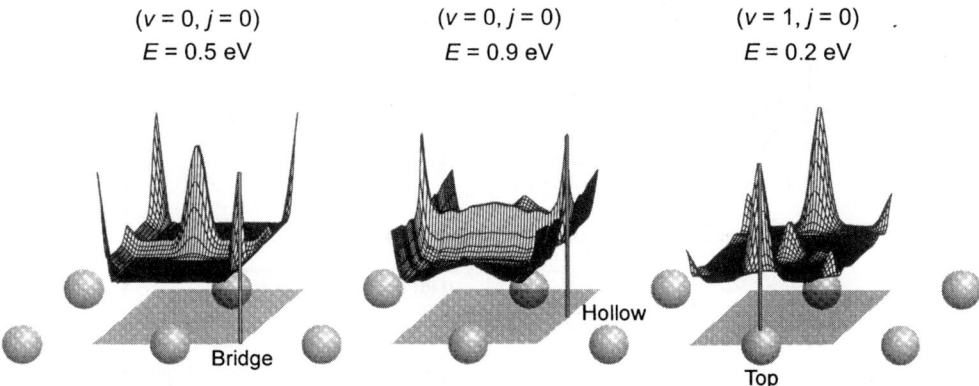

Fig. 3 Reaction probability density as a function of surface site. The plots show the reaction probability density vertically for a single unit cell. The unit cell is shaded and shown relative to the surface Cu atoms. Each plot is labeled with the incident state and collision energy. The scales for the reaction probability density are *not* the same for each plot.

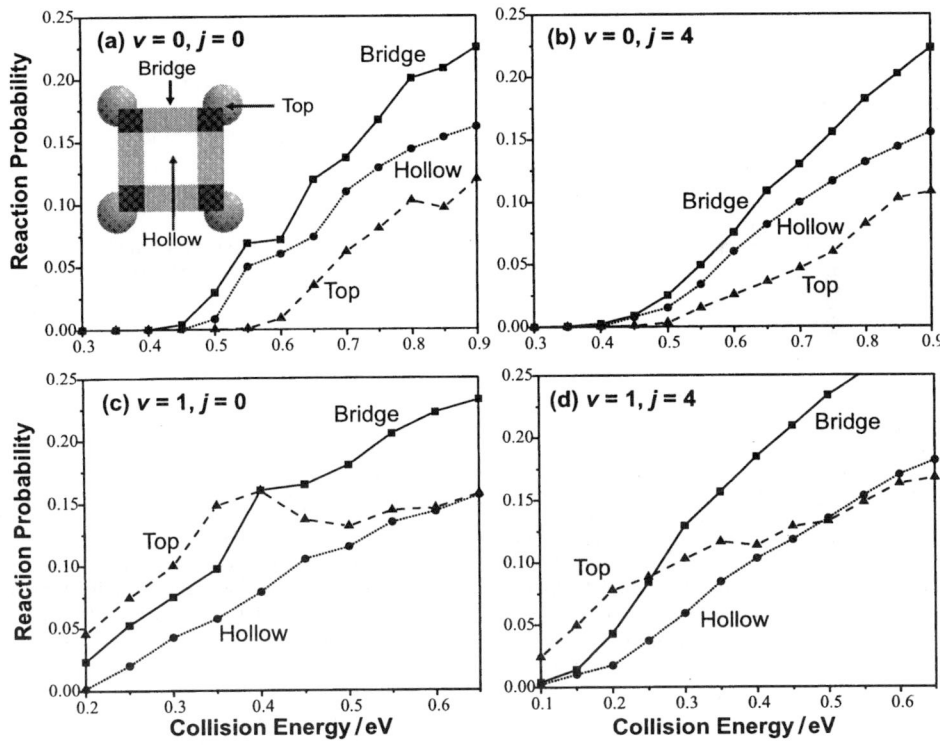

Fig. 4 The reaction probability associated with different regions of the surface unit cell for degeneracy-averaged incident states. The plots are calculated by integrating the reaction probability density over the regions shown in the inset of (a) at each collision energy. Each region is labeled by the nearest high-symmetry site. Plots are shown for (a) the $(v = 0, j = 0)$ incident state; (b) the $(v = 0, j = 4)$ degeneracy-averaged state; (c) the $(v = 1, j = 0)$ state; and (d) the $(v = 1, j = 4)$ degeneracy-averaged state.

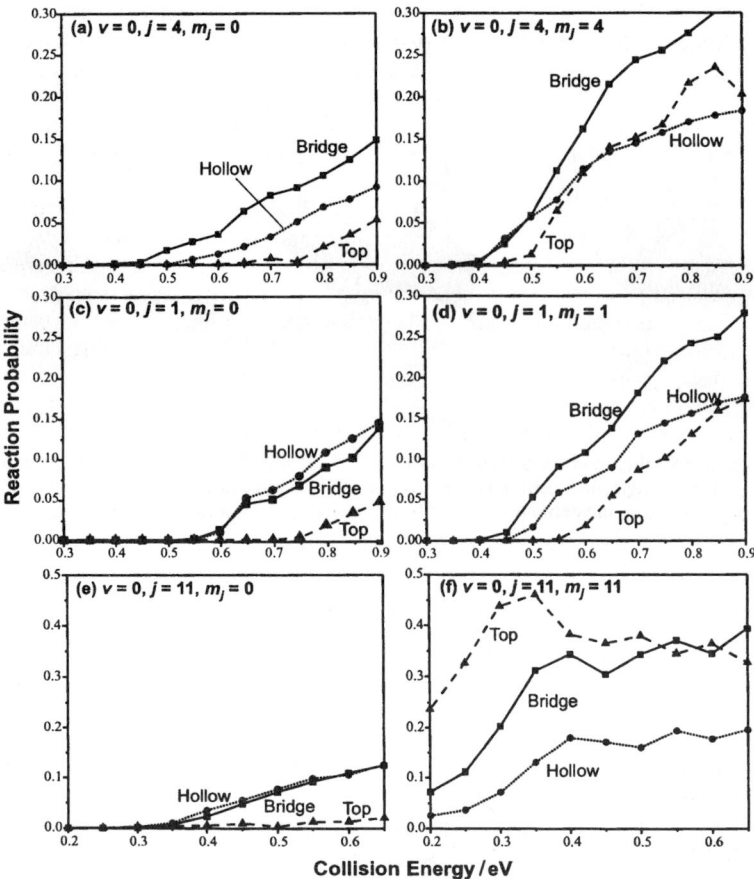

Fig. 5 The portion of reaction probability associated with different regions of the surface unit cell for helicopter and cartwheel incident states. The plots are calculated as in Fig. 4. Results are shown for the cartwheel ($m_j = 0$) and helicopter ($|m_j| = j$) incident states of (a) and (b) ($v = 0, j = 4$); (c) and (d) ($v = 0, j = 1$); and (e) and (f) ($v = 0, j = 11$).

C. Site-specific reaction probabilities

Fig. 3 shows the reaction probability density as a function of surface site for the ($v = 0, j = 0$) and ($v = 1, j = 0$) incident states, at various collision energies. The plot for ($v = 0, j = 0$) at the relatively low collision energy of 0.50 eV shows a strong preference for reaction at the bridge sites, directly between two surface Cu atoms, and very little reaction at any other sites.

The same initial rovibrational state at 0.90 eV collision energy shows a very different dependence on reaction site (Fig. 3), with the hollow sites, in the center of the square surface unit cell, most reactive. At this collision energy, the bridge sites are around half as reactive as the hollow sites. The top sites, directly over a surface Cu atom, show some reaction, though less than the other two high-symmetry sites.

The plot for ($v = 1, j = 0$) at 0.20 eV (Fig. 3), a relatively low collision energy for this state, is different again, with the top sites most reactive. Some reaction occurs at the bridge sites, but very little at the hollow sites.

In order to analyze the dependence of site-reactivity on collision energy, we can integrate the reaction probability density, as given in Fig. 3, over various regions of the unit cell, plotting the results against incidence energy. The integration regions are shown in the inset of Fig. 4(a). This division of the surface unit cell simply corresponds to assigning each point to the nearest high-symmetry site. Each region is labeled with the name of its associated high-symmetry site (*i.e.*, bridge, top, or hollow). Here, when we refer to the 'probability of reaction' for a given region, we mean the probability that a molecule reacts and that it reacts in the given region. For example, a value of 0.10 for the top site is referred to as the 'top-site reaction probability', but actually means that the probability for any incident molecule reacting in the top-site region, not only those incident upon the top site, is 0.10. The probabilities sum to the total reaction probabilities shown in Fig. 1.

For the $(v = 0, j = 0)$ incident state, the bridge site has the highest reaction probability at all collision energies shown (Fig. 4(a)). This may seem to contradict Fig. 3, where for $E_c = 0.9$ eV the hollow site was found to be most reactive. However, there are twice as many bridge sites as there are top or hollow sites. The surface unit cell region associated with the bridge site is thus twice as large as that of the other high-symmetry sites, meaning that even if the actual bridge site is not most-reactive, more reaction can still occur in the region surrounding the bridge site than at either of the other two. Results for $(v = 0, j = 4)$ (Fig. 4(b)) are qualitatively similar to those for $(v = 0, j = 0)$, although the top site is noticeably more reactive at low collision energies for $(v = 0, j = 4)$.

The situation for $v = 1$ is very different. For both the $(v = 1, j = 0)$ and $(v = 1, j = 4)$ incident states, most reaction at low energies occurs at the top sites, as already seen in Fig. 3. At high collision energies, the ordering of site reactivities is more similar to that seen for $v = 0$ states, with the largest portion of reaction occurring at the bridge sites, followed by the hollow and top sites (but note that for $v = 1$ the top site is more reactive than or as reactive as the hollow site, even at higher collision energies, whereas for $v = 0$ the hollow site is generally more reactive).

Fig. 5 shows energy-dependent site reaction probabilities for various incident helicopter and 'cartwheel' (*i.e.*, $m_j = 0$) states. As for the total reaction probability (Fig. 1(d)), the helicopter is generally more reactive than the cartwheel state at all sites. More interesting are the qualitative differences that exist between the two cases. Most significant is the reactivity of the top site, which for the cartwheel states is suppressed relative to that of the hollow and bridge sites when compared with the corresponding plots for $(v = 0, j = 0)$ (Fig. 4(a)). In contrast, top site reaction is enhanced relative to the other sites for the helicopter incident states, as exemplified by the results for $(v = 0, j = 11, m_j = 11)$ (Fig. 5(f)) where the top site is actually most reactive at most energies. Another noteworthy result is that, for most energies, more reaction occurs at the hollow site than at other sites for the cartwheel incident state $(v = 0, j = 1, m_j = 0)$ and $(v = 0, j = 11, m_j = 0)$, but not for the $(v = 0, j = 4, m_j = 0)$ cartwheel state.

IV. Discussion

A. Agreement with experiment

It is encouraging that the extension of the PES instigated here has resulted in overall better agreement with the experimental fits of Michelsen and Auerbach for the total reaction probability.[63] The changes have resulted in small increases in the calculated reaction probabilities, bringing them into better agreement with the experimental findings. The improved fitting of the PES, which is now based on DFT calculations for more orientations than previously, has caused the barriers for some geometries, for which the previous PES was purely interpolated, to decrease somewhat in height (Table 1), thereby boosting reaction. This is particularly true of the bridge site, where the barrier for a molecule in a parallel orientation with $\phi = 45°$ has dropped almost 0.5 eV (Table 1). The barriers for the top and hollow sites are also lower, even for those geometries which are not interpolated, due to differences in the DFT method used (see Section II.A). The remaining discrepancy between experiment and theory could be due to any number of factors, including inaccuracies in the potential fit, the inaccuracy inherent in the DFT approach, and inaccuracies in the experimental data.[67] A detailed discussion of such influences has been given elsewhere.[32]

B. Circumvention of the transition state

An immediate and startling conclusion to be drawn from the site reactivity results of Fig. 3, 4, and 5 is that, even at low collision energies, reaction does not necessarily proceed primarily through the transition state (at the bridge site; here the 'transition state' refers to the lowest point of the potential ridge separating reactants from products). Reaction can actually be dominant at other sites where the barrier to reaction is higher, depending on the initial state. Other aspects of the PES play a role in determining which site is most reactive, and these are outlined in detail below. However, the general conclusion must be that, even at low collision energies, reaction does not necessarily proceed in the vicinity of the minimum energy path; instead, it is possible that most of the reactive flux follows a very different route.

C. Effect of rotation on site-reactivity

The good agreement between experiment and theory for reaction probabilities gives us confidence that our results for site-reactivity are trustworthy. Most of the observed dependences of site-reactivity on initial state can be explained with reference to three quantities: barrier height, position ('lateness') of barrier, and dependence of the PES on molecular orientation. For example, the high reactivity of the bridge site at low collision energies for the incident $(v = 0, j = 0)$ state (Fig. 3) is to be expected, because the system-wide minimum barrier to reaction occurs for the bridge-to-hollow geometry (*i.e.*, when the molecule is positioned above the bridge site with the atoms dissociating into neighboring hollow sites; Fig. 2). However, the PES at the bridge site barrier geometry is also strongly dependent on orientation: Not only are tilted geometries unfavorable for reaction (Fig. 2(b)), but dissociation in the bridge-to-top geometry is also very unfavorable (Fig. 2(c) and (d)). This explains why the hollow site becomes more reactive at high collision energies (Fig. 3): the PES at the hollow site is much more isotropic in front of the barrier, for rotation in both θ and ϕ (Fig. 2(f)), meaning molecules approaching this site will experience a lower barrier *on average*.

The influence of steric effects in molecule–surface reactions has been the subject of several earlier experimental[67–69] and theoretical studies.[10,16,38] Here we will break down these effects into two categories: 'orientational hindrance' and 'rotational hindrance'. The term 'orientational hindrance' has been used elsewhere interchangeably with 'steric hindrance', but here we use it to mean a very specific type of reaction hindrance. In particular, an example of 'orientational hindrance' (Fig. 6(a)) is the suppression of reaction seen at the bridge site, relative to the hollow site, at high collision energies for the $(v = 0, j = 0)$ incident state (Fig. 3). Here, orientational hindrance is a static effect, in the sense that it is related only to the orientation of a molecule, and not to its internal motion.

In contrast to orientational hindrance, rotational hindrance is a dynamical effect related to the rotation of the molecule (Fig. 6(a)). Rotational hindrance results when there is a strong anisotropy of the potential in the direction of rotation. In such a case, the faster the molecule rotates, the more likely it is to encounter a repulsive orientation before it can react. The molecule will only be able to react if it can move through the interaction region and react before it has rotated into an unfavorable geometry, and the faster it rotates the smaller the 'window' of reactive paths gets.[8,10,12]

Fig. 6(a) represents orientational and rotational hindrance diagrammatically. As stated above, it is common to group these two effects under the one umbrella term, but there are good reasons to distinguish between them. We have already seen a case of orientational hindrance: the relative suppression of bridge site reaction at high collision energies for the $(v = 0, j = 0)$ incident state. This is clearly unrelated to rotation, because the molecules are not rotating, at least not initially. To demonstrate rotational hindrance, we consider the well-known suppression of reaction with increasing j seen experimentally[67,68] and in calculations[8,10,38] for the H_2/D_2 reaction on Cu(111). This suppression, which we will refer to as the 'steric hindrance effect' to distinguish it from orientational and rotational hindrance, occurs for j less than about 4. The effect cannot be due simply to orientational effects, because the degeneracy-averaged beams for different js all have the same isotropic orientational distribution initially. As has been demonstrated by Darling and Hollow,[8,10] the effect is actually due to rotational hindrance; as j increases, molecules are more likely to rotate into unfavorable geometries while they are close to the barrier. The hindrance of reaction in these two examples clearly have different roots, and it makes sense to explain them in terms of two different effects.

The qualitative differences between the reactivity of the top site relative to other sites for different incident states (Fig. 5) can be partially attributed to orientational and rotational hindrance.

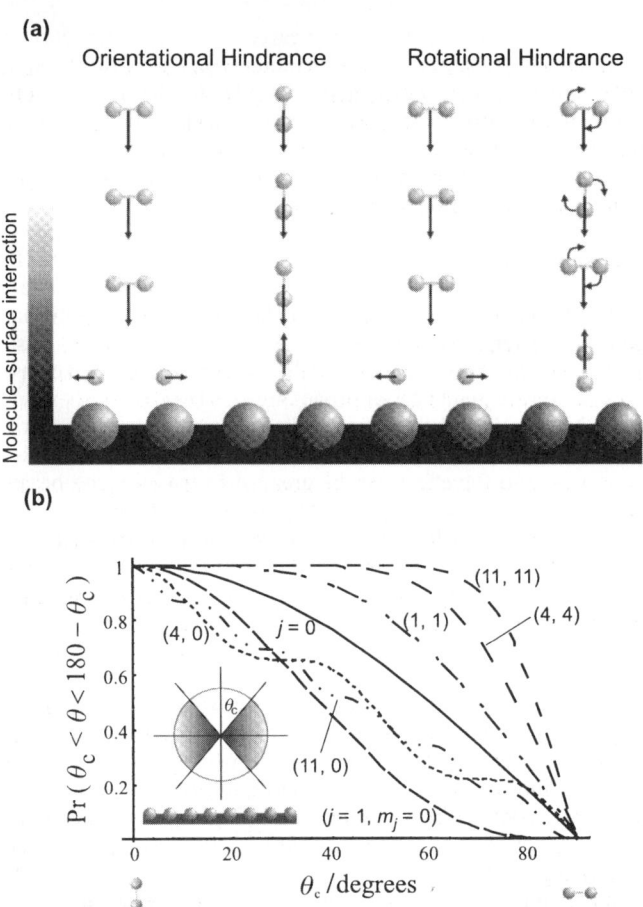

Fig. 6 Orientational and rotational hindrance, and the orientational distribution of rotational states over θ. (a) Diagrammatical representation of orientational and rotational hindrance. The bar on the left with the grayscale gradient represents the strength of the molecule–surface interaction. Orientational hindrance is demonstrated by comparing two different molecules approaching the surface: one parallel to the surface, and the other perpendicular. The poorly-oriented (perpendicular) molecule is unable to react and is back-scattered, while the well-oriented (parallel) molecule dissociates. Two molecules are also shown for rotational hindrance: one is not rotating, and the other is. The frames demonstrate that the rotating molecule can rotate into a poor orientation whilst in the strong interaction region, thus preventing reaction. The faster the molecule rotates, the more likely this is to occur. (b) The probability that a molecule in a given rotational state (j, m_j) is oriented in the range of angles $\theta_c < \theta < 180 - \theta_c$, plotted against θ_c. The inset depicts the range of angles graphically in relation to the surface. Small diagrams below indicate the θ_c values for which the molecule is perpendicular and parallel to the surface. The plot for each different rotational state is labeled by (j, m_j).

Orientational hindrance helps account for the general trend, seen for many of the incident states, that the top site is least reactive of the three: Fig. 2(h) shows that tilted geometries are particularly repulsive for the top site, when compared with other sites (Fig. 2(b) and (f)). The dramatic contrast between helicopter and cartwheel reaction at the top site (Fig. 5) is also partially attributable to orientational hindrance. Fig. 6(b) shows how the helicopter molecules are much more likely to be parallel to the surface than either the isotropically-distributed (*i.e.*, $j = 0$) or cartwheel states. The helicopter states thus suffer less from orientational hindrance, because parallel geometries are much more favorable for reaction at the top site than tilted or perpendicular geometries (Fig. 2).

Rotational hindrance also influences the comparison between helicopter and cartwheel reaction at the top site. The top site is less reactive in proportion to other sites for the incident ($v = 0$,

$j = 11$, $m_j = 0$) state (Fig. 5(e)) than for the lower-j cartwheel states (Fig. 5(a) and (c)). This is true even though the ($j = 11$, $m_j = 0$) state is better oriented for reaction than, for example, the ($j = 1$, $m_j = 0$) state (*i.e.* it is more likely to be close to parallel to the surface; Fig. 6(b)), ruling out orientational hindrance as the cause. Increased rotation increases the likelihood that the incident molecules will encounter a repulsive geometry, suppressing top-site reaction of the ($v = 0$, $j = 11$, $m_j = 0$) in proportion to the slower rotating ($v = 0$, $j = 1$, $m_j = 0$) and ($v = 0$, $j = 4$, $m_j = 0$) states.

So far, we have explained the suppression of top site reaction for the cartwheel states in Fig. 5, but have not yet addressed the question of why the top site is relatively more reactive than other sites for the helicopter states, particularly ($v = 0$, $j = 11$, $m_j = 11$) (Fig. 5(f)). This is largely due to barrier position. Fig. 2(g) shows that the barrier is much 'later' at the top site than at other sites, with the bond significantly extended. This has been shown to lead to rotational enhancement of reaction due to a transfer of energy from rotation to dissociative motion.[10] Basically, the rotational energy has an inverse dependence on bondlength, which leads to a decrease in the rotational energy as the molecule's bond extends on approaching the barrier. With total energy conserved, the excess energy must be transferred to other modes, including the dissociative mode. The high rotational energy of the $j = 11$ molecules means that there can be a considerable liberation of energy as the bond extends, and the effect will be greatest for the top site because it has the latest barrier.

This is demonstrated in Fig. 7, which shows plots of the rotationally-adjusted potential,

$$V_j(r, Z, X, Y, \theta, \phi) = V(r, Z, X, Y, \theta, \phi) + \frac{j(j + 1)}{2\mu r^2}, \tag{28}$$

for parallel orientations and $j = 11$. The barrier at the top site (Fig. 7(f)) for this potential is much lower than at the hollow site (Fig. 7(d)), and slightly lower than at the bridge site (Fig. 7(b)). This partially explains the relatively high reactivity of the top site, but not totally. A second contribution is the very low azimuthal anisotropy of the potential for the top site (Fig. 2(h)) relative to that of the bridge site (Fig. 2(b) and (d)). This reduces rotational hindrance and, together with the lower rotationally-adjusted barrier, gives the top site a significant reactive advantage for the ($v = 0$, $j = 11$, $m_j = 11$) incident state at low collision energies.

Another interesting aspect of ($v = 0$, $j = 11$, $m_j = 11$) state reaction is that the top-site reaction probability exceeds the probability initially associated with the site at nearly all collision energies (Fig. 5(f)). 25% of incident molecules are located in the region assigned to the top site; if no redistribution of molecules parallel to the surface were to occur, the maximum attainable top-site reaction probability would be 0.25. Fig. 5(f) includes probabilities in excess of 0.45. This could only occur if there was a convergence of reacting molecules toward the top site. Interestingly, such a convergence was seen for this same state in classical trajectory calculations performed on an earlier version of the PES, PES III.[39] The reason for this convergence can be seen in the rotationally-adjusted potential for the top site (Fig. 7(e) and (f)). This potential is considerably lower in the entrance channel than the potential at other sites, and actually exhibits a small well. The repulsion of the molecule (*i.e.*, building up of the barrier) begins later at the top site than at other sites (Fig. 2), and the reactant channel actually broadens as the molecule approaches the surface (Fig. 2(g)), giving rise to the well. The resulting potential difference between the top site and other sites lead to forces which act to attract an incident molecule toward the top site.

An interesting feature of the plots for cartwheel states in Fig. 5 involves hollow site reaction. For the even-j ($v = 0$, $j = 4$, $m_j = 0$) state, less reaction occurs at the hollow sites than at the bridge sites (Fig. 5(a)), but the opposite is generally true for the odd-j ($v = 0$, $j = 1$, $m_j = 0$) and ($v = 0$, $j = 11$, $m_j = 0$) states (Fig. 5(c) and (e)). For the ($v = 0$, $j = 1$, $m_j = 0$) state this is probably explained by orientational hindrance: odd-j cartwheel states have a node for parallel configurations, meaning they are more poorly aligned for reaction; Fig. 6(b) shows that this is particularly true for ($j = 1$, $m_j = 0$). Because the PES at the hollow site is more isotropic in the θ direction (Fig. 2(f)) than at the bridge site, bridge-site reaction will likely suffer more from orientational hindrance than hollow-site reaction, leading to the observed effect. It is less clear whether this is also the case for the ($v = 0$, $j = 11$, $m_j = 0$) state, because Fig. 6(b) shows that it is less poorly-aligned. Another possibility is that rotational hindrance is responsible for the effect in the case of the ($v = 0$, $j = 11$, $m_j = 0$) incident state. The greater rotation of this state means it suffers more

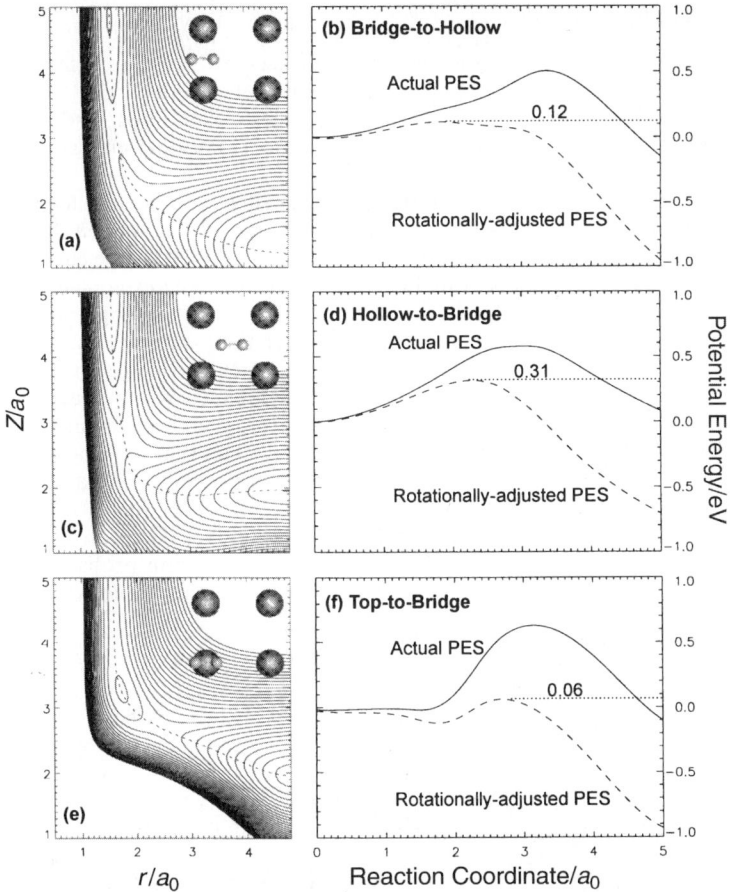

Fig. 7 Rotationally-adjusted potential for $j = 11$. (a), (c) and (e). Contour plots of 2D cuts through $V_{j=11}$ for the bridge-to-hollow, hollow-to-bridge and top-to-bridge configurations, respectively. The insets show the geometry of the molecule relative to the surface unit cell. In all cases, the molecule is parallel to the surface. The contour lines are for potential energies 0.1 eV apart. The reaction path is shown in each plot as a dotted line. (b), (d) and (f). The dependence of V and $V_{j=11}$ on the reaction path coordinate. (Note that the plots for V are the same as in Fig. 2, and are only reproduced here for convenience.) The respective plots are for the geometries given in the corresponding plots on the left. The reaction path coordinate is taken to be zero at $Z = 5.0\ a_0$ and increasing moving toward the surface. For the rotationally-adjusted PES plots, the reaction path of the rotationally-adjusted 2D PESs, as shown in (a), (c) and (e), are used. Plots for the actual PES, V, are shown relative to the reaction paths for the 2D PESs given in Fig. 2. The plots for $V_{j=11}$ have been shifted down 0.87 eV so that the gas-phase minimum potential is zero, in order to better facilitate comparison. Barrier heights (in eV) are indicated on the rotationally-adjusted plots.

from rotational hindrance than the $(v = 0, j = 1. m_j = 0)$ state, and rotational hindrance will be more severe at the bridge site where the θ anisotropy is larger in front of the barrier (Fig. 2(b)).

D. Effect of vibration on site reactivity

Fig. 3 and 4 show that adding vibrational energy to the incident molecules leads to a considerable enhancement of top-site reaction relative to that of other sites, particularly at low collision energies. This is again due to the lateness of the barrier for the top-site PES. The late barrier, and large curvature of the reaction path in the reactant channel which accompanies it (Fig. 2(g)), lead to a strong coupling of the vibrational mode and the dissociative mode,[11,18] allowing much of the vibrational energy to be utilized for barrier crossing.

E. Steric hindrance effect

As already mentioned, experimental and theoretical studies of H_2/D_2 reacting on Cu(111) have demonstrated the existence of a steric hindrance effect for that system.[8,10,38,67,68] Increasing j for j less than approximately 4 suppresses reaction; this is due to rotational hindrance. (Increasing j enhances reaction for j greater than 4 due to the increased rotational energy available; see also Section 4C.)

The question arises as to whether the steric hindrance effect can be seen for H_2 + Cu(100). In our earlier quantum calculations it was not seen,[36] but this could have been due to some irregularities in the PES used.[39] In classical trajectory calculations on PES III the effect was observed;[39] however, insufficient quantum data is available for PES III to make any conclusions regarding the existence or otherwise of the steric hindrance effect. Here we see no clear evidence for the steric hindrance effect in the Cu(100) system (Fig. 1(b)). Neither the $(v = 0, j = 1)$ or the $(v = 0, j = 4)$ degeneracy-averaged reaction probabilities show any clear trend relative to the results for $(v = 0, j = 0)$. At some collision energies there is slight rotational enhancement of reaction, and at others, slight suppression. Reaction probabilities for $v = 1$ states show a clear absence of the effect, with the $j = 4$ incident state more reactive at all collision energies than the $j = 0$ state (Fig. 1(c)).

In order to examine the steric hindrance effect in more detail, we have plotted reaction-probability differences between $j = 0$ and $j = 4$ states for each site (Fig. 8). In this way, enhancement/suppression of reaction with changing j can be considered on a per site basis. Fig. 8(a) shows that the absence of any clear trend in the overall reaction probability carries over to the individual sites. None of the three sites exhibits exclusively either enhancement or suppression

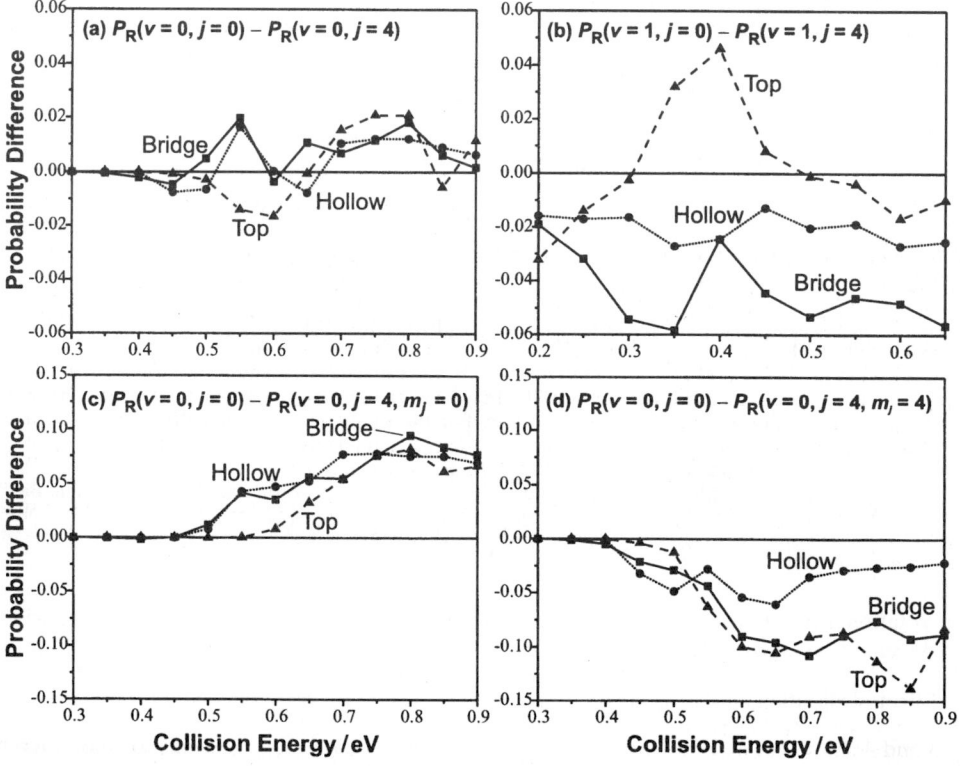

Fig. 8 Differences between the site-associated reaction probabilities of different incident states. (a) The reaction probability associated with the three high-symmetry sites, as shown in Fig. 4, for the $(v = 0, j = 4)$ degeneracy-averaged incident state, subtracted from those of the $(v = 0, j = 0)$ state. (b) As in (a), but for the $(v = 1, j = 4)$ and $(v = 1, j = 0)$ states. (c) As in (a), but for the $(v = 0, j = 4, m_j = 0)$ cartwheel state and the $(v = 0, j = 0)$ state. (d) As in (a), but for the $(v = 0, j = 4, m_j = 4)$ helicopter state and the $(v = 0, j = 0)$ state.

over the entire range of collision energies. For $v = 1$ there is clearly rotational enhancement of reaction at both the hollow and bridge sites; the top site exhibits both enhancement and suppression, dependent upon the collision energy.

The lack of any clear suppression or enhancement of reaction with increased j in the $v = 0$ reaction probabilities (Fig. 1(b)) actually results from a complex cancellation of suppression and enhancement in the individual m_j states of $(v = 0, j = 4)$. For example, at all sites, reaction of the cartwheel state is suppressed and helicopter reaction enhanced relative to $(v = 0, j = 0)$ (Fig. 8(c) and (d)). The steric hindrance effect results from a slight tipping of the balance toward suppression for low j; it is likely to be very sensitive to PES topography. In this regard, it is worth noting that even though full-dimensional calculations for H_2 reacting on Cu(111) do show the presence of a steric hindrance effect,[38] *fixed-site* quantum calculations *for the top site* of Cu(111) do not show any clear effect.[17]

If we assume that our calculations are accurate regarding the absence of a steric hindrance effect, a possible explanation is the part played by the top site in reaction. The top-site barrier is lower relative to that of the bridge site in PES IV (Table 1) than has been found to be the case in DFT calculations for Cu(111);[50] the top site would thus be expected to make up a greater portion of the total reaction probability for Cu(100) than for Cu(111). Although the top site does not exhibit the steric hindrance effect for Cu(111),[17] this does not significantly influence the overall reaction because the other sites contribute the bulk of the reaction probability; however, this is less true of Cu(100). Close examination of Fig. 8(a) reveals that, if the top site contribution to reaction were to be removed, the resultant total reaction probability would show a general suppression with increasing j, because the bridge and hollow sites do exhibit the steric hindrance effect for most energies shown. (In the present calculations, a small rotational enhancement is observed for the bridge and hollow sites at the lowest collision energies, but there was also no clear suppression seen in the Cu(111) calculations at the lowest energies.[17,38]) We thus conclude that the existence of the steric hindrance is very much dependent upon the relationship of the top site barrier to the other barriers in the system.

F. Experimental signature

So far we have made several predictions about site reactivity on the assumption, based on good agreement between theory and experiment for reaction probabilities (Fig. 1), that our PES adequately represents the $H_2 + $ Cu(100) system. Now we will go one step further by making a prediction which can be verified by contemporary experimental techniques. We will provide a signature for our site-reactivity finding that, at low collision energies, $(v = 0, j = 4)$ H_2 reacts primarily at the bridge site, while $(v = 1, j = 4)$ H_2 reacts primarily at the top site.

Experimentally, the molecular beam experiments corresponding to our calculations for dissociative adsorption are difficult to perform. Another approach is to study the reverse reaction, associative desorption,[65,66,68,70–72] and invoke detailed balance, which has been shown to hold for $H_2 + $ Cu(111).[67] Several studies for H_2/D_2 desorbing from Cu(111) have focussed on the orientational distribution of angular momenta in the desorbing beam,[66,70,72] with the most recent resolved with respect to vibrational and rotational state, and desorption energy.[66] These experiments measure the quadrupole alignment of the angular momentum vector, j, for the desorbing molecules:

$$A_0^{(2)} \equiv \langle 3 \cos^2 \Theta - 1 \rangle, \quad (29)$$

where Θ is the angle between j and the surface normal. This quantity is -1 if all desorbing molecules are in the cartwheel state, and 0 if there is an isotropic distribution over m_j states. The quadrupole alignment is positive if there is a preference for helicoptering motion, with the theoretical maximum 2. However, this maximum is unattainable in a quantum system, with the maximum attainable value actually a monotonically increasing function of j.

Our calculations can be compared with the results of associative desorption experiments if detailed balance is invoked. We can calculate the quadrupole alignment of only those molecules in the incident beam that *go on to react*, which is directly related to the quadrupole alignment of desorbing molecules in the associative–desorption experiments. The quadrupole alignment of j can be written in terms of our calculated reaction probabilities as[73]

$$A_0^{(2)}(v, j; E_c) = \frac{\sum_{m_j} P_R^{vjm_j}(E_c)[3m_j^2 - j(j+1)]/[j(j+1)]}{\sum_{m_j} P_R^{vjm_j}(E_c)}, \quad (30)$$

where $P_R^{vjm_j}(E_c)$ is the reaction probability for incident state (v, j, m_j) at collision energy E_c.

The dominance of bridge-site reaction for the $(v = 0, j = 4)$ incident state and top-site reaction for the $(v = 1, j = 4)$ incident state, at relatively low collision energies, leads to differences between the two states in the proportion of reacting molecules that are initially in a helicopter state (Fig. 9(a)). The top site PES is considerably more anisotropic in θ for geometries in front of the barrier (Fig. 2(h)), and this leads to a stronger preference for reaction of helicoptering molecules at the top site compared with at the bridge site. This is also seen in the quadrupole alignment given by eqn. (30) (Fig. 9(b)). The collision-energy scale is different for each state: the scales have been shifted relative to one another such that the total reaction probabilities are comparable at the lowest collision energies shown. The figure shows a qualitative difference in the collision-energy dependence of $A_0^{(2)}$ between the two states, with the $(v = 1, j = 4)$ alignment rising much more abruptly at low energies. This is simply a reflection of the greater preference for helicopter reaction in the $(v = 1, j = 4)$ beam (Fig. 9(a)), at collision energies for which the total reaction probabilities of $(v = 0, j = 4)$ and $(v = 1, j = 4)$ are comparable. This dependence of $A_0^{(2)}$ on v could be used as an experimental signature for our site reactivity findings, because it is a direct result of differences in the potential topology between the preferred reaction sites of $v = 0$ and $v = 1$ H_2.

G. Implications for experiment

The theoretical site-reactivity results presented here could have other important ramifications for experiment. Fig. 10 shows plots for the percentage of reaction occurring at each site as a function of collision energy and initial state. Results for $(v = 0, j = 0)$ show that at low collision energies, more than 90% of reaction occurs in the vicinity of the bridge site (Fig. 10(a)); for $(v = 0, j = 4)$ more than 80% of reaction is at the bridge site at 0.4 eV. The plots for $v = 1$ states show similar high percentages, but instead for top site reaction; at 0.1 eV, nearly 80% of $(v = 1, j = 4)$ reaction occurs at the top site.

Such dominance of reaction by individual surface sites could be used experimentally to probe the surface. Through appropriate selection of initial conditions, experimentalists could choose the site at which they wish reaction to occur; the site chosen does not necessarily have to be the site of minimum barrier. The PES could effectively be used as a filter, allowing reaction at one site but restricting it at another. For example, by vibrationally exciting the incident molecules top-site reaction is enhanced relative to that of other sites. By choosing a relatively low collision energy, reaction is effectively 'switched-off' at all sites except the top site.

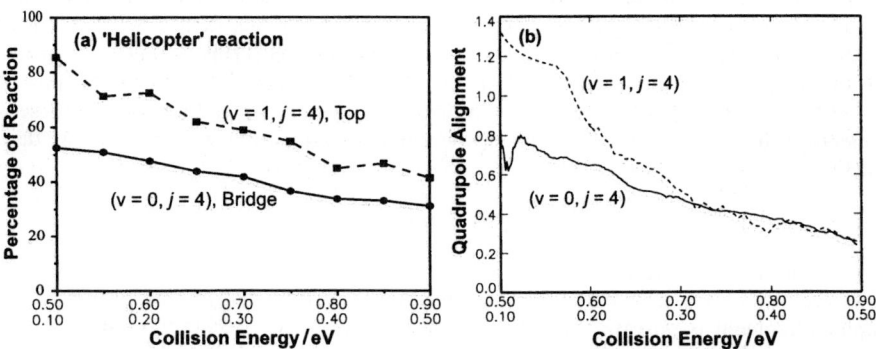

Fig. 9 The percentage of reaction due to helicopter states, and the quadrupole alignment of j for reacting molecules. (a) The percentage of reaction due to helicopter states for the $(v = 1, j = 4)$ degeneracy-averaged state in the vicinity of the top site, and the corresponding quantity for the $(v = 0, j = 4)$ state at the bridge site. Two collision-energy scales are shown: the upper one applies to the $(v = 0, j = 4)$ state, and the lower to the $(v = 1, j = 4)$ state. (b) The quadrupole alignment of j for those molecules which go on to react, for the same states as in (a). The same collision energy scales as in (a) are used.

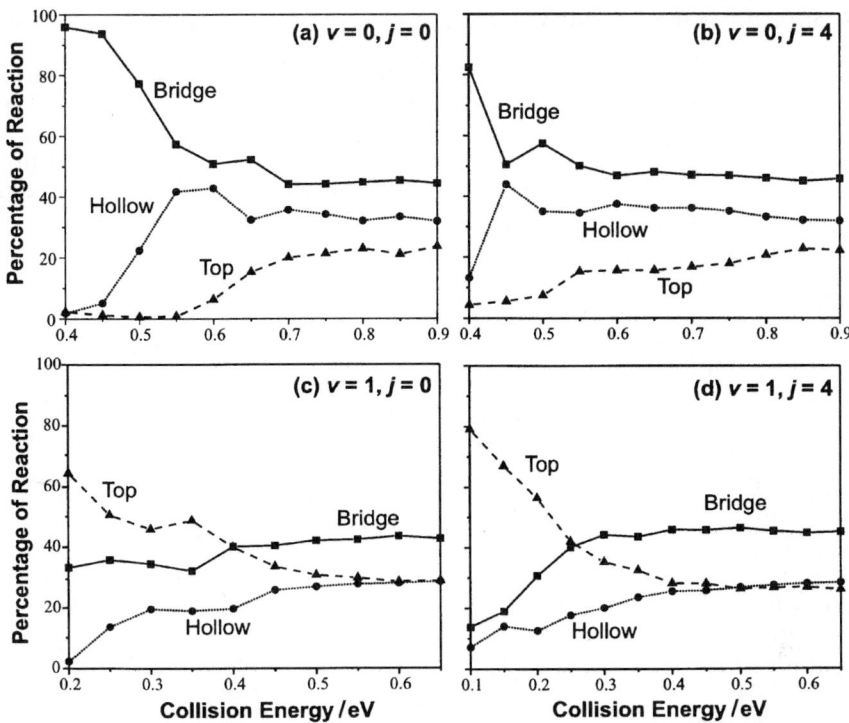

Fig. 10 The percentage of reaction occurring in the vicinity of the three high-symmetry sites, as a function of collision energy, for various (v, j) initial states. Plots are given for (a) the $(v = 0, j = 0)$ incident state; (b) the $(v = 0, j = 4)$ degeneracy-averaged state; (c) the $(v = 1, j = 0)$ state; and (d) the $(v = 1, j = 4)$ degeneracy-averaged state.

This filtering should also apply in the reverse reaction: associative desorption. Rather than aiming for reaction of incident molecules at a particular surface site, in associative desorption, state-selection could be used to select only those molecules originating from a particular site. In fact, this may already have been achieved. Hou et al. have measured the quadrupole alignment of j for D_2 desorbing from Cu(111), as a function of desorption energy, for the $(v = 0, j = 11)$ and $(v = 1, j = 6)$ states.[66] They found a high preference for helicopter reaction at the lowest desorption energies measured, concluding that ". . . the dissociative adsorption of D_2 at a Cu surface shows a strong steric preference for broadside collisions when the molecules strike the surface at low translational energy".[66] Our calculations suggest that this experiment may not have measured a general preference for broadside reaction at low energies, but actually a preference for broadside reaction *at the top site*. The two states in the experiment are both cases for which we would expect nearly all reaction at low translational energies to occur in the region of the top site: in this study, both vibrational and high rotational excitation has led to a preference for top site reaction at low translational energies. The orientational preference for reaction at the bridge site could presumably be probed by performing the same experiment for low-j $v = 0$ states.

V. Conclusions

We have performed 6D quantum wavepacket calculations for various rovibrational initial states of H_2 dissociatively adsorbing on Cu(100) at normal incidence. The PES used, known as PES IV, is improved over those used in earlier studies. Forming the PES involved calculating points using DFT/GGA for various 2D (r, Z) cuts, which were then interpolated using bicubic splines. Interpolation of the remaining coordinates involved a 14-term expansion in symmetry-adapted rotation–diffraction functions, including up to fourth-order spherical harmonics. The most significant difference between the new PES and older ones is that PES IV includes anisotropy in the azimuthal angle (ϕ) at the four-fold top and hollow sites, where earlier PESs were azimuthally isotropic at these sites.

Reaction probabilities calculated here were in better agreement with an experimental fit[63] to molecular-beam and associative-desorption data than those calculated previously, due mostly to improved interpolation in the current PES, which caused barriers for certain geometries to lower. The calculated reaction probabilities for ($v = 0, j = 0$) were still a bit too low relative to the experimental estimate. Agreement between theory and experiment for ($v = 1, j = 0$) was excellent.

We have detailed and utilized a new analysis technique for studying reaction as a function of surface site. This involves calculating the flux, resolved with respect to surface site and collision energy, across a cut in the bondlength coordinate positioned in the product channel. This approach has been used to examine the effect of translational energy, and rovibrational initial state on reaction at specific surface sites.

The results for site-reactivity showed that reaction is not necessarily predominantly at the site with the minimum barrier to reaction. Other factors, including lateness of barrier and orientational dependence of the PES are also very important, and can result in a site with a higher barrier dominating reaction. Molecules approaching in $v = 1$ states, for example, reacted predominantly at the top site at low collision energies. This means that, at low collision energies, $v = 1$ H_2 reacts by circumventing the transition state. Most reaction of ($v = 0, j = 0$) molecules occurred at the bridge site, but the hollow site was more reactive at high collision energies.

Site-reactivity was strongly influenced by the orientation of j. The most dramatic example of this occurred for the top site, where reaction was more suppressed than at other sites for the cartwheel ($v = 0, j = 4, m_j = 0$) incident state, but particularly enhanced for the helicopter ($v = 0, j = 4, m_j = 4$) state. The effect was even more dramatic for ($v = 0, j = 11$) states. These findings were explained with respect to barrier position and orientational/rotational effects.

Vibrationally-exciting the incident molecules had the effect of enhancing top-site reaction in proportion to that at other sites. The top site, which for ($v = 0, j = 0$) was least reactive of the three high-symmetry sites, was the most reactive for the ($v = 1, j = 0$) state at low collision energies. This was again explained with reference to barrier position.

The steric hindrance effect seen in experimental and computational studies for H_2/D_2 reacting on Cu(111),[8,10,38,67,68] in which reaction is hindered with increasing j for low-j, was not seen here. Incident $v = 0$ molecules showed no clear trend of suppression or enhancement with increasing j, while reaction of $v = 1$ molecules was clearly enhanced by increasing j from 0 to 4. It was suggested that the absence of a clear suppression of reaction for $v = 0$ may be due to the fact that the top site barrier is lower relative to other sites in our Cu(100) DFT calculations than it is in DFT calculations for Cu(111).[50] Results here and in a fixed-site study of H_2 + Cu(111)[17] show no steric hindrance effect for reaction at the top site; whether this carries over to the total reaction probability will be sensitive to the relative contribution of each site to reaction, and therefore to differences in barrier height. The present results suggest that the steric hindrance observed for H_2 + Cu(111) is not necessarily a general phenomenon. Instead, it might depend sensitively on the crystal face on which reaction takes place. This issue certainly deserves experimental attention.

The implications of the site-reactivity results for experiment were also examined. The calculated results were used to show that by appropriate selection of initial conditions, experiments could, in principle, be devised, using contemporary techniques, to induce reaction almost solely at a particular surface site. Different conditions could be used to select different sites for reaction. The calculations predicted upward of 80% of reaction could be concentrated in the vicinity of an individual surface site. We also suggested that a recent associative desorption experiment designed to probe the orientational preference for reaction of D_2 on Cu(111),[66] may have actually probed this preference in the region of a single surface site: the top site.

Lastly, and most importantly, we gave a prescription for testing our site-reactivity predictions experimentally. It was established that differences in PES topography at the top and bridge sites could be used as an experimental signature for the findings. The top-site PES has a much stronger dependence on the polar angle than the bridge-site PES, which results in a higher proportion of helicopter state reaction at the top site. Because most $v = 0$ reaction at low energies was found to occur at the bridge site, and most $v = 1$ reaction at the top site, there are more helicoptering molecules that react for the ($v = 1, j = 4$) incident state than for the ($v = 0, j = 4$) state. This preference for helicopter reaction is reflected in the quadrupole alignment of reacting molecules, which can be measured experimentally. Confirmation of the predicted dependence of alignment on incident vibrational state would add significant weight to our predictions for site-reactivity.

Acknowledgements

We acknowledge allocations of computer time by the National Computing Facilities foundation (NCF) and on the VUA beta-cluster, which is part of the Distributed ASCI Supercomputer. Work at NRL was funded by the Office of Naval Research through the NRL, and at VUA by NRSC-Catalysis. The research of R.A.O. is financed by the National Research School Combination "Catalysis Controlled by Chemical Design" (NRSC-Catalysis).

References

1 M. R. Hand and S. Holloway, *J. Chem. Phys.*, 1989, **91**, 7209.
2 M. Hand and S. Holloway, *Surf. Sci.*, 1989, **211**, 940.
3 J. Sheng and J. Z. H. Zhang, *J. Chem. Phys.*, 1992, **97**, 6784.
4 J. Sheng and J. Z. H. Zhang, *J. Chem. Phys.*, 1993, **99**, 1373.
5 T. Burner and W. Brenig, *Surf. Sci.*, 1994, **317**, 303.
6 G. R. Darling and S. Holloway, *J. Chem. Phys.*, 1992, **97**, 5182.
7 G. R. Darling and S. Holloway, *Chem. Phys. Lett.*, 1992, **191**, 396.
8 G. R. Darling and S. Holloway, *Faraday Discuss.*, 1993, **96**, 43.
9 G. R. Darling and S. Holloway, *J. Electron Spectrosc. Relat. Phenom.*, 1993, **64–5**, 571.
10 G. R. Darling and S. Holloway, *J. Chem. Phys.*, 1994, **101**, 3268.
11 G. R. Darling and S. Holloway, *Surf. Sci.*, 1994, **307–409**, 153.
12 G. R. Darling and S. Holloway, *Surf. Sci.*, 1994, **304**, L461.
13 G. R. Darling and S. Holloway, *Faraday Discuss*, 1998, **110**, 253.
14 G. R. Darling, M. Kay and S. Holloway, *Surf. Sci.*, 1998, **400**, 314.
15 J. Q. Dai, J. Sheng and J. Z. H. Zhang, *J. Chem. Phys.*, 1994, **101**, 1555.
16 J. Q. Dai and J. Z. H. Zhang, *Surf. Sci.*, 1994, **319**, 193.
17 J. Q. Dai and J. Z. H. Zhang, *J. Chem. Phys.*, 1995, **102**, 6280.
18 G. J. Kroes, G. Wiesenekker, E. J. Baerends and R. C. Mowrey, *Phys. Rev. B*, 1996, **53**, 10397.
19 G. J. Kroes, G. Wiesenekker, E. J. Baerends, R. C. Mowrey and D. Neuhauser, *J. Chem. Phys.*, 1996, **105**, 5979.
20 R. C. Mowrey, G. J. Kroes, G. Wiesenekker and E. J. Baerends, *J. Chem. Phys.*, 1997, **106**, 4248; R. C. Mowrey, G. J. Kroes, G. Wiesenekker and E. J. Baerends, *J. Chem. Phys.*, 1999, **110**, 2740.
21 R. C. Mowrey, G. J. Kroes and E. J. Baerends, *J. Chem. Phys.*, 1998, **108**, 6906.
22 A. Gross, *J. Chem. Phys.*, 1999, **110**, 8696.
23 G. R. Darling and S. Holloway, *Rep. Prog. Phys.*, 1995, **58**, 1595.
24 A. Gross, S. Wilke and M. Scheffler, *Phys. Rev. Lett.*, 1995, **75**, 2718.
25 A. Gross and M. Scheffler, *Prog. Surf. Sci.*, 1996, **53**, 187.
26 A. Gross and M. Scheffler, *Chem. Phys. Lett.*, 1996, **263**, 567.
27 A. Gross and M. Scheffler, *Chem. Phys. Lett.*, 1996, **256**, 417.
28 A. Gross and M. Scheffler, *Phys. Rev. Lett.*, 1996, **77**, 405.
29 A. Gross, S. Wilke and M. Scheffler, *Surf. Sci.*, 1996, **358**, 614.
30 A. Gross and M. Scheffler, *Phys. Rev., B*, 1998, **57**, 2493.
31 A. Gross, C. M. Wei and M. Scheffler, *Surf. Sci.*, 1998, **416**, L1095.
32 G. J. Kroes, E. J. Baerends and R. C. Mowrey, *J. Chem. Phys.*, 1997, **107**, 3309; G. J. Kroes, E. J. Baerends and R. C. Mowrey, *J. Chem. Phys.*, 1999, **110**, 2738.
33 G. J. Kroes, E. J. Baerends and R. C. Mowrey, *Phys. Rev. Lett.*, 1997, **78**, 3583; G. J. Kroes, E. J. Baerends and R. C. Mowrey, *Phys. Rev. Lett.*, 1998, **81**, 4781.
34 D. A. McCormack, G. J. Kroes, E. J. Baerends and R. C. Mowrey, *Faraday Discuss.*, 1998, **110**, 267.
35 D. A. McCormack, G. J. Kroes, R. A. Olsen, E. J. Baerends and R. C. Mowrey, *Phys. Rev. Lett.*, 1999, **82**, 1410.
36 D. A. McCormack, G. J. Kroes, R. A. Olsen, E. J. Baerends and R. C. Mowrey, *J. Chem. Phys.*, 1999, **110**, 7008.
37 J. Q. Dai and J. C. Light, *J. Chem. Phys.*, 1997, **107**, 1676.
38 J. Q. Dai and J. C. Light, *J. Chem. Phys.*, 1998, **108**, 7816.
39 D. A. McCormack and G. J. Kroes, *Phys. Chem. Chem. Phys.*, 1999, **1**, 1359.
40 G. te Velde, Ph. D. Thesis, Vrije Universiteit, Amsterdam, 1990.
41 G. te Velde and E. J. Baerends, *Phys. Rev. B: Condens Matter*, 1991, **44**, 7888.
42 G. te Velde and E. J. Baerends, *J. Comput. Phys.*, 1992, **99**, 84.
43 P. Hohenberg and W. Kohn., *Phys. Rev. B*, 1964, **136**, 864.
44 W. Kohn and L. J. Sham, *Phys. Rev. A*, 1965, **140**, 1133.

45   F. Herman and S. Skillman, *Atomic Structure Calculations*, Prentice-Hall, Englewood Cliffs, NJ, 1963.
46   G. Wiesenekker, G. te Velde and E. J. Baerends, *J. Phys. C-Solid State Physics*, 1988, **21**, 4263.
47   S. H. Vosko, L. Wilk and M. Nusair, *Can. J. Phys.*, 1980, **58**, 1200.
48   A. D. Becke, *Phys. Rev. A*, 1988, **38**, 3098.
49   J. P. Perdew, *Phys. Rev. B*, 1986, **33**, 8822.
50   B. Hammer, M. Scheffler, K. W. Jacobsen and J. K. Nørskov, *Phys. Rev. Lett.*, 1994, **73**, 1400.
51   G. Wiesenekker, G. J. Kroes, E. J. Baerends and R. C. Mowrey, *J. Chem. Phys.*, 1995, **102**, 3873.
52   G. Wiesenekker, G. J. Kroes and E. J. Baerends, *J. Chem. Phys.*, 1996, **104**, 7344.
53   C. Kittel, *Introduction to Solid State Physics*, Wiley, New York, 1986.
54   A. Messiah, *Quantum Mechanics*, North Holland, Amsterdam, 1961, vol. 1.
55   G. J. Kroes, J. G. Snijders and R. C. Mowrey, *J. Chem. Phys.*, 1995, **103**, 5121.
56   V. A. Mandelshtam and H. S. Taylor, *J. Chem. Phys.*, 1995, **103**, 2903.
57   G. G. Balint-Kurti, R. N. Dixon and C. C. Marston, *J. Chem. Soc., Faraday Trans.*, 1990., **86**, 1741.
58   G. G. Balint-Kurti, R. N. Dixon and C. C. Marston, *Int. Rev. Phys. Chem.*, 1992, **11**, 317.
59   D. H. Neuhauser, M. Baer, R. S. Judson and D. J. Kouri, *Comput Phys. Commun.*, 1991, **63**, 460.
60   D. H. Zhang and J. Z. H. Zhang, *J. Chem. Phys.*, 1994, **101**, 1146.
61   J. Z. H. Zhang, *Theory and Application of Quantum Molecular Dynamics*, World Scientific, Singapore, 1999.
62   R. A. Olsen, P. H. T. Philipsen, E. J. Baerends, G. J. Kroes and O. M. Lovvik, *J. Chem. Phys.*, 1997, **106**, 9286.
63   H. A. Michelsen and D. J. Auerbach, *J. Chem. Phys.*, 1991, **94**, 7502.
64   G. Anger, A. Winkler and K. D. Rendulic, *Surf. Sci.*, 1989, **220**, 1.
65   G. Comsa and R. David, *Surf. Sci.*, 1982, **117**, 77.
66   H. Hou, S. J. Gulding, C. T. Rettner, A. M. Wodtke and D. J. Auerbach, *Science*, 1997, **277**, 80.
67   C. T. Rettner, H. A. Michelsen and D. J. Auerbach, *J. Chem. Phys.*, 1995, **102**, 4625.
68   H. A. Michelsen, C. T. Rettner, D. J. Auerbach and R. N. Zare, *J. Chem. Phys.*, 1993, **98**, 8294.
69   C. T. Rettner, H. A. Michelsen and D. J. Auerbach, *Faraday Discuss.*, 1993, **96**, 17.
70   S. J. Gulding, A. M. Wodtke, H. Hou, C. T. Rettner, H. A. Michelsen and D. J. Auerbach, *J. Chem. Phys.*, 1996, **105**, 9702.
71   H. A. Michelsen, C. T. Rettner and D. J. Auerbach, *Phys. Rev. Lett.*, 1992, **69**, 2678.
72   D. Wetzig, M. Rutkowski, R. David and and H. Zacharias, *Europhys. Lett.*, 1996, **36**, 31.
73   R. N. Zare, *Angular Momentum*, Wiley, New York, 1988.

Energy disposal during desorption of D_2 from the surface and subsurface region of Ni(111)

S. Wright, J. F. Skelly and A. Hodgson*

Surface Science Research Centre, The University of Liverpool, Liverpool, UK L69 3BX. E-mail: andrewh@ssci.liv.ac.uk

Received 13th May 2000
First published as an Advance Article on the web 30th August 2000

The recombination of surface and subsurface D atoms on Ni(111) has been studied using resonance-enhanced multiphoton ionisation (REMPI) to measure the internal state and translational energy distributions of the desorbing product. By detecting D_2 formed during temperature-programmed desorption we were able to examine the reaction between subsurface and surface D atoms, and the recombination of two D atoms chemisorbed on the surface. Translational energy distributions for D_2 formed by recombination of surface D are very sensitive to coverage. Desorption from a low coverage surface produced a translational energy release of 2.6 kT, but a thermal rotational distribution, reflecting an entrance channel barrier to dissociative chemisorption on the clean Ni(111) surface. Sticking probabilities predicted from detailed balance are consistent with molecular beam adsorption measurements. Desorption from D coverages above 0.5 ML resulted in a sub-thermal energy release, desorption being mediated by a molecular precursor state with D_2 dissociation occurring *via* a non-activated, trapping–dissociation channel. In contrast, the reaction of subsurface D produces translationally hot D_2, with a mean energy approaching 8 kT_s at 180 K. This is consistent with the energetics for direct recombination of a chemisorbed D atom with a metastable subsurface D atom, which overcomes an activation barrier to resurface of between 0.35 and 0.47 eV depending on D concentration. The energy release decreases at higher temperature, probably as a result of a reduction in the energy of resurfacing D as the subsurface D concentration drops. This low energy component is attributed to accommodation of resurfacing D which is unable to react directly, followed by slow thermal desorption *via* the high coverage, surface D recombination channel. No internal rotational or vibration excitation was found in D_2 formed by reaction of subsurface D.

1 Introduction

Despite the enormous importance of hydrogenation in synthetic chemistry, the role of subsurface H in Ni catalysed reactions is only slowly becoming clearly appreciated. It has long been suspected that the high efficiency of Raney Ni, which is frequently used as a hydrogenation reagent, may be due to the reactivity of subsurface H atoms present in the catalyst,[1] surface adsorbed H being inactive for a number of reactions which nevertheless occur under high pressure conditions.

Recently several hydrogenation reactions involving subsurface H atoms have been identified and characterised through the use of ultra high vacuum (UHV) techniques.[2–5] In the first of these studies, Ceyer and coworkers[2] demonstrated that CH_3 groups chemisorbed on a Ni(111) surface will react with subsurface H atoms to form CH_4 at 170 K, whereas surface adsorbed H is unreactive even at much higher temperatures. Subsurface H will also react with higher hydrocarbons, hydrogenating ethylene[3] and cyclohexenes[4] and activating the C–C bonds in cyclopentane.[5]

The subsurface H atoms occupy octahedral interstitial sites in the Ni lattice,[6] corresponding to the stable binding sites in nickel hydride. Originally it was suggested that the reaction between adsorbed CH_3 and subsurface H occurs readily because these H atoms can access a sterically favourable transition state which is not available to surface H atoms.[2] The subsurface H atoms would appear to be in the ideal position to approach and react with CH_3 groups which are adsorbed in fcc three-fold hollow adsorption sites directly above the subsurface H. However, calculations[7–9] indicate that this reaction path is unfavourable, since the presence of an adsorbate in the hollow site directly above the subsurface H increases the activation barrier for H to resurface.[7] Instead of being the result of the specific approach geometry, the unusual reactivity of subsurface H atoms is due to the translational energy acquired as H resurfaces from the metastable subsurface site. The minimum energy reaction pathway involves H overcoming a barrier to resurface at the fcc hollow site, producing a hot H atom which then reacts with CH_3 bound at an adjacent hollow site. The activation barrier for the surface reaction H + CH_3 is similar to the energy liberated as H resurfaces[8,9] and so reaction can occur even at 170 K.[2] The double barrier predicted by these calculations of the minimum energy reaction path suggests that the dynamics of the subsurface H + CH_3 reaction will be complex, the lifetime of the hot H atom being determined by energy dissipation to electron–hole pair excitation.[10]

In order to explore the resurfacing of H atoms from the bulk and the reaction dynamics of the hot H formed, we have chosen to simplify the problem by investigating the reaction between subsurface H atoms and H chemisorbed at surface sites. In this case the rotational, vibrational and translational energy distributions of the nascent H_2 molecules can be more readily measured and interpreted, yet the reaction is otherwise analogous to the hydrogenation of CH_3.[7] Although there have been numerous studies concerning H_2 dissociation on Ni surfaces, studies of surface dynamics have traditionally ignored the movement of H atoms to and from the subsurface region. This is probably because formation of subsurface H requires considerable energy, occurring either by absorption of surface H during energetic gas-surface collisions at high pressure[11] or when H atoms are available to populate the bulk sites directly. Exposing a clean Ni surface to H atoms results in efficient uptake, the H atoms dissipating their 2.5 eV binding energy slowly during deep excursions into the bulk.[10] However, as a layer of H accumulates on the surface, direct Eley–Rideal abstraction starts to compete efficiently with H incorporation into the bulk.[12] Other methods of forcing H atoms into subsurface sites include cathodic charging in aqueous solution[13] and ion implantation.[14,15] Ion implantation is particularly convenient to use in UHV because of the relatively high currents that can be supplied by a conventional discharge source. However, in this case the H atoms may be implanted some distance into the bulk (depending on the ion energy), while at energies above 1 keV lattice defects appear to be created which trap H in the bulk.[15]

Here we describe measurements on the translational and internal energy distributions of D_2 formed by recombination of surface and subsurface D atoms at Ni(111). These reactions are followed by first adsorbing D into the surface and subsurface sites on Ni(111) and then heating the surface to thermally desorb D_2. First we describe the energy release obtained following recombination of surface D *via* the β_1 and β_2 desorption channels, contrasting the dynamics of D_2 formation (and dissociation) at high and low coverage. On a low D coverage Ni(111) surface ($\Theta_D < 0.5$ ML) D_2 is formed with an excess translational energy, consistent with a small barrier to dissociative chemisorption of D_2 on Ni(111). As the D coverage increases above 0.5 ML a different mechanism takes over in which D_2 desorbs from a molecular precursor state. This high coverage recombination reaction acts as a baseline with which to compare the reaction of subsurface D as it resurfaces and reacts with a surface chemisorbed D atom. In this case the energy disposal is quite different, an energy of 0.12 ± 0.02 eV being released into translational motion of D_2 away from the surface. We discuss the dynamics of this reaction and compare our results with models developed in recent calculations.[7]

2 Experimental

A Ni crystal (99.999%, Metal Crystals) was mounted by two tantalum heating wires on the liquid nitrogen cooled stub of a UHV manipulator. The crystal could be cooled to 130 K and ramped to 1200 K without significant heating of the surrounding mount. Laué measurements on the crystal showed the front surface to be 1.25° from the (111) plane. The surface was cleaned using repeated cycles of Ar^+ bombardment at a sample temperature of 900–1000 K followed by annealing for 20 min at the same temperature. Carbon and sulfur contamination was removed by heating the sample to 1000 K in a background pressure of 10^{-7} Torr O_2. The thin oxide layer was then reduced by heating the sample in a background pressure of 1×10^{-6} Torr D_2. The resulting surface reproduced the reported temperature-programmed desorption (TPD) spectra for D_2.[16] A surface covered by 1 ML of D could be prepared by exposing the sample to a background pressure of 1×10^{-6} Torr D_2 for 10 min. D atoms were implanted into the bulk of the crystal by exposure either to D atoms from a microwave discharge atomic beam source[17] or to 500 eV D_2^+ ions from an ion gun.

TPD spectra of D_2 were obtained by placing the front face of the sample 20 mm from the aperture of a shroud which surrounded the quadrupole mass spectrometer.[18] Quantum state specific time-of-flight (TOF) information for D_2 desorbing along the Ni(111) surface normal was collected using resonant two-photon ionisation via the $E,F^1\Sigma_g^+$ transition to produce D_2^+ ions. The Ni(111) crystal was held in a sample manipulator at the entrance to a set of parallel plate, ion extract optics (Fig. 1). The detection laser (~0.5 mJ in 5 ns at 202 nm) was focused onto the desorbing D_2 some 12 mm from the front face of the crystal, the other side of a fine repeller grid, which prevented changes in the Ni/H workfunction during desorption from distorting the ion TOF measurements. The D_2^+ ions were extracted axially, along the Ni(111) surface normal, into a short drift tube and detected on a microchannel plate whose output was fed to a digital storage

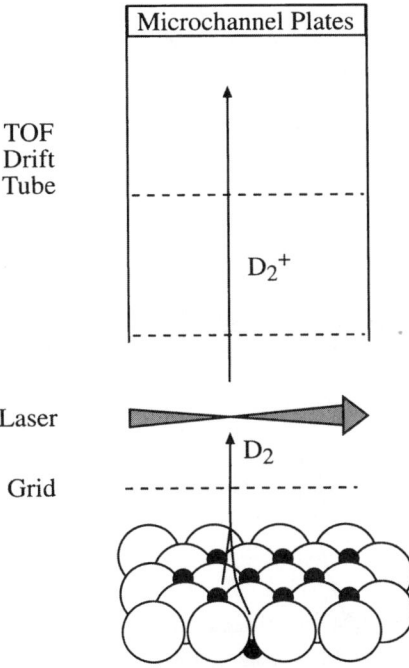

Fig. 1 Schematic showing the detection of D_2 by REMPI following reaction of subsurface D with surface chemisorbed D. A saturated layer of chemisorbed D (solid circles shown adsorbed in hcp hollow sites) reacts with a D atom from the octahedral subsurface site. The desorbing D_2 is ionised 12 mm from the surface at the focus of the probe laser and then extracted into a short ion drift tube.

scope before being discriminated and signal averaged on a microcomputer. The TOF ion detector was operated with a weak extraction field (1.2 V cm^{-1}) so that the arrival time at the microchannel detector was sensitive to the initial velocity of the ions. The TOF data were converted into D_2 translational energy distributions using the detector dispersion obtained from calibration measurements of the 1D thermal distributions from a Knudsen source, as described in detail elsewhere.[19]

During each experiment the surface was loaded with D and then the ion TOF signal recorded continuously while heating the crystal at 1–2 K s^{-1}. The data were collected into bins, each of which represented a temperature interval of 18–20 K. When investigating desorption of D from the surface chemisorption sites, the data from up to 10 experiments had to be averaged to obtain an adequate signal-to-noise ratio. In the case of the subsurface D reaction enough counts could be obtained during a single desorption experiment if at least 10 ML of D was implanted into the bulk.

3 Results and discussion

A Adsorption and thermal desorption of surface and subsurface D

Thermal desorption profiles following molecular D_2 exposure showed the β_1 and β_2 desorption peaks near 320 and 360 K (Fig. 2(a)) characteristic of clean Ni(111).[16,20–22] These peaks appear, with identical intensity and shape, whether one doses D_2, D atoms or D_2^+ ions (Fig. 2(b),(c)). The

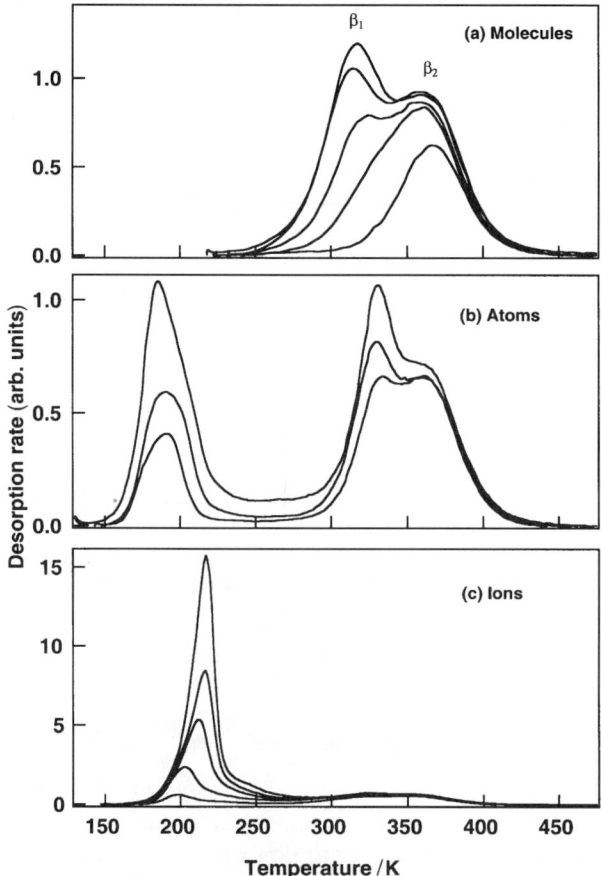

Fig. 2 Thermal desorption of D_2 from Ni(111) following exposure to (a) molecular D_2, (b) a beam of atomic D and (c) 500 eV D_2^+ ions. The heating rate is 1.5 K s^{-1}.

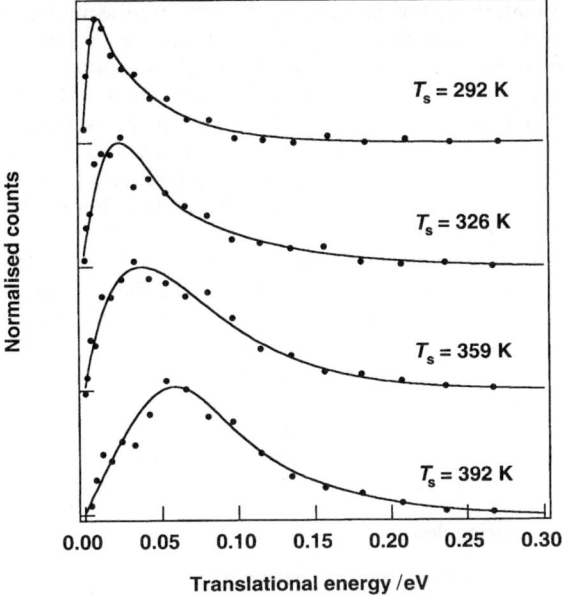

Fig. 3 Translational energy distributions for D_2 desorbing along the Ni(111) surface normal following recombination of D chemisorbed on the surface. The corresponding surface temperatures are indicated.

low coverage β_2 peak near 360 K saturates at a coverage of 0.5 ML, corresponding to an ordered (2 × 2) low-energy electron diffraction (LEED) pattern. The structure of this overlayer has been determined by LEED and shows D adsorbed in threefold hollow sites, the D atoms occupying both fcc and hcp sites in a hexagonal lattice.[21] The (2 × 2) structure is disordered above 270 K, so that D_2 forms at a disordered D overlayer during a TPD experiment.[20] The same order–disorder phase transition is seen in the presence of steps, but the transition temperature increases due to pinning of H adsorption at step edges.[23] Above 0.5 ML, uptake of D_2 becomes slower, finally saturating at a D coverage of around 1 ML.[20] The efficiency of adsorption into the β_1 state is noticeably different between different studies. It appears that adsorption into the β_1 state, from 0.5 ML up to saturation coverage near 1 ML, is dependent on the quality and cleanliness of the surface. Rendulic has shown that the position of the β_1 desorption peak shifts down in temperature on stepped or defective surfaces, suggesting that desorption is mediated by defects.[16,24] This low energy dissociation channel could be blocked by coadsorption of O which made the step sites inactive.

Exposure to a beam of D atoms results in an additional TPD peak near 185 K (Fig. 2(b)). This peak does not appear to saturate, but continues to grow with exposure as octahedral subsurface sites in the Ni are populated.[6] Once the subsurface D has desorbed there is no further desorption until the β_1 surface peak near 320 K. When D is incorporated into the surface using low energy ion implantation the desorption peak shifts up towards 210 K and broadens as the exposure is increased, developing a tail towards high temperature. There is no indication that D uptake becomes saturated, even when more than 20 ML of D has been deposited. As observed previously, the β_1 peak is easier to saturate when the surface is exposed to D atoms rather than D_2.[6]

The activation energy for formation of D_2 by desorption of subsurface D can be obtained from a leading edge analysis of the TPD spectra of Fig. 2. Following implantation of up to 1 ML of D into the subsurface by exposure to the atom beam, the activation energy for reactive desorption was found to be 0.35 ± 0.05 eV. TPD peaks for higher subsurface D loadings, produced by ion dosing, showed a common leading edge, desorption increasing exponentially with temperature irrespective of the subsurface D concentration. This behaviour is characteristic of zero-order desorption kinetics from a reservoir of constant concentration. The leading edge analysis showed an activation energy which increased steadily with D uptake, from 0.41 eV at 1 ML up to 0.47 eV for an uptake of 20 ML. These values are very similar to the activation energy for D diffusion in the bulk of Ni, which is 0.41 eV at low D atom concentrations and increases at higher D concentration due to the strain induced in the Ni lattice.[25] The desorption behaviour is consistent with diffusion of D to the surface being the rate limiting step in D_2 formation from subsurface D.

The TPD spectra of D_2 reported here are noticeably simpler than those reported in the other two studies in which H/D atom sources were used.[6,12] Ceyer and coworkers created H by cracking H_2 on a hot W filament and observed two peaks at 185 and 215 K, which coincide very closely with those we see for neutral atom and ion dosing respectively. Premm et al.[12] passed H_2 through a heated W tube and observed peaks at 185, 210 and 265 K. It is possible that these sources also generate a small number of ions which are responsible for the extra structure seen during TPD. Our neutral atom beam, created in a high pressure environment, contains no ions and only a single peak was seen during TPD. Similarly we did not see any evidence for extra adsorption states, or surface damage, caused by ion exposure.[26] Chorkendorff et al.[15] observed that the subsurface TPD peaks shifted to higher temperatures as the ion energy was increased above 1 keV. This reflects the implantation depth of D into the bulk and the kinetics of diffusion back to the surface, with formation of trapping states becoming important at higher energies. We find the activation barrier to desorption to be very similar following atom or low energy (500 eV) ion dosing, but a high temperature tail develops for high D uptakes (Fig. 2(c)), probably associated with diffusion of D into and then back out of the Ni crystal.

B Energy release into D_2 during desorption of surface D

A sequence of distributions showing the translational energy release along the Ni(111) surface normal is shown in Fig. 3 for various surface temperatures during desorption. These show that the translational energy distributions for D_2 formed by recombination of surface D change during the course of desorption, the distribution becoming considerably hotter as the D coverage drops. The mean energy released into translational motion perpendicular to the surface, $\langle E \rangle$, was determined from the energy distributions and is shown in Fig. 4 for different rotational states of $D_2(v = 0)$. No vibrationally excited D_2 could be detected, nor was there any significant variation in the translational energy release with rotational state. Also shown as a solid line in Fig. 4 is the energy release, $\langle E \rangle = 2kT_s$, which would be obtained if the desorbing molecules were entirely thermalised to the surface temperature and obeyed a Knudsen energy distribution. When desorption starts at around 290 K the energy release is sub-thermal, the distribution peaking at an energy well below

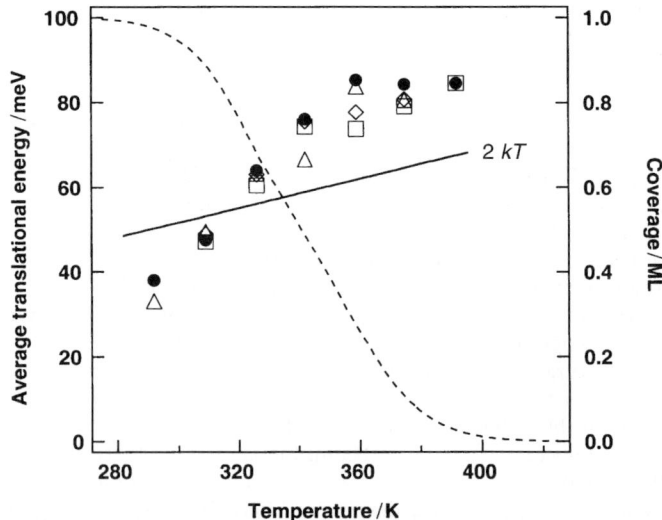

Fig. 4 Mean translational energy release as a function of T_s for recombinative desorption of surface adsorbed D. The energy release (left hand scale) is shown for $J = 0$ ($\diamond$), 1 ($\square$), 2($\bullet$) and 3($\triangle$) while the corresponding D coverage ($\cdots$) is shown on the right-hand scale. The thermal expectation value, $\langle E \rangle = 2 kT_s$, is shown for reference.

the $E = kT_s$ expected for a thermal distribution and having a mean energy just 70% that for a 290 K gas. As desorption proceeds the energy release increases with temperature to reach 81 ± 5 meV (or 2.6 kT_s) above 360 K, corresponding to a translational temperature of 490 K. The rotational temperature was measured by taking the ratio of the populations for two quantum states, $J = 0$ and 2 (Fig. 5) and shows that the rotational temperature follows the same trend as the translational energy release, being subthermal at high Θ_D and approaching the surface temperature at low coverage. Since the desorption behaviour is correlated to the D coverage (Fig. 4), the energy release will be discussed in turn for desorption from the low and high coverage peaks, which show qualitatively different behaviour.

(i) **D_2 desorption and adsorption for $\Theta_D < 0.5$ ML.** For temperatures above 350 K, where desorption is entirely *via* the β_2 channel, the mean energy release, $\langle E \rangle$, approaches 2.6 kT_s. Desorption at coverages less than 0.5 ML is characterised by an excess energy release into translational motion away from the surface, reflecting a repulsion between the Ni(111) surface and D_2 as it desorbs following D recombination. The excess energy release is a consequence of the barrier to D_2 dissociative chemisorption on this surface,[27–29] which acts to accelerate desorbing D_2 away from the surface. Interestingly, there is no evidence for any significant change in the energy release as the D coverage drops from 0.4 to 0.05 ML. This means that the repulsion between the D_2 and the Ni surface does not change significantly with coverage, and that dissociation has a barrier which is independent of coverage. The activation energy for desorption *via* the β_2 channel is also independent of Θ_D,[21] which is again consistent with an adsorption/desorption barrier that is insensitive to coverage below 0.5 ML. This is in contrast to D_2 adsorption and desorption from Pt(111) where the stability of adsorbed D drops and the barrier to dissociation increases as the D coverage increases towards the saturation coverage of 0.8 ML.[30–32]

Using detailed balance we can invert the desorption distributions to obtain the relative sticking probability as a function of translational energy for a given temperature and D coverage.[33] This is shown in Fig. 6(a) for D_2 desorption at 360 K from a Ni(111) surface with a mean D coverage of 0.4 ML. Above 30 meV the sticking probability shows a steady increase with energy, mimicking the results of molecular beam measurements for dissociative chemisorption on a low coverage surface,[27–29] shown as the solid line in Fig. 6. However, at low energy the sticking curves obtained from the desorption data do not fall to zero, indicating a non-activated component to adsorption/desorption. This is not observed in beam experiments, which show a small threshold of 10 to 15 meV[27,28] for dissociation. Rendulic and Winkler have shown that at low energy the sticking probability is sensitive to both the surface quality[16,24] and the presence of steps,[34] dissociation being non-activated at steps and on the more open Ni(100) and (110) surfaces.[27,29] Adsorption of

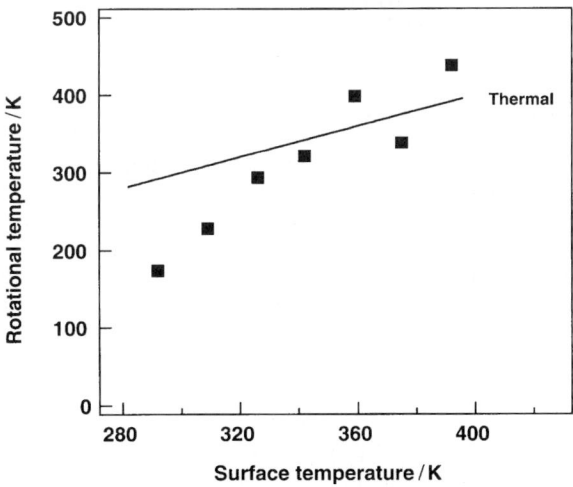

Fig. 5 Rotational temperature of D_2 formed during recombination of the surface adsorbed D. The straight line shows the equilibrium thermal expectation.

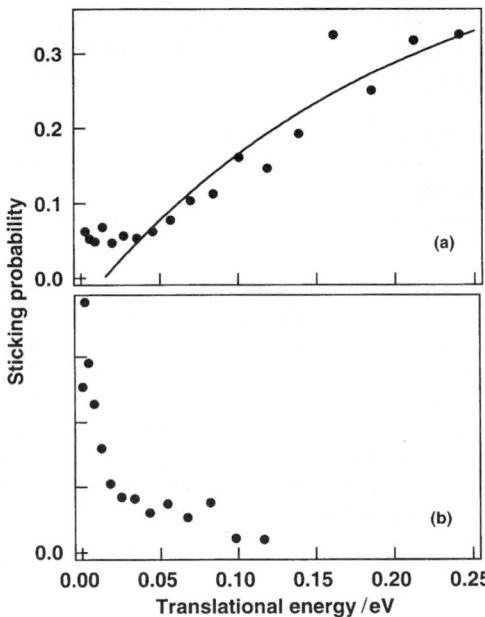

Fig. 6 (a) Sticking function for $D_2(v = 0)$ on a Ni(111) surface covered with 0.27 ML of D, as inferred from the energy distribution of D_2 desorbing from a 360 K surface. This data has been scaled so that it can be compared with molecular beam adsorption measurements on the clean surface (solid line) by Rendulic et al.[29] The form of $S(E)$ remained identical as the coverage dropped towards 0.05 ML at 400 K. (b) Equivalent data for adsorption at high coverage, $T_s = 290$ K and $\theta_D = 0.97$ ML. In this case there is no absolute sticking measurement to which the data can reliably be normalised.

even low coverages of S or O can passivate these step sites[34] and minor changes in the surface quality are probably responsible for the range of sticking values found for thermal adsorption, which vary between 0.02 and 0.07.[22] The Ni(111) crystal used in the present study was miscut by 1.25° from the (111) plane, providing a ready source of step sites at which recombination or dissociation can occur. We believe that this is the cause of the non-activated adsorption seen in Fig. 6(a) at low energy, where the desorption data indicate a higher sticking probability than some of the adsorption measurements. The thermal sticking probability on the clean surface was 0.02, consistent with the values seen previously on Ni(111).[22]

An excess energy release was originally reported in high temperature H_2/D_2 permeation experiments by Comsa et al.[35] who found a mean energy release of $2.9\ kT_s$ for desorption from Ni(111) at 1143 K, while Dabiri et al.[36] obtained the same value following permeation through from a polycrystalline Ni foil. These energy releases are slightly greater than the $2.6\ kT_s$ we obtain from thermal desorption near 360 K, but this is consistent with the reduced weighting expected for the non-activated channel as the temperature is increased. Rendulic and coworkers have examined H_2/D_2 reaction at Ni(111) using detailed balance to relate the angular distributions for adsorption and desorption on the low coverage surface.[22] They found a $\cos^{3.5}\theta$ distribution for adsorption and concluded that this was consistent with the $\cos^{4.5}\theta$ angular distribution of desorbing D_2. They also calculated the translational energy release based on the sticking function and obtained numbers similar to those determined experimentally.[35] In conclusion, adsorption and desorption of D_2 from Ni(111) appears entirely consistent with detailed balance, with a small activation barrier to dissociation on the Ni(111) terraces but non-activated dissociation at steps and defect sites.

(ii) **D_2 desorption and adsorption for $\Theta_D > 0.5$ ML.** As the D coverage increases above 0.5 ML, the coverage associated with the ordered (2 × 2) structure,[21] desorption starts to occur at lower temperatures via the β_1 channel. As this occurs there is an abrupt reduction in the translational energy released into excitation of D_2 (Fig. 3 and 4), the translational energy distributions becom-

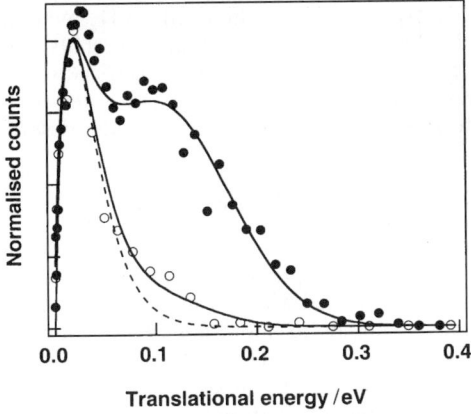

Fig. 7 Translational energy distributions for D_2 desorbing along the Ni(111) surface normal following recombination of subsurface D atoms at 200 K (●) and at 240 K (○) [which corresponds to the tail of the subsurface desorption peak shown in Fig. 2(c)]. The dotted line shows a thermal energy distribution at $T_s = 200$ K for reference.

ing cooler than the surface temperature. Chemisorption of D at coverages above 0.5 ML is associated with a reduction in the sticking probability for D_2.[16] This is consistent with destabilisation of D on the surface[21] and an increase in the activation barrier to direct dissociation on the D covered Ni(111) terraces.[22] However, an increased dissociation barrier suggests that the energy released during D recombination should increase for coverages above 0.5 ML, whereas the opposite occurs. Recombinative desorption shows no evidence for a repulsive D_2–Ni(111) interaction, the energy release being dominated by low translational energies, and detailed balance predicts non-activated adsorption, the sticking probability decreasing with energy as shown in Fig. 6(b).

Low energy, non-activated dissociation occurs on both the Ni(110) and (100) surfaces even at zero coverage, sticking being characterised by a high dissociation probability, $S(E)$, at low energy, followed by a minimum in $S(E)$ before activated dissociation becomes efficient at higher energies.[27,29] A similar behaviour has been observed for H_2/D_2 dissociation on clean Pd(100)[29] and W(100)[37,38] surfaces. The increase in $S(E)$ at low energy may originate from two possible mechanisms which differ in the timescale for dissociation. In the direct model, H_2/D_2 is steered into favourable dissociation sites by an attractive potential, the steering only being efficient when molecules approach the surface with a low energy.[39] Alternatively, the increase in sticking at low energy may originate from a trapping–dissociation mechanism,[38,40] similar to that observed for many heavier molecules where trapping into a physisorption well is efficient at low energy. The shallow physisorption well and large mass mis-match between H_2/D_2 and the metal substrate mean that the efficiency of trapping on a clean surface is expected to be low, favouring steering as a mechanism for the low energy channel.[39]

However, the case of adsorption/desorption from the high coverage ($\Theta_D > 0.5$ ML) Ni(111) surface is somewhat different. The presence of preadsorbed H/D on the surface provides sites with a low effective mass where collisional energy exchange may be more efficient, favouring trapping at high coverages compared to the clean metal surface. Furthermore Russell et al.[41] observed a molecular desorption feature at 100 K after exposure to D_2, but no corresponding H_2 or HD species could be seen. Although this peak was relatively small compared to the β_1 and β_2 features, it displayed first-order desorption kinetics and appeared to be associated with the flat Ni(111) surface.[41] Small Ni/H clusters will also take up molecular H_2,[42] while a stable molecular chemisorption state has been identified by electron energy-loss spectroscopy (EELS) on a stepped Ni(100) surface, H_2 binding above step sites on the H saturated surface.[43,44] This molecular species has a substantially reduced vibrational frequency and desorbs near 125 K. These reports suggest that the H saturated Ni(111) surface supports a molecular chemisorption state which acts as a precursor to desorption or dissociation.[41] The energy release into D_2 at $\Theta_D > 0.5$ ML is characteristic of a trapping–desorption mechanism, in which D_2 molecules cool during desorption from a molecular well.[45-47] Reducing the surface quality shifts the β_1 peak to lower temperature,[16,24] suggesting defect or step sites may mediate dissociation. We observed a similar shift in the β_1 desorption temperature depending on the precise cleaning routine; traces of O left on the surface after O_2 treatment to remove C were sufficient to hinder desorption and shift the β_1 peak to higher temperature.

C Energy release into D_2 during desorption of subsurface D

Subsurface D was loaded into the Ni by ion exposure and the translational energy distributions were measured during recombinative desorption near 200 K (solid points, Fig. 7). These distributions look very different to those associated with the surface peaks, having a pronounced shoulder at high energy, with a tail which extends to about 0.4 eV. A thermal energy distribution at this temperature is shown for reference in Fig. 7. The mean energy release has a maximum value at low temperature (Fig. 8), but drops at higher temperatures as D is depleted from the bulk (open circles, Fig. 7). Varying the D dosage and desorption conditions always produced a decrease in $\langle E \rangle$ with T_s, but it became clear that the surface history was a factor in determining the total energy released into translation along the surface normal. The energy release was always a maximum for a freshly prepared Ni(111) surface when it was exposed to D_2^+ or D for the first time. Repeated loading with D atoms steadily reduced the mean energy release seen during desorption of subsurface D, which could be as high as 0.14 eV during the first TPD, some 5 times higher than expected for a thermal distribution at 180 K. After repeated loading with subsurface D we find that the β_1 and β_2 surface desorption peaks both grow in together, suggesting a defective surface has formed which could not be returned to its original condition even by extensive cleaning. This is probably the result of stress induced in the surface by the lattice mis-match between bulk Ni and the deuteride,[25] repeated loading with D creating dislocations which slowly destroy the surface quality.

Although the surface quality clearly influenced the mean translational excitation, the high energy tail seen in the desorption distributions persists in all the measurements, becoming smaller as desorption proceeds. At 200 K, the energy distributions (Fig. 7) show a distinct shoulder above 0.1 eV and the distribution can be decomposed into two components, a thermal component and a translationally excited component, which could be simulated using a gaussian function, centred near $E = 0.12$ eV. This decomposition is not unique but with suitable weighting these two components will adequately describe all the energy distributions. The rotational temperature of the D_2 product is essentially thermal at the surface temperature, Fig. 9, no excess energy being deposited

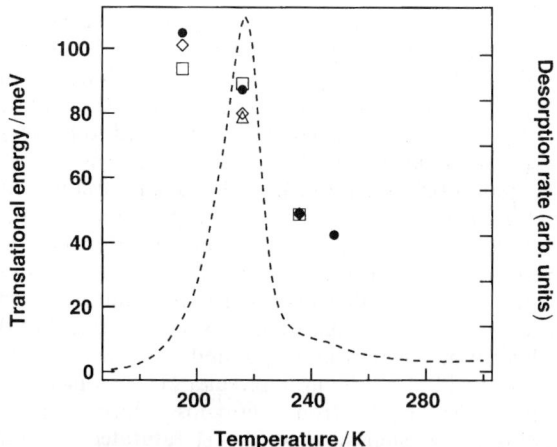

Fig. 8 Mean translational energy as a function of surface temperature for recombinative desorption of subsurface D. The energy release (left-hand scale) is shown for $J = 0$ (◇), 1 (□), 2(●) and 3(△), while the right hand scale shows the corresponding desorption rate.

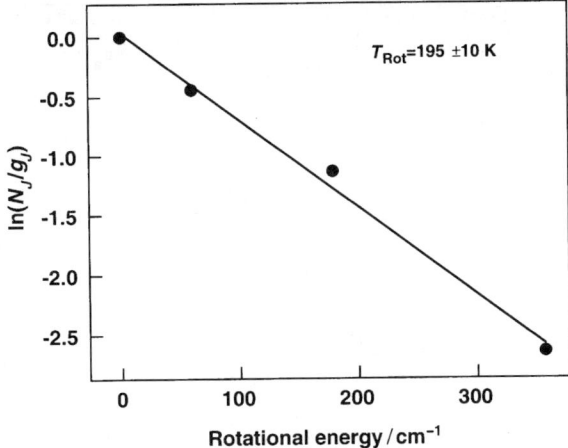

Fig. 9 Rotational energy distribution for D_2 formed during recombinative desorption of subsurface D on a 210($\pm$ 10) K surface.

into product rotation. Similarly we could find no sign of any population in $D_2(v = 1)$ from the reaction of subsurface and surface chemisorbed D.

The reaction between a D atom from the subsurface and one chemisorbed on the surface is energetically very finely balanced. A D atom in a subsurface octahedral site in Ni is metastable by some 0.17 eV with respect to $1/2D_2$ in the gas phase, while the activation barrier for diffusion of D through the bulk is 0.41 eV, increasing at higher D atom concentrations.[25] Assuming that all of this activation energy is present as excitation of D (rather than as a distortion of the Ni lattice) implies that resurfacing will produce a hot D atom on the surface, with an excess energy of 0.58 eV relative to $1/2D_2(g)$. The activation energy for desorption of D_2 from Ni(111) at saturation coverage (1 ML) is 0.43 eV and this must be provided by the hot D atom if it is to react directly with surface chemisorbed D to produce gas phase D_2. This means that we expect an energy release of *ca.* 0.15 eV to be available for excitation of the products, in addition to the thermal energy associated with the inactive coordinates. The average translational energy release obtained at 180 K, the lowest temperature bin, is 0.12 $\pm$ 0.02 eV, or $8kT_s$, some 4 times the thermal expectation and very close to the value estimated above.

The agreement between the D_2 translational energy release and the energy available from reaction of resurfacing D with a surface D atom suggests that this reaction is direct and occurs without dissipation of significant energy from the hot D atom to the surface. The translational energy distribution created is quite unlike those obtained during thermal recombination of surface adsorbed D atoms. Even at 400 K the energy released during repulsion of D_2 from the barrier to adsorption on the bare Ni(111) surface is only some 2/3 that observed for reaction at 180 K. The energy distribution from subsurface desorption has a much greater enhancement of the high energy tail than seen for surface recombination, as is clearly seen if one inverts the desorption energy distributions to obtain the relative probability of D_2 dissociating to create subsurface D. Instead of following either the slightly activated form characteristic of dissociation on bare Ni(111) surface (Fig. 6(a)), or the trapping form seen at high coverages (Fig. 6(b)), the probability for direct dissociation to produce subsurface D increases exponentially with energy (Fig. 10), with little evidence that this channel saturates at energies below 0.4 eV. This is consistent with the observation that beam adsorption measurements do not show any evidence for formation of subsurface H/D during molecular adsorption on Ni and contrasts with the behaviour on Pd where dissociation appears to be able to populate subsurface sites directly.[48]

The total absence of $D_2(v = 1)$ in the reaction products is a consequence of the limited energy release available during reaction and the low surface temperature, which rules out thermal excitation contributing to excitation of vibrational motion ($\Delta E = 0.35$ eV). Although the rotational coordinate has no such tight constraints, it too reveals little evidence of direct reaction dynamics. The rotational temperature is thermal (Fig. 9) and there is no correlation between the rotational state and the translational energy release (Fig. 8). It appears that the rotation of D_2 product

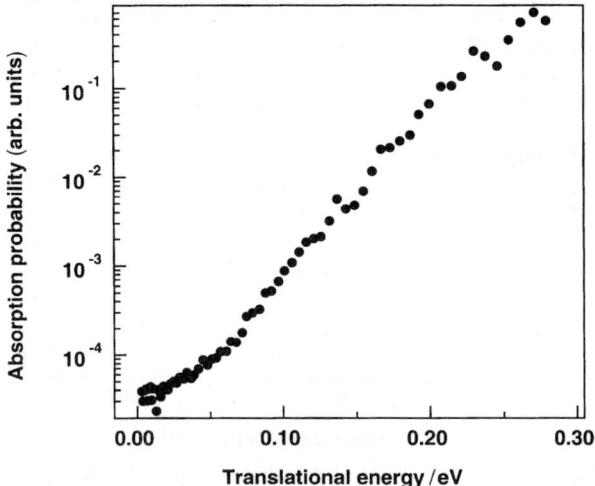

Fig. 10 Relative probability, obtained from detailed balance, for direct formation of subsurface D from molecular D_2, as a function of the translational energy of D_2. The probability increases exponentially towards higher energy, consistent with this process having a low probability at thermal energies (no subsurface H is observed after exposure to 36 000 L H_2 [6]).

desorbing along the Ni(111) surface normal is determined by thermal excitation, without the dynamical excitation of product angular momentum which might have been anticipated.[15]

Baer, Zeiri and Kosloff[7] have performed dynamical calculations on the resurfacing of H, and its reaction with surface H atoms, using an embedded atom model. They found that the H atom resurfaces by tunnelling from the octahedral subsurface site into the fcc threefold hollow site immediately above.[7] The H atom motion is highly correlated with the motion of the Ni atoms, the transition state occurring as H passes through the plane of the surface Ni atoms which involves considerable relaxation of the Ni lattice.[8,9] The subsurface H forms a hot atom, which on the bare Ni(111) surface dissipates its energy to electron–hole pair excitation on a timescale of a few picoseconds.[10] The presence of adsorbates on the surface is known to inhibit resurfacing.[25] Baer et al.[7] found that the barrier to H resurfacing doubled when H was adsorbed in the fcc site directly above the subsurface H, and recent DFT calculations give similar results for blocking by CH_3.[8,9] At 200 K there is insufficient energy for subsurface H to resurface and react directly with an atom adsorbed immediately above it in the fcc hollow site. Instead resurfacing would presumably require displacement of the adsorbed H away from the fcc site, in a similar manner to that found for the H + CH_3 reaction.[8,9] This suggests that reaction at 200 K involves resurfacing into a vacant fcc site followed by reaction of the hot H with H adsorbed in an adjacent hollow site.[7] Since surface chemisorbed H occupies the fcc and hcp threefold hollow sites without distinction, half of the subsurface sites will be blocked and half will have a vacant hollow site immediately above. Chorkendorff et al.[15] pointed out that a H atom resurfacing and reacting with surface H from an adjacent hcp hollow site might be expected to induce rotation in the product. However, we have found no evidence for rotational excitation in D_2 desorbing along the surface normal, only translational excitation from activated resurfacing of D from the octahedral subsurface site.

Although the product translational excitation suggests that the reaction is direct at low temperatures, there are two aspects of the data which suggest an indirect mechanism must operate in parallel with the direct reaction between subsurface and surface D. The first is the observation that formation of a defective surface, by repeated loading with D atoms, steadily reduces the mean energy released into D_2. As this occurs the high energy peak in the translational energy distribution is attenuated and the energy distribution begins to look increasingly thermal. The second factor is the decrease in the energy release at higher temperatures (Fig. 8) as D is depleted from the bulk. As pointed out above, the energetics of this reaction are finely balanced. If a hot D atom has an energy less than 0.43 eV it will not be able to abstract a chemisorbed D atom and desorb. There are several mechanisms which may reduce the energy of the hot D. Resurfacing involves displacement of the Ni atoms surrounding the reaction path by as much as 5%[9] so some of the activation energy may be associated with the Ni lattice rather than the D atom which comes to the surface. Similarly, tunnelling of D may also reduce the energy of the hot D, which will dissipate its energy rapidly if reaction does not occur immediately.[7,10] Repulsion between subsurface D will change the energy of the hot D depending on its concentration,[7] as will the increase in the activation barrier to D resurfacing for large D uptakes (Section 3A). DFT calculations show a barrier of 0.49 eV for H resurfacing from a saturated subsurface layer (1 ML),[8] compared to just 0.18 and 0.15 eV from 1/3 and 1/9 ML of subsurface H, respectively.[9] Although we know the total D loading of the surface, the local subsurface D concentration during desorption is difficult to quantify.[15] Initially, adsorption of D or D_2^+ presumably saturates the octahedral sites near the surface[6] but the local subsurface coverage will drop during TPD as D resurfaces and a concentration gradient is established within the bulk. The fall in the D_2 energy as desorption progresses (Fig. 8) is consistent with a decrease in the energy of the hot D atoms reaching the surface.

If a hot D atom is created or relaxes to an energy below ~ 0.43 eV with respect to 1/2 D_2(g), it will be unable to react directly with a surface chemisorbed D atom to produce molecular D_2(g). Since the surface is saturated with D, an atom resurfacing does not see the deep atomic chemisorption well which is available at surface coverages less than 1 ML. The potential energy surface calculated by Baer *et al.*[7] for the reaction of subsurface H indicates that the fcc hollow remains a favourable binding site even when the neighbouring hcp hollow sites are saturated with D. If the hot D atom cannot react directly it must thermalise, compressing the overlayer and creating an excess local coverage. It is noticeable that atom dosing (Fig. 2) leads to a slightly greater surface D coverage than can be obtained without an enormous dose of molecules. Increasing the D coverage above its saturation value will lead to indirect, thermally activated desorption to maintain the saturation coverage. Desorption from the saturated surface, *via* the β_1 channel, shows a low, sub-thermal energy release and is sensitive to the presence of defects. We find that the energy released by recombination of subsurface and surface D decreases as the surface quality degrades, suggesting a fraction of the D desorbs *via* this indirect route, with a low energy release from a super-saturated surface. Since some D_2 desorbs with a low translational energy (Fig. 7), even from the most ordered, high coverage surfaces at 180 K, the indirect channel appears to operate as a minor route even in the most ideal circumstances.

5 Conclusion

The recombinative desorption of D chemisorbed on Ni(111) is consistent, *via* detailed balance, with previous measurements of weakly activated dissociation. The repulsive energy release of 2.6 kT_s at 380 K is slightly lower than the 2.9 kT_s found in permeation experiments at high temperatures,[35,36] reflecting the greater weighting afforded a non-activated component at 380 K. A non-activated channel was also seen during adsorption on stepped surfaces.[34] As the surface coverage increases above 0.5 ML the D_2 formed becomes much cooler, indicating a change in reaction mechanism. The adsorption/desorption behaviour is consistent with a non-activated trapping–dissociation mechanism, direct dissociation on the D covered Ni(111) surface either having a large activation barrier or being sterically unfavourable. The sensitivity of adsorption to surface quality suggests that this channel involves trapping into a molecular precursor state[41] followed by dissociation at step or defect sites.[16]

Reaction of D resurfacing from a 180 K Ni(111) surface, saturated with chemisorbed D, produced translationally hot D_2. No rotational or vibrational excitation was observed. The translational energy release was consistent with the energetics for direct reaction of subsurface D with D chemisorbed on the surface. As the reaction proceeds there is a decrease in the translational energy of the D_2 formed. This reflects a decrease in the energy of the hot D created as the diffusion barrier and the strain energy in the subsurface layer decrease at lower D concentrations. The D atoms start to resurface with an energy which is too low to react directly to form gas phase D_2 and the translational energy release falls to thermal levels. Recombination from this super-saturated surface occurs *via* the trapping-dissociation channel characteristic of the D covered Ni(111) surface.

References

1. R. J. Kokes and P. H. Emmett, *J. Am. Chem. Soc.*, 1959, **81**, 5032.
2. A. D. Johnson, S. P. Daley, A. L. Utz and S. T. Ceyer, *Science*, 1992, **257**, 223.
3. S. P. Daley, A. L. Utz, T. R. Trautman and S. T. Ceyer, *J. Am. Chem. Soc.*, 1994, **116**, 6001.
4. K. A. Son, M. Mavrikakis and J. L. Gland, *J. Phys. Chem.*, 1995, **99**, 6270.
5. K. A. Son and J. L. Gland, *J. Am. Chem. Soc.*, 1995, **117**, 5415.
6. A. D. Johnson, K. J. Maynard, S. P. Daley, Q. Y. Yang and S. T. Ceyer, *Phys. Rev. Lett.*, 1991, **67**, 927.
7. R. Baer, Y. Zeiri and R. Kosloff, *Phys. Rev. B*, 1997, **55**, 10952.
8. A. Michaelides, P. Hu and A. Alavi, *J. Chem. Phys.*, 1999, **111**, 1343.
9. V. Ledentu, W. Dong and P. Sautet., *J. Am. Chem. Soc.*, 2000, **122**, 1796.
10. R. Baer and R. Kosloff, *J. Chem. Phys.*, 1997, **106**, 8862.
11. K. J. Maynard, A. D. Johnson, S. P. Daley and S. T. Ceyer, *Faraday Discuss. Chem. Soc.*, 1991, **91**, 437.
12. H. Premm, H. Polzl and A. Winkler, *Surf. Sci.*, 1998, **401**, L444.
13. A. Kimura and H. K. Birnbaum, *Acta Metal. Mater.*, 1991, **39**, 295.
14. A. Golchet and J. M. White, *Chem. Phys. Lett.*, 1990, **175**, 143.
15. I. Chorkendorff, J. N. Russel and J. T. Yates, *Surf. Sci.*, 1987, **182**, 375.
16. K. D. Rendulic, A. Winkler and H. P. Steinruck, *Surf. Sci.*, 1987, **185**, 469.
17. M. J. Murphy, J. F. Skelly and A. Hodgson, *Chem. Phys. Lett.*, 1997, **279**, 112.
18. F. Healey, R. N. Carter and A. Hodgson, *Surf. Sci.*, 1995, **328**, 67.
19. M. J. Murphy and A. Hodgson, *Surf. Sci.*, 1997, **390**, 29.
20. K. Christmann, *Surf. Sci. Rep.*, 1988, **9**, 1.
21. K. Christmann, R. J. Behm, G. Ertl, M. A. V. Hove and W. H. Weinburg, *J. Chem. Phys.*, 1979, **70**, 4168.
22. H. P. Steinruck, K. D. Rendulic and A. Winkler, *Surf. Sci.*, 1985, **154**, 99.
23. A. T. Hanbicki, S. B. Darling, D. J. Gaspar and S. J. Sibener, *J. Chem. Phys.*, 1999, **111**, 9053.
24. K. D. Rendulic and A. Winkler, *Int. J. Mod. Phys. B*, 1989, **3**, 941.
25. G. Alefeld and J. Völkl, *Hydrogen in Metals*, Springer, New York, 1978.
26. A. Golchet, G. E. Poirier and J. M. White, *Surf. Sci.*, 1990, **239**, 42.
27. H. J. Robota, W. Vielhaber, M. C. Lin, J. Segner and G. Ertl, *Surf. Sci.*, 1985, **155**, 101.
28. D. O. Hayward and A. O. Taylor, *Chem. Phys. Lett.*, 1986, **124**, 264.
29. K. D. Rendulic, G. Anger and A. Winkler, *Surf. Sci.*, 1989, **208**, 404.
30. K. Christmann, G. Ertl and T. Pignet, *Surf. Sci.*, 1976, **54**, 365.
31. P. R. Norton, J. A. Davies and T. E. Jackson, *Surf. Sci.*, 1982, **121**, 103.
32. A. C. Luntz, J. K. Brown and M. D. Williams, *J. Chem. Phys.*, 1990, **93**, 5240.
33. A. Hodgson, *Prog. Surf. Sci.*, 2000, **63**, 1.
34. A. Winkler and K. D. Rendulic, *Surf. Sci.*, 1982, **118**, 19.
35. G. Comsa, R. David and B. J. Schumacher, *Surf. Sci.*, 1979, **85**, 45.
36. A. E. Dabiri, T. J. Lee and R. E. Stickney, *Surf. Sci.*, 1971, **26**, 522.
37. H. F. Berger, E. Grosslinger, G. Eilmsteiner, C. Resch, A. Winkler and K. D. Rendulic, *Surf. Sci.*, 1992, **275**, L627.
38. D. Butler, B. E. Hayden and J. D. Jones, *Chem. Phys. Lett.*, 1994, **217**, 423.
39. A. Gross, S. Wilke and M. Scheffler, *Phys. Rev. Lett.*, 1995, **75**, 2718.
40. D. Butler and B. E. Hayden, *Surf. Sci.*, 1995, **337**, 67.
41. J. N. Russell, S. M. Gates and J. T. Yates, *J. Chem. Phys.*, 1986, **85**, 6792.
42. L. Zhu, J. Ho, E. K. Parks and S. J. Riley, *Z. Phys. D*, 1993, **26**, 313.
43. A. S. Martensson, C. Nyberg and S. Andersson, *Phys. Rev. Lett.*, 1986, **57**, 2045.
44. A. S. Martensson, C. Nyberg and S. Andersson, *Surf. Sci.*, 1988, **205**, 12.
45. R. R. Cavanagh and D. S. King, *Phys. Rev. Lett.*, 1981, **47**, 1829.
46. D. S. King and R. R. Cavanagh, *J. Chem. Phys.*, 1982, **76**, 5634.
47. J. W. Gadzuk, U. Landman, E. J. Kuster, C. L. Cleveland and R. N. Barnett, *Phys. Rev. Lett.*, 1982, **49**, 426.
48. R. A. Olsen, P. H. T. Philipsen, E. J. Baerends, G. J. Kroes and O. M. Lovvik, *J. Chem. Phys.*, 1997, **106**, 9286.

The role of rotational excitation in the activated dissociative chemisorption of vibrationally excited methane on Ni(100)

Ludo B. F. Juurlink, Richard R. Smith and Arthur L. Utz*

Tufts University, Department of Chemistry, 62 Talbot Avenue, Medford, MA 02155, USA

Received 9th May 2000
First published as an Advance Article on the web 6th October 2000

We have measured the sticking probability of methane excited to $v = 1$ of the v_3 antisymmetric C–H stretching vibration on a clean Ni(100) surface as a function of rotational state ($J = 0, 1, 2$ and 3) and have investigated the effect of Coriolis-mixing on reactivity. The data span a wide range of kinetic energies (9–49 kJ mol^{-1}) and indicate that rotational excitation does not alter reactivity by more than a factor of two, even at low molecular speeds that allow for considerable rotation of the molecule during the interaction with the surface. In addition, rotation-induced Coriolis-splitting of the v_3 mode into F^+, F^0 and F^- states does not significantly affect the reactivity for $J = 1$ at 49 kJ mol^{-1} translational energy, even though the nuclear motions of these states differ. The lack of a pronounced rotational energy effect in methane dissociation on Ni(100) suggests that our previous results for ($v = 1, v_3, J = 2$) are representative of all rovibrational sublevels of this vibrational mode. These experiments shed light on the relative importance of rotational hindering and dynamical steering mechanisms in the dissociative chemisorption on Ni(100) and guide future attempts to accurately model methane dissociation on nickel surfaces.

Introduction

Chemical reaction dynamics seeks to explain macroscopic chemical phenomena through a detailed understanding of how translational, vibrational, rotational and electronic energy influence chemical reactivity at the molecular level.[1,2] Dynamics experiments often are designed to elucidate how these different forms of energy contribute to the transformation of reactants into products. In the study of gas–surface reactivity, scattering experiments have long been used to probe energy transfer in non-reactive systems while studies that probe the reactive channel of the gas–surface encounter reveal how different forms of energy couple to the reaction coordinate.[3] Both types of studies provide complementary and detailed information that helps characterize the potential energy surface (PES) governing the interaction between the gas-phase molecule and the surface.

A common difficulty encountered in studies of gas–surface reactivity is that the thermal distribution of internal quantum states present in a reagent sample can obscure the contribution of a particular internal degree of freedom to the observed reactivity. Diatomic molecules have a single vibrational degree of freedom and the coupling of rotational and vibrational degrees of freedom is generally weak. Consequently, for closed-shell diatomic molecules, separation of internal state effects is experimentally challenging but straightforward. In contrast, polyatomic reagents have a much more complex internal structure. Not only may distinct vibrational modes affect the gas–surface interaction to varying degrees, Fermi resonance interactions and rotation-induced Coriolis mixing can alter the identity of the vibrational degrees of freedom on a state-by-state basis. In a polyatomic molecule, the separation of rotational and vibrational effects becomes much more difficult to study experimentally. State-resolved experimental techniques are required to reveal how specific internal degrees of freedom influence reactivity.

Researchers have used a variety of experimental techniques to investigate internal state effects in gas–surface dynamics. The most detailed of these experiments reveal the behavior of molecules in specific quantum eigenstates.[4] Early studies employing molecular beam methods or permeation techniques have evolved to include sensitive state-resolved detection of scattered molecules using laser-induced fluorescence (LIF) of NO[5–7] and resonance-enhanced multiphoton ionization (REMPI) spectroscopy of N_2.[8] Recently, state-selected preparation techniques have enabled gas–surface dynamics studies of molecules in selected internal states. For example, Gostein et al. studied scattering of H_2 ($v = 1$, $J = 1$) on Cu(110)[9] and Pd(111)[10] using stimulated raman pumping (SRP), and Hou et al. used stimulated emission pumping (SEP) to investigate inelastic scattering[11] and reactivity[12] of highly vibrationally excited NO on (oxygen covered) Cu(111) surfaces. Experiments in Miller's group and in our own laboratory use state-resolved IR excitation to perform such studies on polyatomic reagents. White and Miller use sensitive bolometric detection to probe CH_4 scattered from a LiF surface[13] while we use surface-bound product detection to investigate the reactive channel for CH_4 dissociation on Ni(100).[14,15]

Hydrogen dissociation on copper surfaces is the model system for understanding rotational and vibrational effects in strongly activated dissociative adsorption. Both theoretical[16,17] and experimental studies[18–20] on the H_2/Cu system agree that vibrations strongly enhance reactivity and that rotations may hinder reaction. In the "rotational hindering model",[21,22] rotational excitation inhibits dissociation for low-J states in H_2 by reducing the probability for finding the lowest energy pathway to dissociation. This model is closely related to the "dynamical steering mechanism"[23,24] which is used to explain decreasing reactivity with increasing J state for small values of J in H_2 on Pd(111).[25] In this nearly non-activated system, the gas-phase reactant experiences a torque upon approaching the surface due to corrugation in the PES. At low incident kinetic energies, the molecule is more easily steered into a favorable adsorption geometry and reacts with higher probability if less initial rotational motion is present.

Several studies have suggested that rotational state effects may influence the activated dissociation of methane on nickel surfaces. In general, the reactivity of a gas-phase molecule is expected to vary with rotational state if the PES has a significant dependence on orientation of the gas-phase molecule.[4] Transition state calculations show that the height of the repulsive barrier experienced by a methane molecule approaching a nickel atom in a Ni(111) surface does depend on the orientation of the molecule.[26] Barrier height differences of up to 80 kJ mol^{-1} are found when rotating the molecule relative to the surface.[27] Dynamical calculations also suggest the presence of rotational energy effects in methane dissociation. Jansen and Burghgraef find a reduction of the reactivity of methane on Ni(111) upon including rotational degrees of freedom in a calculation based on a LEPS potential.[28] In another study, Carré and Jackson use an ab-initio PES and wave-packet propagation methods to calculate the rotational state dependence for CH_4 dissociation on Ni(100).[29] Their reduced-dimensionality calculation treats methane as a pseudo-diatomic molecule but includes all three rotational degrees of freedom. They find that reactivity of the vibrational ground state increased significantly with rotational excitation at low kinetic energies, while the opposite trend held for high kinetic energies. In addition, the influence of m_j on reactivity varied with kinetic energy which suggests the importance of steering. In a recent experimental study of methane dissociation on the reconstructed Pt(110)-(1 × 2) surface, dynamical steering was invoked to explain an unusual translational energy dependence of the reaction probability at low incident kinetic energies.[30,31]

In this paper we report on our continued experimental efforts to understand activated dissociation by studying how selectively prepared excited states react on surfaces. We focus our attention on the effects of rotational motion by quantifying the reactivity of individual rotational levels of vibrationally excited CH_4 ($v = 1$, v_3). We also investigate how rotational excitation might affect reactivity by altering the vibrational character of the excited molecule. Our studies probe the low rotational states present in our supersonic molecular beam where dynamical effects such as rotational hindering and dynamical steering are likely to exert their greatest influence on the dissociation dynamics.

Experimental approach

We combine molecular beam techniques with state-selected infrared (IR) excitation to measure the dissociative chemisorption probability of molecules in individual excited quantum states as described previously.[15] Mixtures (2%, 15% and 40%) of CH_4 (99.99%, Northeast Airgas) in H_2 (99.9999%, Northeast Airgas) and 100% CH_4 expand through a 25 µm orifice in an Inconel 600 nozzle. The pressure in the nozzle is 370 kPa and the nozzle temperature is held at 287 ± 1 K. Time-of-flight (TOF) measurements verify that the kinetic energy of CH_4 in the molecular beams ranges from 9 to 49 kJ mol^{-1} with respective $\Delta E/E$ (FWHM) of 35% to 15%. The supersonic molecular beam is triply differentially pumped before entering the ultrahigh vacuum (UHV) portion of the apparatus where it impinges on the Ni(100) crystal at normal incidence. An optical multipass cell[32] and a room-temperature pyroelectric detector (PED) that acts as a bolometer[33] allow us to excite methane molecules in the beam with an IR laser and then verify the extent of excitation. The UHV chamber houses our nickel single crystal, which is cut and polished to $<0.1°$ of the (100) plane and mounted on a manipulator that allows for translation in three dimensions and rotation around the front face of the crystal. The crystal can be cooled to 78 K and heated radiatively or by electron bombardment using a thoriated tungsten filament. For all the experiments described in this paper, the surface is held at 475 ± 1 K to minimize adsorption of hydrogen and carbon oxides during the dose. Cleaning procedures generally involve sputtering with Ar^+ ions at glancing incidence (500 eV, 2 µA cm^{-2}, 3 min) to remove sulfur and annealing at 1100 K (15–30 min) to restore surface order. Two oxidation–reduction cycles remove carbon and oxygen (exposure to 5×10^{-6} Torr s O_2 at room temperature, followed by a temperature ramp to 1000 K and a dose of 2.4×10^{-4} Torr s of H_2). Auger Electron Spectroscopy (AES) confirms that contamination is below the 0.01 ML detection limit for carbon, sulfur, and oxygen. The base pressure in the UHV chamber is 5×10^{-11} Torr and background adsorption of carbon-containing species on the nickel crystal is measured to be 0.01 ML carbon h^{-1}. When trapped with a dry ice–acetone bath, the 100% CH_4 beam may impinge on the crystal for more than 4 h before contaminants interfere with the experiment. We find that we can measure absolute initial sticking probabilities as low as 5×10^{-7}.

IR light from a tunable, continuous wave color-center laser (Burleigh, FCL-20, $\Delta v < 1$ MHz) enters the optical multi-pass cell and excites methane molecules in the molecular beam to the desired rovibrational eigenstate. The optical cell creates 10–12 orthogonal crossings of the laser beam with the molecular beam. The IR light is linearly polarized and its electric field vector crosses the molecular beam at 45° with respect to the Ni(100) surface normal. The color-center laser is frequency stabilized with a home-built, computer-controlled servo-loop that uses a temperature and pressure stabilized 150 MHz Fabry–Perot etalon as its reference.[15] The pyroelectric bolometer intercepts a solid angle of the molecular beam that is centered on, but smaller than the solid angle intercepted by our Ni(100) surface. Once we detect laser-excited molecules impinging on the PED, we can be confident that laser excited molecules will impinge on the Ni(100) surface when we retract the PED from the molecular beam. Although we have demonstrated that the laser frequency can be held stable for several hours with no significant decrease in excitation efficiency,[15] we verify the resonance between the laser and the methane transition of interest every 30 min during long exposures.

We have shown in previous work that the absolute reaction probability of molecules in the laser-excited state is given by the difference in average sticking probability measured with and without laser excitation divided by the total fraction of methane molecules excited by the IR laser.[14] Since physisorbed methane desorbs from the surface promptly at the 475 K surface temperature we use, surface-bound carbon is a signature of methane's dissociative chemisorption. Auger electron spectroscopy measurements that are calibrated with the 0.50 ML carbon coverage resulting from the self-limiting adsorption of C_2H_4 on an 475 K Ni(100) surface provide an absolute measure of the number of dissociated methane molecules.[34] We chose methane exposure times that deposit from 0.08 to 0.12 ML of carbon. The initial sticking probability of methane is known to be constant up to 0.15 ML coverage of carbon.[35] We use the experimentally determined pumping speed and steady-state partial pressure of methane in the UHV chamber during the dose to determine the incident flux of methane on the surface.

We excite methane from $v = 0$ to $v = 1$ of the v_3 antisymmetric C–H stretching vibration and the total fraction of molecules excited in the beam is the product of the $v = 0$, J'' state population and the fractional excitation out of that state. Populations of J'' levels ($v = 0$) in the molecular beam are calculated by scanning the IR laser over $R(J'')$ transitions for all J'' ($v = 0$) levels present in the beam, integrating the absorption signal, and normalizing for transition probability.[36] We determine the fractional excitation using the $R(J'' = 1)$ transition for all kinetic energies using a calibration method described previously.[15] For this calibration we apply a well-understood mathematical model for the dependence of absorption on transit time, transition probability, radiation density, and state degeneracies[2] with no adjustable parameters and obtain an excellent fit to the data.[15] Thus, we know the absolute fraction of molecules excited in R(1) transitions for all kinetic energies. We calibrate the response of our PED with the absolute fractional excitation of the R(1) transition and use its calibrated response to calculate the fractional excitation for other rotational transitions to the v_3 vibrational fundamental.

Our experiments deposit a vertical stripe of carbon on the nickel crystal.[37] This effect is due to the narrow bandwidth of the laser exciting only methane molecules that have very small velocity components perpendicular to the molecular beam axis and parallel to the laser's propagation direction. Molecules hitting the left or right edges of the crystal are sufficiently Doppler shifted (due to their perpendicular velocity), that their transit-time-dominated absorption profile is not resonant with the IR laser. We measure by (AES) the C(272 eV)/Ni(848 eV) ratio across the crystal after exposing it to the molecular beam. We fit this profile to its expected Gaussian form. The amplitude of the Gaussian above baseline (expressed in ML of carbon) is a direct measure of the enhancement in carbon deposition due to laser excitation. The non-zero carbon coverage of the baseline is due to carbon deposition from methane molecules not excited by the laser, trace impurities in the beam, and carbon-containing residual gas in the UHV chamber. Dividing the amount of carbon deposition due to laser excitation by the time-integrated flux of CH_4 molecules (expressed in CH_4/Ni atom) and the total fractional excitation yields the state-resolved sticking probability. Neglecting the sticking probability of the vibrational ground state introduces an error smaller than 2%.[14] This error is insignificant compared to the random and systematic sources of uncertainty in our measurements. These include base line variations in the AES carbon signal and varying amounts of undetectable surface impurities that affect the reactivity. We have determined previously that the nature of these uncertainties limits our ability to measure the sticking probability to a relative error of 23% (one standard deviation).[38]

Results

Table 1 shows the transitions used to assess the effect of rotational motion on the reactivity, the maximum population in the initial J'' states, and the maximum fractional excitation out of each J'' state. We discuss the additional angular momentum labels appearing in Table 1 in the Discussion section. The transitions access $J' = 0$, 1, 2 and 3 in the $v = 1$ (v_3) vibrational state. Our experiments are limited to these J' states, since the supersonic expansion strongly relaxes the rotational population and IR transitions are only allowed for $\Delta J = 0, \pm 1$. We measure the population of

Table 1 Transitions used to probe rotational effect from 9 to 49 kJ mol^{-1}. T = transition, RS = rotational symmetry, and VS = vibrational sublevel

	$v = 0$					Maximum fractional	$v = 1$					Maximum fractional	Lower energy	Transition frequency
T	N''	l''	J''	RS	VS	population	N'	l'	J'	RS	VS	excitation	/cm^{-1}	/cm^{-1}
P(1)	1	0	1	F_1	A_1	0.563	1	1	0	F_2	F^+	0.250	10.4816	3009.0113
R(0)	0	0	0	A_1	A_1	0.313	0	1	1	A_2	F^-	0.750	0	3028.7522
R(1)	1	0	1	F_1	A_1	0.568	1	1	2	F_2	F^-	0.625	10.4816	3038.4985
R(2)	2	0	2	E	A_1	0.125	2	1	3	E	F^-	0.583	31.4421	3048.1692

each J state in the beam using IR absorption spectroscopy and find that rotational cooling becomes increasingly effective with hydrogen content. Nuclear spin statistics in methane result in three distinct sets of rotational levels that are not interconverted by collisions.[39,40] For $\leqslant 15\%$ CH_4/H_2 we find that rotational cooling is essentially complete under room temperature conditions and that thermally populated A, E and F rotational levels cool entirely into the lowest possible states: $J = 0$ for A, $J = 1$ for F and $J = 2$ for E. When cooling is complete $J = 0$, 1 and 2 have relative populations of 5/16, 9/16 and 2/16 respectively.[39] Cooling is not complete in room-temperature expansions of 40% CH_4/H_2 and 100% CH_4. For example, in a 100% CH_4 expansion at 294 K, we find that of all methane molecules with F rotational symmetry $\sim 23\%$ and 3% remain in $J = 2$ and $J = 3$, respectively and do not relax into $J = 1$. In addition to incomplete cooling, we find that the rotational temperature of the molecular beam varies with symmetry of the rotational state. Specifically, methane molecules with F rotational symmetry reach a lower rotational temperature than those with A symmetry. This behavior has been observed before by Luijks *et al.*[41] We agree with their explanation that this is due to missing A-symmetry J states at low values of J.

Employing the measured populations in J'' states and the described calibration for fractional excitation out of these states, we calculate fractional excitations for transitions in Table 1 for 2%, 15%, and 40% CH_4/H_2 and 100% CH_4. In Table 2 we show a matrix of the fractional excitation of all methane molecules in the molecular beam as a function of ro-vibrational transition and molecular velocity. The fractional excitation ranges from 3.7% to 21.3% of all methane molecules in the molecular beam. This large spread reflects the differences in initial state population and maximum fractional excitation out of that state. It is obvious that R(1) and R(0) transitions allow us to excite considerably more molecules than P(1) and R(2) transitions. Consequently, the dose times for experiments using the former transitions are much shorter than for the latter. Since the reactivity of methane on Ni(100) decreases rapidly with decreasing kinetic energy, we cannot measure the state-resolved reactivity of $v = 1$ (v_3), $J = 0$ and 3 for the slower 40% CH_4/H_2 and 100% CH_4 beams.

For all transitions with fractional excitations listed in Table 2 we have measured carbon deposition across the crystal with AES. A typical example of a plot of carbon coverage *vs.* position on the crystal is shown in Fig. 1. The data are collected after a 15 min dose of 15% CH_4/H_2 while exciting the R(0) transition. The individual points are found by integration of AES features for carbon and nickel and the previously described calibration procedure. Random errors in the AES measurements of the carbon feature are accounted for by integrating over slightly varying regions and establishing an error for the carbon integral. A minimum of four integrals per point on the crystal is taken and the error bars in Fig. 1 reflect two standard deviations of the variation in the size of this integral. Since the nickel integral is an order of magnitude larger and almost invariant over all experiments, we do not establish an error for individual measurements of this feature. The continuous line in Fig. 1 is a Gaussian curve, which is fit to the data employing appropriate weighting of individual data points. The curve fitting procedure indicates that ro-vibrationally-enhanced deposition reaches a maximum of 0.096 ± 0.003 ML carbon on top of 0.022 ± 0.001 ML carbon background. The peak is located 0.5 mm to the left of the center of the crystal. The location of the peak is nearly invariant in all experiments reflecting the excellent pointing stability of our laser. We verify that the width of the Gaussian fit agrees with the expected width.[37] In cases where the curve fitting procedure does not converge properly, we constrain the width to the expected value.

Table 2 Total fractional excitation

Kinetic energy /kJ mol^{-1}	Molecular velocity /m s^{-1}	Transition			
		P(1)	R(0)	R(1)	R(2)
49	2.47×10^3	0.037	0.154	0.173	0.044
30	1.93×10^3	0.045	0.154	0.204	0.051
17	1.46×10^3			0.172	0.213
8.9	1.06×10^3			0.189	0.205

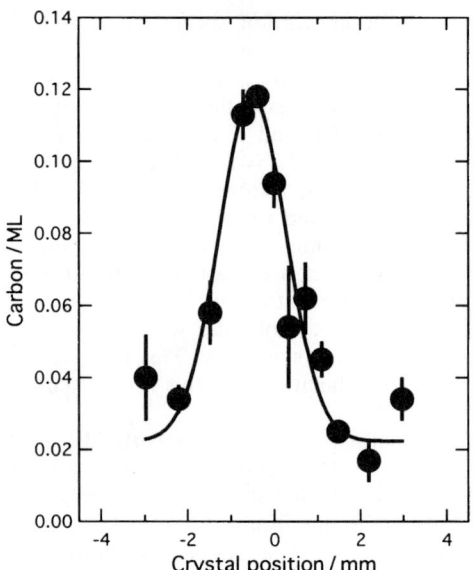

Fig. 1 Carbon deposition (in ML) across the crystal surface. Individual data points were collected by AES after a 15 min dose of 15% CH_4/H_2, while exciting $v = 1$ (v_3), $J = 1$ through the R(0) transition. The curve is a floating Gaussian fitted to the data as indicated in the text.

We calculate the state-resolved sticking probability for all v_3, $v = 1$, J' states from the enhancement in carbon deposition, the total methane flux and the fractional excitation. For the data shown in Fig. 1, the total flux of methane was 2.7×10^3 CH_4 molecules per surface nickel atom, the enhancement in carbon deposition was 0.096 carbon atoms per surface nickel atom and the fractional excitation was 17.2%. The value for $S_0^{v=1,J=1}$ is therefore $0.096/(2.7 \times 10^3 \times 0.172) = 2.1 \times 10^{-4}$. Fig. 2 shows the values we calculate for $S_0^{v=1,J}$ as a function of kinetic energy. The error bars are two standard deviations of the convoluted uncertainties from

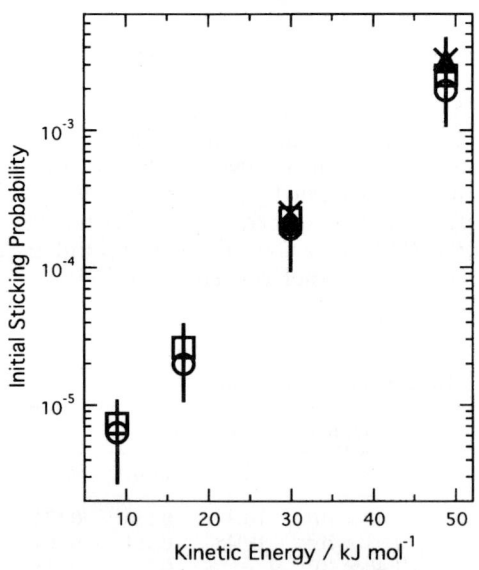

Fig. 2 Initial sticking probability for $v = 1$ (v_3), $J = 0$ (▲), $J = 1$ (□), $J = 2$ (○) and $J = 3$ (×) as a function of kinetic energy. Error bars are two standard deviations of convoluted uncertainties.

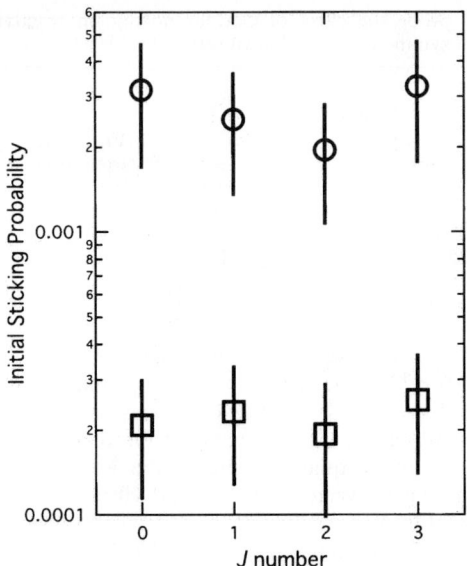

Fig. 3 Initial sticking probability for $v = 1$ (v_3) as a function of J state for $E_{kin} = 49$ kJ mol^{-1} (○) and $E_{kin} = 30$ kJ mol^{-1} (□). Error bars are two standard deviations of convoluted uncertainties.

the individual curve fittings and our estimate of other sources of error. An additional error is introduced by uncertainty in our calibration procedure for the fractional excitation of R(1), which is estimated to be no larger than 20%. It affects all data points in Fig. 2 equally and dominates the uncertainty in the absolute scaling of the y-axis.

In Fig. 3 we plot $S_0^{v=1,J}$ vs. J state to allow for a clearer visualization of possible rotational state effects on the sticking probability. The error bars are the same as those reported in Fig. 2. Fig. 4

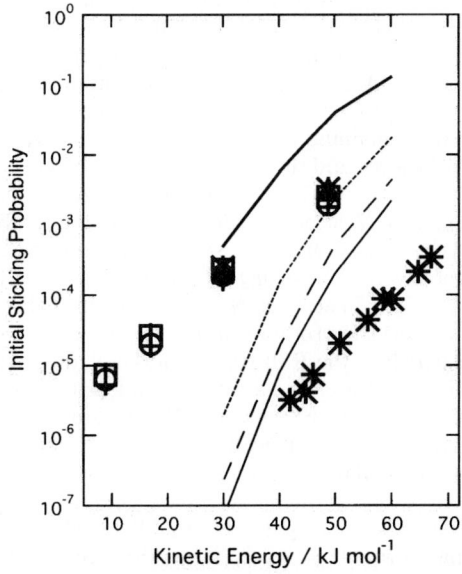

Fig. 4 Comparison of predicted and measured initial sticking probabilities for $v = 0$ and $v = 1$ (v_3). Our data are represented by markers: $v = 0$ (*) and $v = 1$, $J = 0$ (▲), $J = 1$ (□), $J = 2$ (○) and $J = 3$ (×). Data from Carré and Jackson appear as thin lines for $v = 0$ {$J = 0$ (—), $J = 1$ (— —), $J = 2$ (· · ·)} and as a thick line for $v = 1$ (—).

Table 3 Transitions used to probe the effect of Coriolis mixing on reactivity at $E_{trans} = 49$ kJ mol^{-1}. T = transition, RS = rotational symmetry, and VS = vibrational sublevel

T	$v = 0$					$v = 1$					Fractional excitation	Total internal energy /cm^{-1}	Sticking probablity
	N''	l''	J''	RS	Population	N'	l'	J'	RS	VS			
R(0)	0	0	0	A_1	0.313	0	1	1	A_2	F^-	0.154	3028.7522	0.0025 ± 0.0012
Q(1)	1	0	1	F_1	0.563	1	0	1	F_2	F^0	0.110	3029.3058	0.0021 ± 0.0010
P(2)	2	0	2	E	0.125	2	−1	1	E	F^+	0.016	3030.5024	0.0019 ± 0.0013

compares our results to the calculated sticking probabilities for $v = 0$ ($J = 0$, 1 and 2) and $v = 1$ as reported by Carré and Jackson.[29] Our comments regarding the difference in absolute values for the $v = 1$ data have been published elsewhere[14] and we focus here on rotational effects.

Finally, we present our results from measurements of the reactivity for different ro-vibrational states described by the same v' and J' quantum numbers in Table 3. In the Discussion section we will outline how Coriolis mixing between rotational and vibrational motion in methane lifts the vibrational state degeneracy of v_3 and results in three distinct ro-vibrational eigenstates that have different nuclear motions but nearly identical energies. Table 3 shows the transitions of interest, the fractional excitation, the total internal energy and the sticking probability for $v' = 1$, $J' = 1$ as measured at 49 kJ mol^{-1} translational energy. The same methods as described previously have been applied to calculate the reported sticking probability and error.

Discussion

The spherical symmetry of methane and the presence of vibrational angular momentum in the triply degenerate v_3 vibration complicate the angular momentum labeling of the rovibrational states we excite. Simpson et al.[42] and Hougen[43] discuss the role of angular momentum in the infrared excitation of methane. The total angular momentum of the molecule, J, is composed of both pure rotation of the molecule as a whole, denoted by N and the vibrational angular momentum of the v_3 vibration, l. The total angular momentum is given by the vector sum $J = N + l$. The $v = 0$ ground state of methane has $l = 0$, so $J = N$ for $v = 0$, as indicated in Table 1. The $v = 1$ level of v_3 is split by Coriolis mixing into three nearly degenerate sublevels with vibrational angular momentum $l = 0$ and ±1. A C–H stretching excitation polarized parallel to the molecular rotation axis does not experience a Coriolis force and corresponds to the $l = 0$ vibrational level. C–H stretch motions along the remaining two directions perpendicular to the molecular rotation axis are coupled by the Coriolis force and the linear combinations of these motions give rise to the $l = ±1$ levels. The directional nature of the Coriolis force results in l lying parallel or antiparallel to the molecule-fixed rotation axis, N. Selection rules based on the rovibrational symmetries of the states result in P- ($\Delta J = -1$), Q- ($\Delta J = 0$) and R- ($\Delta J = +1$) branch excitation accessing v_3 levels with $l = -1$, 0 and +1 respectively. The strong IR transitions in the fundamental band that we excite ($\Delta v_3 = +1$) have $\Delta N = 0$.[43] Therefore, $N' = N'' = J''$ in Table 1 and we are led to the conclusion that the extent of molecular rotation in the vibrationally excited molecule is given by J'', not J'. Thus, the state excited by the R(0) transition is not physically rotating because it has $J' = 1$ but $N' = 0$. Excitation of P(1) prepares a rotating molecule with $J' = 0$ but with $N' = 1$ and $l = -1$ to give total $J' = 0$. Since both the vibrational and purely rotational angular momentum have identical inertial constants in this spherically symmetric molecule, the *total angular momentum* and the total energy due to angular momentum (to first order) both depend on J'.

Our use of plane-polarized IR laser light also affects the orientation of the oscillating C–H stretch dipole of the v_3 mode. Polarized light preferentially excites an oscillating dipole moment vector of the C–H stretching vibration (μ_{IR}) parallel to the electric field vector of the laser field. Excited vibrational states prepared *via* a Q-branch transition have J nearly parallel to μ_{IR} while P- and R-branch excitation prepares states with J nearly orthogonal to μ_{IR}.[42]

Rovibrational states with $N' > 0$ are rotating as they approach the surface, and we now consider the extent of rotational motion during methane's interaction with the surface. We use the methane rotational constant, $B \approx 5.25$ cm^{-1},[40] and calculate the classical rotational periods for $N' = 1$ and 2 of 2.25×10^{-12}, 1.30×10^{-12} s, respectively. Assuming that the distance over which the surface and the methane molecule strongly interact is at least 0.5 Å from the transition state, we calculate an interaction time based on our measured molecular beam speeds. We then calculate the extent of rotation during the interaction time for each of the beams in Table 2 and report the values in Table 4. Since methane molecules slow as they experience the repulsive portion of the gas–surface potential, these calculations should be treated as lower limits on the extent of rotation. We find that even the fastest molecules rotate by up to 11° during their interaction time in the $N = 2$ state and more slowly moving molecules rotate through even larger angles. These angular displacements are not large, but they may be influential since the potential energy varies so strongly with molecular orientation.[26,27] At a fixed C–Ni distance of 2.34 Å, Kratzer et al. calculate differences in energy of up to 80 kJ mol^{-1} for different high symmetry orientations of the molecule above a nickel atom.[27] Calculations by Yang and Whitten indicate that this difference increases to 350 kJ mol^{-1} for two orientations differing by a 55° rotation at a C–Ni distance of 2.11 Å.[26] Consequently, we believe that the portions of the barrier height distribution sampled by the $N' = 0$, 1 and 2 states are different enough to result in J-dependent sticking probability variations if rotational hindering is important.

We note that, except for the R(0) transition that excites $N' = 0$, the vibrationally excited molecule rotates freely between the excitation region and the surface. Therefore, our data for the $N' = 1$ and 2 states are averaged with respect to the angle between the surface normal and the vibration-induced dipole moment in the molecule. For $N' = 0$, the induced dipole moment orientation is preserved at a 45° angle relative to the surface normal. Since the orientation of the molecule-fixed coordinate system with respect to the nuclear displacement vectors for the v_3 mode in XY$_4$ molecules is not uniquely defined by symmetry[40] our results for all J' values are averaged with respect to the orientation of the molecule relative to the Ni(100) surface normal upon impact.

We consider Fig. 3 to assess whether we find clear evidence for rotational hindering effects in methane dissociation. The trends observed in H$_2$/Cu(111)[20] and D$_2$/Cu(111)[44] show a substantial increase in threshold energy to dissociation upon increasing rotational excitation at low values of J. At larger values of J, the threshold energy drops due to coupling of rotational energy to the reaction coordinate. This trend is observed in various vibrational states, though its magnitude decreases with v.[20,44] At a fixed translational energy, this trend results in an initial drop in sticking probability with J followed by an increase at higher J values. In methane, we consider the variation in reactivity with N, the quantum number for pure rotation of the molecule.

Fig. 3 shows the dissociation probability for methane molecules with $J' = 0, 1, 2$ and 3 and with $N' = 1, 0, 1$ and 2, respectively, measured at fixed translational energies of 30 and 49 kJ mol^{-1}. We note that the different N' states differ in reactivity by, at most, 60%, which is less than the error in our measurement. Since the difference between the lowest energy orientation for dissociation and a highly repulsive orientation can be as little as 55°, one might expect that the rotational states investigated in this study should exhibit a much more significant difference in reactivity were a rotational hindering model important at the kinetic energies investigated. If rotational hindering does influence methane dissociation on Ni(100), its effect on the sticking probability must be less than a factor of two for $E_{\text{trans}} = 49$ kJ mol^{-1} and $N < 2$.

The data in Fig. 2 and 3 also allow us to investigate whether we find evidence for dynamical steering in our data. Calculations by Kratzer et al.,[27] Jansen and Burghgraef[28] and Carré and

Table 4 Estimated interaction time and rotation of methane for various transitions and kinetic energies

Kinetic energy /kJ mol^{-1}	Molecular velocity /m s^{-1}	Time/s	Rotation/degrees		
			$N = 0$	$N = 1$	$N = 2$
49	2.47×10^3	4.1×10^{-14}	0	6	11
30	1.93×10^3	5.2×10^{-14}	0	8	14
17	1.46×10^3	6.9×10^{-14}		11	19
8.9	1.06×10^3	9.5×10^{-14}		15	26

Jackson[29] have all suggested that dynamical steering may be of importance in methane dissociation on nickel surfaces. We find that our measured values of $S_0^{v=1,J=1}$ are consistently larger than $S_0^{v=1,J=2}$ at all four kinetic energies investigated. Both of these reaction probabilities are measured using transitions with large and very similar excitation efficiencies. When one calculates the ratio of these two experimentally measured sticking probabilities, systematic errors in the excitation fraction for each state cancel and the relative error of the ratio is less than that of an individual sticking probability measurement. Our data indicate that the ratio of $S_0^{v=1,J=1}/S_0^{v=1,J=2}$ is relatively constant (1.25 ± 0.2) for E_{trans} ranging from 9 to 49 kJ mol^{-1} and suggest that the $J' = 1$ level may indeed be slightly more reactive than $J' = 2$, although this difference is just at the limit of uncertainty in our measurement.

If the difference in reactivity between the $J' = 1$ and $J' = 2$ state is statistically significant, a dynamical steering mechanism may be causing the differences in $S_0^{v=1,J=1}$ and $S_0^{v=1,J=2}$ that we observe. State-resolved experiments[25] have recently confirmed theoretical predictions[23,24] of dynamical steering in H_2 dissociation on Pd(111). An unusual J-dependence in the reaction probability, in which S_0 initially decreases with increasing J before reaching a minimum and then increasing monotonically with J, was a signature for the dynamical steering mechanism. Since the torque required to "steer" or reorient the molecule varies with the total angular momentum of the molecule, we expect steering in methane to depend on J, not N. Since N and J can vary in methane, it is, in principle, possible to distinguish rotational hindering from dynamical steering in methane dissociation. Our data for $J = 1$ and 2 show a possible decrease in S_0 with increasing J, which mimics the behavior observed for H_2 dissociation on Pd(111).

Translational energy can play an important role in dynamical steering mechanisms since it defines the time frame during which the molecule can be reoriented. Experiments[25] and calculations[23] on the H_2/Pd system exhibit a strong translational energy dependence in which steering effects become much less significant at high translational energies. Molecules incident with different translational energies also sample different portions of the potential energy surface. In the PES illustrated in Fig. 2 of ref. 23, low-energy molecules sample a portion of the H_2/Pd potential in which a significant gradient along the molecular orientation coordinate leads to a reorienting torque on the molecule. At higher translational energies, molecules sample a portion of the potential that is highly repulsive with respect to z, but has a minimal gradient with respect to molecular orientation.

Carré and Jackson investigated the effect of m_j on methane reactivity and found reactivity to be independent of m_j up to a kinetic energy of about 70 kJ mol^{-1}.[29] Above that energy they saw a significant difference in the reactivity of different m_j states; those states with the active C–H bond aligned parallel to the surface were much more reactive than those with the active C–H bond aligned along the surface normal. They attributed this trend to dynamical steering. At lower translational energies, gradients in the PES with respect to molecular orientation were able to exert a torque on the methane molecule and reorient incident molecules with unfavorable m_j into a favorable orientation for reaction. At high translational energies, molecules with unfavorable m_j orientations had too little time to reorient and experienced a much higher barrier to reactivity.

Our results for $J = 1$ and 2 do not exhibit a strong translational energy dependence over the range of $E_{\text{trans}} = 9$ to 49 kJ mol^{-1}. In fact, the ratio of $S_0^{v=1,J=1}/S_0^{v=1,J=2}$ remains almost constant for the entire range of translational energies we investigate. All of these data are collected at kinetic energies less than 70 kJ mol^{-1} where Carré and Jackson predict that dynamical steering is operative.[29] If our data do reflect the presence of dynamical steering in methane dissociation, they suggest that steering may influence methane dissociation at translational energies well above those reported in the study of H_2 dissociation on Pd(111). Calculations by Yang and Whitten[26] and Kratzer et al.[27] suggest why this may be true. They find evidence for gradients along the molecular orientation coordinate that increase in slope as the methane molecule moves closer to the surface. Therefore, at lower translational energies, slower methane molecules feel a weaker orienting force for a longer period of time, while faster methane molecules approach closer to the surface and are oriented by the stronger forces for a shorter time. Thus, the relationship between interaction time and reorientation force may result in steering effects persisting for CH_4/Ni(100) to much higher translational energies than was observed for the H_2/Pd(100) system. At translational energies higher than those we discuss here, steering effects likely become less important. We note that in our studies of v_3, $J = 2$ dissociation, we observe that S_0 increases less rapidly at translational energies above 45 kJ mol^{-1}.[14,38] This behavior may reflect a transition from a translational energy regime where steering facilitates dissociation to one where the molecular speed is too great to permit significant reorientation of the methane molecule during its interaction with the surface.

Experiments that probe $N = 0$ and $N > 0$ differ in the extent to which the orientation of the vibrational dipole is averaged with respect to the surface normal. Our linearly polarized laser prepares $N = 0$ molecules via the R(0) transition with a vibrational dipole preferentially oriented at 135° relative to the surface normal, which is close to the 118° calculated orientation of the active C–H bond at the transition state. In contrast, the vibrational dipole in molecules with $N > 0$ (e.g. the $N = 1$ state prepared by R(1) excitation) is tumbling with respect to this orientation. The reactivity of vibrationally excited molecules in the beam must be averaged over a full range of incident orientations of the C–H bond extension. The vibrationally excited molecules that we detect react with significantly enhanced probability relative to molecules in $v = 0$, which suggests that the vibrational dipole of the C–H stretching state that we excite projects along the R–H coordinate at the transition state. Therefore, we might expect that the highly favorable orientation of the laser-excited $N = 0$, v_3, $v = 1$ molecules would lead to a significantly enhanced reaction probability relative to the orientationally averaged molecules in $N = 1$ and 2. We find instead that the $J = 1$ level prepared with $N = 0$ is only very slightly more reactive than the $N = 1$ level prepared via R(1) excitation. This result is consistent with either an extremely modest orientational effect, or with a dynamical steering mechanism that reorients the tumbling $N = 1$ and 2 molecules into a favorable orientations similar to that of the non-rotating $N = 0$ molecules.

Taken together, these results provide tentative support for the role of dynamical steering in CH_4 dissociation on Ni(100). While many aspects of the data are consistent with a dynamical steering mechanism for CH_4 dissociation on Ni(100), the uncertainty in the measurements presented prevents us from conclusively establishing the importance of steering. We are planning further experimental studies that will better characterize the role of dynamical steering in methane dissociation on Ni(100).

Our data also identify the magnitude of the rotational enhancement for CH_4 dissociation on Ni(100). In Fig. 4 we compare our data to results from dynamical calculations by Carré and Jackson, who treat methane as a pseudo-diatomic with one vibrational and three rotational degrees of freedom.[29] Their calculations predict a strong enhancement in reactivity with H–CH$_3$ vibrational excitation with which they model the asymmetric stretch of CH_4. We have pointed out previously that their calculated absolute reaction probabilities for the vibrationally excited state agree well for $E_{trans} < 40$ kJ mol^{-1}, but significantly overestimate S_0 for $E_{trans} > 40$ kJ mol^{-1}.[14]

In their calculations on the vibrational ground state, Carré and Jackson observe a strong dependence of the reaction probability on rotational state.[29] At 40 kJ mol^{-1}, they calculate that $S_0^{v=0}$, averaged over incident orientations, increases 200-fold from $J = 0$ to $J = 2$. At kinetic energies above ~ 110 kJ mol^{-1} this trend is reversed and lower-J states react more readily. Our results indicate that this predicted strong rotational dependence is not observed for $v = 1$, v_3. At most, the reactivities of $J = 0$ and 2 differ by a factor of 2. Although our experiment probes the reactivity of J levels in a vibrational state other than $v = 0$, we would expect the $v = 1$ level of v_3 to exhibit at least as strong a J-level dependence as the $v = 0$ level. In order for energy in the v_3 coordinate to promote reactivity most efficiently, the C–H stretch vibration must be polarized along an orientation similar to that at the transition state. This constraint places more, not fewer restrictions on the orientational anisotropy of the PES. Since we see at most a small rotational state effect in the $v = 1$ level of v_3, we do not expect that $v = 0$ would exhibit a significantly larger difference in reactivity between $J = 0$ and 2. Thus, we believe that our results are not consistent with the sizeable rotational enhancement predicted in ref. 29. We expect that a major contribution to the discrepancy between the results is the treatment of methane as a pseudo-diatomic molecule in the calculations. We agree with Carré and Jackson that constraining the reaction to one of the four C–H bonds likely overestimates the effect of rotation. Due to the high symmetry of the real molecule and the nature of the C–H stretching eigenstates, rotation over small angles allows different C–H bonds to become the reactive oscillator.

Next, we address the effect of Coriolis mixing in the v_3 antisymmetric C–H stretching mode. In Fig. 5 we show representations of this triply degenerate, F_2, mode that are consistent with the

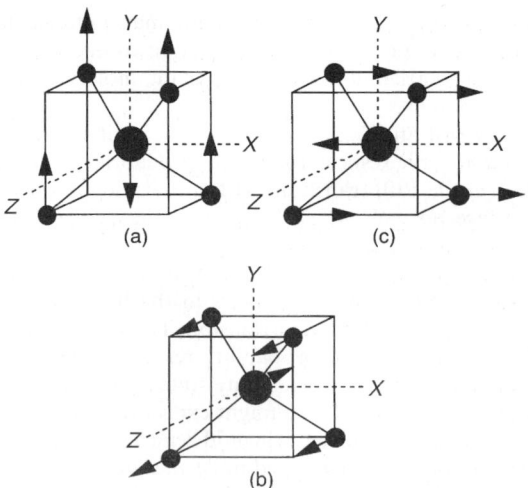

Fig. 5 Representations of the asymmetric stretch mode (v_3) in XY_4 molecules (adapted from Herzberg[40]). Nuclear displacement coordinates indicate direction only.

symmetry of the molecule and that help illustrate the origin of Coriolis mixing in methane (adapted from Herzberg[40]). Rotation along the z-axis couples the v_{3a} and v_{3c} modes through Coriolis forces. Consequently, the vibrational eigenstates we excite are more properly described by v_{3b} and by linear combinations of the v_{3a} and v_{3c} zero-order states in Fig. 5. These linear combinations are the clockwise and counterclockwise circular rotations of the nuclei about their equilibrium position in the x–y plane. The vibrational angular momentum vector resulting from this vibrational motion is aligned parallel or antiparallel with the rotational angular momentum vector. Thus, the triply degenerate v_3 mode is split into three eigenstates: the unperturbed F^0 state (v_{3b}), and the Coriolis-mixed F^+ and F^- states, which are derived from the symmetric and antisymmetric linear combination of v_{3a} and v_{3c}. The nuclear displacement coordinates for the F^+ and F^- states are similar, but they differ significantly from the linear harmonic displacement of nuclei in the unperturbed F^0 state. Nonetheless, the vibrational energies for the three states differ by less than 2 cm^{-1} (see Table 3). The three vibrational sublevels of v_3 are accessed through different rovibrational transitions; Q branch excitation accesses the F^0 state, while the P and R branch excitations access the F^+ and F^- states, respectively. Dennison points out that the exact forms of the normal displacement coordinates for an XY_4 molecule are not uniquely determined by symmetry,[45] but also depend on potential constants.[45] In the case of CH_4, the displacement coordinates for the zero-order v_3 sublevels are more nearly polarized along the C–H bond axes than those illustrated in Fig. 5 and Coriolis mixing leads to an elliptical motion of the H atoms rather than a circular motion.

The measured reactivities for the three vibrational sublevels with identical total angular momentum $J = 1$ state are shown in Table 3. We find that the F^+ and F^- states have the same reactivity as the F^0 state within our experimental error. This result sheds light on the nature of the reaction coordinate in the region of the transition state. The three vibrational eigenstates that we excite are coherent excitations of all four C–H oscillators in the molecule. The three vibrational sublevels of v_3 differ considerably in the relative motion of one C–H oscillator relative to the others. We find that changing the motion of the non-reactive C–H oscillators relative to the motion of the active oscillator has little effect on reactivity. This result is consistent with the notion that a localized C–H bond extension is crucial in driving methane dissociation on Ni(100). The three sublevels of v_3 share similar reactivity because each can be thought of as a localized C–H excitation coupled, with a very specific phase, to the other (non-reactive) C–H oscillators in the molecule. This picture would suggest that the v_1 symmetric C–H stretching state should also share a similar efficacy for vibrational activation because the localized motion of its C–H oscillators is very similar to that of the F^0 level of v_3. It is the phase of the non-reactive C–H oscillators relative to the reactive oscillator that differs. We suggest that the enhanced reactivity of thermally excited C–H stretching eigenstates in CH_4, all of which involve the concerted motion of four localized C–H oscillators, is chiefly derived from the projection of this concerted motion onto a reaction coordinate in which a single extended C–H bond has a favorable orientation and impact parameter with respect to the surface. The relative motion of the remaining non-reactive C–H oscillators in these states appears to have little impact on the efficacy of vibrational excitation. We do not rule out the possibility that bending excitation may also facilitate methane dissociation, as has been suggested in previous experimental[46] and theoretical work.[27] In fact, a comparison of our measured reaction probabilities for $v = 1$ of v_3 with values of S_0 averaged over a thermal distribution of vibrational states indicates that modes other than the pure C–H stretching eigenstates must dominate the thermally averaged reaction rate constants.[14]

Finally, we comment on the relevance of this work to the reactivity of methane in a high-temperature thermalized sample. Rotational cooling in our supersonic molecular beam source limits our studies to low-lying rotational levels of the v_3 vibration. These levels are the ones that are most susceptible to dynamical effects such as dynamical steering, but they do not span a wide enough range of rotational energy content to allow us to predict the reactivity of highly excited rotational states. We calculate that at 1000 K, the peak of the rotational state distribution is near $J = 12$, with significant population extending to rotational states as high as $J = 30$. If some rotational energy couples to the reaction coordinate, we would expect these highly excited rotational levels to be more reactive than those discussed here. We note that our studies do characterize the rotational levels that dominate the population in beam-surface studies. The excellent agreement between the beam-surface data of Holmblad et al.[35] and Luntz,[47] in which the rotational state distribution is partially cooled, and the thermal bulb studies of Ølgaard-Nielsen et al.,[48] in which the rotational distribution is thermalized, suggests that highly excited rotational states are not dramatically more reactive than those present in a molecular beam. Therefore, we believe that the v_3 antisymmetric C–H stretching mode is well characterized by the results shown here, and that the conclusions published previously for $v = 1$ (v_3), $J = 2$[14] are representative for the vibrational mode in general.

Summary

We have measured the reactivity of various rotational states of vibrationally excited methane ($v = 1$, v_3) on Ni(100) and find that: (1) the reactivity of methane in different J states at a fixed total translational energy differs by less than a factor of two; (2) the reactivity of methane in the $J = 1$ rotational level is consistently greater than the reactivity of CH_4 in $J = 2$; (3) that the orientational averaging inherent to our studies of levels with $N > 0$ does not result in significantly less reactivity than we measure for $N = 0$ molecules oriented at a near-optimal approach angle; and (4) that Coriolis mixing of the degenerate sublevels of the v_3 vibration affects the reactivity of the sublevels by less than a factor of two. Observation (1) allows us to conclude that, if rotational hindering does influence methane dissociation on Ni(100), its effect on the sticking probability must be less than a factor of two for $E_{trans} = 49$ kJ mol^{-1} and $N < 3$. Observations (2) and (3) suggest that dynamical steering may play a role in methane dissociation on Ni(100), but further studies are needed to confirm this interpretation of our data. Observation (4) suggests that the enhanced reactivity of the v_3 mode is due to the elongated C–H bond length in the one oscillator that is positioned to become the dissociating C–H bond. This interpretation also suggests that the reactivity of v_1 and v_3 may be very similar, since both involve the concerted excitation of similar localized C–H bond modes. Finally, these results taken together point to the modest effect of rotational excitation and rotation-induced vibrational mode mixing on methane reactivity and indicate that our detailed studies of the v_3 ($v = 1$), $J = 2$ state are representative of all sublevels of the v_3 vibration.

Acknowledgements

This material is based upon work supported by the National Science Foundation under Grant No. 9703392 and the W. M. Keck Foundation. We thank Prof. Terry Haas and Prof. Bret Jackson for helpful discussions.

References

1. R. D. Levine and R. B. Bernstein, *Molecular Reaction Dynamics and Chemical Reactivity*, Oxford University Press, Oxford, 1987.
2. R. B. Bernstein, *Chemical Dynamics via Molecular Beam and Laser Techniques*, Oxford University Press, Oxford, 1982.
3. C. T. Rettner, D. J. Auerbach, J. C. Tully and A. W. Kleyn, *J. Phys. Chem.*, 1996, **100**, 13021.
4. D. C. Jacobs, *J. Phys. Condens. Matter*, 1995, **7**, 1023.
5. F. Frenkel, J. Hager, W. Krieger, H. Walther, C. T. Campbell, G. Ertl, H. Kuiper and J. Segner, *Phys. Rev. Lett.*, 1981, **46**, 152.
6. A. W. Kleyn, A. C. Luntz and D. J. Auerbach, *Phys. Rev. Lett.*, 1981, **47**, 1169.
7. G. M. McClelland, G. D. Kubiak, H. G. Rennagel and R. N. Zare, *Phys. Rev. Lett.*, 1981, **46**, 831.
8. G. O. Sitz, A. C. Kummel and R. N. Zare, *J. Chem. Phys.*, 1987, **87**, 3247.
9. M. Gostein, H. Parhikhteh and G. O. Sitz, *Phys. Rev. Lett.*, 1995, **75**, 342.
10. M. Gostein, E. Watts and G. O. Sitz, *Phys. Rev. Lett.*, 1997, **79**, 2891.
11. H. Hou, Y. Huang, S. J. Gulding, C. T. Rettner, D. J. Auerbach and A. M. Wodtke, *J. Chem. Phys.*, 1999, **110**, 10660.
12. H. Hou, Y. Huang, S. J. Guiding, C. T. Rettner, D. J. Auerbach and A. M. Wodtke, *Science*, 1999, **284**, 1647.
13. A. C. Wight and R. E. Miller, *J. Chem. Phys.*, 1998, **109**, 1976.
14. L. B. Juurlink, P. R. McCabe, R. R. Smith, C. L. DiCologero and A. L. Utz, *Phys. Rev. Lett.*, 1999, **83**, 868.
15. P. R. McCabe, L. B. F. Juurlink and A. L. Utz, *Rev. Sci. Instrum.*, 2000, **71**, 42.
16. G. R. Darling and S. Holloway, *Rep. Prog. Phys.*, 1995, **58**, 1595.
17. G. J. Kroes, *Prog. Surf. Sci.*, 1999, **60**, 1.
18. B. E. Hayden, in *Dynamics of Gas–Surface Reactions*, ed. C. T. Rettner and M. R. Ashfold, The Royal Society of Chemistry, Cambridge, 1991.
19. H. A. Michelsen, C. T. Rettner and D. J. Auerbach, in *Surface Reactions*, ed. R. J. Madix, Springer, Berlin, 1994.
20. C. T. Rettner, H. A. Michelsen and D. J. Auerbach, *J. Chem. Phys.*, 1995, **102**, 4625.
21. D. O. Hayward and A. O. Taylor, *Chem. Phys. Lett.*, 1986, **124**, 264.
22. H. A. Michelsen, C. T. Rettner and D. J. Auerbach, *Phys. Rev. Lett.*, 1992, **69**, 2678.
23. A. Gross, S. Wilke and M. Scheffler, *Phys. Rev. Lett.*, 1995, **75**, 2718.
24. M. Kay, G. R. Darling, S. Holloway, J. A. White and D. M. Bird, *Chem. Phys. Lett.*, 1995, **245**, 311.
25. M. Gostein and G. O. Sitz, *J. Chem. Phys.*, 1997, **106**, 7378.
26. H. Yang and J. L. Whitten, *J. Chem. Phys.*, 1992, **96**, 5529.
27. P. Kratzer, B. Hammer and J. K. Nørskov, *J. Chem. Phys.*, 1996, **105**, 5595.
28. A. P. J. Jansen and H. Burghgraef, *Surf. Sci.*, 1995, **344**, 149.
29. M.-N. Carré and B. Jackson, *J. Chem. Phys.*, 1998, **108**, 3722.
30. A. V. Walker and D. A. King, *J. Chem. Phys.*, 2000, **112**, 4739.
31. A. V. Walker and D. A. King, *Phys. Rev. Lett.*, 1999, **82**, 5156.
32. T. E. Gough, D. Gravel and R. E. Miller, *Rev. Sci. Instrum.*, 1981, **52**, 802.
33. R. E. Miller, *Rev. Sci. Instrum.*, 1982, **53**, 1719.
34. C. Klink, L. Olesen, F. Besenbacher, I. Stensgaard, E. Laegsgaard and N. D. Lang, *Phys. Rev. Lett.*, 1993, **71**, 4350.
35. P. M. Holmblad, J. Wambach and I. Chorkendorff, *J. Chem. Phys.*, 1995, **102**, 8255.
36. L. S. Rothman, C. P. Rinsland, A. Goldman, S. T. Massie, D. P. Edwards, J. M. Flaud, A. Perrin, C. Camy-Peyret, V. Dana, J. Y. Mandin, J. Schroeder, A. McCann, R. R. Gamache, R. B. Wattson, K. Yoshino, K. V. Chance, K. W. Jucks, L. R. Brown, V. Nemtchinov and P. Varanasi, *J. Quant. Spectrosc. Radiat. Transfer.*, 1998, **60**, 665.
37. L. B. F. Juurlink, R. R. Smith and A. L. Utz, *J. Phys. Chem. B*, 1999, **104**, 3327.
38. L. B. F. Juurlink, R. R. Smith, P. R. McCabe, C. D. DiCologero and A. L. Utz, in preparation.
39. E. B. Wilson Jr., *J. Chem. Phys.*, 1935, **3**, 276.
40. G. Herzberg, *Molecular Spectra and Molecular Structure II. Infrared and Raman Spectra of Polyatomic Molecules*, Van Nostrand Reinhold, New York, 1945.
41. G. Luijks, S. Stolte and J. Reuss, *Chem. Phys.*, 1981, **62**, 217.
42. W. R. Simpson, T. P. Rakitzis, S. A. Kandel, A. J. Orr-Ewing and R. N. Zare, *J. Chem. Phys.*, 1995, **103**, 7313.
43. J. T. Hougen, in *Physical Chemistry*, ed. D. A. Ramsay, Butterworths, Boston, vol. 2, 1976.
44. H. A. Michelsen, C. T. Rettner, D. J. Auerbach and R. N. Zare, *J. Chem. Phys.*, 1993, **98**, 8294.
45. D. M. Dennison, *Rev. Mod. Phys.*, 1940, **12**, 175.
46. M. B. Lee, Q. Y. Yang and S. T. Ceyer, *J. Chem. Phys.*, 1987, **87**, 2724.
47. A. C. Luntz, *J. Chem. Phys.*, 1995, **102**, 8264.
48. B. Ølgaard Nielsen, A. C. Luntz, P. M. Holmblad and I. Chorkendorff, *Catal. Lett.*, 1995, **32**, 15.

General Discussion

Prof. Wolf opened the discussion of Dr Saalfrank's paper: I have a question regarding the physical mechanism behind the temperature dependence of the excited state lifetime assumed in eqn. (19) of the paper. I agree that temperature plays a role in STM induced desorption of H/Si (i) *via* the Boltzmann population in the ground state (but as you pointed out the level spacing is quite large for H/Si) and (ii) by vibrational energy relaxation mediated by phonon excitation in the substrate. However, I don't see the mechanism behind eqn. (19) which leads to a temperature dependence of the electronic lifetime?

Dr Saalfrank responded: Our modelling of the (admittedly surprisingly strong) temperature-dependence of the lifetime of the electronically excited state, which we considered for the hole-resonance state, is based on earlier work of Thirstrup *et al.*[1] They assume that the hole resonance width increases substantially with temperature, and that this effect is more important than the temperature-dependence of the ground state vibrational lifetime when it comes to the interpretation of their experimental data for the STM-induced desorption at a bias of -7 V. We have used their parameterization whereby a simple picture of the temperature-dependence of the excited state lifetime is due to the strong exponential dependence of the lifetime on the adsorbate–surface distance which in turn is (periodically) affected by temperature-mediated motion of the substrate atoms.

A more sophisticated explanation of the temperature effect is given in Thirstrup *et al.*'s paper.[1] Prof. Stokbro, could you give us a few details?

1 C. Thirstrup, M. Sakurai, T. Nakayama and K. Stokbro, *Surf. Sci.*, 1999, **424**, L329.

Prof. Stokbro replied: Let me give a brief summary of why we in our paper[1] concluded that there must be a temperature dependence of the electronic lifetime. We measured the temperature dependence of the STM-induced H desorption rate, R, from Si(100) at negative sample bias, V_b. At $V_b = -7$ V we find that R decreases by a factor of 200 for a temperature increase of 80 K, whilst it only decreases by a factor of 3 at -5 V upon the same temperature change. The factor of 3 is what you would expect taking into account the temperature dependence of the vibrational lifetime, as in Foley *et al.*,[2] while another mechanism must account for the strong effect at $V_b = -7$ V. In the paper[1] we discuss a range of different effects, and find that the only effect that can account for the experimental data is a temperature dependence of the electronic lifetime. A temperature dependent electronic lifetime gives a strong bias dependent temperature effect in excellent agreement with the experimental data.

1 C. Thirstrup, M. Sakurai, T. Nakayama and K. Stokbro, *Surf. Sci.*, 1999, **424**, L329.
2 E. T. Foley, A. F. Kam and J. W. Lyding, *Phys. Rev. Lett.*, 1998, **80**, 1336.

Prof. Wolf commented: I understand that you obtain a good fit assuming the described temperature dependence of the excited state lifetime, which you associate with coupling of a surface phonon to the hole in the SiH bond. But I still don't see the mechanism. Are you saying that the hole can only be filled if you get momentum scattering from the phonon?

Prof. Stokbro answered: No! An electron tunnels from the substrate to the tip and leaves a hole behind. This hole state decays with a certain lifetime, and the lifetime depends on the substrate temperature. In the paper we relate the origin of the temperature dependence to a coupling between the surface resonance and the substrate phonons. It will be too lengthy here to go through all the details and I refer to the paper which is already published.[1] Let me just note that a temperature dependence of the electronic lifetime due to electron–phonon coupling is not something new that we invented, but has been measured directly for silicon bulk eigenstates, and interpreted in a series of papers by Cardona (for instance ref. 2).

1 C. Thirstrup, M. Sakurai, T. Nakayama and K. Stokbro, *Surf. Sci.*, **424**, L329.
2 P. Lautenschlager, M. Carriga, L. Vina and M. Cardona, *Phys. Rev. B*, 1987, **36**, 4821.

Prof. Wolf commented further: My reservation to the proposed temperature dependent lifetime according to eqn. (19) in Dr Saalfrank's paper originates because the underlying mechanism remains unclear. A possible interpretation maybe related to the temperature dependence of the Si band structure, but this will likely be a minor effect. The observed very pronounced temperature dependence should be related to a mechanism where for the filling of the hole the presence of a phonon is very critical and I don't see an explanation for that.

Prof. Stokbro responded: The temperature dependence is four times larger than the temperature dependent lifetime of Si eigenstates measured by Cardona. The relevant electronic state is in our case a surface resonance, so it does not seem to be an unreasonable strong temperature dependence. In the above mentioned paper we list some factors that could give rise to the additional temperature dependence. Let me also note that Prof. Seideman finds evidence for a temperature dependent electronic lifetime in the case of benzene on Si(100).

Prof. Seideman addressed Dr Saalfrank: You describe calculations of the STM-induced desorption of H from Si(100) and state in the conclusion that it would be important for future theoretical studies to take into account also the STM excitation process. I'd like to show you the result of recent work where we study the H/Si(100) problem[1] in the same (single electronic transition) regime explored in your paper, accounting for the STM-excitation dynamics within the framework outlined in our Discussion paper[2] and applied there to the STM-induced desorption of benzene from a silicon surface. The H/Si(100) desorption serves in this case as a test study. By contrast to the systems studied in ref. 2, for this system the nuclear dynamics and the substrate electronic density of states are relatively simple and a wealth of detailed experimental data, taken by several groups,[1,3] is available.

We treat the lifetime as a parameter and extract it from a least-squares fit of the experimental data to the theoretical bias dependence. In Fig. 1 the circles show the data of ref. 1 at 11 and 300 K and the curves show the corresponding calculated yield curves. The lifetime determined from

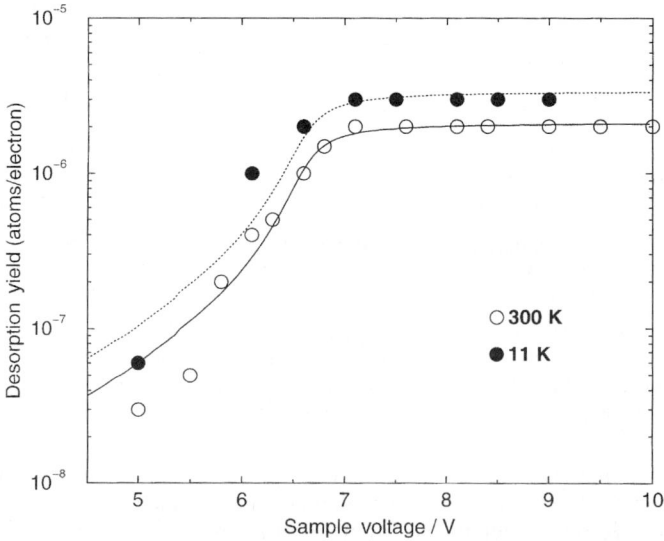

Fig. 1 STM-induced desorption of H from Si(100) in the single electronic transition regime. The symbols show the data of Foley *et al.*[1] at 11 and 300 K and the curves are the results of a least squares fit.

the least squares fit, about 1.2 fs, is close to the value of 0.75 fs estimated in your Discussion paper based on the isotope effect of ref. 1. (Probably both your result and ours should be reported as $\tau \approx 1$ fs, in agreement with the value of 1–3 fs estimated in ref. 1 based on the isotope effect.) This agreement may not suffice to suggest validity of the potential energy curves and the assumptions introduced in your dynamical modelling. It is worth noting, however, that while estimation of the lifetime based on the isotope ratio is subject to accuracy of the potential energy surfaces and the method of computing the nuclear dynamics, its determination from fit of the data to the theoretical bias dependence is essentially independent of the potential surfaces and the nuclear dynamics but depends on the substrate and tip densities of electronic states. In addition it relies on different experimental data (bias dependence of the yield, rather than comparison between two systems).

1 T.-C. Shen, C. Wang, G. C. Abeln, J. R. Tucker, J. W. Lyding, Ph. Avouris and R. E. Walkup, *Science*, 1995, **268**, 1590; Ph. Avouris, R. E. Walkup, A. R. Rossi, H. C. Akpati, P. Nordlander, T.-C. Shen, G. C. Abeln and J. W. Lyding, *Surf. Sci.*, 1996, **363**, 368; Ph. Avouris, R. E. Walkup, A. R. Rossi, T.-C. Shen, G. C. Abeln, J. R. Tucker and J. W. Lyding, *Chem. Phys. Lett.*, 1996, **257**, 148; T.-C. Shen and Ph. Avouris, *Surf. Sci.*, 1997, **390**, 35; E. T. Foley, A. F. Kam, J. W. Lyding and Ph. Avouris, *Phys. Rev. Lett.*, 1998, **80**, 1336.
2 S. Alavi, R. Rousseau, G. P. Lopinski, R. A. Wolkow and T. Seideman, *Faraday Discuss.*, 2000, **117**, 213.
3 D. P. Adams, T. M. Mayer and B. S. Swartzentruber, *J. Vac. Sci. Technol.*, 1996, **14**, 1642; M. Schwartzkopff, P. Radojkovic, M. Enachescu, E. Harmann and F. Koch, *J. Vac. Technol. B*, 1996, **14**, 1336.

Dr Saalfrank responded: I am pleased to learn that Prof. Seideman's estimated lifetime for the $\sigma \rightarrow \sigma^*$ resonance (MGR-A in our notation), is close to what we get based on the isotope effect. However, it should also be said that the isotope effect $I_{des} = P_{des}(H)/P_{des}(D)$ depends quite dramatically on the lifetime τ, once $\tau < 1$ fs.[1] Hence, it seems only fair to say that $\tau < 1$ fs and it is, to my opinion, neither possible nor useful to extract τ to a too high accuracy. This is even more so since Avouris in his STM-experiments, and Zhu in his photodesorption experiments, find somewhat different I_{des} (see our paper), although both should be in the DIET limit. Also the value of τ based on I_{des} depends on whether coordinate-dependence of the lifetime is accounted for or not.

1 See refs. 10, 11 and 33, and Fig. 5b in our paper.

Prof. Stokbro commented: For the hole resonance you model the desorption process as a DIET process. Experiments clearly show a power law dependence between the current and the desorption rate indicating a DIMET process. Our toy model where parameters are obtained from first principles,[1] gives excellent agreement with the experimental data. I find that the main difference between this study and your study is a difference in the excited state potentials, your potential has a slope which is 4 times larger that the potential we used. Our excited state potential was constructed on the basis of the changes in the PDOS of the 5σ resonance as a function of the Si–H distance. How was your potential constructed?

1 K. Stokbro, C. Thirstrup, M. Sakurai, U. Quaade, Ben Yu-Kuang Hu, F. Perez-Murano and F. Grey, *Phys. Rev. Lett.*, 1998, **80**, 2618.

Dr Saalfrank answered: It is true that we actually model a (so far hypothetical) DIET process *via* a hole resonance state, not a DIMET one as you did. Our construction of the excited state is indeed very approximate and uses an idea of Jennison *et al.*[1] who exploited the analogy of the Si–H/Si–H$^+$ and the H_2/H_2^+ systems to construct a hole resonance state of "H desorption in the bulk". Details are given in our paper. I agree that our potential is only a model potential.

1 D. R. Jennison, J. P. Sullivan, P. A. Schultz, M. P. Sears and E. B. Stechel, *Surf. Sci.*, 1997, **390**, 112.

Prof. Gadzuk asked: (1) What descriptive name do you use for the alternatives to "toy models"? What relevant physics do these alternatives contain that are omitted in the "toy models"?

(2) I would also like to hear your thoughts on the issue of temporal control in surface femtochemistry/DIET/DIMET by optimal pulse shaping procedures. In free standing gas phase intra-molecular dynamics, the electronic ground and excited states are directly coupled by the time-dependent dipole coupling, $V_{eg} = -\mu_{eg}(Z)F(t)$, [eqn. (5) of your paper] where there is a one-to-one correspondence between the time dependence of the incident laser pulse and the time

dependence of the driving force acting on the molecular degree of freedom being excited. This now-standard direct coupling has permitted a certain type of analysis involving optimal control models which have formed the basis for a continuing run of symposia and publications on coherent control in chemical dynamics (see ref. 1 for a nice introduction). In much of the actual work in surface femto/photochemistry, the excitation at the reaction scene is indirect, that is the time-dependent laser pulse excites substrate electrons and it is these hot electrons which in turn provide the equivalent of V_{eg}. With this in mind, what are your views on the relevance and immediate applicability of the time dependence of the incident laser pulse on the time dependence of the actual excitation acting upon the surface molecular system under study?

1   R. J. Gordon, L. Zhu and T. Seideman, *Acc. Chem. Res.*, 1999, **32**, 1007.

Dr Saalfrank answered: (1) I use the term "toy models" for one-dimensional two-state models. I don't have any descriptive name for more complicated alternatives. The "toy models" were and are extremely useful in elucidating the basic physical mechanisms leading to DIET and DIMET. It is also true, however, that for several specific systems multi-dimensionality is required (*e.g.* for NH_3 at various substrates,[1] or that more than one excited state is of importance, perhaps for STM-induced desorption of H from Si(100)).

(2) I fully agree that UV/Vis laser excitation of adsorbates at metal surfaces (and possibly many others as well) is dominantly substrate-mediated.[2] However, for H/Si(100) the CW-laser experiments by Zhu *et al.*[3] show, through measurements using polarized light, that a direct coupling mechanism highly analogous to the gas phase examples you mentioned should prevail in this case. Hence the response of the adsorbate to the laser field is "easy" and direct in this case, and that is why we think that this system is a good candidate to demonstrate "controllability" of photochemistry at surfaces.

1   X.-Y. Zhu and J. M. White, *Phys. Rev. Lett.*, 1992, **68**, 3359; see also ref. 14 in our paper.
2   F. Weik, A. De Meijere and E. Hasselbrink, *J. Chem. Phys.*, 1993, **99**, 91.
3   T. Vondrak and X.-Y. Zhu, *Phys. Rev. Lett.*, 1999, **82**, 1967.

Prof. Polanyi commented: I should like to draw a parallel (not, perhaps, immediately evident) between the successive papers by Saalfrank *et al.*[1] and Giorgi *et al.*[2] In the Saalfrank *et al.* paper a prominent example of what is termed DIET ('desorption induced by electronic transitions') is the non-adiabatic cleavage of an Si–H bond at a surface when an electron hops from a nearby metal atom, M, of an STM probe to that surface. The electron-hop is induced electro-magnetically. The separation between M and the silicon hydride is in the region of 6 Å (the tip-to-surface distance). Turning to the Giorgi *et al.* paper one finds a metal atom (Na) stimulated by an electromagnetic disturbance (light) hopping to a nearby Cl–H with resultant cleavage of the Cl–H bond.

The difference between these two events lies not so much in the phenomena, I believe, as in the language used to describe them. Saalfrank *et al.*[1] describe a physical process of bond cleavage due to dissociative attachment of an electron (or hole) at the Si–H. The metal atom is not specified in the discussion of their STM-induced 'DIET,' since it is regarded merely as a source of electrons (or holes) infinitely distant from the site of dissociation. This is in marked contrast to the vocabulary of Giorgi *et al.*[2] who regard the non-adiabatic 'harpooning' of their Cl–H as being profoundly affected by the nearby presence of an electronically-excited metal atom or cluster, $(Na)_n$. Theirs, they believe, is a chemical process.

An equivalent statement is that in the literature of DIET the dissociation denoted by the D is into a vacuum, whereas in the discussion of 'harpooning,' in the Giorgi *et al.* paper, dissociation is to form a new bond at the metal. The former would indeed be pure dissociation, whereas the latter is chemical reaction. The notable thing about chemical reaction is that the energy-barrier separating the reagent (a hydride here) from the product can be markedly reduced by the formation of the new chemical bond concurrently with the dissociation of the old one. The charge that binds the original molecule is (so to speak) siphoned into the low-energy space separating the atoms of the new molecule.

This discussion is preparatory to a question. Given the fact that in STM-induced DIET one of the atoms of the molecule undergoing dissociation is within a chemical-bonding distance of the STM tip, ought one not keep open the possibility that the dissociation is assisted by bond formation at the metal (or otherwise reactive) tip? In this case the STM-induced process is better described as a chemical reaction with the tip—specifically an exchange reaction in which an atom (or radical) is transferred to the tip.

We shall know whether this is a better description when we have data that allow us to compare the threshold energy for dissociation induced by charge transfer from (or to) the tip, using tips having different chemical properties. Given the small tip-to-surface separation I would put my money on the chemical rather than the physical interpretation.

1 P. Saalfrank, G. Boendgen, C. Corriol and T. Nakajima, *Faraday Discuss.*, 2000, **117**, 65.
2 J. B. Giorgi, T. G. Lee, F. Y. Naumkin, J. C. Polanyi, S. A. Raspopov and J. Wang, *Faraday Discuss.*, 2000, **117**, 85.

Dr Saalfrank replied: It is in fact true that there is a close link between STM- and light-induced bond cleavage at surfaces. Based on previous work by others, we in fact use exactly corresponding models (at least in the 'DIET' single-excitation limit) for both laser- and STM-induced desorption.

It is also true that the 'desorbing' atom/molecule in STM-induced processes will often end up at the STM tip, forming a new bond there. (This process is sometimes reversible and leads to the construction of so-called 'atomic switches'.[1]) However, I don't believe that for the STM-processes we modelled the H–Si desorption rate, for example, will depend very much on the chemical forces exerted by the nearby STM tip. Actually, we have to distinguish between two ways in which a tip may influence the desorption rate: (1) by creating a double-minimum potential [with potential minima for reactant (Si–H···tip) and product configurations (Si···H–tip)] in which the activation barrier and to a lesser extent the whole reactant well are modified by the nearby tip. This is what Prof. Polanyi suggests. (2) Another possible difference of the double-minimum (rather than the Morse potential) description comes from the fact that in DIMET, for example, which proceeds by ladder climbing in the ground state, tunnelling through the barrier may become important thus increasing the reaction probability by a dynamical effect.

As for this second point, it has been argued on the basis of quantum mechanical calculations that tunnelling through the barrier usually plays no significant role.[2] Concerning the first point which has to do with the reduction of the barrier by the nearby tip I would argue as follows. For a strong chemical bond such as the one between Si and H or the one between H and W, the forces are 'chemical' and therefore short-ranged, and a tip 6 Å away from the Si–H should lead to a double-well potential with two isolated, well-separated wells for the reactants and products, respectively. The situation may be different when the adsorbate is bound by ionic or van der Waals type forces which are more long-ranged. I agree, however, that details of the tip may influence the desorption process in certain cases—this was in fact so far neglected in our work.

1 D. M. Eigler, C. P. Lutz and W. E. Rudge, *Nature*, 1991, **352**, 600.
2 R. E. Walkup, D. M. Newns and Ph. Avouris, *J. Electron. Spectrosc. Relat. Phenom.*, 1993, **64/65**, 523.

Prof. Wolf added: Transfer of atoms and molecules to the tip has been investigated for a number of systems. Here I show an example for breaking of the CO/Cu(III) bond by tunnelling electrons accompanied by a transfer of the CO molecules to the STM tip (work by Rieder and co-workers[1,2]). However, for the double-minimum potential for this process between the surface and the tip one would expect very little perturbation in the barrier region compared to the situation of CO absorbed on Cu (without the STM tip).

1 L. Bartels, G. Meyer and K.-H. Rieder, *Appl. Phys. Lett.*, 1997, **71**, 213.
2 L. Bartels, M. Wolf, G. Meyer and K.-H. Rieder, *Chem. Phys. Lett.*, 1998, **291**, 573.

Prof. Kleyn opened the discussion of Prof. Polanyi's paper: Avouris *et al.*[1] found that rearrangements of the solid following an electronic excitation can be extremely rapid. Did you take this possible rearrangement of the LiF substrate into account in the calculations? Could the vibrational excitation observed be related to excitation of a LiCl or perhaps a LiF bond?

1 R. E. Walkup and Ph. Avouris, *Phys. Rev. Lett.*, 1986, **56**, 524.

Prof. Polanyi replied: In response to Prof. Kleyn's first question, we could detect no change in the observed reaction with time, suggestive of a gradual change in the LiF substrate. We don't think, therefore, that photoinduced reaction of Na* with the underlying LiF crystal is important. We know, however, from our gas-phase studies of the harpooning of hydrogen halides in Na···XH complexes, that the reaction has a large cross-section when a complex is excited by visible light to its lowest electronically-excited state. Much the most likely cause of the observed pulse of atomic H(g) in the present experiments would seem to be the electronic excitation of (Na···XH)(ad) (and similar complexes) at the surface, leading to harpooning with the release of H(g); i.e., (Na···XH)(ad) → (Na*···XH)(ad) → (Na$^+$···H$^-$)(ad) → Na$^+$X$^-$(ad) + H(g).

The vibrational structure observed in the ejected H(g) is significantly wider than that associated with alkali halides, so it seems improbable that we are leaving behind a LiF or LiCl product due to the photoinduced reaction of the substrate with the adsorbed halide. It should perhaps be noted that both Na and HCl must be present at the surface in order that the light pulse shall yield H(g). The yield of H(g) no longer increases if the number of HCl(ad) exceeds the Na(ad). The reaction of Na* with HCl, in the adsorbed state, appears to be implicated.

If the proposal is that the departing H(g) deposits a variable number of quanta of vibration as phonon excitation in LiF, this seems improbable. From the Debye temperature of LiF, $\theta_D = 730$ K, the phonon spacing in the bulk is 0.06 eV. It will be even less (0.045 eV) near the surface. By contrast we observed a vibrational spacing of 0.25 eV. Our best guess as to the meaning of this is that we are seeing an H-atom oscillating to-and-fro in a hydrogen-bonded residue, NaCl···HCl, formed by harpooning in the complex Na···(HCl)$_2$. This is substantiated by *ab initio* calculation.

Dr Hodgson said: Although the overall energy distribution of H does not change as the detection angle is varied from 0 to 35°, the position of the "structure" does shift. If this is associated with the vibrational spacing of an intermediate it should not shift in energy with θ, why do these features shift in view of the overall similarity of the H energy distribution?

Prof. Polanyi answered: We asked ourselves the same question and came up with a speculative answer, namely that the observed two peaks in the H(g) angular distribution at low coverage are due to the presence of two different geometries of complex at the surface. We went so far as to hazard a guess as to what these complexes might be. The reader is referred to eqn. (3) of the paper in which Na···(HCl)$_2$(ad) is drawn with an L shape for (HCl)$_2$ yielding a freely-escaping H(g) that comes off the crystal along the normal. By contrast in eqn. (4) of the paper for (Na)$_2$···(HCl)$_2$(ad), with Na···Na in the crystal plane, the end-H-atom is the one that can escape freely, and the direction of its escape would be expected to be away from the normal.

Dr Darling communicated: You attribute the oscillations in the H atom energy distributions to the excitation of vibrational modes predominantly involving H–Cl motion. Have you tried isotopically substituting D for H and would you expect this to give useful additional insight?

Prof. Polanyi communicated in response: We really should try substituting D for H in the HCl (as well as in the HF analogue we are currently studying). Apart from the cost, we are deterred by the unfortunate fact that the lower translational energy (E'_T) of the departing D(g) will make the smaller vibrational spacing in the deuterated residue especially difficult to detect (lower E'_T is associated with larger error bars).

There is, however, a compelling argument for doing this difficult experiment, namely that (for gaseous Na···FH) Kosloff *et al.*[1] have predicted a large increase in the rate of the reactive as opposed to the non-reactive pathway for the deuterated halide, in a quantum scattering calculation. This interesting effect merits investigation.

1 Y. Zeiri, G. Katz, R. Kosloff, M. Topaler, D. Truhlar and J. Polanyi, *Chem. Phys. Lett.*, 1999, **300**, 523; Y. Zeiri and R. Kosloff, private communication.

Prof. Kosloff asked: (1) What is the relative branching ratio between the two reaction channels and how can it be influenced?

(2) The structure in the observed time of flight for the ejected hydrogen atom can be also interpreted as a diffraction effect. This can happen if the ejected hydrogen can be temporally trapped between the Cl atom and the surface resulting in the hydrogen wavefunction interfering with itself.[1,2]

1   R. Kosloff and Y. Zeiri, *J. Chem. Phys.*, 1992, **97**, 1719.
2   V. J. Barclay, J. C. Polanyi, Y. Zeiri and R. Kosloff, *J. Chem. Phys.*, 1993, **98**, 9185.

Prof. Polanyi answered: I wish we knew the branching ratio between the reactive product of photoinduced harpooning, namely H(g), and the non-reactive, HCl(g). The reason we don't know it is that the two products had to be detected by different techniques on different machines. The relative sensitivity of the Rydberg-atom TOF used for H(g) and the REMPI used for HCl(g) is not known (nor were the geometries in the two machines identical). All we can say at present is that the absolute yield (certainly), and the relative yield (perhaps), of H(g) increased with the use of increasingly energetic radiation to excite the alkali-metal plus halide complexes.

As regards Prof. Kosloff's second point, it would be delightful if we were seeing the effect of H-atom matter-waves bouncing off and under the neighbouring adsorbate molecules. I am sceptical. The approximate 0.25 eV spacing, with clear evidence of anharmonicity (see the vertical lines in Fig. 4 of the paper), is suggestive of vibrational excitation in a residue left behind at the surface.

Dr Shluger commented: It is known that interaction of Li vapour with LiF would lead to so-called "additive colouration". After some time the crystal becomes coloured due to creation of F-centres at the surface and their diffusion inside the sample. I guess that a similar effect should take place when you deposit Na on LiF and suggest that this should be considered in your model. Another point concerns surface steps, which are always present on cleaved surfaces. In other experiments, it has been observed that metal tends to cluster first at steps, which can be the case at low coverages in your experiments too. I think that this also should perhaps be taken into account in your model.

Prof. Polanyi responded: In our temperature range for Na adsorption on LiF, namely 40–50 K, we would not expect to see F-centres being created by reaction between the Na adlayer and the substrate. As a matter of fact our LiF crystals as delivered to us are yellow due to the manufacturer introducing F-centres to harden the crystals. We can anneal out these F-centres at 700 K (1 h), whereupon the crystals are clear. There is no evidence of yellowing due to the multiple laser shots on the Na-coated surface, nor do we see our results change due to the gradual alteration of our substrate. The notion that the Na remains predominantly on the surface rather than diffusing into it gains credence from Prof. Träger's observations (see ref. 44 of our paper) that with increasing temperature in the range approximately 90–110 K the Na(ad) diffuses across the surface to form larger clusters.

Dr Shluger's point about the effect of defects, which in our case will I imagine be predominantly step edges, is a good one. Our low coverage (which is, of course, the most susceptible to defects) is 0.05 ML. One would not expect an annealed LiF crystal (heated to 700 K for 1 h, with noticeable effect) to retain quite this high a concentration of defects. Our results are reproducible, so the variable balance between reaction at the terraces and step-edges does not appear to be affecting the outcome.

Dr Saalfrank opened the discussion of Prof. Groß's paper: In the fewest switches surface hopping algorithm you rescale the particle velocities after a surface hopping has occurred, to account for total energy conservation for a single trajectory. My question is why that has to be done, and why not simple Franck–Condon transitions are used instead. Shouldn't energy conservation be demanded for the whole ensemble of trajectories only, and couldn't this in fact be used as a criterion for the convergence of the calculation with respect to the number of trajectories?

To give an analogy: an open-system Liouville-von Neumann equation can be solved by stochastic wavepacket methods, *e.g.* the Monte Carlo wavepacket method.[1] The correct loss of energy in this case will not be resembled by a single quantum trajectory, but has to be recovered as the ensemble average only.

1   K. Mølmer, Y. Castin and J. Dalibard, *J. Opt. Soc. Am. B*, 1993, **10**, 524.

Prof. Groß replied: The fewest switches algorithm has been especially designed for closed systems in which the energy is conserved. Hence the energy adjustment and a jump is an integral part of the algorithm and actually ensures the time-reversibility of this method.

In the case of Franck–Condon transitions it is generally assumed that the whole energy difference between the two potential energy surfaces that are involved in the jump is taken up by the electronic system. In such a situation the electrons act as a bath that leads to dissipation. Thus the theoretical description involving Franck–Condon transitions corresponds to the treatment of an open system where time-reversibility is no longer valid. Now it is true that many electronic transitions in reactions at surfaces are of the Franck–Condon type. The fewest switches algorithm has to be adjusted to be able to describe such reactions. This is not a fundamental problem and can be done for example along the lines already proposed by Sholl and Tully.[1] The easiest approach would be just to omit the energy adjustment, as suggested by Dr Saalfrank, and only allow for one jump. This would correspond to the usual treatment of, *e.g.*, DIET processes (desorption induced by electronic transitions). However, the position of the jump would not be determined by assumed lifetimes but by the coupling between the electronic states.

1   D. S. Sholl and J. C. Tully, *J. Chem. Phys.*, 1998, **109**, 7702.

Prof. Gadzuk said: If you include things like surface corrugation in your model, presumably this would lead to a spatial distribution of curve crossings points (in the surface normal direction) as the impact parameter of the incident I_2 beam ranges over the unit cell. Thus a distribution of time intervals or trajectory times between in and out crossings would result. Do you think that this averaging would be sufficient to wash out the Stückelberg oscillations that you have discussed here?

Prof. Groß answered: Definitely this averaging due to, *e.g.*, corrugation will wash out the Stückelberg oscillations. As you point out, a distribution of curve crossing points causes a distribution in the path lengths between in and out crossings that suppresses the quantum phase coherence. Indeed this is exactly what happens when we include molecular rotations in our calculations.

Prof. Stokbro asked: What are the system sizes you can treat with full quantum mechanical calculations, and what are the system sizes you can treat within the semi-classical approach?

Prof. Groß replied: If no electronic transitions are taken into account and hydrogen dissociative adsorption is considered, then full quantum mechanical calculations can be performed in which all six molecular degrees of freedom are explicitly treated dynamically.[1,2] However, already for heavier molecules such as NO typically only up to three-dimensional simulations in restricted geometry are possible.[3] The computational bottleneck is not the CPU time itself but the high memory requirement of the quantum methods. If open systems with electronic transitions are treated quantum mechanically, then at most two molecular degrees of freedom can be taken into account at the moment.[4]

In the semi-classical approach there is in principle no fixed limit for the system size that can be treated. The fewest switches algorithm, for example, requires the solution of a system of coupled differential equations which needs usually very small computer memory. Of course the computational effort still increases with system size, and it might not be feasible to run a sufficiently large number of semi-classical trajectories in order to satisfy statistical requirements for the evaluation of reaction probabilities, in particular since the numerical integration of the time-dependent electronic Schrödinger equation makes a rather small time step necessary. Still semi-classical calculations with several tens of atoms might be possible.

1   A. Groß, S. Wilke and M. Scheffler, *Phys. Rev. Lett.*, 1995, **75**, 2718.
2   G. J. Kroes, E. J. Baerends, R. C. Mowrey, *Phys. Rev. Lett.*, 1997, **78**, 3583.
3   D. Lemoine and T. Duhoo, *Chem. Phys.*, 1998, **238**, 59.
4   H. Guo, P. Saalfrank and T. Seideman, *Prog. Surf. Sci.*, 1999, **62**, 239.

Dr Dujardin commented: You mention in your paper that "in the experiment the diamond surface has been hydrogen-covered". However, in the following discussion you consider only the electronic properties of clean diamond. This may be a problem since hydrogen on the diamond surface is known to dramatically change its electronic properties. For example the positive electron affinity is turned into negative electron affinity after hydrogenation. New surface states appear in the bulk band gap and also the diamond, which is an insulator, becomes conductive upon hydrogenation.

Prof. Groß responded: It is certainly true that covering the diamond surface with hydrogen will influence the electronic properties of the surface. However, it is not true that our construction of the potential energy surface is only based on the electronic properties of clean diamond. It is rather adjusted to reproduce the basic features of the scattering experiment.[1] For example, from the difference between the surface work function of clean diamond and the electron affinity of molecular iodine a threshold of 3.5 eV in the ionization probability is expected, however, in the experiment it is found to be close to 3.0 eV.[1] We have therefore taken this experimental value of 3.0 eV. Furthermore, our choice of a shallow physisorption well of 0.1 eV for the adsorption of neutral iodine molecules was also motivated by the assumption of a hydrogen passivated diamond surface.

Unfortunately, there is little information on the microscopic details of the interaction in the iodine/diamond system. Hence our particular choice of the potential energy surface should be regarded as an educated guess that is based on the results of the scattering experiment and not on an analysis of the electronic structure of the considered system.

1   A. Danon and A. Amirav, *Phys. Rev. Lett.*, 1988, **61**, 2961.

Prof. Kosloff commented: Fig. 2 shows a modification of the potential used by Prof. Groß for the I_2 diamond system. The general topology is very similar to the original one. In the table in parentheses are the original values of the potential where they have been modified. What can be observed in the insert is an inner crossing point at 11 eV. The smaller value of the non-adiabatic coupling constant emphasizes the dynamics close to the inner turning point.

Fig. 3 shows the result of a two channel wavepacket calculation on the above potential. Snapshots of the wavefunction are shown for every 25 fs. After 75 fs the wavepacket hits the inner repulsive wall where most of the non-adiabatic transfer takes place (the amplitudes of wavepacket

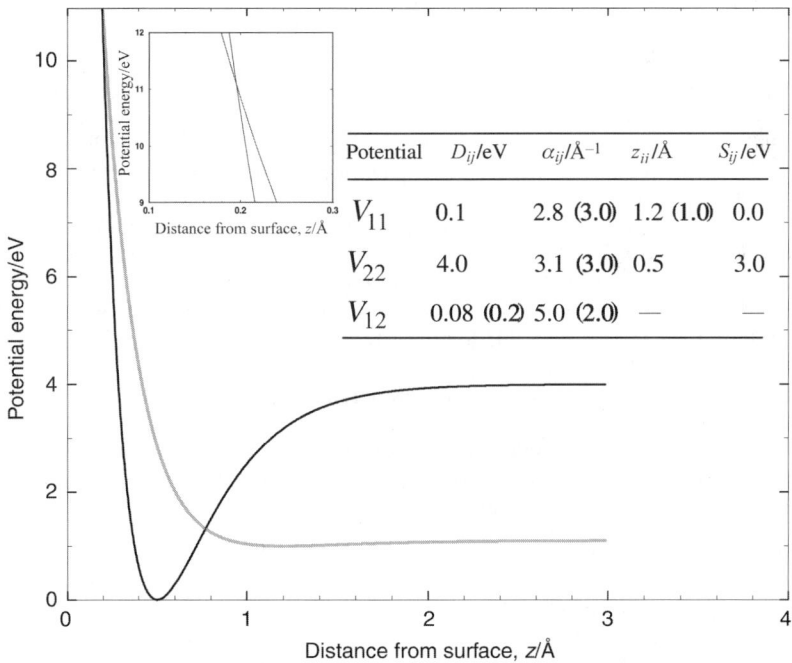

Fig. 2 Modification of the potential for the I_2 diamond system.

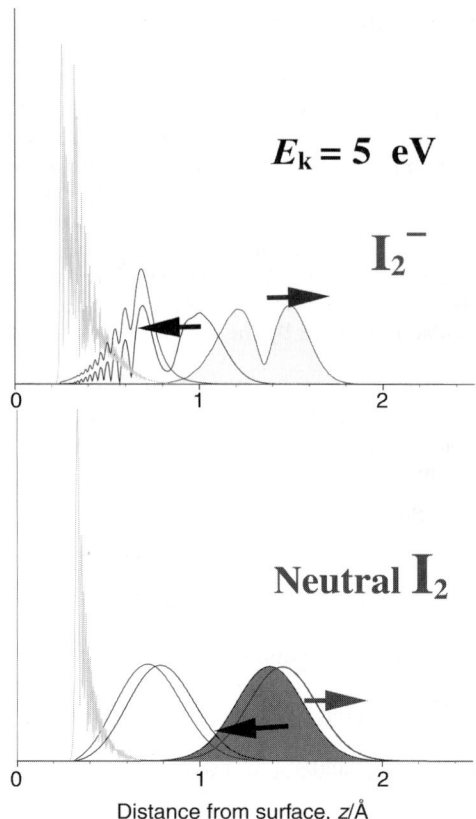

Fig. 3 Two channel wavepacket calculations on the potential for the I_2 diamond system.

on the neutral and ionic states are not in the same scale). Notice that for the collision energy presented (5 eV) the turning point on the upper ionic potential is located at a smaller distance from the surface. The calculations were performed using a Fourier representation of 1024 points with grid spacing of delta z of 0.006 a_0. The propagation in time was performed using the Chebychev scheme with a time step of 1000 $\hbar/E_h$ and a polynomial order of $N = 1007$.

Fig. 4 shows the negative ion formation as a function of incident kinetic energy. The general trend is a large increase in the ion formation as a function of kinetic energy. The reason is that the dynamics are due to tunnelling from the inner turning point of the neutral potential to the inner turning point of the ionic potential. For energies above the crossing point this trend will saturate. If the outer crossing point is responsible for the dynamics we expect the kinetic energy to hinder the ion formation probability.

As stated correctly by Prof. Groß quantum interference oscillations will be wiped out in multi-dimensional dynamics.

Another issue is the ability of semiclassical method to describe even qualitatively multidimensional surface crossing dynamics. We found in a recent paper[1] that semiclassical methods have qualitative discrepancies in non-adiabatic multi-dimensional problems.

1 Y. Zeiri, G. Katz, R. Kosloff, M. S. Topaler, D. G. Truhlar and J. C. Polanyi, *Chem. Phys. Lett.*, 1999, **300**, 523.

Prof. Gadzuk addressed Prof. Kosloff: In the I_2(neutral) → I_2^-(negative ion) scattering, from what states do the harpooning electrons originate, valence band states or within-gap occupied defect and/or surface states or something else?

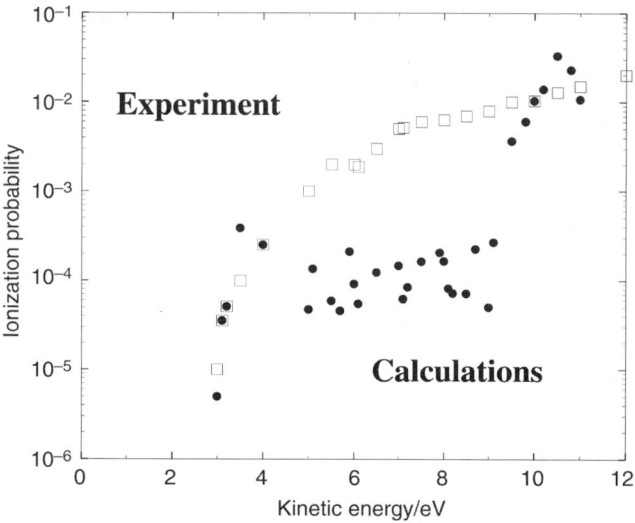

Fig. 4 Negative ion formation as a function of incident kinetic energy.

Prof. Kosloff responded: The parameters show that the inner turning point is the most important in the dynamics. The experiment was performed with hyperthermal energies of up to 10 eV of incident kinetic energy of the impinging I_2. This energy is above the band gap of diamond of ~ 4.5 eV. It is therefore reasonable that the charge comes from the valence band and is not a minority effect. The compression of the diamond surface due to the heavy I_2 is the source of the non-adiabatic charge transfer.

Prof. Groß added: In fact the existence of an inner crossing point between the two diabatic potential energy surfaces in the region of the exponentially increasing repulsive wall of the potential will have a significant influence on the ionization probability. However, the exact location of such a crossing point will strongly depend on the impact point and orientation of the incoming molecule. Hence it really requires high-dimensional simulations in order to determine the true consequences of the existence of inner crossing points.

Alternatively, the exponential increase in the ionization probability can also be obtained in one-dimensional two-state calculations by increasing the coupling between the two diabatic potentials by a factor of about ten with respect to our choice. This, however, appears unrealistically large to us. We believe that we have a rather robust explanation for the observed increase in the ionization probability, namely the energy loss to other degrees of freedom which suppresses the ionization probability dramatically, in particular at energies close to the ionization threshold. This explanation is actually supported by the measured broad distribution in kinetic energy of the scattered molecules.[1]

Still it is fair to say that no particular electronic coupling scenario can be ruled out as long as there are no reliable theoretical or experimental studies to determine the relevant potential energy surfaces.

1 A. Danon and A. Amirav, *Phys. Rev. Lett.*, 1988, **61**, 2961.

Prof. Kleyn asked: How do you know that the curve crossing between "diamond"–I_2 and "diamond$^+$"–I_2^- is in the attractive part of the potential. The difference between work function (~ 6 eV) and vertical electron affinity (~ 1 eV) does not necessarily imply this.

Prof. Groß replied: In fact we do not "know" that the curve crossing between the two different charge states is in the attractive part of the potential. This is just an assumption. However, the ionization threshold is not at 5 eV, as you imply in your question, but at approximately 3 eV in the experiment. This makes it reasonable that the curve crossing is in the attractive part.

However, we have also checked what happens if we moved the curve crossing point closer to the surface, *i.e.* we shifted the whole potential of the negative ion state towards the surface and determined the Landau–Zener ionization probability. In fact the ionization probability only changed by less than 30% for all possible curve crossing points that are closer to the surface if we left all other potential parameters unaltered. Hence it seems that within our parametrization the exact location of the curve crossing point is not too critical.

Dr Kroes added: I think that if there is a crossing in the inner turning point region, there should be a crossing in the outer turning point region as well. If the I_2^- + metal potential is above the I_2 + metal potential at large molecule–surface distance, and the I_2^- + metal potential is also the highest of the two close to the surface, then it can only be the lowest of the two in the intermediate region if there is a crossing with the I_2 + metal potential in the inner turning point region as well as in the outer turning point region.

Prof. Gadzuk commented: Some fifteen years ago Prof. Holloway and I looked at the related I_2/insulator dissociative scattering problem in which dissociation resulting from the excitation due to temporary negative ion existence between the inward and outward Laudau–Zener transitions at the curve crossings was considered[1] in the spirit of your model. We were able to account for the energy dependence of the dissociation probability for I_2 incident upon MgO reported by Kolodney and Amirav[2] using the charge transfer model as an alternative to a dissociative-scattering-due-to-rotational-excitation model advanced by Gerber and Elber.[3] This old success seems quite related to what you have just presented and could add some additional support to your current proposal.

1 J. W. Gadzuk and S. Holloway, *J. Chem. Phys.*, 1986, **84**, 3502.
2 E. Kolodney and A. Amirav, *J. Chem. Phys.*, 1983, **79**, 4648.
3 R. B. Gerber and R. Elber, *Chem. Phys. Lett.*, 1984, **107**, 141.

Prof. Groß answered: Indeed we have in the meantime also taken into account the molecular vibration and dissociation channel in our semiclassical simulations.[1] For the sake of simplicity we have assumed the same interatomic potential for the neutral and the ionic molecular state. Hence no vibrational excitation in the spirit of your model due to a temporary changed interatomic potential could occur in our calculations. Still we were able to reproduce the measured large dissociation probability in the scattering of molecular iodine from diamond which rises up to 60% at an incident kinetic energy of 10 eV.[2] Our analysis of the dissociation events shows that the dissociation of iodine indeed proceeds according to the centrifugal model of Gerber and Elber,[3] *i.e.*, the dissociation is a consequence of the strong rotational excitation during the scattering.

Actually for our particular choice of the coupling between the neutral and the ionic state the large dissociation probability can not be caused by the formation of a temporary negative ion. Our scattering scenario corresponds to the diabatic limit, *i.e.*, most scattered molecules stay on the diabatic potential energy surface during the scattering event.

But again we have to emphasize that we can not rule out the transient negative ion mechanism as a possible explanation for vibrational excitation and molecular dissociation since our potential just represents an educated guess. Still it is our feeling that this mechanism is not too probable for the iodine/diamond system. In order to account for the high dissociation probability the majority of scattered molecules have to become temporarily ionized. We consider this rather unlikely to happen at a diamond surface because of its large band gap. And it is also not necessary to invoke this mechanism since the centrifugal model alone is sufficient to explain the high dissociation probability.

1 C. Bach and A. Groß, *J. Chem. Phys.*, submitted.
2 E. Kolodney, private communication.
3 R. B. Gerber and R. Elber, *Chem. Phys. Lett.*, 1984, **107**, 141.

Prof. Hasselbrink commented: It is certainly impressive how the "toy models" are able to semi-quantitatively reproduce experimental findings. That certainly helps to rationalize these observations. In that sense they are, as Prof. Gadzuk has called to strive for, intuition boosters. However, there is always the problem of ambiguity, *i.e.* the questions remains whether there is more than one toy model explaining the same result. Thus, to enable progress it is necessary that the models allow predictions which differentiate them and which then can be tested experimentally. Can you please comment on this in view of your results?

Prof. Groß responded: I fully agree with your statement. As long as different toy models do reproduce the same results, there is of course no way to tell which model is closer to reality. If there are two different models, it is particularly desirable to find aspects of a system where the two models yield different results. Hence it is important that the toy models can be extended to yield results on additional features of the studied systems which either have already been determined or which can be tested.

For the particular system of our study, apart from the ionization probability also the dissociation probability, the angular distribution and the kinetic energy distribution of scattered ionized molecular iodine have been measured. Consequently, high-dimensional simulations that include the molecular rotation and vibration and the surface corrugation should be performed within the same framework because it lends more credibility to a model if it is able to reproduce several aspects of an experiment. In the meantime we have also included molecular vibration in our model. This extension does not only confirm our energy dissipation argument to explain the dependence of the ionization probability on incident kinetic energy, but it also yields results for the dissociation probability in good agreement with the experiment.

Dr Saalfrank said: I have my doubts as to whether the "three-step-approach" (1. calculation of excited state potential energy surfaces, 2. calculation of all couplings between them, 3. dynamics on coupled surfaces), proposed by Prof. Groß is feasible, in particular for metal surfaces. Not only may the number of excited states become prohibitively large, but also many potential energy surfaces of areas of configuration space will be unimportant for the non-adiabatic process of interest. It would appear as a waste of effort to calculate all this unimportant information, apart from being largely impractical in most cases.

Prof. Groß replied: You raised a very important point. I can understand your doubts, and in fact I do share them, at least partially. At metallic surfaces there is a continuum of electronically excited states, and it would indeed be a waste of effort to calculate many, most probably rather similar potential energy surfaces. As far as the electron–hole pair excitation at metallic surfaces is concerned, treating these excitations as a friction force will be more appropriate, as proposed by Head-Gordon and Tully.[1] In this scheme the friction coefficient can be calculated by electronic structure calculations which was demonstrated for the interaction of CO with Cu(100).[2]

In the treatment of DIET processes at surfaces, molecular excited states or negative ion resonances play a crucial role. There may still be a large number of them. However, they might have similar shapes so that it makes sense to determine a representative excited state potential energy surface. Such an approach was followed for the treatment of laser-induced desorption of NO from NiO(100).[3] This is still better than to rely entirely on guess work. The determination of the excited state surfaces does furthermore not necessarily have to be on an *ab initio* level. Semi-empirical schemes like for example a Newns–Anderson approach might also be rather helpful.

As far as the treatment of reactions with electronic transitions at surfaces is concerned, theoretical studies have been so far on a rather qualitative level. In my opinion there is a strong need for a more quantitative approach. A full quantitative determination of all relevant excites states and couplings might indeed hardly be possible. Still it is desirable that more input from electronic structure calculations will be included in dynamical simulations. This certainly requires a clever choice of the particular excited states that are computed. But I do not see any alternative for progress in this theoretical field.

1 M. Head-Gordon and J. C. Tully, *J. Chem. Phys.*, 1995, **103**, 10137.
2 J. T. Kindt, J. C. Tully, M. Head-Gordon and M. A. Gomez, *J. Chem. Phys.*, 1998, **109**, 3629.
3 T. Kluener, H.-J. Freund, V. Staemmler and R. Kosloff, *Phys. Rev. Lett.*, 1998, **80**, 5208.

Prof. Holloway addressed Prof. Clary: In the long history of applying semiclassical methods to problems in the gas phase, are we at a point to review the success of such methods?

Prof. Clary replied: Faraday Discussions 110 on Chemical Reaction Theory had several papers that describe applying semiclassical theories to reactions in the gas phase. The semiclassical initial value representation described by Miller in his Spiers Memorial Lecture[1] is a particularly promising approximation for complex systems. In addition, the full multiple-spawning method developed by Martinez et al.[2] has allowed calculations to be performed on non-adiabatic transitions in biomolecules as complex as retinal. A problem in the case of computations on non-adiabatic transitions is that there are only very few exact quantum dynamics calculations on reactions in three dimensions that have been performed to test the accuracy of semiclassical techniques (see ref. 3 for a summary).

1. W. H. Miller, *Faraday Discuss.*, 1998, **110**, 1.
2. M. Ben-Nun, F. Molnar, H. Lui, J. C. Phillips, T. J. Martinez and K. Schulten, *Faraday Discuss.*, 1998, **110**, 447.
3. D. G. Truhlar, *Faraday Discuss.*, 1998, **110**, 521.

Prof. Groß added: It is true that there is a long history of applying semiclassical models to gas phase problems. However, unfortunately the semiclassical schemes have often only be tested for low-dimensional model potentials by comparing them to full quantum calculations. Typically one-dimensional two-state problems have been used for these comparisons with only few exceptions.[1,2]

Consequently, there are in my opinion still some open questions with respect to the semiclassical schemes that can only be answered by comparing them to higher-dimensional quantum calculations. One of them is the problem of the energy adjustment in surface-hopping schemes, another one is the question whether quantum coherence is suppressed in higher-dimensional simulations. Hence there is certainly still room for further development of semiclassical methods.

1. M. S. Topaler, T. C. Allison, D. W. Schwenke and D. G. Truhlar, *J. Chem. Phys.*, 1998, **109**, 3321.
2. Y. Zeiri, G. Katz, R. Kosloff, M. S. Topaler, D. G. Truhlar and J. C. Polanyi, *Chem. Phys. Lett.*, 1999, **300**, 523.

Prof. Kosloff responded: The issue is the ability of semiclassical methods to describe even qualitatively multi-dimensional surface crossing dynamics. We found in a recent paper that semiclassical methods have qualitative discrepancies in non-adiabatic multi-dimensional problems.[1]

The problem arises when the nuclear and electronic motion cannot be separated.

1. Y. Zeiri, G. Katz, R. Kosloff, M. S. Topaler, D. G. Truhlar and J. C. Polanyi, *Chem. Phys. Lett.*, 1999, **300**, 523.

Prof. Gadzuk commented: With regards the issue of "beyond one dimension", sometimes study of a well chosen two-dimensional problem in which one degree of freedom is the one you are interested in (such as Miller's "reaction path", ref. 1) and the other one is *the rest of the world*, all lumped into an "effective coordinate", can lead to significant insights and a better intuitive understanding of the full problem, which of course is what we ultimately wish to achieve. This 2D system will probably contain stochastic electronic and phonon heat baths, time-dependent potentials/forces, and extreme anharmonic potentials and couplings leading to non-linear (at least classical) dynamics, which opens up all the issues under the chaos umbrella. While this may not be the general answer to the non-adiabatic multi-dimensional problem, it does suggest that there are some useful alternative stopgap measures that can be taken prior to the arrival of the ultimate and complete answer.

1. W. H. Miller, *J. Chem. Phys.*, 1980, **72**, 99.

Dr Kolasinski opened the discussion on Dr Hodgson's paper: In the desorption of molecular hydrogen from silicon, significant lattice distortions accompany the reaction. Because of this, little of the energy from the desorption activation barrier is deposited in the desorbing molecule. Instead, the energy remains in the Si surface.[1-3] The diffusion of hydrogen atoms out of the bulk is also associated with large scale lattice distortions; hence, we might anticipate that the energy of the bulk diffusion barrier is not transferred to the H atoms but rather to the lattice. The presence of a long-lived hot H atom state after diffusion out of the bulk would, of course, also have implications for permeation/desorption experiments of the type performed by Zare et al.[4]

Have you considered that the excess energy observed in desorbed molecular hydrogen may result from repulsive interactions between atoms in the adsorbed phase? These repulsive interactions arise because the chemisorbed layer is saturated, at least locally. At high subsurface loading, diffusion out of the bulk keeps a large fraction of the surface saturated with chemisorbed H atoms. A hydrogen atom that diffuses out of the bulk then finds itself in a supersaturated layer from which desorption as H_2 is exothermic. The exothermicity then leads to hot desorbed H_2. If the H atom diffuses out of the bulk into a region with below saturation coverage, thermalized desorption occurs. This would explain how the two channels (hot and thermal) can be seen in desorption and why the hot component becomes less important as the subsurface region becomes depleted and is unable to maintain large (local) H coverage.

1 K. W. Kolasinski, W. Nessler, A. de Meijere and E. Hasselbrink, *Phys. Rev. Lett.*, 1994, **72**, 1356.
2 K. W. Kolasinski, *Int. J. Mod. Phys. B*, 1995, **9**, 2753.
3 U. Höfer, *Appl. Phys. A: Mater. Sci. Processes*, 1996, **63**, 533.
4 S. F. Shane, H. A. Michelsen and R. N. Zare, in *Laser Spectroscopy and Photochemistry on Metal Surfaces*, ed. H. L. Dai and W. Ho, World Scientific, Singapore, 1995, p. 977.

Dr Hodgson responded: The experiment described here is quite different to the permeation/desorption experiments referred to on Cu and Pd.[1] There the crystal is held at a high temperature (typically 900 K for Cu) to allow H or D to permeate through the bulk. Since a saturated monolayer desorbs from Cu at around 300 K the resulting equilibrium coverage of D is very low. This ensures that a hot D atom, formed by permeation to the surface, has a sufficiently long lifetime (determined by the temperature and permeation rate) to lose its excess energy and thermalise long before it reacts with a second chemisorbed D atom. The recombination reaction is therefore of Langmuir–Hinschelwood type, the D atoms having lost all knowledge of how they were originally formed. This is supported by the equivalence of permeation and atom dosing measurements of the product energy release at the same surface temperature.

Resurfacing of D is certainly assisted by phonons, as indicated by calculations on the tunnelling rate for D through the bulk of Ni by Baer and Kosloff.[2] Several recent DFT calculations have also looked at how the energy barrier to D resurfacing is effected by lattice distortion during the reaction with surface adsorbed CH_3. These calculations indicate that the Ni atoms surrounding a D atom resurfacing from the subsurface octahedral site are displaced outwards slightly (*ca.* 5%) and we should certainly expect that the energy associated with relaxation of the Ni lattice back to its optimum geometry will remain in the lattice rather than being partitioned into the D atom. The D excitation energy is associated with repulsion of D from the planar Ni_3D transition into the surface threefold hollow site. Both phonon effects and coupling to electron–hole pair excitation are expected to reduce and broaden the D energy distribution, so that estimates based on the activation barrier to resurfacing and the energy of the sub-surface D, see Fig. 5, provide only an estimate of the maximum energy of hot D formed.

We considered carefully whether it was possible to understand the D_2 energy distribution in terms of reaction from a surface supersaturated with thermalised D atoms, *i.e.* without invoking direct reaction of the hot D atoms. However, we could not find a consistent model that accounts for the experimental observations on this basis and neither does the scheme suggested in the question. While recombination of thermalised D from a supersaturated region might be used to explain the hot D_2 desorption, D which has thermalised in a region with less than saturation coverage would not desorb, since at 200 K we are well below the thermal desorption peak for surface D. In practice these atoms would remain on the surface, replacing those which have reacted with hot D. Secondly, there is no evidence that the surface coverage of chemisorbed D depends on the amount of subsurface D as suggested. On the contrary, the D coverage remaining on the surface is entirely independent of the amount of subsurface D which has been desorbed, while the D_2 energy release drops as the concentration of subsurface D drops, consistent with a change in the energy of the hot D atoms created.

In our model, D resurfacing from the bulk finds a Ni(111) surface which is saturated with chemisorbed D. As indicated by the question, the hot D atom, although bound to the surface, does not feel the deep chemisorption well associated with the stable threefold hollow site at low

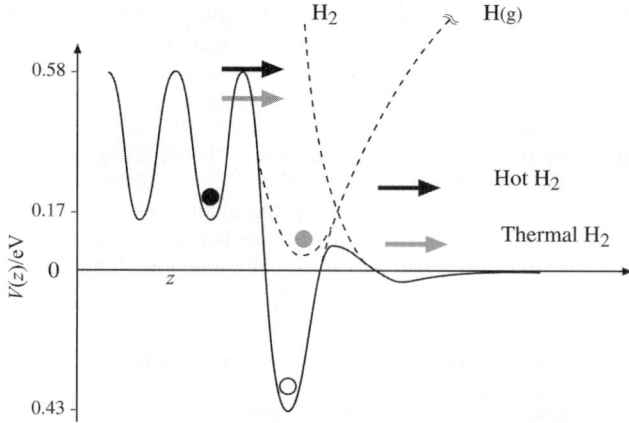

Fig. 5 Schematic showing the energy of a D atom in the bulk (●) and chemisorbed with a coverage of 1 monolayer on the surface (○). When an additional atom resurfaces it must overcome an activation barrier of between 0.35 and 0.47 eV, depending on the concentration of subsurface D, sufficient to create a hot atom which can react directly to form hot D_2. However if the energy of the species formed by permeation drops below 0.43 eV it will be unable to react with surface D and will thermalise to give a supersaturated surface. This state, indicated schematically by the dashed well and shaded circle, will destabilise local D adsorption sites and has sufficient energy to desorb slowly, forming cold D_2 and maintaining the surface saturation coverage.

coverages. Since the surface is saturated with D atoms, the energy of the additional adsorbate must be above that of $1/2H_2$, so that these species can recombine to maintain the saturation coverage. This is illustrated in Fig. 5, where the energy of the subsurface and surface species is shown schematically. The energy distribution of the hot D atoms created by permeation depends on the concentration of subsurface D, as shown by the change in the activation barrier to desorption with D loading (see section 3A of the paper). Provided the D atom has an energy of more than 0.43 eV (with respect to $1/2H_2$) it will be able to abstract a chemisorbed D atom directly. The energy of the hot D formed from a saturated layer of subsurface D is estimated to be as much as 0.64 eV, consistent with the energy released into the hot D_2 channel. If a hot D atom relaxes so that its energy is below 0.43 eV then it will be unable to abstract a chemisorbed D atom to form D_2. Once this has happened the D atom will thermalise, desorption occurring slowly by thermal recombination of pairs of these "extra" D atoms to maintain the D saturation coverage. We know that desorption from the D saturated surface gives rise to cold D_2, probably by recombination at defect or step sites, and this is consistent with the low energy channel present during desorption of subsurface D.

This mechanism is also consistent with the observation by Ceyer and co-workers that D from the bulk would react with adsorbed CH_3 in the presence of chemisorbed D but no CH_4 product was seen. This means that reaction occurs not from an excess concentration of chemisorbed H/D (which would be expected to exchange efficiently and give rise to CH_4 as well as CH_3D) but by means of direct reaction with the hot D before this thermalises. The reaction between subsurface H and CH_3 or H seem to be very similar but in this case we were not able to use isotopes to check for thermalisation of the subsurface species prior to reaction.

1 S. F. Shane, H. A. Michelsen and R. N. Zane, in *Laser Spectroscopy and Photochemistry on Metal Surfaces*, ed. H. L. Dai and W. Ho, World Scientific, Singapore, 1995, p. 977.
2 R. Baer and R. Kosloff, *J. Chem. Phys.*, 1997, **106**, 8862.

Prof. Kosloff commented: Comparing the potentials to the experiment of Dr Hodgson (Figs. 6 and 7), the direct recombination route is severely restricted by the large potential barrier. The other possibility seems to be geometrically too restrictive. This means that a possible mechanism is for the hot resurfacing hydrogen to diffuse on the surface and recombine with a surface hydrogen before it had time to loss its energy (Fig. 8). Together with Dr Baer we have calculated the

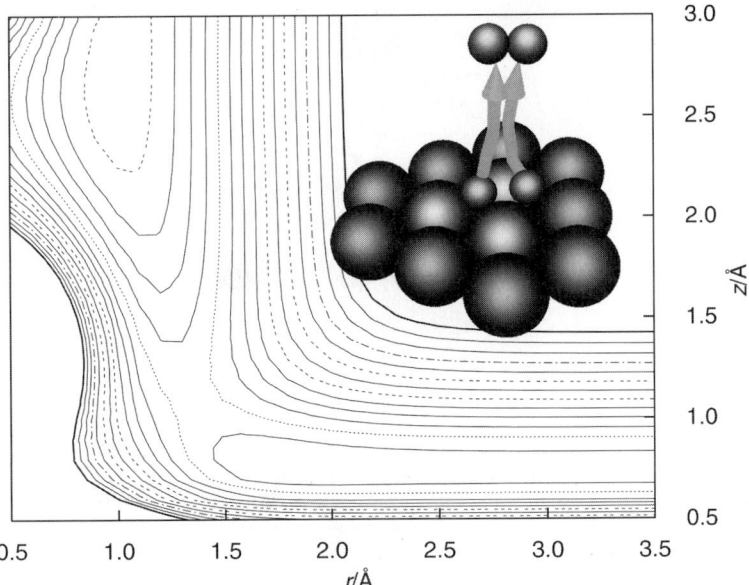

Fig. 6 Shows a potential energy surface of hydrogen recombination on a Ni(111) surface (the distances are in Å, each contour is 0.2 eV). The well depth of the H_2 molecule is 4.72 eV and the double desorption well is 5.2 eV, the barrier to dissociation is 0.17 eV.

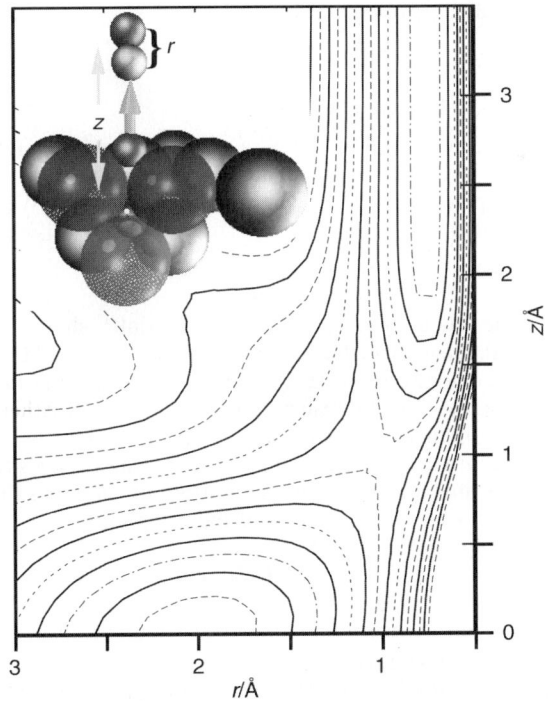

Fig. 7 Shows the potential for subsurface–surface hydrogen recombination where the surface hydrogen atom is in the three-fold hollow site and the subsurface hydrogen is directly underneath. The barrier for recombination is 1.2 eV. The distance between contour lines is 0.2 eV.

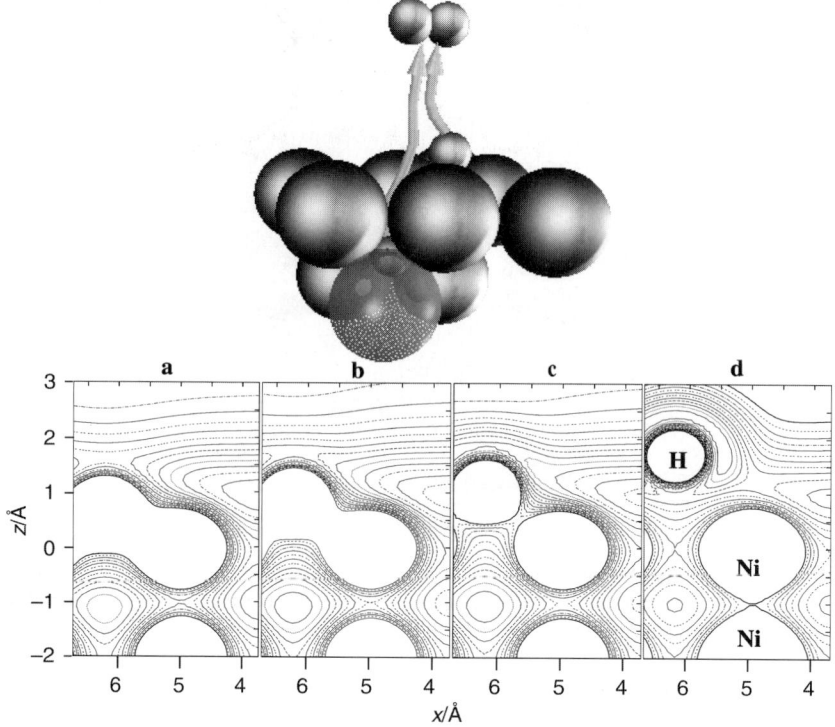

Fig. 8 Shows a series of potentials for subsurface–surface hydrogen recombination where the distance of the surface hydrogen is varied. The distance of the surface hydrogen atom is fixed at (a) 0.7 Å (b) 0.9 Å, (c) 1.1 Å and (d) 1.7 Å above the surface site. Notice in (d) the appearance of the H_2 potential energy well. The difference between contour lines is 0.1 eV.

lifetime of the hot surface hydrogen and we found a value of 1 ps.[1] The relaxation mechanism is due to electron–hole pairs and we used the surrogate Hamiltonian method.

The Ni atoms do participate in the tunnelling of the hydrogen. We found the tunnelling dynamics to be highly correlated.

1 R. Baer and R. Kosloff, *J. Chem. Phys.*, 1997, **106**, 8862.

Dr Hodgson replied: The calculations of Baer and Kosloff[1] show that the D atom must resurface from an octahedral subsurface site into an empty fcc three-fold hollow site, since if the site is occupied by D the energy barrier to resurfacing doubles, making it inaccessible at 160 K. Previously it had been suggested that the resurfacing D atom would either react directly with D adsorbed immediately above it in the fcc site (which we now know is not viable as discussed above) or it would directly abstract a neighbouring D atom from an hcp site (see Fig. 9). We have looked for the signature of a direct abstraction reaction by investigating the angular distribution of the desorbing D_2. If the D atom abstracts one of the neighbouring D atoms we would expect the D_2 distribution to retain some memory of the site at which the D resurfaced. This means that we might anticipate an angular distribution which is *not* peaked along the surface normal, showing instead a lobular form with a three-fold azimuthal symmetry. We have looked at this carefully but find that this is not the case, the angular distribution does not appear to reflect the three-fold symmetry of the resurfacing site and is lobular about the surface normal. We believe that this is strong evidence that the hot D atom does not immediately react with an adjacent D atom but instead travels some distance across the surface as a hot precursor, losing all memory of its initial creation site. Baer and Kosloff[1] calculated a lifetime of around a picosecond for relaxation of the hot D atoms, which is sufficient for a hot D atom to travel of the order of 50 Å.

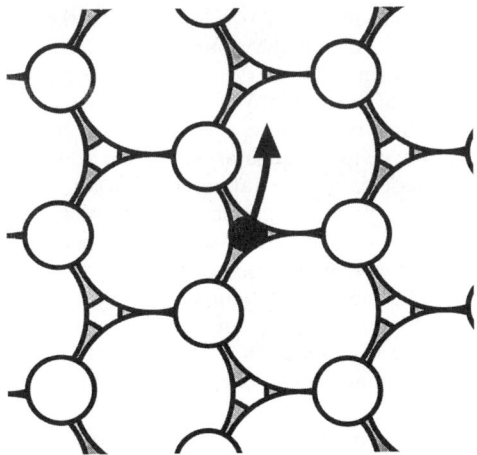

Fig. 9 Abstraction of a neighbouring D atom from an hcp site.

Looking at the potentials shown in Fig. 8, it is evident that the interaction of D with a chemisorbed D atom is repulsive unless the other adsorbate is displaced some distance above the surface. This suggests that the D atom may be steered away from these sites towards the Ni atoms, travelling across the D saturated surface until it relaxes or reaction occurs.

1 R. Baer and R. Kosloff, *J. Chem. Phys.*, 1997, **106**, 882.

Dr Shluger said: I would like to ask Dr Hodgson and Prof. Kosloff about the role of quantum diffusion of H in Ni. It is known that quantum effects play an important role in H diffusion in metals. What is the zero point energy of H in Ni, how does it depend on the hydrogen position with respect to the surface, and how does this affect the model of the desorption in the experiments presented by Dr Hodgson?

Prof. Kosloff responded: The calculations of hydrogen resurfacing and diffusion dynamics were fully quantum mechanical. They were based on the surrogate Hamiltonian method where the bath modes and electron–hole pairs are represented by additional degrees of freedom.[1]

The effect of zero point motion along the reaction path is included automatically. At low temperature the diffusion dynamics is dominated by tunnelling. This leads to a large isotope effect and an almost temperature independent diffusion constant.

The model predicts that hot hydrogen will resurface with an excess energy of ~ 0.4 eV. This hot hydrogen will pick up another hydrogen in a neighbouring site. The lifetime of this excited hydrogen is estimated to be 1 ps.

We are not able to do a quantum calculation of subsurface surface hydrogen recombination. The number of participating degrees of freedom is too large and is therefore beyond the current computation capability.

1 R. Baer, Y. Zeiri and R. Kosloff, *Phys. Rev. B*, 1997, **55**, 10952; R. Baer, Y. Zeiri and R. Kosloff, *Surf. Sci. Lett.*, 1998, **411**, L783.

Dr Kroes commented: No rotational cooling was observed in the "ordinary" desorption channel and this was related to the barrier to reaction being early. Nevertheless, rotational cooling was predicted theoretically and observed experimentally for non-activated dissociations. Why does it not occur for early barrier systems, and is there experimental evidence for other early barrier systems that there is no rotational cooling in associative desorption.

Dr Hodgson replied: Rotational cooling during desorption seems to occur either during desorption from a stable molecular well (by coupling of rotation to the desorption coordinate) or due to steric constraints during recombination, where rotation reduces the adsorption/desorption probability by preventing a molecule following the optimum path. An example of the latter is the non-activated pathway to dissociation of D_2 at Pd, where rotation inhibits adsorption/desorption by reducing the ability of a molecule to steer into the ideal geometry. However, for Ni(111) there is an activation barrier to dissociation on the low coverage ($\Theta_H < 0.5$ ML) surface and steering is less important than on an attractive surface such as Pd. When adsorption is activated rotation of the molecule can either inhibit dissociation, by taking a dissociating molecule away from its ideal orientation, or it may contribute to overcoming a barrier by allowing coupling to lower (hindered) rotational levels at the transition state. These two conflicting mechanisms make interpreting the role of rotation ambiguous, the balance depending on the exact form of the PES and the localisation of the transition state. Only when the transition state has an extended bond does rotation clearly favour dissociation, e.g. D_2 on Cu(111), but even here some rotational cooling is seen at low J. It is therefore unclear to me whether early barrier systems will generally show a rotational cooling or not and I am not aware of any other experimental evidence either way.

Prof. Utz said: You describe two methods for preparing subsurface D: ion implantation and dosing of D atoms. Do you observe any differences in translational energy release for the subsurface D desorption feature that depends on the method of subsurface D preparation?

Dr Hodgson responded: Because of the need to obtain adequate statistics we measured the translational energy release only for subsurface D prepared by ion implantation, where several monolayers of D_2 can be obtained in a single desorption experiment.

Dr Darling commented: You employ microscopic reversibility to extract a sticking curve for the precursor mediated reaction at high coverage. Could you comment on the validity of this in the light of Harris' suggestions[1] that adsorption and desorption may explore different regions of phase space?

1 J. Harris, *Faraday Discuss.*, 1993, **96**, 1.

Dr Hodgson replied: Harris presented an argument which relied on microscopic reversibility to suggest that the adsorption and desorption processes were qualitatively different. His argument was that adsorption started with a flat surface, the molecule exciting phonons in the surface and leaving a "dent" where the molecule struck, whereas during desorption the surface provided energy to the adsorbate and remained excited after the molecule had departed. This picture suggests that adsorption and desorption are entirely different processes. However, naïve arguments based on microscopic reversibility do not deal with the proper statistical weighting of adsorption and desorption channels, whereas detailed balance arguments (which is what we use) relate adsorption and desorption quantitatively at a particular surface temperature. Harris's microscopic reversibility argument implicitly assumes that phonon excitation does not aid adsorption, an assumption which does not seem to be justified experimentally. For example, when we apply detailed balance quantitatively to the activated dissociation of H_2 on Cu we see that desorption is biased towards low energy, thermal adsorption channels which are strongly dependant on the surface temperature. As the surface temperature is decreased the adsorption behaviour inferred from detailed balance tends to that observed experimentally at 100 K. Similarly the adsorption of H_2 on Ni(111) at low coverage (see section 3B of our paper) reproduces the energy dependence seen in adsorption and detailed balance arguments appear to hold in all cases where they have been properly tested.

In some cases adsorption and desorption experiments may well explore very different regions of phase space, for example where the adsorption reaction is highly activated and requires extreme conditions or where the two processes occur *via* different mechanisms in the different temperature regimes of the experiments. The energy range probed by desorption reflects the region of phase space which is important for thermal adsorption from a gas at the surface temperature, even if this only occurs with a vanishingly small probability! In such cases it may not be possible to learn anything about sticking of a high energy beam, simply because this energy range is not represented in the desorption data. In this case microscopic reversibility arguments will become misleading, but it does not invalidate detailed balance arguments. An example of this is H_2 on Si where it was originally suggested that detailed balance was contravened. In some cases increasing the surface temperature may allow high energy channels to be observed in desorption, even when thermal desorption is normally dominated by channels which have a very low probability in adsorption, as seen for H_2 on Ag(111) for example.

Prof. Polanyi opened the discussion on Prof. Utz's paper: Prof. Utz and his co-workers have measured the reactive cross section (they also refer to it as the sticking probability) for the dissociative attachment of vibrationally-excited methane at a clean nickel surface. They lay particular stress on the measurement of the J-dependence (rotational quantum number), at various collision energies, and report that, though the dependence on J is modest, $J = 1$ is consistently more reactive than $J = 2$ and 3. The change in rate over this small range of J is by no more than a factor of two.

The surprising thing to me is that both the sign and magnitude of the J-dependence at low-J is in accord with what was found some years ago[1] in a gas-phase study of the reaction between vibrationally-excited hydrogen halides and atomic sodium (and earlier work from the same laboratory cited there). It was observed that for all v-levels, $v = 1-4$, the reactive cross section at thermal collision energies fell by approximately a factor of two for the early stages of rotational excitation and then increased with substantial increase in J. This was ascribed tentatively (a) to the disadvantageous effect of rotational disalignment at low J, later (b) compensated by the advantageous effect of rotation-induced bond-stretching (rotation–vibration, R–V, interaction, especially important for 'late-barrier' reactions), or rotation-induced collision-energy enhancement (in effect rotation–translation, R–T, interaction, most important for 'early barrier' reactions).

To a rather surprising degree these simple notions garnered from studies of exchange-reactions in gases seem to be of use in understanding reaction at a surface (see, for example, refs. 20 and 44 of Utz *et al.*'s paper[2]). Prof. Utz has clearly thought about this. Would he care to comment on the utility and limitations of these gas phase insights when applied to reaction at a surface?

1   B. Blackwell, J. Polanyi and J. Sloan, *Chem. Phys.*, 1978, **30**, 299.
2   L. B. F. Juurlink, R. R. Smith and A. L. Utz, *Faraday Discuss.*, 2000, **117**, 147.

Prof. Utz replied: Highly detailed dynamical studies of gas-phase reactions have spawned a host of useful paradigms for understanding chemical reactivity. Many of these ideas have shaped and will continue to influence our understanding of gas–surface reactivity. At the same time, there are fundamental differences between a surface and a gas-phase reagent that can account for new or substantially different dynamics in a gas–surface encounter.

In many cases, gas-phase and gas–surface reactions share a propensity for an orientational anisotropy in their reactive cross section. This anisotropy leads to lower energy reaction paths for favourably oriented reagents. When rotational motion is relatively fast compared to the timescale for chemical reaction, rotations can move reactants out of a favourable and into a more repulsive orientation before the reactive system accesses the transition state. In this way, molecular rotation may be thought to "hinder" the reaction by reducing the probability for following the lowest energy pathway to reaction. In their studies of D_2 desorption from a Cu(111) surface, Michelsen *et al.* observe a non-monotonic trend in translational energy release as a function of D_2 rotational state. They interpret this trend within the context of such a "rotational hindering" model.[1,2] Thus, both the rotational disalignment effect for gas-phase reactions that Prof. Polanyi cites and the "rotational hindering" model for gas–surface interactions share a common physical basis.

Several important differences between gas-phase and gas-surface reactions highlight how these two classes of reactions might differ in their rotational state dependence. The large steric bulk of a surface may restrict the range of energetically accessible orientations that lead to reaction. The presence of a surface, with its essentially infinite inertial moment, also opens up the possibility for significant reorientation, or "steering" of only one reactant as it approaches the stationary surface. In the gas phase, molecular collision partners with similar masses are both reoriented by angular gradients in the intermolecular potential. Another effect arises from the large area and significant effective mass of the surface. All incident gas-phase molecules eventually slow as they approach the repulsive wall of the gas–surface interaction potential, regardless of impact parameter. This effect alters both the interaction time between reactants as well as the time available for rotational hindering or steering to influence reactivity. In the gas phase, long-lived collision complexes can significantly enhance the probability for a reactive encounter. Formation of these complexes hinges on the availability of efficient channels for dissipating the reagents' translational energy. Surfaces, with their many vibrational and electronic degrees of freedom, provide a highly efficient sink for dissipating collision energy and trapping an incident molecule in a precursor state. Such trapped molecules can then sample a variety of potentially reactive surface sites and orientations while tumbling and diffusing on the surface. While the results we present in our paper are unlikely to be dominated by long-lived methane precursors, the potential for precursor-mediated dynamics is in general an extremely important feature of gas–surface reactivity.

The J-state selected studies of methane dissociation presented here indicate that rotational state effects in methane's dissociative chemisorption are modest. Unfortunately, the spherical symmetry of the methane molecule coupled with the signal level of our experiment does not allow us to attribute the effects we see to a particular mechanism. Thus, we are currently unable to attribute our results to a specific dynamical mechanism. We are planning to carry out further experiments that better probe the origin of the effects we observe.

1 H. A. Michelsen, C. T. Rettner and D. J. Auerbach, *Phys. Rev. Lett.*, 1992, **69**, 2678.
2 H. A. Michelsen, C. T. Rettner, D. J. Auerbach and R. N. Zare, *J. Chem. Phys.*, 1993, **98**, 8294.

Dr McCoustra communicated: In your paper you allude to the use of your dynamical data in the estimation of high temperature thermal sticking coefficients of methane on transition metals. In a previous Faraday Discussion,[1] we reported on just such a calculation and comparison for the sticking of ethane on the Pt(111) surface. There we found that estimates based on our dynamical studies with appropriate thermal averaging were approximately two orders of magnitude larger than those measured by Rodriguez and Goodman.[2] At the time, we could not offer a suitable explanation for this large discrepancy other than, perhaps, rotational effects.

Could you comment on the comparison of estimates of thermal sticking coefficients for methane on Ni(100) based on your dynamical studies and measurements of such if they exist? Would you expect thermal averaging of the dynamical data to over- or underestimate the thermal sticking coefficient?

1 H. E. Newell, D. J. Oakes, F. J. M. Rutten, M. R. S. McCoustra and M. A. Chesters, *Faraday Discuss.*, 1996, **105**, 193.
2 J. A. Rodriguez and D. W. Goodman, *J. Phys. Chem.*, 1990, **94**, 5342.

Prof. Utz communicated in response: Comparisons of beam-surface data and thermally averaged bulb data for dissociative sticking are frequently complicated by difficulties in modelling the internal state distribution in the impinging gas sample. Even bulb studies carried out at modest pressures may suffer from an incomplete equilibration of vibrational degrees of freedom in the gas sample, particularly when the sample container and the surface are held at different temperatures. The vibrational state distribution in the molecular beam may also differ from the thermal distribution generally assumed in such models.

Several recent experimental measurements of methane dissociation on Ni(100) have attempted to circumvent these issues and compare dynamical data for methane dissociation with thermal data from bulb experiments.[1-3] To compare the data, the experimenters used a model assigning all vibrational activation to thermally excited C–H stretch oscillators in methane. This model yields excellent agreement between the thermally averaged bulb measurements and molecular beam data collected at a variety of nozzle temperatures and translational energies. While this agreement is encouraging, it is based on a highly simplified model of internal energy effects. By focusing on the C–H stretching modes, the model omits contributions to reactivity by the vast majority of thermally populated vibrational states.

Our own measurements reveal the efficacy of energy deposited into specific coordinates for promoting dissociative chemisorption. We find that vibrational energy in the ν_3 antisymmetric C–H stretching vibration has an efficacy nearly identical to that of translational energy directed along the surface normal. Since we have only quantified the reactivity of a single vibrational mode, we are unable to calculate thermal sticking probabilities properly averaged over all vibrational states as would be desired for a direct comparison with bulb data. We can, however, assess whether the efficacy for energy deposited in the v_3 vibration exceeds the average efficacy for energy in all thermally populated vibrational states. To do this, we assign the measured efficacy of the v_3 state to all vibrationally excited states in the beam and weight the reactivity of each state by its fractional population at the beam source temperature. Summing over all states yields a sticking probability. We compare this quantity with the "hot nozzle" experiments of Holmblad et al.[2] Details of this calculation will appear in an upcoming publication.[4]

We find that sticking probabilities calculated in this way overestimate the beam data at high nozzle temperatures ($T > 450$ K). This result has at least three possible explanations. First, the efficacy of energy in the v_3 mode may be somewhat higher than that of the "average" vibrational state in methane. This idea is consistent with electronic structure calculations cited in our discussion which show an elongated C–H bond at the transition state. Our planned studies of the efficacy of energy in other vibrational coordinates will allow us to test this idea. Second, while vibrational cooling in supersonic expansions is generally not very efficient, local cooling among nearly degenerate vibrational states may alter the vibrational state populations in a beam-surface experiment from a thermal distribution. If states differ in their efficacy for reaction, then a comparison of state-resolved data with "hot-nozzle" beam data may be complicated by the non-equilibrium distribution of vibrational states in the molecular beam gas sample. Third, such simulations require an estimate for the shape of the sticking probability function as well as its high-energy asymptote. Our own measurements of CH_4 (v_3) reactivity show a marked decrease in the efficacy for translational energy at high translational energies where the methane reaction probability is still much less than unity. In the absence of state-resolved measurements, the exact form of the sticking functions cannot be modelled with certainty.

The data presented in our Discussion paper do not suggest a dramatic rotational effect for methane dissociation, although we cannot rule out enhanced reactivity for high-J states that are not experimentally accessible in our experiments. Further studies that reveal the reactivity of other vibrational states in methane will undoubtedly shed more light on this important question.

1 B. Ølgaard Nielsen, A. C. Luntz, P. M. Holmblad and I. Chorkendorff, *Catal. Lett.*, 1995, **32**, 15.
2 P. M. Holmblad, J. Wambach and I. Chorkendorff, *J. Chem. Phys.*, 1995, **102**, 8255.
3 A. C. Luntz, *J. Chem. Phys.*, 1995, **102**, 8264.
4 L. B. F. Juurlink, R. R. Smith, P. R. McCabe, C. D. DiCologero and A. L. Utz, in preparation.

Prof. Clary commented: There are other interesting vibrational modes in methane such as the triply degenerate bending mode around 1300 cm^{-1}. A linear combination of these 3 modes gives an umbrella mode which is directed along a C–H dissociation coordinate. Calculations on reactions of CH_4 with atoms in the gas phase suggests that excitation of this bending mode can lead to a significant enhancement of reactivity.[1] By heating the CH_4 it should be possible to access the excited bending modes experimentally. Is this possible in your experiment?

1 D. C. Clary, *Phys. Chem. Chem. Phys.*, 1999, **1**, 1173.

Prof. Utz responded: Prof. Clary has identified an important question in methane dissociation dynamics. Ceyer and co-workers highlighted the importance of bending excitation in their analysis of beam–surface data for methane dissociation[1] and collision induced dissociation of physisorbed methane[2] on Ni(111).

Heating the molecular beam nozzle source increases the population of these bending modes in our molecular beam, but it also increases the population of other C–H stretching and stretch–bend combination vibrations. Beam–surface experiments that rely on thermal excitation to prepare vibrationally excited polyatomic reactants are generally incapable of distinguishing the contribution of distinct vibrational modes to the thermally averaged reaction probabilities that are measured.

Our state-resolved laser excitation technique overcomes this limitation. Any ro-vibrational eigenstate possessing a sufficiently strong infrared excitation probability can be prepared for such state-resolved experiments. As a result, our approach permits a comparative study of how energy deposited into different vibrational coordinates may differ in their efficacy for promoting dissociative chemisorption. Such a comparison permits a direct test of a statistical mechanism for reactivity and reveals how the internal motions of the gas-phase reagent couple to the reaction coordinate. We are currently investigating the reactivity of methane excited to the $v = 3$ level of v_4, the triply degenerate bending mode to which Prof. Clary referred. We anticipate publishing the results of those studies within the coming year.[3]

1   M. B. Lee, Q. Y. Yang and S. T. Ceyer, *J. Chem. Phys.*, 1987, **87**, 2724.
2   J. D. Beckerle, A. D. Johnson, Q. Y. Yang and S. T. Ceyer, *J. Chem. Phys.*, 1989, **91**, 5756.
3   L. B. F. Juurlink, R. R. Smith and A. L. Utz, in preparation.

Dr Shluger said: I would like to ask your opinion on whether the following effect could be relevant for interpretation of your experiments. Non-fully symmetric vibrations of the molecule will induce a dynamical dipole moment. If the projection of this dipole moment onto the surface is non-zero, this will induce an image interaction which will tend to increase the tilt of the molecule further. I wonder whether this effect could be checked by, for example, selective excitation of CH_3D along the C–D bond?

Prof. Utz responded: Dr Shluger raises an interesting point. Such reorientation effects affect not only vibrationally excited molecules, but also molecules in the vibrational ground state due to the zero-point motion associated with antisymmetric vibrations. To assess the relative importance of this reorientation force relative to forces arising from chemical anisotropies in the methane–surface potential, we must compare the relative magnitude of the two forces.

The presence of a dynamic dipole in a gas-phase molecule approaching the surface produces an induced image dipole in the metal. One result of this dipole-induced dipole interaction is a net attraction of the molecule toward the surface. This attractive force contributes to the depth of the molecular physisorption well.

If the projection of the dynamic dipole onto the surface is non-zero, then the end of the dipole moment nearer the surface will experience a stronger attraction relative to the more distant end of the dynamic dipole. The *difference* in attraction between the two ends of the dipole moment leads to a reorienting torque that further tilts the dipole toward the surface normal. Since the reorientation force arises from a difference in attraction experienced by opposite ends of the molecule, it is necessarily no larger in magnitude than the total attractive force resulting from the dipole-induced dipole interaction. The well depth for methane physisorption on several transition metal surfaces ranges from 10–25 kJ mol^{-1}.[1]

The methane–nickel interaction potential is characterized by a highly repulsive potential energy barrier whose barrier height can vary by 100 kJ mol^{-1} or more depending on the orientation of the incident molecule. The anisotropy in this potential is far greater than the total magnitude of the dipole-induced dipole interaction that Dr Shluger described. Thus, we expect the angular anisotropy in the gas–surface potential that we describe in our Discussion paper will be the chief contributor to reorientation effects for methane approaching the nickel surface.

We note that an incident methane molecule experiences the gas–surface potential anisotropy upon encountering the repulsive wall of the gas–surface potential. The dipole-induced dipole interaction will be most important in the long-range attractive portion of the gas–surface potential. Therefore, it is possible that dipole-induced dipole forces may begin to reorient the incident molecule at long range before it experiences the highly anisotropic repulsive wall.

Experiments in which the C–H (or C–D) stretch excitation is localized within the molecular frame have the potential to address orientation effects in methane dissociation. Experiments such as ours that probe the reactive channel of the gas–surface interaction are sensitive only to the net reorientation of the incident molecule. Experiments that probe the rotationally inelastically scattered methane molecules would be better suited to distinguishing these two mechanisms for reorientation.

1   D. C. Seets, M. C. Wheeler and C. B. Mullins, *J. Chem. Phys.*, 1997, **107**, 3986.

Prof. Groß opened the discussion of Dr Kroes' paper: I have two questions with respect to the vibrational effects in H_2 adsorption.

(1) You state that the vibrational effects are due to the large curvature of the reaction path in the reactant channel. However, the appropriate curvature relevant for vibrational effects is determined in mass-weighted coordinates, in which the Z-axis is stretched by a factor of two. If you

plot your potential energy surfaces in mass-weighted coordinates, the curvature will be much smaller. Are the vibrational effects not in fact caused by the lowering of the vibrational frequency at the barrier position, *i.e.* by vibrationally adiabatic effects?[1]

(2) In your rotationally adjusted potential the barrier position has moved to the entrance channel, which means that the vibrational efficacy in adsorption should be reduced. Have you checked this effect?

1   A. Groß and M. Scheffler, *Chem. Phys. Lett.*, 1996, **256**, 417.

Dr Kroes responded: (1) Your first question is whether the vibrational effect we see (the preference for top site reaction for $v = 1$ at an energy where the reaction probability is still low) is not in fact caused by the lowering of the vibrational frequency at the barrier position, *i.e.* whether it is a vibrationally adiabatic effect. The answer to that question is a very definite no. This can actually be seen by comparing Figs. 4b and 4d of our paper, which show the site-specific top-site reaction probability for $(v = 0, j = 4)$ and $(v = 1, j = 4)$, respectively. For $(v = 1, j = 4)$, the onset of the site-specific reaction probability (which can be operationally defined as the energy where the site-specific reaction probability first becomes 0.025) is lower than the onset of the site-specific reaction probability for $(v = 0, j = 4)$ by a full quantum of vibrational energy (at 0.1 eV for $v = 0, j = 4$), compared to 0.6 eV for $(v = 1, j = 4)$). This corresponds to a vibrational efficacy of one, which can only be obtained if the mechanism is vibrationally non-adiabatic (the molecule gives up all of its excess energy to react). This conclusion is also born out by calculating vibrationally adiabatic potentials for bridge-to-hollow and top-to-bridge dissociation, in which the energy of vibrational motion perpendicular to the reaction path is added to the potential along the reaction path, for $v = 1$. As shown in Fig. 4 of ref. 1, the $v = 1$ vibrationally adiabatic top-site potential has a higher barrier than the $v = 1$ vibrationally adiabatic bridge site potential, so that a vibrationally adiabatic mechanism could not explain why top-site reaction should be preferred over bridge-site reaction for $v = 1$. This preference can only be explained by a vibrationally non-adiabatic mechanism in which the molecule gives up *all* of its vibrational energy to react. This can occur if the potential has a large curvature in front of an especially late barrier, and these features are present in the top site PES.[2] Such a PES can also give rise to vibrational excitation, and we have previously shown that scattering from the top site in fact causes very efficient vibrational excitation.[3,4]

(2) Concerning your second question of whether the vibrational efficacy in adsorption is reduced for higher j due to a shift of the barrier position in to the entrance channel: we have not yet checked this effect, but my guess would be that it could be quite important for $j > 4$. However, this is presently the highest j for which we have m_j-averaged results for both $v = 0$ and $v = 1$. We will certainly investigate the effect you suggest once we have results for higher j.

1   D. A. McCormack, G. J. Kroes, R. A. Olsen, J. A. Groeneveld, J. N. P. van Stralen, E. J. Baerends and R. C. Mowrey, *Chem. Phys. Lett.*, 2000, **328**, 317.
2   G. R. Darling and S. Holloway, *Surf. Sci.*, 1994, **307–309**, 153.
3   G. J. Kroes, G. Wiesenekker, E. J. Baerends and R. C. Mowrey, *Phys. Rev. B*, 1996, **53**, 10397.
4   G. J. Kroes, G. Wiesenekker, E. J. Baerends, R. C. Mowrey and D. Neuhauser, *J. Chem. Phys.*, 1996, **105**, 5979.

Prof. Seideman commented: You point out in your paper that probing experimentally the predicted anisotropies by selecting the reagents with respect to magnetic components would be difficult (although optical preparation of an anisotropic distribution of magnetic components has been reported in this meeting) and note, in addition, that rotational effects would typically prevent molecular beams of selected rotational levels from probing the potential energy surface.

A different method for producing an aligned beam of reagents, which circumvents the problem of rotational hindrance, produces a reagent that is much better spatially-defined than molecules in a single quantum state, and is experimentally practical, is resonant intense-laser-alignment.[1,2] Molecular alignment in an intense laser field[1–5] is similar in concept to traditional alignment in a strong DC field[6–8] in that it produces a broad superposition of total angular momentum states (which is what allows the system to be well-defined in the conjugate angle space). The advantages of a laser field over a DC field are that alignment can be preserved after turn off of the laser pulse,[2] and hence the surface reaction can take place under field free conditions, and that much sharper alignment is achievable, making practical the alignment of nonpolar and light molecules. Intense field alignment can be achieved at both near-[1] and non-resonant[3] frequencies. For H_2, however, due to its light mass and low polarizability anisotropy, only the near-resonance route is practical.

1 T. Seideman, *J. Chem. Phys.*, 1995, **103**, 7887.
2 T. Seideman, *Phys. Rev. Lett.*, 1999, **83**, 4971.
3 B. Friedrich and D. R. Herschbach, *Phys. Rev. Lett.*, 1995, **74**, 4623.
4 J. J. Larsen, I. Wendt-Larsen and H. Stapelfeldt, *Phys. Rev. Lett.*, 1999, **83**, 1123.
5 J. J. Larsen, K. Hald, N. Bjerre, H. Stapelfeldt and T. Seideman, *Phys. Rev. Lett.*, 2000, **85**, 2470.
6 See, for instance, B. Friedrich, D. R. Herschbach, J.-M. Rost, H.-G. Rubahn, M. Renger and M. Verbeek, *J. Chem. Soc., Faraday Trans.*, 1993, **89**, 1539.
7 See, for instance, H. J. Loesch and J. Möller, *Phys. Chem.*, 1993, **97**, 2158.
8 See, for instance, M. Wu, R. J. Bemish and R. E. Miller, *J. Chem. Phys.*, 1994, **101**, 9447.

Dr Kroes answered: I would say that the primary goal of our paper is to demonstrate that vibrationally excited ($v = 1$) H_2 in low j-states reacts at a different site (the top site) than ($v = 0$) H_2 (which reacts at the bridge site). We are saying that to confirm this effect experimentally, one could actually take advantage of orientational hindrance, by performing energy-resolved associative desorption experiments in which the rotational quadrupole alignment of (v, j) H_2 is measured for $v = 0$ and $v = 1$.

We know that H_2 can be prepared in a single m_j state with the use of a magnetic field and techniques pioneered in the late 1960s.[1] Of course the question is whether the technique allows enough population to be created in such a state to be useful for performing measurements of the reaction probability using molecular beams. We also know that stimulated Raman pumping can be used to produce H_2 in a particular (v, j) state with a non-statistical distribution over the m_j-states; in fact, for $j = 2$ the population of a single m_j-state ($m_j = 0$) in a molecular beam should be achievable.[2] Another road towards experimental confirmation of site-specific reaction is then perhaps to show that the reaction probability of (v, j) H_2 changes in a specific way if the incident m_j-distribution is changed with the aid of this technique, and that the change can be explained from the topology of the PES of the dominant reaction site for the (v, j) state investigated. This is one avenue we will explore in the future, but for now we note that molecular beam experiments on scattering of (v, j) H_2 with a non-statistical distribution over the m_j incident states have yet to be performed. In contrast, the feasibility of the associative desorption experiment we suggest has already been demonstrated.[3]

Of course the intense laser field techniques that you suggest are potentially even more interesting, also for investigating site-reactivity, and we will certainly attempt to derive predictions for the experiments you suggest. It would be very nice if experimentalists would attempt such experiments on the reaction of maximally aligned H_2 with Cu: such experiments would produce even more detailed information on this paradigm for activated dissociation.

1 H. Moerkerken, M. Prior and J. Reuss, *Physica*, 1970, **50**, 499.
2 G. O. Sitz and R. L. Farrow, *J. Chem. Phys.*, 1994, **101**, 4682.
3 H. Hou, S. J. Gulding, C. T. Rettner, A. M. Wodtke and D. J. Auerbach, *Science*, 1997, **277**, 80.

Prof. Sitz added: production of anisotropic spatial distributions of angular momenta has been demonstrated using stimulated Raman scattering (SRS).[1–3] This approach produces alignment[1,2] or orientation[3] of the molecular rotational angular momentum. This means that a nonstatistical population distribution among magnetic sublevels is created for a specific value of the rotational quantum number J (in favorable cases a single $| J, m \rangle$ state can be prepared). So far, this technique has only been applied to gas-phase scattering experiments on acetylene[2,3] and N_2,[1] however in preliminary work in my laboratory, we have applied SRS to production of alignment in H_2 for surface scattering experiments and it appears quite promising. In terms of comparison with theoretical calculations of the type reported here by Kroes, scattering experiments utilizing beams of molecules prepared in this manner offer an appealing approach to measuring geometrical aspects of reactive gas–surface interactions.

1 G. O. Sitz and R. L. Farrow, *J. Chem. Phys.*, 1994, **101**, 4682.
2 J. B. Halpern, R. Dopheide and H. Zacharias, *J. Chem. Phys.*, 1995, **99**, 13611.

3 A. D. Rudert, J. Martin, W. B. Gao, J. B. Halpern and H. Zacharias, *J. Chem. Phys.*, 1999, **111**, 9549.

Prof. Bird commented: The process of energy dissipation to the substrate *via* phonon and electron–hole pair excitation has long been recognised as being significant in gas-surface dynamics, but it is often neglected in theoretical work. How can we quantify the effects of energy dissipation in adsorption, dissociation, reaction, adsorbate vibration and diffusion, *etc.*? Some time ago, Hellsing and Persson[1] showed that in the limit of a slow perturbation, the effects of electron–hole pair creation can be described by an effective friction coefficient that can be calculated using the eigenstates and self-consistent potential of density functional theory. We have recently started a project to calculate this coefficient for hydrogen atoms moving above the Cu(111) surface. The key result to emerge is that the friction coefficient for motion perpendicular to the surface is strongly peaked when the hydrogen atom is 2.5 Å from the surface—the value here being about ten times larger than for the equilibrium height of 1 Å. This peak correlates with the affinity level of the hydrogen atom (the minority spin 1s state) passing through the Fermi level. Simple, one-dimensional, classical dynamical simulations based on the calculated potential energy surface and friction coefficient show that the rate of energy loss into electron–hole pairs is significant. For a one-dimensional potential well the sticking probability changes from one at low incident energy to zero at a particular incident energy.

Preliminary results show that this switch occurs at 0.42 eV for H atoms and 0.28 eV for D. These results indicate that energy loss to electron–hole pairs could be of considerable importance in gas–surface dynamics, particular for light atoms and molecules such as H and H_2.

1 B. Hellsing and M. Persson, *Phys. Scr.*, 1984, **29**, 360.

Dr Kroes replied: In a previous Faraday Discussion, I have commented in great detail on the possible influence of phonons[1] and electron–hole pair excitations[2] on dissociative chemisorption of H_2 on a Cu surface. To estimate the effect of electron–hole pair excitations on dissociation of H_2 on a metal surface, one cannot simply say that the effect will be similar in size to that found for sticking of H on a metal surface. Of course, H_2 is a closed-shell molecule, which only starts to interact chemically with the surface close to the barrier to dissociation, which occurs between 1 and 1.5 Å from the surface. As discussed in ref. 2 there is little charge on H_2 close to the barrier, which suggests that electron–hole pair excitations should not be so important.

The influence of phonons and electron–hole pair excitations on dissociative chemisorption remains of course of interest, but also one should note that, for now, we obtain reaction probability curves which agree with experiment to within experimental uncertainty (in the range 0.1–0.2 eV for the H_2 + Cu(100) system we investigated[3]). The inclusion of phonons and electron–hole pair excitations would really become interesting if reaction probability curves can be determined with enough accuracy experimentally and theoretically to state with certainty that theoretical results obtained within an electronically adiabatic, static surface treatment are significantly different from the experimental results, for a cold surface. This requires improvements in experimental technology (although much more accurate results are available for H_2 + Cu(111)[4] than for H_2 + Cu(100)[5]) as well as in electronic structure theory (more accurate density functionals are required). One could then check whether a treatment of the phonons and/or electron–hole pair excitations would yield better agreement with experiment, keeping in mind that an exact treatment of these degrees of freedom is however not possible and probably never will be. Nevertheless, it would be interesting to use the charges computed on H_2 interacting with Cu(100)[6] to estimate the possible effect of electron–hole pair excitations on the dissociation, using for instance the surrogate Hamiltonian method.[7,8]

1 G. J. Kroes, *Faraday Discuss.*, 1998, **110**, 347 and 356.
2 G. J. Kroes, *Faraday Discuss.*, 1998, **110**, 375.
3 D. J. Auerbach, private communication.
4 C. T. Rettner, H. A. Michelsen and D. J. Auerbach, *J. Chem. Phys.*, 1995, **102**, 4625.
5 H. A. Michelsen and D. J. Auerbach, *J. Chem. Phys.*, 1991, **94**, 7502.
6 P. Vernooys and E. J. Baerends, unpublished results.
7 R. Baer and R. Kosloff, *J. Chem. Phys.*, 1997, **106**, 8862.
8 R. Baer and R. Kosloff, *Phys. Rev. B*, 1997, **55**, 10952.

Prof. Sitz asked Prof. Bird: What is the approximate limit for one H-atom round trip in the atom–surface well, and how does this time compare to the 1 ps H-atom relaxation time (quoted earlier in this discussion by Prof. Kosloff)?

Prof. Bird replied: For the incident energy when an incident hydrogen atom just escapes from the one-dimensional surface well, the round-trip time (for travelling from a position 4 Å away from the surface back to 4 Å) is 0.32 ps.

The vibrational quantum in the atom–surface well is about 130 meV and, for an atom near the bottom of the well, we estimate that this amount of energy is lost in around 0.8 ps, which is consistent with Prof. Kosloff's figure. At first sight it might seem strange that an H atom that scatters can lose around 400 meV in 0.3 ps when an adsorbed atom takes 0.8 ps to lose 140 meV. The reason for this is that the scattered H atom has a higher speed and that it experiences the region where the friction due to electron–hole pair excitation is strongly peaked.

Dr Saalfrank commented: We have examined the vibrational relaxation of H_2 ($v = 1$) molecules scattering from Cu surfaces[1] to model experiments by Sitz et al.[2] We used a two-dimensional-open-system density matrix model, and assumed that the electronic friction coefficient increases smoothly with decreasing molecule–surface distance. Under these assumptions we found that electronic friction was not able to largely account for the amount of vibrational relaxation found in ref. 2. However, the situation may be different in a multi-dimensional model (including rotation) and/or if non-monotonic behaviour of the friction coefficient (as suggested by Prof. Bird for H/Cu) is accounted for.

1 L. Pesce and P. Saalfrank, *J. Chem. Phys.* 1998, **108**, 3045.
2 M. Gostein, H. Parhikhteh and G. O. Sitz, *Phys. Rev. Lett.*, 1995, **75**, 342.

Prof. Stokbro addressed Prof. Bird: The vibrational lifetime of H on Si is in the ns range. On metal surfaces the lifetime is much shorter due to electron–hole pair creation, but still in the ps range. That seems to be inconsistent with your calculation that show that the H atom can loose 0.4 eV in 0.1 ps.

Prof. Bird replied: There is some confusion here about the time scales. As stated in my response to Prof. Sitz, the round-trip time for losing 400 meV is around 0.3 ps.

Prof. Wolf commented: One of the questions discussed has addressed the potential role of electron–hole pair excitation during dissociative adsorption of H_2 on metal surfaces. We have performed complementary experiments on associative desorption of H_2 from Ru(001) induced by fs laser pulses. We find that we can induce the recombination of atomic hydrogen by a purely electronic mechanism, suggesting that electron–hole pair excitation might be of relevance also in the reverse process of dissociation.

Prof. Gadzuk said: This question is directed to Prof. Bird (and probably Prof. Persson). You showed us your calculated results of electronic friction/viscosity as a function of H_2 distance from a surface in which the friction increases as the H_2–surface separation decreases from infinity, rising quickly to a noteworthy maximum around ~2.5 Å from the surface and then dropping significantly for closer separations. You suggested that the reason for the special behaviour at ~2.5 Å was that this is where the electron affinity "level" crosses the Fermi level, hence where an electron tunnels or "harpoons" into the incoming H_2 converting it into a negative molecular ion. Throughout the 70's, Suhl, Schaich, Kumar, Nourtier, and others studied Fokker–Planck dynamics in which electronic friction between a charged particle outside a surface and the dissipative electron–hole pair heat sink/source was the crucial element of their model. Considerable work was done by them (and others such as Ritchie and his protégé who were involved in questions of stopping powers) in which friction was described in terms of some *non-local* dielectric response function, often obtained within the random phase approximation which stressed the role of charge density fluctuations (electron–hole pairs as well as bulk and surface plasmons) on and within the surface which were driven by external charged particles. Since you have placed special emphasis on the viscosity near 2.5 Å where the level crossing leads to charging of the incident H_2, it would seem that the frictional loss mechanisms that they considered should also be relevant to your problem. To what extent are these loss processes (probably expressible in terms of the imaginary part of some surface dielectric response function) included within your (*local*?) density-functional-theory realisation of the Persson/Persson/Hellsing prescription for calculating electronic friction coefficients which might be relevant to the H_2 scattering problem under consideration here?

Prof. Persson replied: Our prescription of calculating electronic friction to electron–hole pairs is based on the low-frequency limit of time-dependent density functional theory[1] and can be expressed in terms of the imaginary part of the density response function. The early work that you refer to was based on the time-dependent Hartree approximation using density response functions based on the homogeneous electron gas and infinite-barrier-like models for the surface but they kept full frequency dependence. However, we are primarily interested in slowly moving particles which can only excite low-energy electron–hole pairs for which the low-frequency limit of the density response function is appropriate. Thus for a slowly moving particle our method includes all the electron–hole pair loss processes of their work. Finally, we would like to stress that in our prescription we give a proper account of the ground state electronic structure of the surface in the presence of the particle that was completely lacking in the early models.

1   B. Hellsing and M. Persson, *Phys. Scr.*, 1984, **29**, 360.

Prof. Sitz asked Dr Kroes: How does the resolution of the structure in the spatial maps of reactivity compare to the De Broglie wavelength of the incident H_2? That is, the site selectivity close to being 'quantum limited'?

Dr Kroes answered: An accurate answer to your question, of whether the site selectivity is close to being 'quantum limited', requires very detailed investigations, which we have not performed. In the calculation of site-specific reaction probability, the area we integrate over is 2.4 by 2.4 a_0. A back-of-the-envelope calculation shows that a De Broglie wavelength of 2.4 a_0 corresponds to an energy of 25 meV. So the De Broglie wavelength of the incident H_2 is certainly small enough to compute site-specific reaction probabilities at the incidence energies we consider (in excess of 0.1 eV). Note also that in computing the site-specific reaction probability we simply make use of the standard, probabilistic interpretation of the wavefunction, we are certainly not doing anything that violates Heisenberg's uncertainty principle.

Dr Hodgson commented: There is of course a very strong temperature dependence to associative desorption, and hence to adsorption, when the translational energy is anywhere below threshold. What the mechanism for this may be is not yet clear. Bearing this in mind, I would like to ask Dr Kroes about the effect of surface temperature on his prediction of different alignments for H_2 ($v = 0$) and ($v = 1$) from Cu(100). In practice this must be measured with a hot surface (900 K or so), will the difference in alignment persist?

Dr Kroes responded: The site-specific effects that we discuss are for collision energies above threshold, where the total reaction probability is of the order of a few percent. Surface temperature has an effect in this energy regime but the effect is not as large as in the energy regime below threshold, for which experiments[1] have shown large effects. I would not expect the effects of surface temperature to affect our main conclusions for the collision energies we discuss, although the site-specific reaction probabilities might change by a few percent. I expect the site-specific effects to remain visible in the alignments that would be measured in associative desorption experiments; I do not expect them to be washed out by the effect of the high surface temperature required in these experiments.

1   M. J. Murphy and A. Hodgson, *J. Chem. Phys.*, 1998, **108**, 4199.

Structure and dynamics of excited electronic states at the adsorbate/metal interface: $C_6F_6/Cu(111)$

C. Gahl, K. Ishioka,† Q. Zhong, A. Hotzel and M. Wolf*

Fritz-Haber-Institut der MPG Faradayweg 4–6, D-14195 Berlin, Germany.
Fax: +49 30 8413 5106

Received 16th May 2000
First published as an Advance Article on the web 17th October 2000

Excited state electron transfer at the adsorbate/metal interface represents a key step in molecular electronic devices. The dynamics of such processes are governed by ultrafast energy relaxation which can be probed directly by time-resolved two-photon photoemission (2PPE). Using 2PPE spectroscopy we investigate the energetics and lifetimes of the unoccupied electronic states of C_6F_6 adsorbed on Cu(111) as a model system for electron transfer at organic/metal interfaces. With increasing C_6F_6 layer thickness we find a pronounced decrease in the energetic position of the lowest unoccupied state, which is accompanied by a strong increase in its lifetime as well as a decrease in the effective electron mass. The frequently employed dielectric continuum model which describes delocalized (quantum well) states within adsorbate layers does not give a consistent explanation of these findings. By adsorption of Xe overlayers onto $C_6F_6/Cu(111)$ we can show that, even for one monolayer of C_6F_6, the excited state must be localized predominantly inside the C_6F_6 layer and thus originates from a molecular state (presumably an antibonding σ* orbital). With increasing coverage this state becomes more delocalized within the adsorbate layer, which reduces the coupling to the metal substrate and thus enhances the excited state lifetime.

1. Introduction

Electron transfer across interfaces is of fundamental importance in many areas of physics, chemistry and biology ranging from electronic devices to surface reactions and photosynthesis.[1,2] In the rapid development of semiconductor electronics the continuing miniaturization will soon approach the physical limits. Attractive candidates in the search to overcome such limitations are molecular devices, *e.g.*, room-temperature transistors using a single carbon nanotube have already been demonstrated.[3] An alternative to nanotubes are organic molecules, which allow easy incorporation of functional groups and the use of self-assembly strategies for parallel fabrication of a large number of molecular electronic devices.[4,5] Organic semiconductors are also widely used for light emitting diodes in large area displays.[6] A very critical issue in this field of molecular electronics is the contact resistance and efficiency of carrier transport across the interface between the metal electrodes and the organic layer.[1,5] On a microscopic level the dynamics of this charge transfer across the organic/metal interface is governed by the strength of the electronic coupling between the molecule and the underlying metal substrate.[7] This problem is thus closely related to

† Present address: National Research Institute for Metals, 1-2-1 Sengen,Tsukuba, 305 Japan.

DOI: 10.1039/b003308l

surface photochemistry, which is often believed to be driven by hot electron transfer to the reactant molecules.[8] From a surface science perspective this requires a systematic investigation of the dynamics of electron transfer at organic/metal interfaces using well chosen model systems.

Our approach here is to study small organic molecules with a high electron affinity adsorbed on a single crystal metal surface, namely hexafluorobenzene (C_6F_6) and variants (like $C_6H_2F_4$) adsorbed on Cu(111).[9] We use time-resolved 2PPE spectroscopy to investigate the symmetry, binding energy and lifetimes of unoccupied electronic states.[10] Previous studies of the dynamics of excited states at adsorbate/metal interfaces have focused on physisorbed adsorbates (like Xe or N_2)[11–13] and chemisorbed species (like CO or Cs).[14–16] The nature of the electronic states investigated in these studies can be classified into two limiting cases: (i) image potential states, where the electrons are localized in the direction normal to the surface, but fully delocalized (i.e. free-electron-like) parallel to the surface, and (ii) electronic states derived from molecular orbitals or molecular resonances which have essentially localized character. As will be shown below, the system of C_6F_6/Cu(111) lies between these limiting cases and exhibits a gradual transition from a more localized to a delocalized character with increasing coverage.

Fig. 1 (left) illustrates schematically the principle of time-resolved 2PPE spectroscopy. A femtosecond pump pulse (with a photon energy hv_1) excites an electron from below the Fermi level, E_F, to a normally unoccupied electronic state. By a second time-delayed (probe) pulse (with photon energy hv_2) the electron is subsequently lifted to a final state above the vacuum level where its kinetic energy, E_{kin}, is analyzed. Varying the time-delay between the pump and probe pulses allows a determination of the lifetime of the excited state.[10,11] The efficiency of the first excitation step depends critically on the overlap between the wavefunction of the excited state and the electronic bulk states of the metal substrate. Likewise, the relaxation rate out of the excited state will be faster the larger this overlap is, i.e. the stronger the coupling to the metal substrate via the tail of the excited state wavefunction. For example, an (anti-)correlation between the relaxation rate (lifetime) and the part of the probability density $|\Psi|^2$ inside the substrate has been demonstrated for image potential states on clean and adsorbate covered metal surfaces.[11,12,17–19]

In general, both elastic channels (like excited state electron transfer to isoenergetic bulk states)

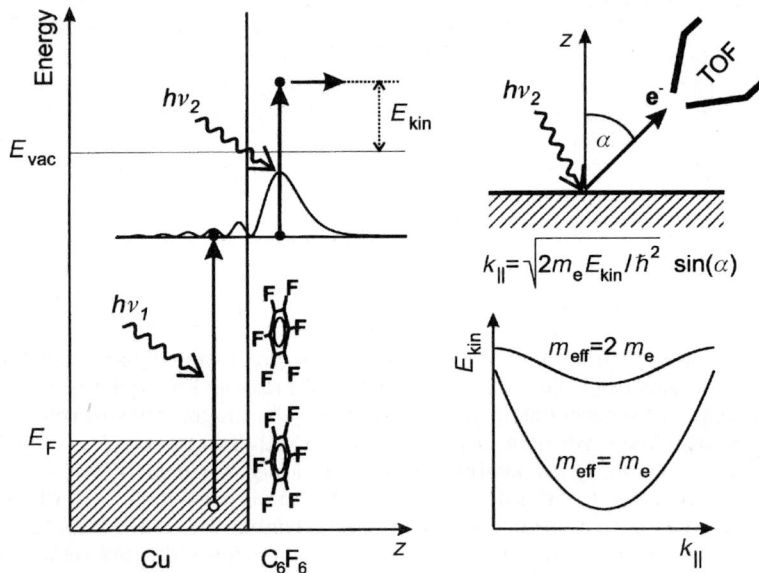

Fig. 1 Angle-resolved 2PPE spectroscopy and dispersion of electronic states: An electron is excited by a femtosecond laser pulse (photon energy hv_1) to a normally unoccupied state (shown is the probability density $|\Psi|^2$ of an adsorbate induced state or image potential state, which is localized at the surface). After excitation with a second time-delayed fs probe pulse (hv_2) the photoemitted electron is detected in a time-of-flight detector at an angle α with respect to the surface normal (right). Measurement of the dispersion of the electron kinetic energy, E_{kin}, as a function of the parallel momentum, $k_\parallel$, allows one to determine the effective mass, m_{eff}, of the excited state.

as well as inelastic processes (leading to electron–hole pair or phonon excitation in the substrate) may contribute to the relaxation of the excited state. Note that on Cu(111) the existence of a lower lying partially unoccupied surface state opens additional decay channels at the surface, which may significantly reduce the excited state lifetime.[20]

Analysis of the dispersion of electronic states provides information about the degree of localization of their wavefunctions. In angle-resolved 2PPE spectroscopy (see Fig. 1 right) a measurement of the photoelectron kinetic energy, E_{kin}, as a function of the emission angle, α, provides information on the electron momentum in the directions parallel to the surface, $k_\parallel = (2m_e E_{kin})^{1/2} \sin(\alpha)/\hbar$, where m_e is the free electron mass. For small values of $k_\parallel$ the effective mass, m_{eff}, of an electronic state is determined from a fit of its binding energy, E_B, as a function of $k_\parallel$ according to $E_B(k_\parallel) = \hbar^2 k_\parallel^2 / 2m_{eff} + E_B(k_\parallel = 0)$. In the case of negligible corrugation of the electronic potential in the directions parallel to the surface the electronic state will exhibit a free-electron-like dispersion ($m_{eff} = m_e$). This is typical for image potential states, which are located mainly in front of the surface.[21] On the other hand, lateral corrugation of the potential can lead to a larger degree of localization of the wavefunction parallel to the surface, which implies a larger value of the effective mass ($m_{eff} > m_e$). This is typical for adsorbate band structures derived from molecular orbitals.[22] A simple Kronig–Penney model can qualitatively explain this increase in the effective mass due to an increased localization of the wavefunction in a periodic array of square well potentials.[23,24] Fig. 1 shows an example for a calculated dispersion curve, which is characterized by $m_{eff} = 2m_e$ near the bottom of the band. In this model the increased localization is accompanied by an upward shift of the binding energy compared to quasi-free electrons ($m_{eff} = m_e$). In general, changes in the localization of the wavefunction parallel to the surface will affect the measured binding energies due to the differences in localization energy.

The use of angle-resolved 2PPE spectroscopy allows one to analyze the dispersion of normally unoccupied electronic states. In an elegant example Harris and coworkers were able to time-resolve the transition from a delocalized image potential state to a localized (*i.e.* non-dispersing) state in alkane overlayers on Ag(111).[25] This was attributed to the formation of a "small" polaron. In our study of C_6F_6/Cu(111) we present a systematic measurement of the effective mass of the lowest unoccupied state as a function of coverage. With increasing coverage we observe a gradual decrease in the binding energy as well as in the effective mass parallel to the surface. To analyze the degree of localization in the direction normal to the surface we have adsorbed Xe overlayers on top of C_6F_6/Cu(111) and find that the excited state wavefunction is localized predominantly inside the first C_6F_6 layer. With increasing thickness of the C_6F_6 layers the wavefunction of this state becomes more delocalized inside the adsorbate layer, accompanied by a reduced coupling to the metal electrons in the substrate, leading to a pronounced increase of the excited state lifetime.

2. Experimental

The experiments were performed in a UHV chamber equipped with an electron time-of-flight spectrometer for 2PPE spectroscopy as described elsewhere.[26] Femtosecond laser pulses are generated by a 200 kHz Ti : Sapphire laser system which pumps an optical parametric amplifier (OPA). The visible OPA output with photon energies between $h\nu = 1.7$ and 2.7 eV is frequency doubled to generate ultraviolet pulses with a duration <80 fs. The visible and frequency doubled pulses are delayed with respect to each other and focused on the Cu(111) crystal. All 2PPE spectra are recorded for emission along the surface normal, except for dispersion measurements. The Cu(111) crystal was cleaned by standard sputter–annealing cycles. Hexafluorobenzene (C_6F_6) was purified by freeze–pump–thaw cycles and dosed through a pinhole doser at a temperature chosen to prevent the adsorption of subsequent layers (155 K for 1 ML C_6F_6/Cu(111), 140 K for 2 ML and 120 K for higher coverages, respectively). The C_6F_6 coverage was determined from thermal desorption spectra (TDS), which were routinely taken after the 2PPE measurements.[7,27] Similar procedures were used to prepare C_6HF_5 and $C_6H_2F_4$ layers on Cu(111). The Xe overlayers were prepared by metered dosing at 50 K after preadsorption of C_6F_6; the coverages of Xe and C_6F_6 were again checked with TDS.

During the 2PPE measurements on C_6F_6/Cu(111) we observed a slight increase in the work function, $\Delta\Phi$, induced by exposure to the ultraviolet pump pulses. This shift is most likely caused by photodissociation of a small amount of C_6F_6 and accumulation of fragments (presumably fluorine) at the surface. However, with the very low laser fluence used in the experiments $<1\ \mu J\ cm^{-2}$) the shift is only of the order of $\Delta\Phi = 5$ meV min^{-1}. Compared to inverse photoemission measurements where dissociation induced by the electron bombardment leads to $\Delta\Phi \sim 1$ eV min^{-1} the effect is much smaller in 2PPE.[28] All spectra were taken within an acquisition time where no influence on the peak positions could be observed within our error bars of 20 meV.

3. Results and discussion

3.1 2PPE spectroscopy of C_6F_6/Cu(111)

Fig. 2 depicts a series of 2PPE spectra of clean and C_6F_6 covered Cu(111) for coverages up to 5 monolayers (ML). The spectra are recorded with photon energies of $hv_1 = 4.26$ eV and $hv_2 = 2.13$ eV and are plotted on the bottom axis as a function of final state energy, $E_{final} - E_F = E_{kin} + \Phi$, and on the top axis vs. intermediate state energy, $E - E_F = E_{final} - E_F - hv_2$. The work function, Φ, which is determined from the low-energy cut-off of the spectra, decreases linearly from 4.89 ± 0.02 eV on clean Cu(111) to 4.5 ± 0.02 eV for 1 ML and finally to 4.47 ± 0.02 eV at and above 2 ML. The two peaks around 6 eV final state energy originate from the occupied surface state (SS) and image potential states (IS), as discussed elsewhere.[27]

For coverages above 0.3 ML a pronounced peak (A) around 5.2 eV final state energy is observed in the 2PPE spectra which originates from an unoccupied electronic state located at 3.14 ± 0.02 eV above E_F for 1 ML C_6F_6/Cu(111). This state shifts to lower energies upon adsorption of each layer and remains at a constant energy of $E - E_F = 2.84 \pm 0.02$ eV for coverages above 4 ML. The asymmetric peak shape of state A may be attributed to inelastic scattering processes of the electrons in the final state. The observation of a double peak structure for non-integer coverages is a signature of island formation on top of complete adsorbate layers and confirms a layer by layer growth of C_6F_6/Cu(111) up to at least 3 ML. Similar observations were

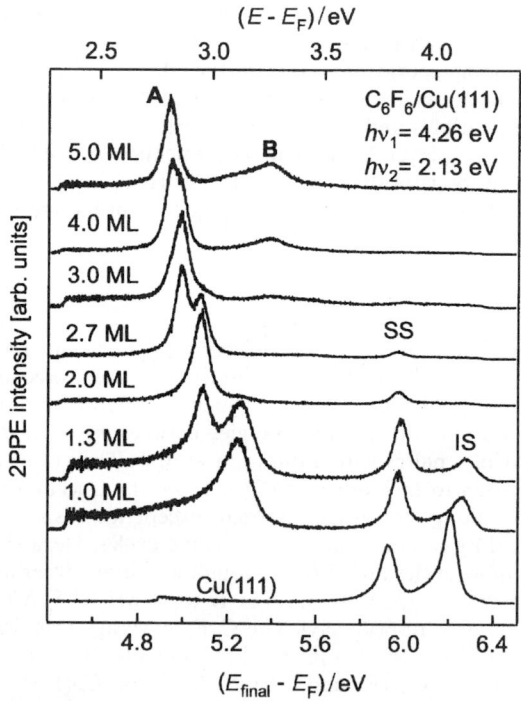

Fig. 2 2PPE spectra of clean Cu(111) and 1 to 5 ML C_6F_6/Cu(111) recorded with $hv_2 = 2.13$ eV and $hv_1 = 4.26$ eV photons. The bottom (top) axis refers to the final (intermediate) state energy in the 2PPE process, as described in the text. The spectra show peaks from the occupied surface state (SS) and image potential states (IS), as well as adsorbate induced states labeled A and B.

made previously for image potential state on Ag/Pd(111).[29] The energetic position of state A agrees well with a 2PPE study by Vondrak and Zhu,[7] where it was assigned to a σ* molecular resonance of hexafluorobenzene based on inverse photoemission[28] and photo-detachment data.[30] In a preliminary report of this work, it has been argued, that the observed downward shift of peak A suggests the formation of molecular quantum well states, which extend over the total thickness of the adsorbate layer.[31] This interpretation needs revisions, as we will show below by modeling and additional overlayer experiments.

For coverages above 3 ML a new peak (B) appears in the 2PPE spectra which arises from an unoccupied state at 3.25 ± 0.03 eV above the Fermi level. At higher coverages the intensity of peak B becomes more pronounced and its maximum remains at constant energy, while an asymmetric tail develops on the low energy side. This suggests that peak B may actually consist of several states. Peak B is visible even after prolonged annealing of the adsorbate layer, which indicates that it does not arise from defects in the C_6F_6 multilayer but must be attributed to intrinsic states of the adsorbate layer. A polarization analysis shows that all peaks observed in the 2PPE spectra of Fig. 2 can be probed only with p-polarized light and are thus totally symmetric with respect to the symmetry group of the C_6F_6/Cu(111) surface (see below).

Fig. 3 displays measurements of the dispersion of state A. The left panel displays the intermediate state energy as a function of $k_\parallel$ for coverages between 1 and 5 ML. State A forms a dispersive band parallel to the surface with an effective mass of $m_{eff} = (1.95 \pm 0.2)\,m_e$ for 1 ML C_6F_6/Cu(111). With increasing coverage the downshift to lower energies is accompanied by a strong decrease in the effective mass which approaches the free electron value above 4 ML. The higher effective mass obtained for the monolayer coverage indicates that the electronic potential is strongly modulated parallel to the surface. This can be rationalized by the fact that we observe an ordered (3 × 3) low-energy electron diffraction (LEED) structure for 1 ML C_6F_6/Cu(111), which suggests a flat adsorption geometry of C_6F_6 on the surface, similar to benzene/Cu(111).[32] For this adsorption geometry an electron moving parallel to the surface will experience a lateral modulation of the potential due to the difference in charge density across a C_6F_6 molecule. On the other hand, the higher effective mass could also be explained by the assumption that the state A originates from a molecular orbital of C_6F_6 (i.e. from a state with localized character) which develops into a band due to the lateral interactions between these orbitals. A difference in adsorption geometry between the first and subsequent layers would imply different lateral interactions and thus a change in the effective mass. Note, however, that the effective mass does not drop abruptly to the free electron value, but rather decreases continuously with increasing coverage. For peak B we have observed a weak upward dispersion, however, it is not possible to quantify the effective mass due to the asymmetric lineshape and the possibility of several overlapping states.

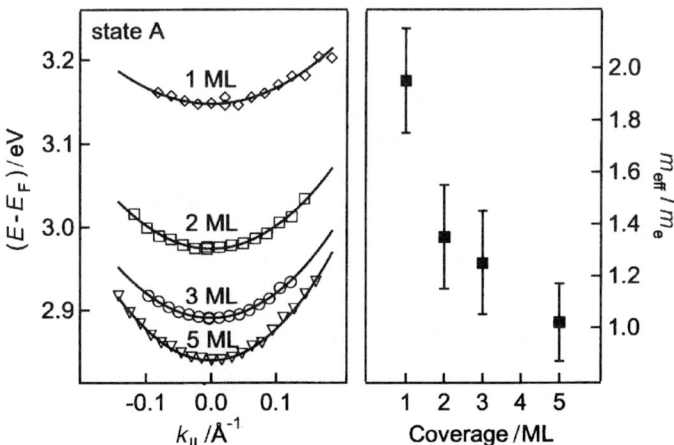

Fig. 3 Left panel: Dispersion of the intermediate state energy of state A as a function of the electron momentum $k_\parallel$ parallel to the surface for C_6F_6 coverages between 1 and 5 ML. Right panel: Effective electron mass m_{eff}/m_e for state A as a function of coverage.

To obtain insight into the coupling strength of the excited states with the underlying metal substrate we use time-resolved 2PPE to measure the lifetimes of the states A and B as a function of coverage. Fig. 4 shows cross correlation curves for state A taken with 4.26 eV pump and 2.13 eV probe pulses. The time zero and width of the cross correlation between the pump and probe pulses are determined from the signal of the occupied surface state (SS) on the clean Cu(111) surface, as discussed previously.[11] Analysis of the cross correlation curves of peak A shows that, above 2 ML, the data cannot be represented by an instantaneous population by the pump pulse followed by a single exponential decay. In order to obtain a satisfactory fit a delayed population from energetic higher lying states must be included, whereby the temporal change of the electron population is described by a rate equation including a finite rise and decay term with time constants $\tau_{rise}(A)$ and $\tau_{decay}(A)$, respectively. The lifetime of state A, $\tau_{decay}(A)$, extracted from this fit shows a pronounced increase from 7 fs for 1 ML to 32 fs for 5 ML C_6F_6. For the resulting finite rise time we find a value of $\tau_{rise}(A) = 14$ fs for 3 ML which increases to 23 fs for 5 ML. Two different mechanisms could be responsible for the delayed rise of the population of states A: (i) interband decay from energetically higher lying states (like peak B or image potential states) or (ii) intraband decay from states with higher $k_\parallel$ within the upward dispersing band of state A. The latter process of intraband decay has been found to play an important role in the dynamics of image potential states on Xe/Ag(111)[12] and N_2/Xe/Cu(111).[33] We note, that the limited time resolution in our experiment makes it difficult to distinguish from other mechanisms, however, the observed trends for the lifetimes will be robust against different assumptions in the fitting procedure. A lifetime analysis of peak B gives values for $\tau_{decay}(B)$ between 8 fs and 14 fs. These values are similar to the rise times $\tau_{rise}(A)$ of peak A within the experimental errors, suggesting that the state B decays efficiently into the energetically lower lying state A. Energy conservation required in this decay process can be achieved either by electron–hole pair excitation in the substrate or by phonon excitation in the adlayer. The short lifetimes of peak B would only be consistent with an electronic mechanism. A similar scenario was proposed to explain the very short lifetimes of image potential states in the C_6F_6/Cu(111) system.[27]

In a further set of experiments we have studied C_6HF_5 and $C_6H_2F_4$ on Cu(111) to investigate the influence of the adsorbate electron affinity on the electron dynamics. For pentafluorobenzene

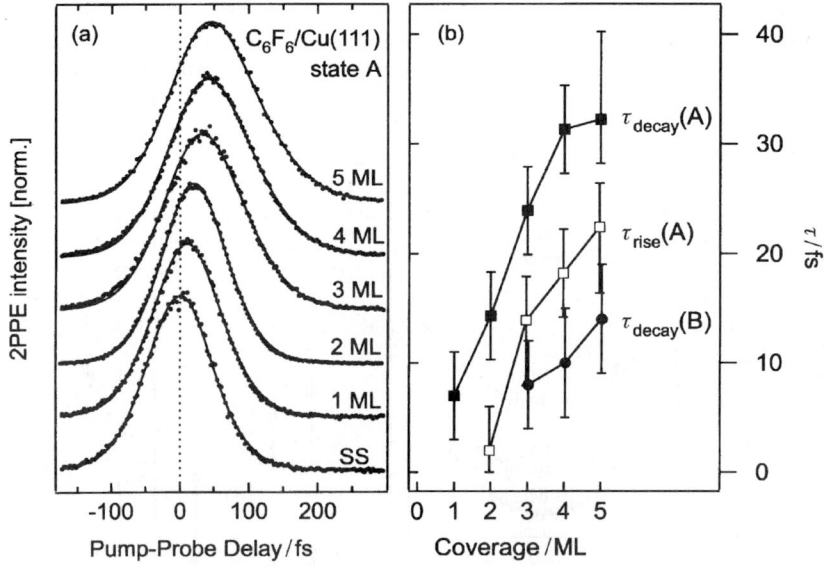

Fig. 4 Time-resolved 2PPE spectroscopy of C_6F_6/Cu(111): (a) Cross-correlation curves of state A on C_6F_6/Cu(111) recorded with $h\nu_1 = 4.26$ eV pump and $h\nu_2 = 2.13$ eV probe pulses. The dotted line indicates zero delay time determined from the cross-correlation signal of the occupied SS on clean Cu(111). (b) Lifetimes (population decay τ_{decay}) for states A and B as a function of coverage. For state A a finite rise time has to be used to obtain a satisfactory fit.

we find no signature of any state equivalent to peaks A or B. We speculate that this may be due to a different adsorption geometry of C_6HF_5 in the first layer compared to the flat lying C_6F_6. For 1 ML tetrafluorobenzene (1–4-$C_6H_2F_4$) we observe an unoccupied electronic state located at $E - E_F = 3.78 \pm 0.02$ eV (not shown). This state exhibits a very short lifetime, which could not be resolved due to our limited time resolution and the interference with the hot electron background. Compared with state A located at $E - E_F = 3.14$ eV for 1 ML C_6F_6/Cu(111) the higher binding energy observed for $C_6H_2F_4$ is expected from the energy shift of the lowest σ^* orbital between C_6F_6 and $C_6H_2F_4$, which amounts to ~ 1.2 eV in the gas phase.[34] For a coverage of 2 ML we find binding energies of 3.89 ± 0.04 eV for $C_6H_2F_4$ compared to 2.97 eV for C_6F_6, *i.e.* the difference becomes even closer to the gas phase value.

3.2 Simulation within the dielectric continuum model

Many weakly bound adsorbates on metal surfaces exhibit unoccupied electronic states which are delocalized parallel to the surface and can be regarded as derived from the image potential states of the bare surface (although binding energies and lifetimes can be changed significantly). For rare gases and N_2 as well as for C_6H_6/Ag(111) these changes have been successfully described by the dielectric continuum model (DCM).[13,35,36] This model can be regarded as a "standard model" in the discussion of the electron dynamics at adsorbate/metal interfaces and we will thus test if the C_6F_6/Cu(111) system can be successfully described within this framework. The DCM treats the adsorbate layer as a homogeneous dielectric slab characterized by its relative permittivity ε and its electron affinity E_{ea}. The band structure of the adsorbate in the direction perpendicular to the surface can be parametrized by an effective mass $m_\perp$ while the dispersion parallel to the surface is taken to be free-electron-like so that the problem can be reduced to one dimension. At the interfaces the classical potential exhibits unphysical divergences which are eliminated by a cut-off at a distance of the Thomas–Fermi screening length in front of the metal and a linear interpolation of the potential over a distance b around the adsorbate/vacuum interface.[37] The vacuum energy, E_{vac}, is obtained from the experimental work function. For states inside the projected band gap of Cu(111) (along the Γ–L direction) trial wavefunctions are calculated inside the substrate within the two-band-approximation of a nearly free electron gas and propagated numerically through the dielectric slab to the vacuum. The binding energies for which the wavefunctions vanish far away from the surface are determined iteratively. Further details of the calculation are described elsewhere.[13]

For adsorbates with a small positive electron affinity like Xe ($E_{ea} = 0.5$ eV), the dielectric screening in front of the metal surface and the formation of an additional potential well in front of the adsorbate layer lead to a decrease in the binding energies and an increase in the lifetimes of the image potential states. This is caused by a shift of the probability density $|\Psi|^2$ towards the adsorbate/vacuum interface and the resulting decrease in overlap with the substrate. The situation is different for an adsorbate like C_6F_6 which has a high electron affinity of about 1.3 eV in the condensed phase.[38] In this case the adsorbate layer forms a potential well which is declined towards the metal surface. Fig. 5(a) shows the model potentials for 1 ML (solid line) and 3 ML (dotted line) C_6F_6 with $\varepsilon = 2.029$[39] and $E_{ea} = 1.3$ eV.[38] The dashed lines indicate the assumed layer thickness of 2.5 Å for the first layer (corresponding to flat lying molecules) plus 5.0 Å for each additional layer. The latter value is close to the average distance between neighboring molecules in a solid C_6F_6 crystal.[40] Because the potential well becomes wider with increasing layer thickness, all states shift towards the bottom of the well as observed for state A in the 2PPE experiments. Fig. 5(b) shows the measured binding energies (solid symbols) together with the results of the dielectric continuum model (open circles connected by solid lines) calculated for $m_\perp = 2\, m_e$, which equals the experimentally obtained value of m_{eff} parallel to the surface for 1 ML.

The binding energies of state A as a function of coverage are well reproduced by the lowest ($n = 1$) state of the model. Also the other peaks observed in 2PPE can be correlated with states obtained from the model. The results of the DCM simulation suggest that the two states around $E - E_F = 4.15$ and 4.35 eV, discussed as image potential states in a previous publication,[27] have a different number of nodes at different coverages and should be identified with the $n = 2$ and $n = 3$ states for 1 ML C_6F_6/Cu(111) and with the $n = 3$ and $n = 4$ states for 2 ML. For coverages above

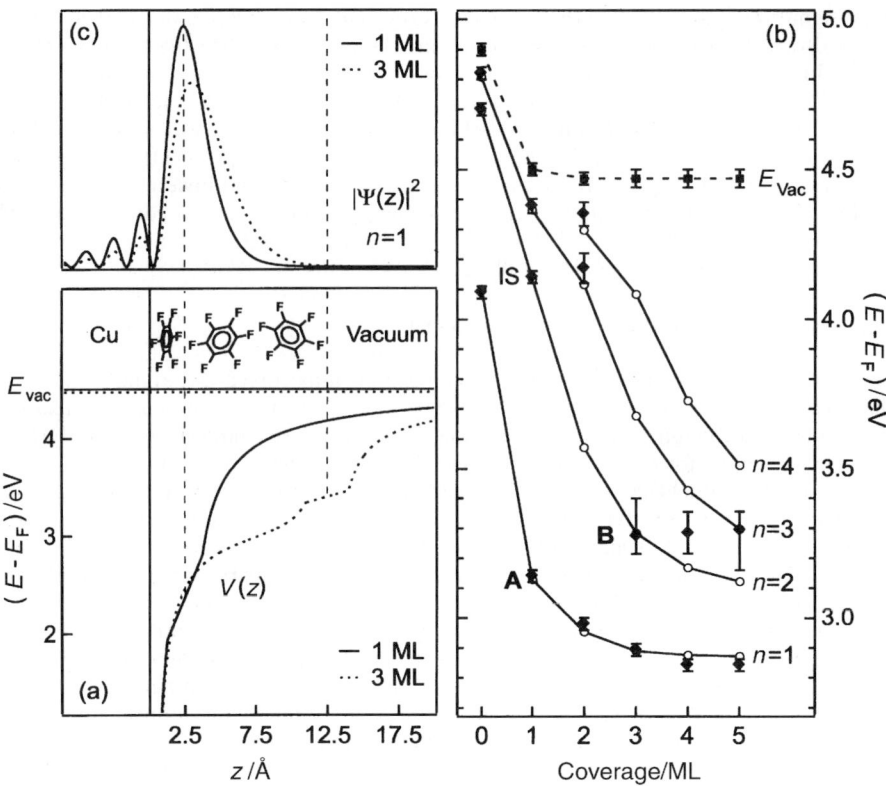

Fig. 5 Simulation within the DCM: (a) and (c) depict the modified image potential $V(z)$ and the calculated probability density $|\Psi|^2$ of the electron wavefunction for 1 ML (——) and 3 ML (···) C_6F_6/Cu(111), respectively. The vertical dashed lines indicate the adsorbate vacuum interface. (b) compares the results of the DCM calculation (○ connected by solid lines) with the experimentally observed binding energies (♦) of the unoccupied states obtained from 2PPE spectra at various photon energies. For peak B the error bars indicate the asymmetric shape of the peak, rather than the uncertainty of its position. Also shown is the measured position of the vacuum level, E_{vac} (■ connected by a dashed line), which enters the simulation as a parameter. The solutions of the stationary states in the potential $V(z)$ are labeled by $n = 1, 2, \ldots$ to indicate the number of nodes of the wavefunction in front of the surface.

3 ML peak B is observed in the energy range where the states $n = 2$ and $n = 3$ are expected from the model. A possible explanation for the constant energy of the peak maximum could be the existence of islands with a local coverage of 3 ML and 5 ML due to an inhomogeneous layer growth above 3 ML. The development of the asymmetric tail on the low energy side of the peak agrees well with the downward shift of the states within the model.

In contrast to the energetically higher lying states the calculated wavefunction of the $n = 1$ state is mainly concentrated in the region close to the substrate as depicted in Fig. 5(c). While for 1 ML the probability density $|\Psi|^2$ extends significantly into the vacuum, it is located completely inside the adsorbate layer at 3 ML. Further adsorption has almost no influence on this state. This localization close to the surface is a result of the attractive potential in front of the metal.

Besides the calculation of binding energies it is also possible to estimate the lifetimes of electronic states from the overlap of their wavefunctions with the metal substrate. Assuming that the population decays mainly into bulk states the lifetimes should be approximately inversely proportional to the fraction of the probability density inside the substrate, defined as the wavefunction overlap.[13] For our model, we estimate the lifetime using a prefactor of 3.85 fs multiplied by the inverse of the wavefunction overlap.[19,33] In this approximation the model predicts lifetimes between 30 and 65 fs for the $n = 1$ state of the C_6F_6-covered surface, which reproduces the experimentally observed trend for $\tau_{decay}(A)$ with increasing coverage but overestimates the absolute values by a factor of two. The calculated wavefunctions of the $n = 2$ and $n = 3$ states in the energy region of peak B have an even smaller overlap with the substrate. Lifetimes between 70 and 175 fs would be expected, one order of magnitude longer than experimentally observed. This supports our interpretation that the energetically lower lying state A acts as an additional decay channel within the adsorbate layer.

3.3 Failure of the dielectric continuum model

At first sight the results of the DCM simulation presented in Fig. 5 seem to demonstrate almost perfect agreement between the model and experiments. The calculation can reproduce quantitatively the coverage dependence of the binding energy for state A as well as qualitatively the trend of its lifetime. The model also provides reasonable values for the binding energies of image potential states and peak B. However, there are at least two major shortcomings: (i) Above a coverage of 1 ML the calculation predicts several unoccupied states between 3.4 and 3.8 eV which are not observed in the experiment (Fig. 5(b)). The calculated wavefunction overlap of these states with the substrate is smaller, but still comparable to that of the $n = 1$ state and thus there is no reason why these states could not be populated by the pump pulse and subsequently be detected in 2PPE. (ii) The one-dimensional DCM model implicitly assumes that the lateral variation of the potential in the adlayer can be neglected. However, as shown in Fig. 4 the effective mass of state A decreases continuously from $m_{eff} = 1.95\ m_e$ for 1 ML to the free electron mass above 4 ML. This implies that the lateral corrugation of the potential will also decrease with coverage and that the resulting release of localization energy has to be accounted for in the analysis of the coverage dependent binding energies (see Fig. 1). Calculations using the Kronig–Penney model show that this contribution can easily amount to several 100 meV. Taken together we conclude that the good fit obtained with the dielectric continuum model does not provide a consistent description of all experimental findings, in particular with respect to the changes of the effective mass with coverage.

3.4 Adsorption of Xe overlayers: localization in the first layer

In order to gain further insight into the nature of the excited states of $C_6F_6/Cu(111)$ we have performed additional experiments using xenon overlayers. In particular we want to characterize the degree of localization of the wavefunction in the direction normal to the surface. The inset in Fig. 6 shows the probability density $|\Psi|^2$ of the $n = 1$ state calculated with the dielectric continuum model for 1 ML Xe adsorbed on top of 1 ML $C_6F_6/Cu(111)$.[33] Note, that the DCM calculations predict that a significant part of $|\Psi|^2$ is located outside the first C_6F_6 layer and must strongly overlap with adsorbates in the second layer. Using adsorbate overlayers with different electronic properties allows therefore to test how much the wavefunction is actually localized inside the first layer. We choose physisorbed Xe, because Xe has an electron affinity of 0.5 eV and thus exhibits a band gap in the energy range of the peaks A and B on $C_6F_6/Cu(111)$. Adsorption of a Xe overlayer is therefore expected to 'push' the $n = 1$ wavefunction towards the metal causing an upward shift in energy due to increased localization. At the same time a decrease in the lifetime is expected due to a stronger coupling with the metal substrate. DCM calculations for this adsorbate heterostructure confirm this picture and predict an upward shift of ~ 300 meV for the $n = 1$ state for 1 ML Xe/1 ML $C_6F_6/Cu(111)$.

Fig. 6 summarizes the results of the overlayer experiments by comparing 2PPE spectra for 1, 2 and 4 ML $C_6F_6/Cu(111)$ with the corresponding spectra taken after adsorption of 1 ML Xe on top of the preadsorbed C_6F_6 layers. Due to Xe adsorption the intensity of peak A is decreased, but there is only a negligible downward shift of its energetic position (~ 20 meV), in clear contrast to the prediction of the DCM calculation. In time-resolved 2PPE measurements we also find that the lifetime of state A does not change upon Xe adsorption within the experimental error of ± 5 fs (not shown). The large spillout of the wavefunction into the second and third layer, as predicted by the DCM, is incompatible with the observed insensitivity of the binding energy and lifetime of state A to Xe adsorption. Even quite different parameter sets in the DCM calculation cannot give a satisfactory agreement. Our findings can thus be only rationalized if we conclude that the observed unoccupied electronic state A is localized predominantly inside the C_6F_6 layer even for

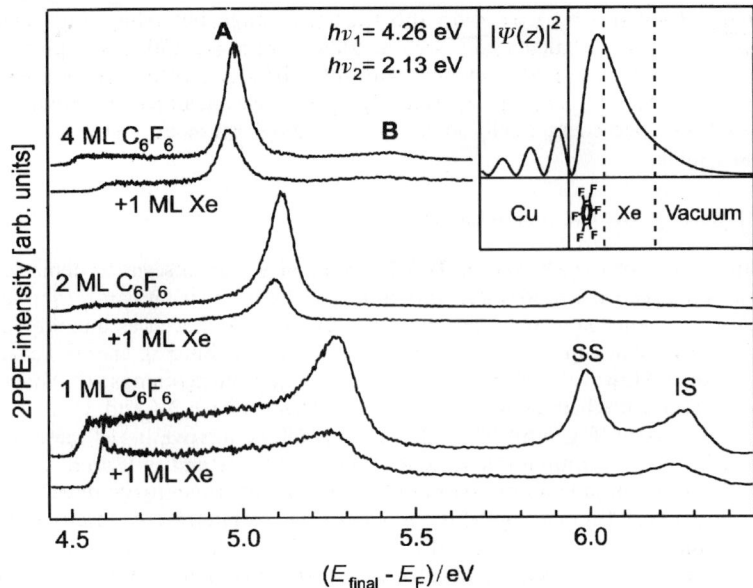

Fig. 6 Comparison between 2PPE spectra of C_6F_6/Cu(111) for coverages of 1, 2 and 4 ML and after adsorption of 1 ML xenon on top of the preadsorbed C_6F_6 layers. The inset illustrates schematically the layer thickness for 1 ML Xe adsorbed on top of 1 ML C_6F_6/Cu(111). The wavefunction of the $n = 1$ state calculated with the DCM model for this heterolayer system would exhibit a significant overlap with the Xe overlayers. Note, that adsorption of the Xe overlayer reduces the intensity of peak A, but results only in minor shifts of its binding energy (<20 meV).

1 ML C_6F_6/Cu(111). Recall that for 1 ML the measurements of the effective mass also indicate an increased localization of the wavefunction parallel to the surface. This localization parallel and perpendicular to the surface strongly suggests that peak A must originate from a molecular state of the C_6F_6 molecules.

For the free hexafluorobenzene molecule several possibilities have been discussed for the lowest unoccupied molecular orbital (LUMO), namely the $\pi^*(e_{2u})$-orbital (which represents the LUMO of benzene) as well as the lowest antibonding σ(C–F)-orbital, which is most likely $\sigma^*(e_{1u})$ or $\sigma^*(e_{2g})$.[34,41] On the other hand the energetically low lying states of the $C_6F_6^-$ radical anion are $\pi^*(e_{2u})$ and $\sigma^*(a_{1g})$.[42] The fact that in 2PPE spectra recorded in normal emission the peaks A and B are not observed with s-polarized probe pulses is consistent with the selection rules of the totally symmetric orbital $\sigma^*(a_{1g})$.[22] Only if the symmetry is lowered to C_{2v} or C_2 (as it is expected for the ground state of the radical anion due to a Jahn–Teller coupling of the $\pi^*(e_{2u})$ and $\pi^*(a_{1g})$ orbitals[42]) also peaks arising from the $\pi^*(e_{2u})$ or $\sigma^*(e_{2g})$ orbital would exhibit the observed polarization dependence, whereas the $\sigma^*(e_{1u})$ orbital can be ruled out. Although from our experiment a definite assignment is not possible, the observed photochemical decomposition of C_6F_6 by the UV pump pulses indicates the transient population of an antibonding orbital with respect to the C–F bond, which suggests that state A originates from a σ^* orbital. For coverages above 1 ML the molecules probably adsorb in a tilted geometry, so that photoemission with s-polarized light should no longer be forbidden by the selection rules for any of the mentioned orbitals. However, because state A is predominantly localized within the first layer molecules in the outer layers may have only minor influence on the polarization dependence.

Finally we note, that in Fig. 6 peak B remains visible after Xe adsorption and does not change its energetic position. This supports our assignment of peak B to an unoccupied state inside the C_6F_6 layer. It is known that many aromatic molecules form excimers.[43] One might thus speculate that peak B arises from excimers in the C_6F_6 layer.[44] This would be consistent with the observation of a broad and weakly dispersing band, but the ultrashort lifetimes seem to contradict this interpretation. As a further test of the interaction between the electronic states of the C_6F_6 layer and the underlying metal substrate we have also tried to vary this coupling using alkane spacer layers. Unfortunately we found that C_6F_6 does not wet octane and pentane layers, as becomes evident by the absence of an energy shift of image potential states in the 2PPE spectra. This is different for the Xe overlayers adsorbed on top of C_6F_6 (Fig. 6), where the observed changes in the work function and the binding energies of the image potential states demonstrate that Xe indeed wets the C_6F_6 surface.

4. Conclusions

In summary, we have used time-resolved 2PPE spectroscopy to systematically investigate the binding energies, dispersion and lifetimes of the unoccupied electronic states of C_6F_6 adsorbed on Cu(111) as a function of coverage. When the coverage is raised between 1 ML and 5 ML C_6F_6/Cu(111) we observe for the lowest unoccupied state a downward shift of the energetic position from 3.14 to 2.84 eV above E_F, which is accompanied by a strong increase in the excited state lifetime from 7 to 32 fs and a decrease in the effective mass parallel to the surface from $m_{eff} = 1.95\ m_e$ to the free electron value. By adsorption of Xe overlayers on top of C_6F_6/Cu(111) we find that, even for 1 ML, the excited state wavefunction must be localized predominantly inside the first C_6F_6 layer. A simulation within the dielectric continuum model reproduces the coverage dependence of the binding energy and the lifetimes, however, it fails to describe the localized character of the wavefunction.

Our interpretation is the following: For 1 ML C_6F_6/Cu(111) the excited state has mainly localized character originating from molecular orbitals (presumably the antibonding σ^* orbital of C_6F_6) which interact *via* the substrate and form an adsorbate band structure with a pronounced dispersion parallel to the surface ($m_{eff} \sim 2\ m_e$). The short lifetime of only 7 fs for 1 ML is a result of this strong coupling to the metal substrate *via* the tail of the excited state wavefunction. With increasing coverage the excited state becomes more delocalized within the adsorbate layer, both parallel as well as normal to the surface, and the electron can move freely in the directions parallel to the surface ($m_{eff} \sim m_e$). Thus for monolayer coverage the situation may be viewed in a tight binding picture of interacting molecular orbitals, which gradually develops at higher coverages to a situation where the picture of delocalized quantum well states is more appropriate. This delocalization is accompanied by a shift of the probability density $|\Psi|^2$ away from the metal, which reduces the coupling with the continuum of electronic excitations in the substrate and thus leads to an increase in the excited state lifetime. The C_6F_6/Cu(111) system provides a nice example of how the transition from a more localized (molecular) state to a delocalized (quasi-free electron) state affects not only the energetics but also the dynamics of the excited state due to the changes in the coupling strength with the substrate. For a complete understanding the three-dimensional potential including chemical interactions between the molecules of the adlayer has to be taken into account while a simple one-dimensional description within the dielectric continuum model must fail.

Acknowledgements

The authors would like to thank G. Ertl for continuous support and X.-Y. Zhu and H. Petek for valuable discussions and important comments. This work was supported in part by the Deutsche Forschungsgemeinschaft through SPP 1093.

References

1. R. J. D. Miller, G. L. McLendon, A. J. Nozik, W. Schmickler and F. Willig, *Surface Electron Transfer Processes*, VCH, New York, 1995.
2. M. Bixon and J. Jortner, *Adv. Chem. Phys.*, 1999, **106**, 35.
3. S. T. Tans, A. R. M. Verschueren and C. Dekker, *Nature*, 1998, **393**, 49.
4. C. A. Merkin and M. A. Ratner, *Annu. Rev. Phys. Chem.*, 1992, **43**, 719.
5. M. A. Fox, *Acc. Chem. Res.*, 1999, **32**, 201.
6. H. Sirringhaus, N. Tessler and R. H. Friend, *Science*, 1998, **280**, 1741.
7. T. Vondrak, H. Wang, P. Winget, C. J. Cramer and X.-Y. Zhu, *J. Am. Chem. Soc.*, 1999, **122**, 4700.
8. E. Hasselbrink, *Appl. Surf. Sci.*, 1994, **97/80**, 34.

9 T. Vondrak and X.-Y. Zhu, *J. Phys. Chem. B*, 1999, **103**, 3349.
10 H. Petek and S. Ogawa, *Prog. Surf. Sci.*, 1998, **56**, 239.
11 M. Wolf, E. Knoesel and T. Hertel, *Phys. Rev. B*, 1996, **54**, R5295.
12 C. M. Wong, J. D. McNeill, K. J. Gaffney, N.-H. Ge, A. D. Miller, S. H. Liu and C. B. Harris, *J. Phys. Chem. B*, 1999, **103**, 282.
13 A. Hotzel, K. Ishioka, G. Moos, M. Wolf and G. Ertl, *Appl. Phys. B*, 1999, **68**, 615.
14 L. Bartels, G. Meyer, H. Rieder, D. Velic, E. Knoesel, A. Hotzel, M. Wolf and G. Ertl, *Phys. Rev. Lett.*, 1998, **80**, 2004.
15 M. Bauer, S. Pawlik and M. Aeschlimann, *Phys. Rev. B*, 1997, **55**, 10040.
16 S. Ogawa, H. Nagano and H. Petek, *Appl. Phys.B*, 1999, **68**, 611.
17 R. L. Lingle, N. H. Ge, R. E. Jordan, J. D. McNeill and C. B. Harris, *Chem. Phys.*, 1996, **205**, 191.
18 E. Knoesel, A. Hotzel and M. Wolf, *J. Electron Spectrosc. Relat. Phenom.*, 1998, **88-91**, 577.
19 E. V. Chulkov, I. Sarria, V. M. Silkin, J. M. Pitarke and P. M. Echenique, *Phys. Rev. Lett.*, 1998, **80**, 4947.
20 J. Osma, I. Sarria, E. V. Chulkov, J. M. Pitarke and P. M. Echenique, *Phys. Rev. B*, 1999, **59**, 10591.
21 T. Fauster and W. Steinmann, in *Electromagnetic Waves: Recent Development in Research*, ed. P. Halevi, Elsevier, Amsterdam, 1995, p. 350.
22 H. P. Steinrück, *J. Phys.: Condens. Matter*, 1996, **8**, 6465.
23 C. Kittel, *Introduction to Solid State Physics*, Wiley, New York, 1996.
24 D. Velic, A. Hotzel, M. Wolf and G. Ertl, *J. Chem. Phys.*, 1998, **109**, 9155.
25 N.-H. Ge, C. M. Wong, R. L. L. Jr., J. D. McNeill, K. J. Gaffney and C. B. Harris, *Science*, 1998, **279**, 202.
26 E. Knoesel, A. Hotzel, M. Wolf and G. Ertl, *Phys. Rev. B*, 1998, **57**, 12812.
27 K. Ishioka, C. Gahl and M. Wolf, *Surf. Sci.*, 2000, **454**, 73.
28 R. Dudde, B. Reihl and A. Otto, *J. Chem. Phys.*, 1990, **92**, 3930.
29 R. Fischer, S. Schuppler, N. Fischer, T. Fauster and W. Steinmann, *Phys. Rev. Lett.*, 1993, **70**, 654.
30 L. G. Christophorou, P. G. Datskos and H. Faidas, *J. Chem. Phys.*, 1994, **101**, 6728.
31 X.-Y. Zhu, T. Vondrak, H. Wang, C. Gahl, K. Ishioka and M. Wolf, *Surf. Sci.*, 2000, **451**, 244.
32 M. Xi, M. X. Yang, S. K. Jo and B. E. Bent, *J. Chem. Phys.*, 1994, **101**, 9122.
33 A. Hotzel, M. Wolf and J. P. Gauyacq, *J. Phys. Chem.*, in press.
34 A. P. Hitchcock, P. Fischer, A. Gedanken and M. B. Robin, *J. Phys. Chem.*, 1987, **91**, 531.
35 J. D. McNeill, J. R. L. Lingle, R. E. Jordan, D. F. Padowitz and C. B. Harris, *J. Chem. Phys.*, 1996, **105**, 3883.
36 K. J. Gaffney, C. M. Wong, S. H. Liu, A. D. Miller, J. D. McNeill and C. B. Harris, *Chem. Phys.*, 2000, **251**, 99.
37 In the calculations for C_6F_6/Cu(111) the parameter b is set to 2.5 Å for 1 ML and 3.0 Å for higher coverages.
38 H. Faidas, L.G.Christophorou and D. L. McCorkle, *Chem. Phys. Lett.*, 1992, **193**, 487.
39 D. R. Lide and H. P. R. Frederikse, *CRC Handbook of Chemistry and Physics*, CRC Press, Boca Raton, FL, 1995.
40 N. Boden, P. P. Davis, C. H. Stam and G. A. Wesselink, *Mol. Phys.*, 1973, **25**, 81.
41 L. N. Shchegoleva, I. I. Bilkis and P. V. Schastnev, *Chem. Phys.*, 1983, **82**, 343.
42 L. N. Shchegoleva, I. V. Beregovaya and P. V. Schastnev, *Chem. Phys. Lett.*, 1999, **312**, 325.
43 J. B. Birks, *Organic Molecular Photophysics*, Wiley, London, 1973, vol. 1.
44 K. Hiraoka, S. Mizuse and S. Yamabe, *J. Phys. Chem.*, 1990, **94**, 3689.

Excitation mechanisms and photochemistry of adsorbates with spherical symmetry

Kazuya Watanabe and Yoshiyasu Matsumoto*

Department of Photoscience, SOKEN-DAI, The Graduate University for Advanced Studies, Hayama, Kanagawa 240-0193, Japan. E-mail: matsumoto@soken.ac.jp; Fax: +81 468 58 1544

Received 10th April 2000
First published as an Advance Article on the web 28th September 2000

By comparing the photo-stimulated desorption of Xe from an oxidized Si(100) surface with the photochemistry of methane on metal surfaces, we try to deduce a common concept in the excited state and the excitation mechanism responsible for the photo-induced processes. Xe atoms are desorbed from the oxidized Si(100) surface by the irradiation of photons in the range 1.16–6.43 eV. Two velocity components with average kinetic energy 0.85 and 0.25 eV are observed in the time-of-flight distributions. The fast component appears only if the photon energy exceeds ~3 eV, but the slow component is present over the entire photon energy range. By analogy with the photochemistry of methane on metal surfaces, the excitation mechanism responsible for the fast component is postulated to be a transition from the 5p state of Xe to the excited state originating from strong hybridization between the 6s state of Xe and the dangling bond at a surface silicon atom bonded to oxygen inserted in the dimer bond. In this scheme an excited electron is transferred from the adsorbate to the substrate, which is the reverse direction to the substrate-mediated excitation frequently assumed in surface photochemistry.

1. Introduction

Electron transfer (ET) is an extremely important elementary process in many branches of chemistry and biology. Chemistry at solid surfaces is no exception. Electrochemistry is an example in which ET plays a central role. Furthermore, ET is a fundamental concept for understanding the nature of chemical bonding at solid surfaces. Adsorption of a molecule at a surface is accompanied by ET and electron redistribution between the adsorbate and the substrate, resulting in the formation of a covalent or ionic bond. Although the concept of ET does not include explicitly the transfer of an atom and/or a group of atoms, ET is intimately related to the dynamic processes at surfaces such as desorption and dissociation of the adsorbate.

It has been well established that ET processes induced by photon irradiation are a major excitation pathway in surface photochemistry.[1] Photon absorption by the substrate produces electron–hole pairs in the bulk. The hot electrons are transferred and attached to an adsorbate or adsorbate–substrate unoccupied orbital, forming a transient negative ion species. Then, the transient species gives rise to desorption or dissociation of the adsorbate. In most of the adsorbate–substrate systems studied in the past using photons with ≤7 eV, the electronic excitation of the adsorbate is explained by this substrate-mediated mechanism. In this excitation mechanism the non-thermal processes are initiated by ET from the substrate to the adsorbate. However, the opposite direction of ET can also be induced by photo-irradiation, as in the case of electron injection into semiconductor electrodes from adsorbed dye molecules.[2]

In this paper, we discuss photo-induced ET in the reverse direction, i.e., ET from the adsorbate to the substrate, which plays an important role in the photochemistry of adsorbates with spherical or pseudo-spherical symmetry. Two adsorption systems are discussed: Xe on an oxidized Si(100) surface and methane on metal surfaces. Although the electronic structures of the substrates discussed here are quite different, there are some similarities between Xe and methane. First they both have a closed-shell structure in the ground state, so that they are very stable and in general physisorbed on solid surfaces. Xe has true spherical symmetry. The lowest unoccupied orbital of Xe is 6s. The excitation from 5p to 6s occurs at 8.4 eV and the ionization potential, E_i is 12 eV. Methane is a spherical-top molecule and also has high structural symmetry (T_d). The first excited state of methane ($^1T_2$) is located at ~ 10 eV and $E_i = 13$ eV. This excited state is antibonding with respect to a C–H bond, but mixes heavily with Rydberg states, so that the wavefunction has an extensive 3s-Rydberg character. Thus, the lowest unoccupied molecular orbital of methane has similar symmetry to that of Xe.

The first example described in detail is the photo-stimulated desorption (PSD) of Xe from the Si(100) surface oxidized by one monolayer of oxygen. We have recently reported on the non-thermal PSD of Xe and Kr from a clean Si(100) surface by irradiation with photons between the near-IR and UV.[3] Here we show that oxidation of the first layer of silicon atoms drastically changes the PSD features. The second example is the photochemistry of methane on metal surfaces. We briefly summarize the major results and the plausible excitation mechanism of this case, since these have been described in detail elsewhere.[4–7] Then, we discuss common features in the excitation mechanisms for the two systems.

2. Photodesorption of Xe from oxidized Si(100)

All the PSD activity of rare gas (Rg) atoms from solid surfaces has been observed only when the photon energy exceeds ~ 7 eV. The excitation involved in the PSD processes is believed to be either due to an excitonic or ionic transition. Below 7 eV, Rg atoms on metal surfaces do not show any non-thermal PSD. Thus, Rg, particularly Xe, atoms are frequently used for forming a space layer between the adsorbate and the substrate to separate the electronic systems.

In contrast, we have recently found that Rg atoms (Xe and Kr) adsorbed on a clean Si(100) surface are desorbed by irradiation with near IR–UV ns laser pulses.[3] The desorption can occur as a result of the direct energy flow to the desorption channel from hot surface phonons generated by charge recombination *via* surface states before they decay into the bulk. Although the desorption is non-thermal in nature, this mechanism does not include any electronic excitation of the adsorbate itself. However, when the surface silicon atoms are oxidized or mono-deuterated, electronic excitation of the adsorbate–substrate complex is feasible and the features of PSD are remarkably changed, as described in the following.

2.1. Experimental

Temperature-programmed desorption (TPD) and time-of-flight (TOF) measurements of desorbed species were performed in a manner similar to that reported elsewhere under the UHV condition ($<1.3 \times 10^{-8}$ Pa).[4,8] A Si(100) sample (Sb-doped, 0.005–0.01 Ω cm, 8×20 mm^2) was mounted on a copper block and could be cooled down to 45 K with an He refrigerator and heated above 1150 K by resistive heating. The surface was initially cleaned by repeated cycles of Ar$^+$-sputtering and annealing followed by flashing to 1150 K which give sharp (2×1) and dim c (4×2) low-energy electron diffraction (LEED) patterns revealing two domains below 100 K. The cleanliness of the sample was checked by X-ray photoelectron spectroscopy (XPS). The sample was cleaned by flashing to 1150 K after each measurement. The excitation light sources were an excimer laser (15 ns, ArF: 193 nm, and XeF: 351 nm), a Nd^{3+}: YAG laser (7 ns, fundamental: 1064 nm, and 2nd harmonics: 532 nm), and a YAG-laser pumped dye laser (7 ns, 440 nm). Laser beams were directed to the surface through a CaF$_2$ window. The angle of incidence was 45° to the surface normal. In this paper, unless indicated, desorption species along the surface normal were detected in the TOF measurements. The surface was oxidized by exposing the clean Si(100) surface at 45 K to 4.0 L

N$_2$O and annealing subsequently to 400 K. This procedure gave ~1 monolayer (ML) (1 ML = 6.8 × 10^{14} cm^{-2}) oxygen atoms on the Si(100) surface.[9] A mono-deuterated phase was prepared by dosing D$_2$ gas over a hot tungsten filament placed in front of the clean sample. During the D atom dosing the sample temperature was kept at 630 K to obtain the mono-deuterated phase.[10] Saturation of the D atom adsorption was confirmed by monitoring the integrated intensity of the D$_2$ associative desorption peak at 750 K. Rare gas dosing and laser irradiation were carried out at a surface temperature of 45–50 K.

2.2. Results

Fig. 1 shows the TPD results of Xe from Si(100) for various surface conditions. Four desorption peaks at 85, 72, 61 and 56 K are observed from the clean surface. The peaks at 72 and 85 K are assigned to desorption from pedestal and cave sites, respectively.[11] The first layer is completed by further adsorption at bridge and/or valley sites, yielding only one desorption peak at 61 K.[3] Desorption from the second layer gives a zero-order peak at 56 K. When the surface is oxidized, the distinction between these adsorption sites is smeared out, and Xe TPD shows a broad peak around 68 K and a sharp peak at 56 K. On the mono-deuterated surface, the interaction between Xe and the substrate seems to be weakened, and a zero-order desorption peak at 62 K and a peak at 53 K are observed. By comparing the TPD results of the modified surfaces with those of the clean surface, we assign the peaks at >60 K to desorption from the first layer and the peaks at <60 K to desorption from the second layer. In this work, we tentatively define 1 ML coverage of Xe when the desorption peak from the first layer is saturated on each surface. Other coverages are determined by taking the ratio of TPD signal intensity to that at 1 ML on each surface.

Xe is desorbed from the various surfaces by photon irradiation. The TOF distributions of Xe are shown in Fig. 2. The TOF distribution from the clean surface is well represented by a Maxwellian with a translational temperature of 380 K. However, when the surface silicon atoms are oxidized, the desorption yield along the surface normal is increased by a factor of 100. In addition, the TOF distribution is remarkably changed and shows two peaks at 200 and 380 μs. Hereafter we denote these non-Maxwellian components as α and β, respectively. The mean kinetic energy $\langle E_K \rangle$ of the component α is 0.85 eV and that of β is 0.25 eV. Note that the α component has a narrow width of 0.22 eV. When the clean surface is mono-deuterated, only a single peak with $\langle E_K \rangle$ = 0.17 eV appears. This component is denoted as γ. The desorption yields measured at 2.33 and 1.16 eV excitation energies are proportional to the laser fluence. Thus, desorption due to two-photon absorption is not likely to occur under the experimental conditions.

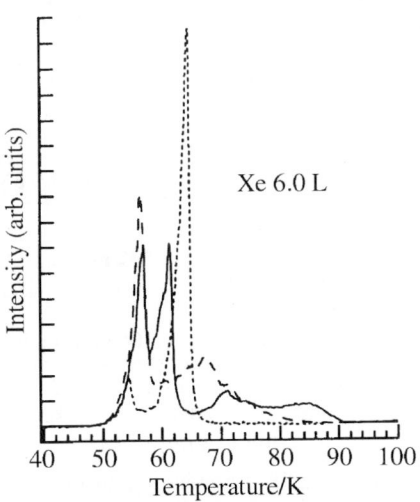

Fig. 1 Xe TPD profiles from clean and modified Si(100) surfaces: clean (———), 1 ML oxidized (– –), and mono-deuterated (···). Xe exposure is 6.0 L and the heating rate is 4 K s^{-1} in each measurement.

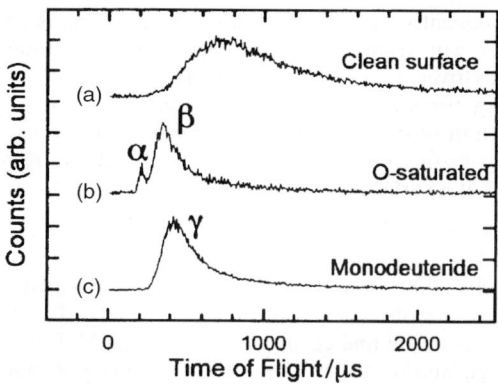

Fig. 2 Xe TOF distributions obtained from various surface conditions. The excitation wavelength is 351 nm and the laser fluence is 3.5 mJ cm^{-2}. Initial Xe exposures (coverage) and total photon numbers irradiated are: (a) 6 L (1.55 ML), 1.5×10^{19} photons cm^{-2}, (b) 0.75 L (0.1 ML), 2.5×10^{18} photons cm^{-2} and (c) 6 L (1.1 ML), 1.2×10^{19} photons cm^{-2}, respectively.

We examined how the TOF distributions depend on the excitation photon energy from 1.16 eV (1064 nm) to 6.43 eV (193 nm) for the oxidized Si(100) surface as shown in Fig. 3. The β component is persistently observed over the entire photon energy range used and its TOF distribution is almost independent of the photon energy. On the other hand, the α component shows remarkable photon energy dependence. This component is observed above 3.5 eV but not at 2.33 or 1.16 eV. Thus, it is clear that different excited states are responsible for the α and β components. The photon energy dependence of the α component suggests that its excited state is located at $\geqslant 3$ eV above the ground state.

2.3. The excitation mechanism

The TOF distributions are insensitive to the laser fluence in the range 0.5–6 mJ cm^{-2}, indicating that laser heating is not responsible for the PSD. As for the PSD from the clean surface, hot phonons localized near the surface may induce desorption.[3] However, the quite large kinetic energy of the desorbed species from the oxidized surface ($\geqslant 0.2$ eV) cannot be accounted for by energy transfer from the hot surface phonon to the adsorbate. Thus, electronic excitation should be responsible for the desorption. Since the wavelength dependence of the α component differs significantly from that of the β component, the excited states responsible for the two components are different. The β component is readily observed over a wide range, 1.16 and 6.43 eV and the

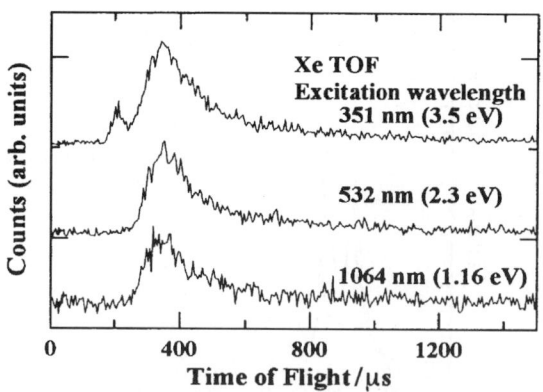

Fig. 3 Wavelength dependence of the Xe TOF distribution from the oxidized Si(100) surface. Excitation wavelengths are indicated in the figure. The laser fluence and photon numbers irradiated are: 3.5 mJ cm^{-2} and 2.8×10^{18} photons cm^{-2} at 351 nm, 3.5 mJ cm^{-2} and 7.7×10^{18} photons cm^{-2} at 532 nm, and 6.5 mJ cm^{-2} and 6.9×10^{19} photons cm^{-2} at 1064 nm. Initial Xe coverage is 0.1 ML in all cases.

wavelength dependence of its cross-section does not mimic that of the absorbance of the substrate. Thus, not only hot carriers but also thermalized carriers are responsible for the β component. In addition, the desorption yield of this component depends on the Xe coverage and the surface temperature in a complicated manner. We will describe these points in greater detail elsewhere[12] and focus only on the α component in this paper.

Since the α component is observed only for the 1 ML oxidized Si(100) surface, the surface structure of the oxidized surface plays a key role in the excitation. There are three different types of adsorption site for atomic oxygen on Si(100): a dimer bond (DB) and two different back bond (BB) sites of dimer atoms. The early first-principles calculations[13] suggested that insertion into the DB site is energetically the most stable. However, Uchiyama and Tsukada[14] found that the BB site of the down atom is the most stable if the silicon atoms of the substrate as well as the silicon dimer atoms are all relaxed. When the oxidation proceeds further, oxygen atoms are inserted into the other sites also. Kato et al.[15] showed that the adsorption energy is largest when one oxygen atom is inserted into the BB site of the down atom and the other into the DB site. Therefore, we adapt this geometry as the structure of the oxidized Si(100) surface with 1 ML oxygen atoms.

Since the α component is observed at submonolayer coverage, the excited state responsible for this component should be formed between a Xe atom and surface atoms in contact with each other. The angular distribution of the desorption yield of this component is sharply peaked along the surface normal ($\cos^4 \theta$). The narrow angular distribution and the very high kinetic energy (0.85 eV) suggest that the excited state is strongly bound, in contrast to the ground state. Since no spectroscopic and theoretical studies for this system are available at the moment, we are forced to speculate on the origin of the excited state.

It is worth considering the electronic structure of Rg monohalides.[16] A Rg atom in the ground electronic state interacts with a halogen atom (X) very weakly. The potential energy curves of $X\,^2\Sigma^+$ and $1\,^2\Pi$ composed of $Rg(^1S) + X(^2P)$ in the separated atom limit are mostly repulsive as a result of the overlap of the pσ orbitals of the two atoms. In contrast, the excited states $2\,^2\Sigma^+$ and $2\,^2\Pi$ composed of $Rg^+(^2S) + X^-(^1S)$ in the separated atom limit are strongly bound because of the Coulombic attraction between Rg^+ and X^- ions. Thus, the electronic configuration of the excited state of Rg monohalides is regarded as an ET state from the Rg atom to the counter halogen atom. The excited state of the Rg atom is considerably lowered by such complex formation. For example, XeF emits a photon of ~ 3.5 eV by the $2\,^2\Sigma^+ \to X\,^2\Sigma^+$ transition, whereas excitation to the first excited state of Xe needs as much as 8.4 eV. Here we note that the large electronegativity of the halogen atom plays an important role in complex formation and its singly-occupied p orbital acts as an electron acceptor.

Similar complex formation between Xe and surface atoms at the oxidized Si(100) surface may be possible in analogy with the Rg monohalides. The dangling bonds at the clean Si(100) surface form a π bond. Although there is some charge transfer from the down atom to the up atom in the asymmetric dimer, the α component is missing in the TOF distribution of Xe from the clean surface. This indicates that the oxygen atoms inserted into the BB and the DB sites are necessary for the appearance of the α component. Since the oxygen atom is inserted into the DB site to make Si–O–Si bonds, the two dangling bonds can no longer form a π bond. Furthermore, the electron density of the dangling bonds decreases because of the large electronegativity of the oxygen atom inserted. Thus, the dangling bond at the surface silicon atom is electron deficient and may play a similar role in ET as does the p orbital of the halogen atom in the formation of the ionic excited state of the Rg monohalides. The absence of the α component from the monodeuterated surface also confirms that the dangling bond is involved in ET in the excited state. Consequently, we propose that the excited state responsible for the α component is an ET state between Xe and oxidized surface silicon atoms in which the ET takes place from the rare gas to the substrate silicon atom. Note that the direction of ET in the excited state is opposite to the substrate-mediated excitation mechanism in which a hot electron is transferred from the substrate to an unoccupied orbital of adsorbate.

2.4. PSD dynamics

Keeping this excitation mechanism in mind, we discuss the desorption dynamics of Xe after electronic excitation. As in the case of the Rg monohalides, the Xe atom is strongly bound to the surface in the excited state by coulombic attraction. Fig. 4 shows a schematic drawing of adiabatic potential energy curves (PEC) as a function of distance between Xe and the oxide surface, z, in the ground and the excited states. The ground-state PEC is constructed by using the Lennard-Jones model potential with the following parameters used in the literature[17] for the Xe/clean silicon surface system: $\varepsilon_{Xe-Si} = 1.6$ kJ mol^{-1}, $\sigma_{Xe-Si} = 0.39$ nm. In the excited state PEC, the coulombic attractive potential between Xe$^+$ and $-e$ charge located at $z = 0$ is simply superimposed on the ground-state PEC. In the case of the charge transfer complex in the gas phase, the energy of the excited-state PEC should be raised relative to the ground-state PEC by $E_i - E_{ea}$ at infinite distance, where E_i is the ionization potential of the electron donor and E_{ea} is the electron affinity of the electron acceptor. In Fig. 4, the ionization potential of Xe (12 eV) and the work function of silicon (4.6 eV) are used for E_i and E_{ea}, respectively. The potential gradient, u, of the excited state at the Franck–Condon region around z_0, is about 13 eV nm^{-1}.

Because the PECs are very crude, we discuss the desorption dynamics qualitatively with the one-dimensional PECs along the surface normal. We follow the desorption model proposed by Antoniewicz.[18] The Xe atom physisorbed in the weakly bound ground state is optically excited to the excited-state PEC. Since the potential gradient in the Franck–Condon region in the excited state is attractive, the Xe atom is attracted toward the surface and then the excited state is quenched back to the ground state. This electronic quenching brings Xe back to the ground-state PEC. There is a critical time, t_c, determined by the shapes of the ground- and excited-state PECs. The Xe atom can desorb if the lifetime on the excited-state PEC, $t_r > t_c$, whereas the Xe atom remains on the surface if $t_r < t_c$. The distribution of lifetimes determines the kinetic energy distribution of desorbed species. Zimmermann and Ho[19] showed that the kinetic energy distribution of desorbed species is well represented by a Maxwellian with the temperature higher than that of the surface if $(\langle t_r \rangle - t_c) \ll t_c$ (the strong quenching limit). Most adsorption systems on metal surfaces are in this limit. However, the α component observed in the current system is far from the Maxwellian distribution.

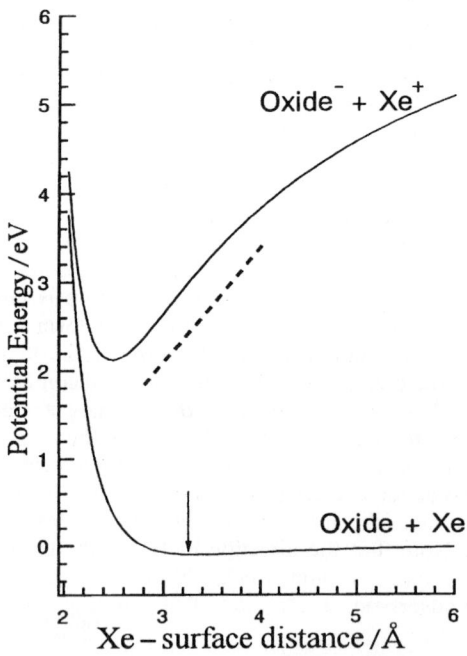

Fig. 4 Schematic PECs relevant to the Xe PSD from the oxidized Si(100) surface: the ground state (the lower curve) and the charge-transfer state (the upper curve). The arrow denotes the equilibrium distance in the ground state (0.33 nm), z_0. The broken line indicates the slope of 13 eV nm^{-1} estimated by the model potential for the Franck–Condon region at z_0.

The longer the lifetime, the more the kinetic energy distribution deviates from a Maxwellian. If $\langle t_r \rangle$ becomes comparable to the period of the vibration motion between Xe and the surface in the excited state, the kinetic energy distribution will show a sharp peak originating in a "translational rainbow".[19] This singularity occurs because the adsorbate reaches the inward classical turning point on the excited-state PEC. A Franck–Condon transition at the turning point brings the adsorbate to the highly repulsive part of the ground-state PEC. Thus, this "translational rainbow" could be responsible for the narrow energy distribution of the α component.

So far the quenching has been assumed to occur at a constant rate. This is a fair approximation in the strong quenching limit. However, when the lifetime becomes longer, the adsorbate moves closer to the surface. Thus, it is reasonable to assume that the quenching rate (Γ) is dependent on the atom–surface distance, *i.e.*, Γ increases with decrease in z such that $\Gamma = \Gamma_0 \exp[-a(z - z_0)]$. Since the population in the excited state monotonically decreases with z, the lifetime distribution has a maximum at τ corresponding to a certain value of z. Suppose that the lifetimes are distributed around τ with a relatively narrow width. Since the ground-state PEC is almost flat in the range of z relevant to the PSD dynamics, thanks to the weak physisorption interaction, the energy gained by the excited-state PEC is approximately equal to the kinetic energy of the desorbed Xe. Thus, $E_K \approx (u\tau)^2/(2m) = 0.85$ eV, where m is the mass of Xe. This simple picture gives $u\tau = 1.5$ eV ps nm^{-1}. It is not possible to determine u and τ independently. However, if $u = 13$ eV nm^{-1} is used, τ is estimated to be ~ 100 fs. This is much longer than that typically assumed in PSD or ESD on metal surfaces (10 fs or less).[19] Thus, this z-dependent quenching rate could also be a possible explanation for the non-Maxwellian distribution of the α component, but it is difficult to determine its true origin without accurate PECs. At any rate, the features of the α component discussed here imply that the lifetime of the excited state responsible for the component is substantially longer than that in the limit of strong quenching. This long lifetime is not unreasonable because the decay channel to the electron–hole pair excitation in semiconductors is much less effective on surfaces than in metals.

3. Photochemistry of methane on metal surfaces

A C–H bond of methane is dissociated on metal surfaces, including Pt(111),[4,5] Pd(111)[6] and Cu(111),[7] by irradiation with 6.4 eV photons. The cleavage of the C–H bond is confirmed by post-irradiation TPD,[4–7] XPS[4] and IR absorption spectroscopy.[20] Here, we focus on the excitation mechanism for methane on Pt(111) as a prototype.

We first examine hot electron transfer to the negative ion states of methane. Transient negative ion states of methane are known to occur in the gas phase and also in a thin film on Pt(111). The resonance features around 9 eV are assigned to the lowest Feshbach resonance, resulting in dissociation to CH$_3$ and H$^-$.[21] The resonance energy shifts from 10.8 to 9.1 eV as the methane film thickness is reduced from multilayers to a monolayer.[22] This energy shift can be well accounted for by the image force stabilization. Since the work function of Pt(111) is 6.0 eV and the photon energy utilized is 6.4 eV, it is unlikely that hot electrons produced by the photon irradiation access the Feshbach resonance states.

Information concerning low-lying negative ion states of methane is less clear. Let us assume that a negative ion state exists below 2 eV with respect to the vacuum. Since methane is physisorbed, this state is likely pinned to the vacuum level. Thus, it can be accessed by the hot electrons as a result of stabilization by the image force, *i.e.*, 1.57 eV at $z = 0.2$ nm (the van der Waals distance). Furthermore, it can be pulled down by alkali metal adsorption, resulting in an increase in the probability of ET from the bulk. In order to examine whether this negative ion state leads to dissociation we have recently measured how the cross-section of the photochemistry depends on the coverage of coadsorbed Cs.[23] We found that the photochemistry is quenched by Cs coadsorption and the cross-section at a Cs coverage of 0.05 ML is 1/4 that of the clean surface. This strong quenching of the photochemistry by Cs coadsorption is in sharp contrast to the photochemistry of Mo(CO)$_6$ on Si(111) 7 × 7.[24,25] In this system a new dissociation pathway *via* dissociative electron attachment is opened in addition to that *via* the direct excitation of the adsorbate. Therefore, the strong quenching of the methane photochemistry by Cs coadsorption suggests that the low-lying transient negative ion state of methane (if any) is not responsible for the photochemistry.

A key finding for the excitation mechanism is obtained from the polarization and incident angle dependence of photochemical cross-sections.[5] The measurements indicate that the excitation cannot be explained just by the substrate-mediated mechanism; the excitation is accompanied by a transition dipole moment perpendicular to the surface. This experimental result strongly suggests that direct excitation of methane is responsible for the photochemistry. As cited in the Introduction, the first excited state of methane in the gas phase is located at 10 eV and is composed of the C–H σ^* state strongly mixed with Rydberg states, so that the wavefuction has an extensive 3s-Rydberg character. We have proposed that this excited state is strongly hybridized with unoccupied metal s,p-states.[4] As a result of the hybridization the excited state is broadened and shifted to the lower energy that can be reached by 6.4 eV photons.

Ab initio calculations on methane–Pt_n ($n = 1$ to 10) clusters support this excitation scheme.[26] The excited state is stabilized to 7 eV from 10 eV by the hybridization. In the excited state, the electron density is shifted from methane to Pt clusters, so that methane is positively charged. Thus, a proton-like hydrogen atom easily forms a chemical bond with Pt. If an electron is completely transferred to the bulk, the required adiabatic energy for the excitation is estimated to be 7 eV [$= 13$ eV (E_i of methane) -6 eV (the work function of Pt(111))]. Thus, approximately 1 eV of the stabilization energy of the charge-transfer state makes it possible to realize excitation from the highest occupied molecular orbital (HOMO) of methane by the 6.4 eV photon. This amount of stabilization energy is feasible from the image force and specific chemical interactions between CH_4^+ and surface platinum atoms. Therefore, it is likely that the methane photochemistry is induced by the transition from the HOMO of methane to the excited state of methane–metal complexes where the electron is transferred from the adsorbate to the substrate.

4. Electron transfer from adsorbate to substrate

We have discussed the possibility of excitation from the electronic state localized at the adsorbate to the charge-transfer excited state in the two systems: Xe on the oxidized Si(100) surface and methane on the metal surfaces. Both of the systems are weakly bound to the substrates in the ground state, but we have proposed that the excited states are strongly hybridized with the unoccupied substrate states. By optical transition from the HOMO to the excited state an electron is transferred from the adsorbates with spherical symmetry to the substrate.

Mixing of the excited state of Rg atoms with metal substrates has been discussed for a long time. The lowest unoccupied s-state of the Rg atom is broadened owing to mixing with empty states of the substrate and is occupied to some extent because a tail extends below the Fermi level. This is called the s-resonance model.[27,28] Interaction through the s-resonance is believed to be the origin of a chemical contribution to the adsorption of Rg atoms on metal surfaces. The optical transition from an occupied state of the adsorbate to the mixed excited state has been readily observed in Ar/Ag(110) by Nilsson *et al.*[29] Using core level X-ray absorption spectroscopy they showed that the 4s-level of Ar is broadened and shifted owing to the s-resonance. The spectral features of the autoionization spectra accompanied by the excitation from the 2p-level of Ar to the continuum are very similar to those obtained with excitation to the absorption threshold and to the 4s-resonance state. This indicates that the electron excited to the 4s level is no longer localized at the adsorbate; rather it is transferred to the substrate and completely delocalized. According to the common nature of the excited states between the Rg atoms and methane, it is reasonable to assume that the optical transition likely occurs from the HOMO of methane to the charge-transfer state due to 's-resonance', similar to the case of Ar/Ag(110).

We have considered four weakly bound systems where the optical transition (may) take place to the charge-transfer excited state: Rg monohalides, Xe on an oxidized Si(100) surface, methane on metal surfaces, and Ar on Ag(110). The electron acceptors in these systems are different from each other. However, all the electron acceptors have low-lying empty or partially filled electronic states: the 2p orbital of a fluorine atom in the case of Rg–monohalide excimers, the dangling bond at a silicon atom bonded with oxygen inserted in the DB site in the case of the Xe PSD from oxidized Si(100), the s,p-states above the Fermi level in the case of the methane photochemistry and the transition from the core level in the case of Ar/Ag(110). On the other hand, the electron donor is common for all the cases, *i.e.*, the s- or s-like states characteristic of an atom (or molecule) with spherical symmetry.

5. Conclusion

We have shown that Xe atoms are desorbed from Si(100) surfaces modified with oxygen and deuterium by the irradiation of photons in the near-IR to UV. The desorbed Xe is very energetic and the energy distribution is very non-thermal. At least two different electronic excited states are involved in the PSD of Xe from the oxidized Si(100) surface. We speculated that the fast component with the average kinetic energy of 0.85 eV originates from the transition to the excited state where the electron is transferred from Xe to the dangling bond at a silicon atom bonded with oxygen inserted in the dimer bond. The lifetime of the excited state could be quite longer than that generally assumed on metal surfaces.

We have discussed some common features in the excitation mechanism of the Xe PSD and the photochemistry of methane on metal surfaces. The electronic structures of the substrates are quite different. However, we pointed out a common concept in the excited states responsible for the photochemistry, *i.e.*, the excited state of the adsorbates with spherical symmetry such as methane and Rg atoms could strongly hybridize with electron deficient or empty states of the substrate. In this excited state, the electron is transferred from the adsorbate to the substrate in contrast to the hot-electron transfer usually assumed in the substrate-mediated excitation in surface photochemistry. Of course, the arguments in this paper are still speculative and we need to await further experimental and theoretical identification of the excited states.

Acknowledgements

This research was supported in part by Grants-in-Aid for Scientific Research (10640570 and 11304041) from the Ministry of Education, Science, Sports and Culture of Japan and by NEDO International Joint Research Grant. We are grateful to the contributions of H. Kato in the early stage of this work.

References

1. W. Ho, in *Laser Spectroscopy and Photochemistry on Metal Surfaces*, ed. H.-L. Dai and W. Ho, World Scientific, Singapore, 1995, p. 1047.
2. F. Willig, in *Surface Electron Transfer Processes*, VCH, New York, 1995, p. 167.
3. K. Watanabe, H. Kato and Y. Matsumoto, *Surf. Sci. Lett.*, 2000, **446**, L134.
4. Y. Matsumoto, Y. A. Gruzdkov, K. Watanabe and K. Sawabe, *J. Chem. Phys.*, 1996, **105**, 4775.
5. K. Watanabe, K. Sawabe and Y. Matsumoto., *Phys. Rev. Lett.*, 1996, **76**, 1751.
6. K. Watanabe and Y. Matsumoto, *Surf. Sci.*, 1997, **390**, 250.
7. K. Watanabe and Y. Matsumoto, *Surf. Sci.*, 2000, **454–456**, 262.
8. J. Lee, H. Kato, K. Sawabe and Y. Matsumoto, *Chem. Phys. Lett.*, 1995, **240**, 417.
9. H. Kato, K. Sawabe and Y. Matsumoto, *Surf. Sci.*, 1996, **351**, 43.
10. C. C. Cheng and J. T. Yates Jr., *Phys. Rev. B*, 1991, **43**, 4041.
11. E. G. Michel, P. Pervan, G. R. Castro, R. Miranda and K. Wandelt, *Phys. Rev. B*, 1992, **45**, 11811.
12. K. Watanabe and Y. Matsumoto, in preparation.
13. Y. Miyamoto, *Phys. Rev. B*, 1992, **46**, 12473.
14. T. Uchiyama and M. Tsukada, *Phys. Rev. B*, 1996, **53**, 7917.
15. K. Kato, T. Uda and K. Terakura, *Phys. Rev. Lett.*, 1998, **80**, 2000.
16. T. H. Dunning Jr. and P. J. Hay, *J. Chem. Phys.*, 1978, **69**, 134.
17. R. Ramírez and L. Utrera, *Phys. Rev. B*, 1993, **47**, 4555.
18. P. R. Antoniewicz, *Phys. Rev. B*, 1980, **21**, 3811.
19. F. M. Zimmermann and W. Ho, *Surf. Sci. Rep.*, 1995, **22**, 127.
20. J. Yoshinobu, H. Ogasawara and M. Kawai, *Phys. Rev. Lett.*, 1994, **227**, 243.
21. M. B. Robin, in *Higher Excited States of Polyatomic Molecules*, Academic Press, London, 1985, vol. 3, p. 82.
22. P. Rowntree, L. Parenteau and L. Sanche, *J. Phys. Chem.*, 1991, **95**, 4902.
23. T. Anazawa and Y. Matsumoto, in preparation.
24. Z. C. Ying and W. Ho., *Phys. Rev. Lett.*, 1990, **65**, 741.
25. Z. C. Ying and W. Ho, *J. Chem. Phys.*, 1991, **94**, 5701.
26. Y. Akinaga, T. Taketsugu and K. Hirao, *J. Chem. Phys.*, 1997, **107**, 415.
27. K. Wandelt and B. Gumhalter, *Surf. Sci.*, 1984, **140**, 355.
28. B. Gumhalter and K. Wandelt, *Phys. Rev. Lett.*, 1986, **57**, 2318.
29. A. Nilsson, O. Björneholm, B. Hernnäs, A. Sandell and N. Mårtensson, *Surf. Sci. Lett.*, 1993, **293**, L835.

Controlling organic reactions on silicon surfaces with a scanning tunneling microscope: Theoretical and experimental studies of resonance-mediated desorption

Saman Alavi, Roger Rousseau, Gregory P. Lopinski, Robert A. Wolkow and Tamar Seideman*

Steacie Institute for Molecular Sciences, National Research Council of Canada, Ottawa, Ontario, K1A 0R6, Canada. E-mail: tamar.seideman@nrc.ca

Received 22nd May 2000
First published as an Advance Article on the web 14th August 2000

The dynamics of tip-induced, resonance-mediated bond-breaking in complex organic adsorbates is studied theoretically and experimentally. Desorption of benzene from a Si(100) surface is found to be efficient and sensitive to voltage, the measured yield rising from below 10^{-10} to *ca.* 10^{-6} per electron within a *ca.* 0.8 V range at low (<100 pA) current. A theoretical model, based upon first principles electronic structure calculations and quantum mechanical wavepacket simulations, traces these observations to multi-mode dynamics triggered by a transition into a cationic resonance. The model is generalized to provide understanding of, and suggest a means of control over, the behaviour of different classes of organic adsorbates under tunneling current.

I. Introduction

The scanning tunneling microscope (STM) has been used in the past decade both as a tool for studying substrate and substrate–adsorbate systems with atomic precision and as a tool for introducing local modifications of surface structure and adsorbate dynamics.[1,2] Included in the first category is the traditional application of the STM to image surfaces and adsorbates[1] and the newly demonstrated possibilities of performing site-specific vibrational spectroscopy,[3] imaging with chemical sensitivity[4] and monitoring the transfer of energy between adsorbate modes[5] with a STM. The second category includes modification of surface processes by local heating,[6] atom sliding on surfaces,[7] dissociation[8] and desorption[9–19] of individual adsorbates, a new approach[20] to localized atomic reactions on surfaces[21] and the exciting possibility of switching on a surface chemical reaction by bringing a single atom into contact with a single adsorbed molecule.[22] On the technological side, STM-triggered desorption has served as the initial step in a new form of lithography[23] and current–voltage curves of substrate–molecule–tip structures were used to characterize molecular wires.[24]

Here we consider two applications which could be categorized under either heading, namely use of the STM environment to study the problem of desorption induced by electronic transitions (DIET) in complex adsorbates[25] and, related, its use to study the response of molecule-based devices to tunneling current. Central to both, as well as to several of the above listed STM-manipulation schemes, are STM-selected ionic states of substrate–adsorbate complexes, as elaborated in the following sections. We are specifically interested in the dynamics of large organic adsorbates on silicon surfaces. The reason is three-fold. First, the properties of organic molecules can be systematically tailored by use of functional group substitutions while silicon is well-studied and experimentally convenient. Second, current electronic structure methods can be used to compute quantitative potential energy surfaces for neutral as well as ionic states of adsorbate/silicon systems. Adsorbate/metal complexes are more difficult to deal with, primarily since cluster models generally fail to mimic the surface environment and since vibrational lifetimes of metal-adsorbates are short and difficult to determine *ab initio*.[26] Third, there is currently considerable interest in the prospect of incorporating organic molecules into silicon-based devices, with the goal of enhancing silicon technology.[27] It is thus germane to investigate carefully the response of organic-molecule/silicon systems to the tunable current of a STM. In the present work we study a single and rather simple current-induced surface process, but one which exhibits several effects which we expect to be general. Namely, the desorption of organic adsorbates from a silicon surface, mediated by STM-triggered ionic states.

In the next section we motivate the use of a STM as a laboratory to better understand the problem of DIET and outline the theory. In order to connect the present contribution with other discussion papers in this volume, we focus on the analogy and distinctions of STM-induced DIET from the extensively studied substrate-mediated counterpart. Section III contains a brief outline of our numerical (A) and experimental (B) methods of determining the observable. In Section IV we use the tools developed in Sections II and III to first (A) study in detail one desorption problem, namely that of benzene/Si(100), and next (B) briefly generalize the discussion to other organic-molecule/silicon systems. With that we hope to gain qualitative insight into the behaviour of different organic functional groups when subjected to tunneling current. The final section concludes with an outlook to future research. Throughout we focus on the theory and the numerical results, using experimental data to motivate the discussion and substantiate the results.

II. STM-induced, ion-mediated desorption: Preliminaries and theory

The problem of desorption induced by an electronic transition has been extensively studied in the context of substrate-mediated desorption from metals, stimulated by either a laser beam or a beam of electrons.[25,28–34] Our interest in DIET of complex molecular systems arises partially from the realization that a similar break up mechanism plays a major role not only in a form of nanometer-scale lithography[23] but also in the radiation damage of cells[35] and in the interaction of light with condensed phases.[36] In the case of substrate-mediated DIET from metal surfaces, the nuclear dynamics is initiated by excitation of hot substrate carriers which may attach to an adsorbate orbital creating an excited transient. Electronic lifetimes of adsorbate/metal-substrate complexes are short but, provided the equilibrium configurations of the initial and resonance states are displaced, brief evolution in the excited state allows for conversion of electronic energy into vibrational excitation. The energy localized in the nuclear vibrations subsequent to electronic relaxation may suffice to surmount the desorption barrier.[30,31]

The general physics of DIET is currently well-understood, owing predominantly to the work of Gadzuk.[28–30] Quantitative modeling of specific chemically interesting systems, however, remains challenging. The main reasons are the difficulty of computing excited state potential energy surfaces and coupling matrix elements for adsorbate/metal-substrate systems, the likelihood of several excited electronic states and several adsorbate modes participating in the dynamics and the difficulty of accounting numerically for the rapid vibrational relaxation which competes with desorption. Rather analogous dynamics can be triggered by the tunneling electrons of a STM. (We distinguish between this desorption mode, where a single electronic transition to an ionic state serves to channel energy into nuclear modes, and the well-studied STM-desorption mechanisms of sequential vibrational heating,[10,13–15] electric field induced desorption[16] and desorption on an excited repulsive potential energy surface.[10,12]) In the STM environment, however, the flux of incident electrons is tunable and measurable, by contrast to a typical photon-stimulated process, and attention can be safely confined to one neutral and one ionic state. Since the experiment can be performed on a nonmetallic surface, vibrational relaxation is typically slow with respect to desorption. Furthermore, the calculation of potential energy surfaces for both neutral and ionic states is significantly simplified by the fact that the processes in mind are localized and hence cluster models of numerically practical size can be expected to approximate well the surface environment (*vide infra*). Hence, the STM route to DIET potentially opens the way to the study of chemically complex adsorbates, since molecular properties can be quantitatively accounted for. The observables of a STM-stimulated DIET experiment can thus be formulated by analogy to those of the substrate-mediated counterpart taking into account, however, that the mode of preparation of the resonant transient is qualitatively different.

The STM-stimulated desorption rate is written as

$$w = \frac{2\pi}{\hbar} \sum_{kk'} \sum_n \int d\varepsilon f_i(\varepsilon_k)[1 - f_f(\varepsilon_{k'})] |T(\varepsilon n, k'|v_0, k)|^2 \delta(\varepsilon_k - \varepsilon - \varepsilon_{k'}), \quad (1)$$

where k, k' are electronic labels, v_0 denotes collectively the vibrational indices of the initial bound state, n are the internal indices of the (nuclear) scattering state and ε is the energy transferred from the electronic to the vibrational system, in terms of which the total scattering energy is $\varepsilon - E_B$, E_B being the binding energy. In eqn. (1) $f_{i(f)}(E)$ is a Fermi Dirac distribution function referred to the Fermi energy, $E_F^{i(f)}$, of the electrode from (to) which current flows and $T(\varepsilon n, k'|v_0, k)$ is a transition matrix element.

Currently we are unable to solve simultaneously for the electronic and nuclear dynamics, at least for the complex systems considered, and hence a crude approximation is inevitable. Elsewhere we show that by separating the transition matrix element into an electronic and a nuclear components eqn. (1) can be expressed as

$$w = \frac{2\pi}{\hbar} \sum_n \int d\varepsilon P_{des}(\varepsilon, n) \sum_{kk'} f_i(\varepsilon_k)[1 - f_f(\varepsilon_{k'})] |T(k'|k)|^2 \delta(\varepsilon_k - \varepsilon - \varepsilon_{k'})$$

$$= \sum_n \int d\varepsilon P_{des}(\varepsilon, n) W_{exc}(\varepsilon) \quad (2)$$

where $T(k'|k) = \langle k'|T|k \rangle = \langle k'|V|k^+ \rangle$ is computed in the electronic basis. Eqn. (2) amounts to separating the vibronic process into two sequential steps; a rapid inelastic electron tunneling event, taking place at a fixed molecular geometry and depositing an excitation ε into the vibrating system, and a subsequent, much slower response of the nuclear system to the sudden excitation. The electronic component, $W_{exc}(\varepsilon)$, has the same physical content as the resonant, inelastic component of the conductance discussed in the context of electron tunneling in tip–adsorbate–substrate systems.[37] In the problems considered in Section IV the STM-bias is negative, accessing a cationic resonance. In this case $f_{i(f)}(E) = f_{s(t)}(E)$, $E_F^f = E_F^t = E_F - |eV_b|$, e being the electron charge and V_b the bias voltage, and W_{exc} describes an inelastic, resonance-mediated hole scattering event.

In the Breit–Wigner limit

$$W_{exc}(\varepsilon) = \frac{1}{2\pi\hbar} \int d\varepsilon_k f_s(\varepsilon_k)[1 - f_t(\varepsilon_k - \varepsilon)] \frac{\Gamma_t(\varepsilon_k - \varepsilon)\Gamma_s(\varepsilon_k)}{(\varepsilon_k - \varepsilon_r)^2 + \Gamma^2/4} \quad (3)$$

where ε_r is the (shifted) resonance center, $\Gamma = \Gamma_s + \Gamma_t = \hbar/\tau$ is the resonance width and Γ_s (Γ_t) is the partial width due to interaction with the sample (tip). We consider the case where the molecule is adsorbed onto the surface and hence Γ_s dominates over Γ_t.

At low temperatures $f(E) \to \Theta(E_F - E)$, Θ being the step function, and eqn. (3) reduces to,

$$W_{exc}(\varepsilon, V_b, \tau) \sim \frac{1}{2\pi\hbar} \int_{E_F + \varepsilon + eV_b}^{E_F} d\varepsilon_k \frac{\Gamma_t(\varepsilon_k - \varepsilon)\Gamma_s(\varepsilon_k)}{(\varepsilon_k - \varepsilon_r)^2 + \Gamma^2/4} \Theta(|eV_b| - \varepsilon) \quad (4)$$

where the dependence of W_{exc} on the resonance life (τ) and the bias voltage (V_b) is explicit on both sides. If energy dependence of the Γ is neglected the ε_k-integration can be carried out analytically and one has

$$w \sim \frac{\Gamma_t \Gamma_s}{\pi\hbar\Gamma} \int d\varepsilon P_{des}(\varepsilon, \tau) \left[\tan^{-1}\left(\frac{E_F - \varepsilon_r}{\Gamma/2}\right) - \tan^{-1}\left(\frac{E_F + \varepsilon + eV_b - \varepsilon_r}{\Gamma/2}\right) \right] \quad (5)$$

where we indicated the dependence of the desorption probability on the resonance lifetime and $P_{des}(\varepsilon, \tau) = \sum_n P_{des}(\varepsilon, n, \tau)$. In case $P_{des}(\varepsilon, \tau)$ is a narrowly-distributed function of ε, as found, *e.g.*, for the NH$_3$/Cu[34b] and C$_6$H$_6$/Si (Section IV A) systems, eqn. (5) can be further approximated as $w \sim P_{des}(\tau)W_{exc}(\varepsilon_0, V_b, \tau)$, where $P_{des}(\tau) = \int d\varepsilon P_{des}(\varepsilon, \tau)$ and ε_0 is the peak of the $P_{des}(\varepsilon, \tau)$ distribution. The physical interpretation of the formalism is discussed elsewhere.[38]

Several notes are worth making in connection with the electronic component in eqn. (2)–(5). First, the sigmoidal structure of W_{exc} [see eqn. (5)] rationalizes the experimentally determined bias dependence of the desorption yield of ref. 10 in the linear regime (the regime where a single electronic transition induces the nuclear dynamics), see also the voltage dependence of the tip-induced rotation rate measured in ref. 39 in that regime. Second, the sensitivity of eqn. (3)–(5) to the resonance width suggests the possibility of extracting Γ by comparing experimental voltage-dependent data to calculations. Third, W_{exc} decreases with decreasing resonance width; in physical terms, the longer the lifetime the less probable tunneling into and out of the resonance. Nonetheless, longer resonance lifetimes give rise to larger desorption probabilities, P_{des}, which may compensate or more than compensate for the decreasing excitation probability. Fourth, while eqn. (5) provides useful insight into the gross features of the bias-dependence of the rate, it entails several approximations and should be regarded as a guideline. Depending on the system it may be possible and necessary to go beyond one or more of these approximations at the price of losing the analytical simplicity. For the case of a silicon surface considered below, the ε_k-dependence of the substrate electronic density of states often needs to be taken into account.

The second component of the rate, P_{des}, contains all the details of the neutral state Hamiltonian and the desorption dynamics, which we now proceed to explore. Our numerical and experimental methods of determining the observable are outlined in the next section. We note here only that, for the complex systems of interest for the present study, both the calculations and the experiments are inevitably approximate.

III. Methods

A. Numerical

Electronic structure calculations of the substrate/organic adsorbate complex are performed using both a surface slab approach and a cluster model of the Si surface. The former calculations employ the Vienna *Ab Initio* software package (VASP),[40] along with the generalized gradient approximations (GGA)[41] to describe exchange and correlation. The core electrons are modeled *via* an ultrasoft pseudopotential[42] and the valence electrons are expanded in a basis set of plane waves with a cutoff energy of 16 E_h. The Si surface is modeled *via* a five atomic layer slab of Si atoms in an orthorhombic box subject to periodic boundary conditions. Hydrogen atoms are employed to saturate the dangling bonds on the Si atoms at the bottom of the slab. Only the coordinates of the bottom two layers of Si atoms and their saturating hydrogens are fixed in the geometry optimizations and no further symmetry constraints are applied.

The cluster calculations employ the B3LYP hybrid HF-DFT nonlocal approach of Becke,[43] as implemented within the GAUSSIAN 98 program.[44] In this approach we include both the core and the valence electrons without the introduction of either pseudopotentials or frozen core approximations. Basis set superposition is not accounted for in calculations employing localized basis sets.

While the surface approach is more reliable, the cluster model simplifies significantly the calculation of ionic states and the construction of full potential energy surfaces. STM-induced, resonance-mediated desorption from a nonmetallic surface is a localized process and hence it is expected that a cluster of moderate size would mimic well the surface environment. The Si cluster must be tailored, however, so as to account for the nature of the ionic state considered. It is also necessary to test the accuracy of the cluster model. To that end we first compute the neutral state structure and energetics within the surface slab approach and ensure that the structure and the energetics are reproduced by the cluster model to within satisfactory accuracy.[45]

Having constructed and tested a case-specific cluster model, we proceed to use it in the calculation of potential energy surfaces for the neutral and ionic states. With the surface modes fixed, the desorption dynamics is $3N$-dimensional, N being the number of adsorbate atoms ($N = 12$ for the adsorbate studied in Section IV A and larger for several of those discussed in Section IV B). Drastic simplification of the $3N$-dimensional problem is thus inevitable. Since the processes in mind are fast, and since high-frequency modes are not expected to couple efficiently with the desorption coordinate, it is not inconceivable that few (carefully identified) modes would dominate the dynamics.

As a first step toward constructing a set of orthogonal modes that collectively capture the essential physics, we compute the geometries, energies and force fields at all stationary state configurations on the two potential energy surfaces in $3N$-dimensional space. Within the same methodology[46] it is possible to construct a minimum energy path connecting the neutral state global minimum with the desorption continuum. Investigation of the structural differences between the neutral and ionic states equilibria identifies the collective mode (modes) dominating the motion in the ionic state. Study of the structural changes along the minimum energy pathway suggests the set of modes dominating the neutral state dynamics. Some indication as to the feasibility of energy transfer between the desorption and internal modes is available from the harmonic force fields. Clearly, the numerical problem is significantly simplified in cases where the system possesses symmetry, in particular if the point group of the adsorbate/substrate complex is conserved throughout. These considerations are illustrated in Section IV A, where the STM-induced desorption of benzene from Si(100) is decomposed into a desorption coordinate and a collective low-frequency mode, describing bending of the benzene ring accompanied by relaxation of the carbon backbone.[38]

With reduced-dimensionality potential energy surfaces for the neutral and ionic states computed and fitted to appropriate functional forms, we proceed to study the desorption dynamics. To that end we adopt a similar approximation to that extensively used in the study of substrate-mediated desorption from metal surfaces.[30,31] Within the general framework of the MGR model,[47] implementing the lifetime averaging scheme of Gadzuk[29] in a quantum mechanical wavepacket simulation, one describes the resonant scattering event (Section II) as an instantaneous transition to the excited (here cationic) state, propagation for a residence time τ_R subject to the excited state Hamiltonian, followed by a second instantaneous transition to the initial state and time evolution subject to the ground state molecular Hamiltonian. The continuous nature of the relaxation process is accounted for by weighing all observables by an exponentially decaying function and averaging over τ_R[29]

$$O(\tau) = \frac{1}{\tau} \int_0^\infty \exp(-\tau_R/\tau) O(\tau_R) \, d\tau_R. \qquad (6)$$

The method is well documented in the literature and not detailed here. Applications to the problem of substrate-mediated DIET are given elsewhere in this volume. Several means of going beyond the instantaneous transitions/lifetime averaging model have been developed in recent years and applied to the problem of substrate-mediated desorption from metals.[25,32,33] For our present purpose, as will become evident in Section IV, the simple approach offers an advantage.

B. Experimental

A direct method of studying electron-induced molecular desorption is by positioning the tip over a subject adsorbate and varying the bias voltage while monitoring the current and/or the tip "height". Under constant height conditions a desorption event is signaled by a current change: The current rises if the clean surface has a larger local density of states than the adsorbate covered site. Conversely, if the adsorbate appears as a topographic maximum, desorption leads to a sudden current decrease. Under feedback-controlled, constant current conditions, desorption leads to the tip moving away from the surface in the case of an adsorbate that appears as a depression and toward the surface for an adsorbate that is imaged as a protrusion.

In practice, it is difficult to place and maintain the tip over a single adsorbate. An alternative approach, taken here, involves scanning the tip over an area of the surface, first at benign imaging conditions, then at the "active" voltage and current setting under investigation and finally once again at benign conditions. Desorption events are then identified by comparison of the images prior and subsequent to the active scan. In order to determine the desorption yield (probability per electron) from the measurements one requires the dose of electrons received by an adsorbate. Here we use the result that the majority of tunneling electrons pass between a narrowly defined tip apex and a point on the surface, to approximate the dose per adsorbate as the product of the effective molecular area (taken directly from the STM image) by the current, divided by the scan rate. To obtain the cross section [and hence the rate of eqn. (1)] the yield is divided by the surface adsorbate density.

IV. Examples

In this section we apply the methods outlined in Section II to study the response of organic molecules adsorbed on a silicon surface to STM-voltage-bias. We begin, in Section IV A, with a detailed study of the desorption dynamics of C_6H_6/Si as induced by STM-excitation of a transient cation. In Section IV B we briefly generalize the discussion to other organic adsorbates on silicon.

A. STM-induced desorption of benzene from Si(100)

1. Observations. The equilibrium configuration of C_6H_6/Si(100) is well characterized.[48,49] At substrate temperatures below 200 K benzene adsorbs onto silicon in only one configuration, the so called "butterfly" structure, where the ring bridges the two atoms of a Si dimer with two of the carbon atoms forming direct Si–C bonds. The "butterfly" configuration, as determined within the surface slab model of Section III A, is shown in Fig. 1 along with a STM image taken under benign conditions, where the single-dimer-bound benzene molecules appear as bright featureless protrusions.

The inset of Fig. 2 shows a STM image taken at a bias voltage of −2.3 V, sufficient to induce bond breaking. Both normal and "truncated" (labeled T) molecule-derived protrusions are seen, the latter resulting when the molecule leaves the surface during the scanning process. In about 80% of the displacement events a truncated image of the adsorbate is seen, indicating that the tip is in the immediate vicinity of the adsorbate when bond breaking occurs. Some of the affected molecules are displaced laterally and readsorb in the single dimer configuration while others desorb permanently. No evidence of C–H or C–C bond breaking is observed. Studies between room temperature and 20 K show that the yield does not increase with temperature, as expected for a thermally activated bond breaking mechanisms, suggesting dissociation *via* an electronic excitation.[50] An approximately linear dependence of the yield on tunneling current indicates that the Si–C bond breaking is a single-electron driven process.

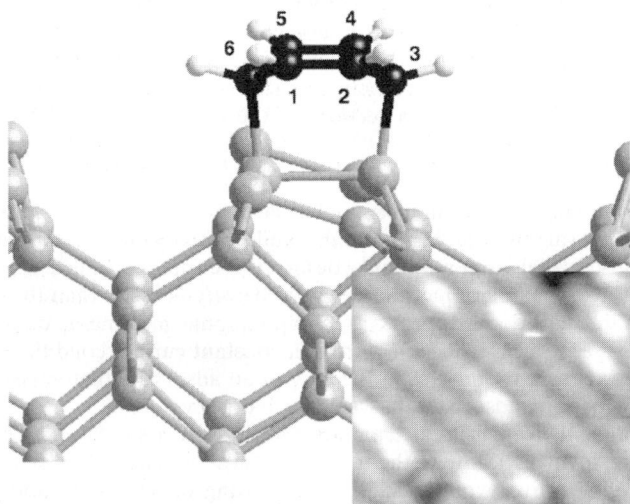

Fig. 1 Optimized geometry of the "butterfly" configuration of benzene on the Si(100) surface as obtained from a first principles surface calculation, Section III A. The Si atoms are represented by large grey spheres, H by small grey spheres and C by black spheres. The inset shows a STM image taken at room temperature, a constant current of 50 pA and bias voltage of −1.5 V. The single dimer bound benzene molecules appear as bright protrusions.

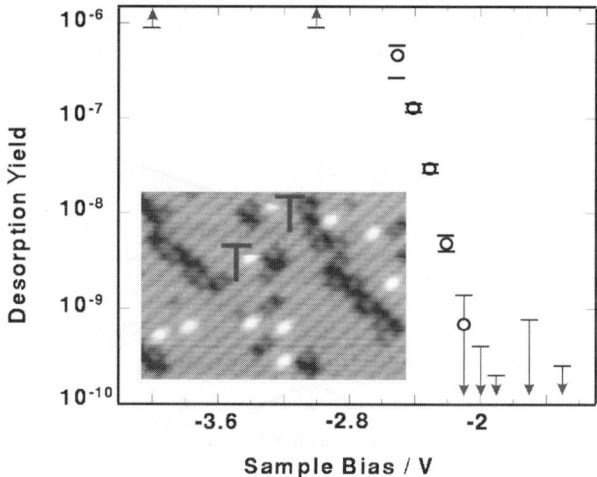

Fig. 2 Inset: a STM image taken at an active (−2.3 V) voltage bias. Molecules that appear truncated (marked "T") desorbed during the scan. Main frame: measured desorption yield for benzene/Si(100) as a function of sample bias at 22 K.

The main frame of Fig. 2 shows the experimental desorption yield *vs.* the STM voltage bias. The bias-dependence of the yield follows nicely the "step-like" functional form predicted by eqn. (3)–(5). The efficiency of the desorption and the steepness of the threshold are nevertheless intriguing. The yield is up to five orders of magnitude larger than that reported in ref. 17, where desorption of CO from a copper surface *via* a single electronic transition mechanism was reported. In the −2 V to −2.5 V range the yield increases by three orders of magnitude. The observed threshold in the yield *vs.* voltage curve coincides approximately with the binding energy of a feature observed in photoemission studies of benzene/Si(100) and attributed to the highest occupied molecular orbital of the adsorption complex.[48]

The large yield of Fig. 2 may come as a surprise given that at negative bias the excitation probability [see eqn. (2)] is expected to be lower than at positive bias.[13,15] The long lifetime of the cationic resonance, suggested by comparison of Fig. 2 with eqn. (4), is consistent with the picture of a rather poorly coupled resonance. An interesting question is thus what features of the C_6H_6/Si(100) system make the desorption probability in eqn. (2) larger than that expected based on previous experience with photon- and STM-stimulated DIET, and to what extent might one expect other organic adsorbates to behave similarly. In what follows we proceed to investigate the structure and the dynamics of the desorption process from first principles.

2. Electronic structure. Upon adsorption to the Si(100) surface, the π-structure of free benzene is severely disrupted. As shown in Fig. 1, the adsorbate assumes a 1,4-cyclohexene-type structure which deviates strongly from planarity and is rather stressed, since the benzene ring size does not precisely match the Si dimer distance. A quantitative measure of the deviation from planarity is the angle α, which the plane of carbons 1, 6 and 5 (or 2, 3, and 4) makes with the plane of the four non-silicon bonded carbons, 1, 2, 4, and 5, *cf.* Fig. 1. Ab initio calculations of benzene on a silicon slab (Section III A) show that at equilibrium this angle is $\alpha_{eq} = 32°$. A second collective mode which characterizes the desorption is the distance, z of the center of the plane of carbons 1, 2, 4, and 5 from the Si-dimer center, calculated to be $z_{eq} = 2.42$ Å at the equilibrium configuration.

Interestingly, the ionic state triggered in the course of the electron tunneling process differs in geometry from the neutral state in only one essential feature. While the C_{2v} symmetry of the neutral state equilibrium is conserved in the ionization process and z increases by only 0.01 Å, the deviation from planarity is markedly reduced, from $\alpha_{eq} = 32°$ in the neutral to $\alpha_{eq}^+ = 20°$ in the ionic structure. The two equilibrium structures are compared in Fig. 3. Two additional stationary state configurations play a role in determining the desorption outcome. A transition state, where

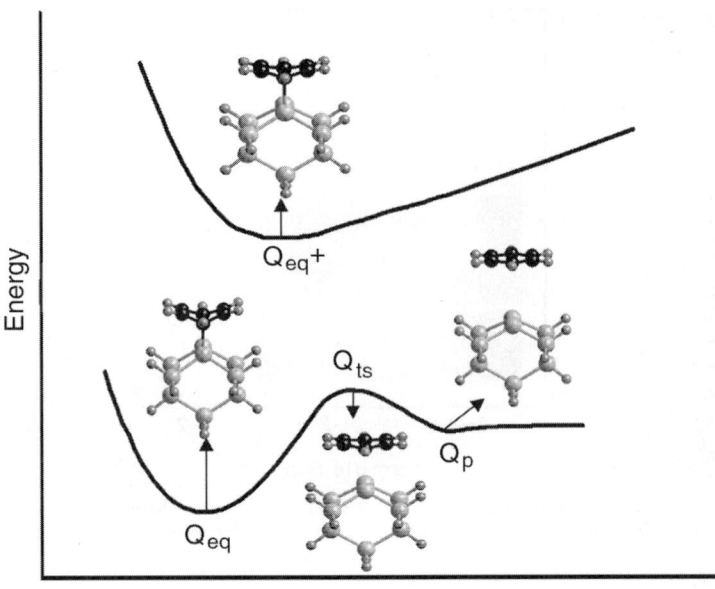

Fig. 3 A schematic illustration of the neutral and ionic potential energy of benzene/Si along the neutral state desorption path. Superimposed are the structures of the stationary states on both surfaces as calculated within a benzene–Si_9H_{12} cluster model. Q_{eq} is the neutral state equilibrium configuration, Q_{ts} is the transition state configuration, Q_p is the physisorbed configuration and Q_{eq}^+ is the equilibrium configuration of the cation. The binding energy is *ca.* 0.9 eV, the transition state barrier is *ca.* 0.2 eV above the desorption continuum and the physisorption well is close to the accuracy of the calculation.[45] The Si atoms are represented by large grey spheres, H by small grey spheres and C by black spheres.

the deviation from planarity is reduced to $\alpha_{ts} = 11°$, and a weak physisorption site, where the ring is flat ($\alpha_p = 0$) and nearly detached from the surface, $z_p = 3.88$ Å. The latter structure shares the C_{2v} symmetry of the two equilibria while the former deviates marginally from this symmetry.

The physical picture which emerges from analysis of the neutral and ion stationary state configurations is summarized in Fig. 3 along with the relative energies at the potential extrema. The neutral state "butterfly" configuration, denoted Q_{eq}, is ionized into a cation whose equilibrium configuration, denoted Q_{eq}^+, is significantly displaced with respect to that of the neutral, primarily along a collective ring-bending mode. The nonstationary superposition produced is accelerated toward the ionic state minimum, converting potential into kinetic energy of the nuclei. We note that the difference between the ionic state energy at Q_{eq} and Q_{eq}^+ is substantial, roughly 0.4 eV. Upon relaxation to the neutral state, the vibrational energy gained, initially deposited for the most part in the ring mode, is rapidly transferred to the desorption mode due to efficient vibrational coupling (*vide infra*). With sufficient energy the system surmounts a transition state barrier, located about 1.1 eV above the neutral state equilibrium configuration, roughly 0.2 eV above the asymptotic free benzene configuration. Finally, both Si–C bonds are ruptured in a concerted fashion. Depending on the kinetic energy gained by the nuclear system in the course of the inelastic resonance tunneling process, the desorption dynamics may be sensitive to the shallow physisorption well, whose configuration is denoted Q_p in Fig. 3.

As discussed in Section III A, an important aspect in a numerical study of large complex adsorbates is the construction of a coordinate system that drastically reduces the dimensionality of the problem but is nevertheless capable of describing the dynamics. Ideally one seeks a set of two or three orthogonal coordinates, to each of which can be attached a physical significance, at least in a qualitative sense. Since the processes in mind are fast, it is conceivable that such a set of strongly dominating modes will often exist.

For the problem at hand, Fig. 3 and a detailed study of the neutral state stationary state configurations suggest that desorption takes place along a C_{2v} pathway and is strongly dominated by two modes; a desorption mode, measuring the distance of the ring from the supporting silicon dimer and a ring-bending mode, describing the deviation of the ring from planarity. We define a pair of unitless orthogonal coordinates as,

$$Z = \frac{z - z_{eq}}{z_p - z_{eq}}, \quad (7)$$

$$X = \frac{(\mathbf{Q}' - \mathbf{Q}_{eq})(\mathbf{Q}'_p - \mathbf{Q}_{eq})}{|\mathbf{Q}'_p - \mathbf{Q}_{eq}|^2} \quad (8)$$

where $\mathbf{Q}$ are $3N$-dimensional vectors which specify the nuclear coordinates at the corresponding configurations (see Fig. 3) and $\mathbf{Q}'$ are the same vectors with the z coordinates of all absorbate atoms replaced by zeros. *Ab initio* potential energy surfaces for the neutral and ionic states of benzene/Si, computed as a function of Z and X are shown as contour plots in Fig. 4. These potential energy surfaces are used in the next subsection to study the nuclear dynamics triggered by the resonance tunneling process.

3. Desorption dynamics. Several features of the benzene/Si(100) system suggest that the essential aspects of the desorption dynamics can be captured within a simple model, following closely the model extensively employed to describe substrate-mediated desorption from metal surfaces. These include the presence of a transition state on the desorption pathway, which ensures that a wavepacket would naturally separate into a desorbed and a chemisorbed portion within a finite propagation time (*vide infra*), the efficient coupling between the X and Z modes and the strict two electronic states nature of the STM-triggered process. In what follows we implement the sudden transitions/lifetime averaging scheme of Gadzuk in a quantum mechanical wavepacket simulation, as outlined briefly in Section III A (see refs. 30 and 31 for details).

The essentials of the approximation are summarized schematically in Fig. 4, where we show two snapshots of the wavepacket, depicting the initial stage of the evolution upon a Franck–Condon transition to the ionic state (Fig. 4(b)) and upon neutralization at $\tau_R = 25$ fs (Fig. 4(a)). Fig. 5 describes the subsequent evolution in the neutral state through plots of the expectation values of X and Z in the wavepacket. Desorption is initiated by large amplitude vibrational motion across the chemisorption well. Energy initially deposited in the X mode shuttles periodically between the coupled modes for only a brief period. The oscillation amplitude of $\langle X \rangle$ decays within *ca.* 400 fs, at the end of which desorption becomes noticeable, as indicated by the subsequent rapid increase of $\langle Z \rangle$. Since a major portion of the wavepacket remains chemisorbed, $\langle X \rangle$ is asymptotically well

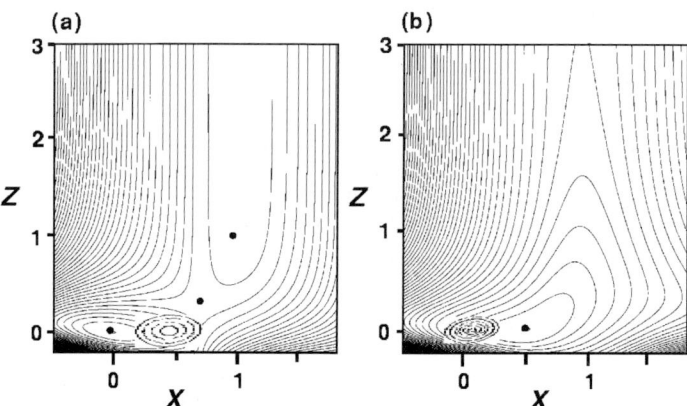

Fig. 4 Calculated potential energy surfaces *vs.* the dimensionless coordinates X and Z: (a) neutral state (b) ionic state. (●) Indicate the location of stationary points. Superimposed on the potential contours is a schematic summary of the instantaneous transitions model, showing the wavepacket upon Franck–Condon ionization (b) and upon Franck–Condon neutralization (a) subsequent to 25 fs evolution subject to the ionic potential energy surface.

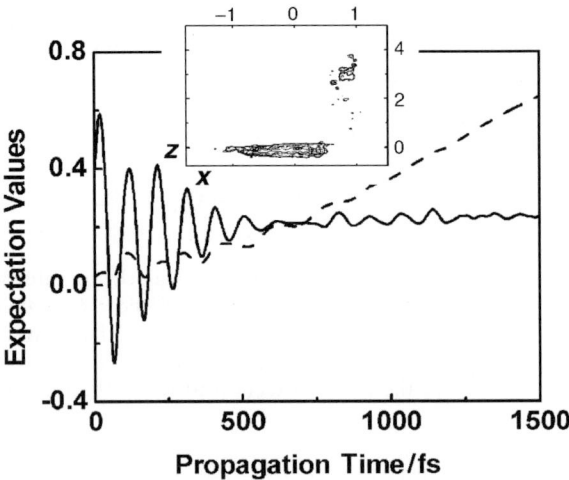

Fig. 5 Expectation values of X (———) and Z (— —) in the evolving wavepacket, subsequent to a residence time of 25 fs in the ionic state. The inset shows a snapshot of the wavepacket 750 fs after relaxation, illustrating highly vibrationally mixed motion in the chemisorption well, some probability density in the physisorption site and a portion of the wavepacket propagating along the exit valley.

below the exit valley configuration of $X = 1$. The inset of Fig. 5 shows a snapshot of the wavepacket at 750 fs, illustrating highly vibrationally mixed motion in the chemisorption well, some probability density in the physisorption site and a portion of the wavepacket propagating along the exit valley. The corresponding desorption probability is $P_{des}(\tau_R = 25 \text{ fs}) = 0.11$.

Fig. 6 and 7 contrast the desorption dynamics subsequent to the 25 fs residence time with that following shorter (20 fs) and longer (30 fs) residence times. In the former ($\tau_R = 20$ fs) case the tunneling event deposits vibrational energy into the nuclear system, giving rise to interesting "bound-state-like" dynamics as the wavepacket dephases and revives and energy is transferred between the adsorbate modes. Desorption, however, is negligible; $\langle Z \rangle(t)$ remains within the chemisorption well and $P_{des}(\tau_R = 20 \text{ fs}) = 0.004$. The inset of Fig. 6 shows that 400 fs subsequent to

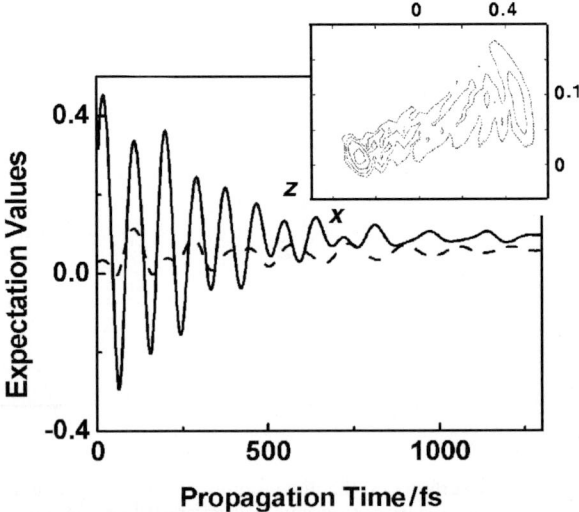

Fig. 6 Expectation values of X (———) and Z (— —) subsequent to a residence time of 20 fs in the ionic state. The inset shows a snapshot of the wavepacket 400 fs after relaxation. The wavepacket undergoes dephasing and revivals and energy is transfered between the two modes but desorption is negligible.

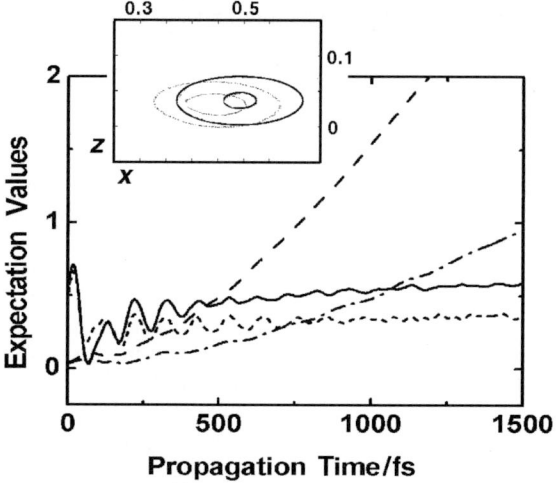

Fig. 7 Expectation values of X and Z subsequent to a residence time of 30 fs in the ionic state. (———) and (— —) give $\langle X \rangle$ and $\langle Z \rangle$, respectively, in a benzene/Si wavepacket. ($\cdots$) and ($\cdot$–$\cdot$) give $\langle X \rangle$ and $\langle Z \rangle$, respectively, in a staggered p-xylene/Si wavepacket. The inset shows snapshots of the benzene/Si (solid contours) and p-xylene/Si (dashed contours) at the neutralization instant, $t = 0$.

neutralization the wavepacket is localized in the well to within eyeball accuracy. In the latter ($\tau_R = 30$ fs) case the system explores highly repulsive regions of the neutral potential energy surface and hence the amplitude of the short-time vibrational motion in the ring mode (solid curve in Fig. 7) is large as compared to the 25 fs counterpart. Within ca. 50 fs, $\langle Z \rangle(t)$ (dashed curve in Fig. 7) has evolved past the transition state and by 400 fs the reaction is complete, with $P_{des}(\tau_R = 30 \text{ fs}) = 0.49$. The dotted and dotdashed curves of Fig. 7 are discussed below.

The solid curve of Fig. 8(a) summarizes the τ_R-dependent dynamics of Fig. 5–7 through a plot of P_{des} vs. the residence time. Comparison with Fig. 4 traces the sharp increase of $P_{des}(\tau_R)$ in the 20–30 fs range to the rapid variation of the ionic and neutral potential energy surfaces near the ionic state equilibrium configuration. The decrease of P_{des} with τ_R beyond ca. 60 fs in Fig. 8(a) corresponds to reversal of the wavepacket direction in the ionic state. We note, however, that P_{des} is not periodic in τ_R, as was found in studies of one-dimensional desorption, due to coupling between X and Z in the ionic potential energy. The large τ_R part of Fig. 8(a) is of no practical interest but is included for reference in the next subsection. The sharp increase of P_{des} with τ_R in the physically relevant region translates, as shown in Fig. 8(b), into a rapid increase of the physical desorption probability, $P_{des}(\tau)$ [see eqn. (6)] with the resonance lifetime. Along with the long lifetime inferred from comparison of eqn. (4) with Fig. 2, Fig. 8 rationalizes the remarkable efficiency of the desorption process. The energy-resolved desorption probability $P_{des}(\varepsilon, \tau)$ is found to be a narrowly-distributed function, peaking close to the barrier height, ca. 0.3 eV above the

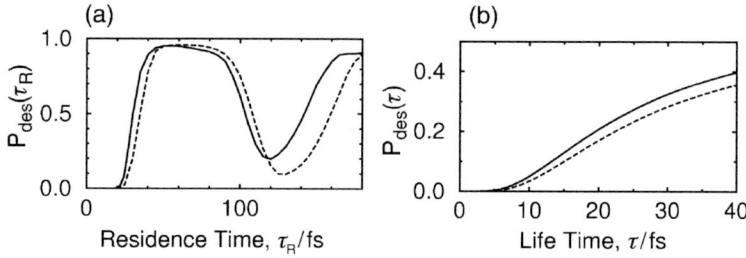

Fig. 8 (a) Residence-time-dependent desorption probability, $P_{des}(\tau_R)$ for benzene/Si (———) and staggered p-xylene/Si ($\cdots$). (b) Life-time-averaged desorption probability, $P_{des}(\tau)$ for benzene/Si (———) and staggered p-xylene/Si ($\cdots$).

desorption continuum. More details of the structure and the desorption dynamics of C_6H_6/Si are given elsewhere.[38]

B. Generalization and implications

In this subsection we proceed to investigate in how far could the qualitative conclusions drawn from the benzene/Si study be generalized to other organic adsorbates on silicon. Our main motivation is the increasingly active interest in the possibility of combining organic molecules with silicon in future molecular-scale devices[27] and in DIET-based nanolithography.[23]

One of the classes of systems widely investigated as potential candidates for molecular wires are π-bonded organic molecules that retain double bonds upon adsorption.[24,51] Such systems are expected and were found to offer resonance-enhanced conductance. Fig. 9 illustrates schematically the binding configuration of two such systems, 1,3-cyclohexadiene/Si (Fig. 9(a)) and styrene/Si (Fig. 9(b)). In both systems two of the carbon atoms form bonds with silicon surface atoms, similar to benzene/Si, but both have larger binding energies, *ca.* 2.3 eV as compared to *ca.* 0.9 eV in benzene/Si. The difference is due primarily to the aromaticity of free benzene which translates into greater stability of the desorbed state. STM images of 1,3-cyclohexadiene show an onset of tip-electrons-induced desorption as the bias is decreased in the −2 to −2.5 V range. Styrene is found to exhibit a similar desorption onset in the −2.4 to −3 V range. The room-temperature desorption yield of both systems is lower than that of benzene/Si by a factor of *ca.* 20, consistent with their larger binding energies.

In contrast, no tip-induced desorption or movements are observed under comparable conditions for adsorbed ethylene (Fig. 9(c)), propylene (Fig. 9(d)), *cis*- or *trans*-but-2-ene (Fig. 9(e) and (f)) or 1*S*(+)-3-carene (Fig. 9(g)). The molecules shown in Fig. 9(c)–(g) adsorb by converting two sp^2 carbon atoms to sp^3 to form a pair of Si–C bonds to a single dimer, similar to adsorbed benzene, 1,3-cyclohexadiene and styrene. They do not, however, retain double bonds in the adsorbed state.

These results are readily understood within the model of Section IV A. Adsorbates possessing π-bonds have low-lying ionic states which σ-bonded systems generally lack. Removal or addition of an electron that is localized in the π-system of small organic adsorbates brings about geometric changes in the nuclear backbone since all of the π-orbitals have a large contribution to the overall bonding of the system. Such equilibrium displacement allows, as discussed above, for conversion of electron energy into vibrational excitation and results in significant bond-breaking probabilities.

The above discussion appears to suggest that organic-molecule/silicon systems that offer resonance-enhanced conductance would typically break under bias. The model extends, however, to suggest π-bonded organic adsorbates where current induced bond-breaking would be unlikely: In systems where only non-bonding electrons are removed by ionization, at most minor rearrangement of the nuclei is expected. This is often the case, for instance, in large aromatic systems, where the energetically available electrons are delocalized over a large number of nuclei.

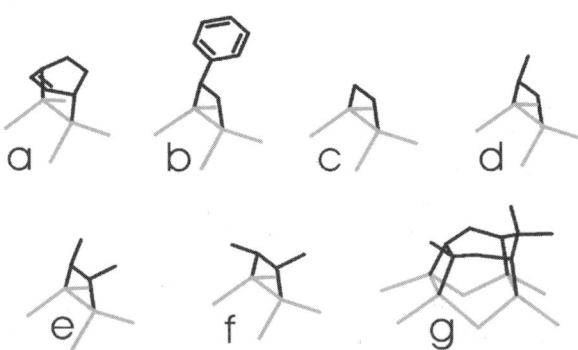

Fig. 9 Schematic illustration of adsorbed (a) 1,3-cyclohexadiene, (b) styrene, (c) ethylene, (d) propylene, (e) *cis*-but-2-ene, (f) *trans*-but-2-ene and (g) carene. Bonds between silicon atoms of the top two substrate layers are shown in grey.

Another example is systems where the π-bond is isolated *via* intermediate saturated hydrocarbons from Si–C bonds formed with the surface.

A different example of the same general effect is illustrated in Fig. 10, where we consider briefly the behaviour of pyridine/Si subject to tunneling electrons current. The stable configuration of pyridine/Si, as determined within a Si-cluster approximation, is shown in Fig. 10(a). The lone pair of the nitrogen atom introduces the possibility of binding through one end of the molecule, forming a single Si–N bond. The global minimum is thus one where the aromatic nature of the free pyridine ring is essentially preserved. Addition of an electron to the lowest-lying empty π state of this configuration yields the negative ion state shown in Fig. 10(b). Only marginal geometric changes ensue upon electron attachment since the orbital involved is delocalized over all six atoms in the ring. Qualitative analysis of the energy gradients dominating the desorption outcome suggests that the desorption yield would be significantly below that of benzene/Si.

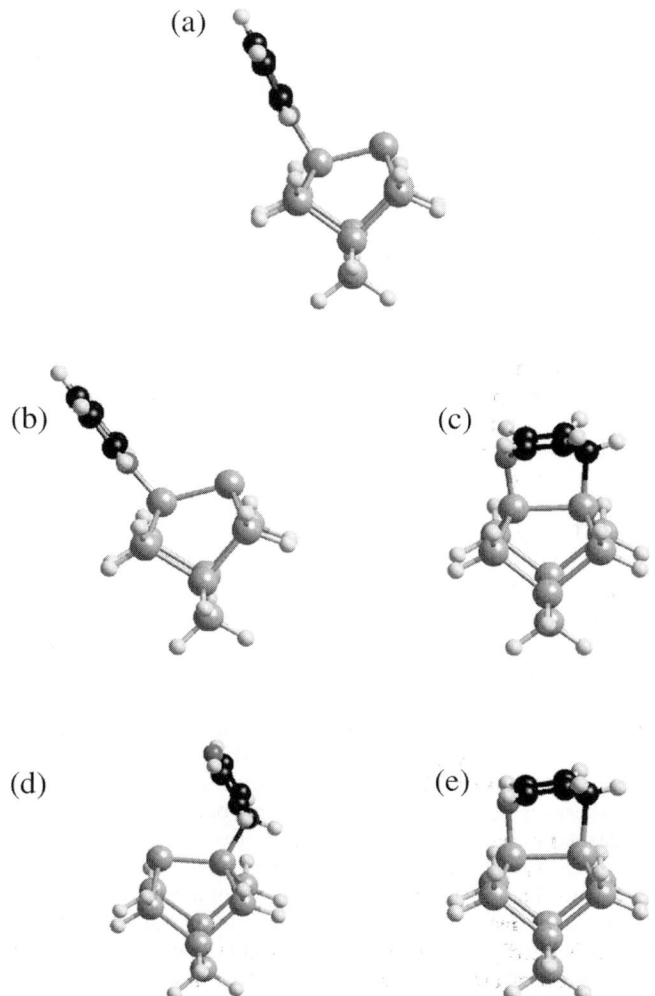

Fig. 10 Selected stationary points on the pyridine/Si potential energy surface. (a) The stable equilibrium configuration. (b) The equilibrium configuration obtained upon electron attachment to the configuration of panel (a). (c) The equilibrium structure of a "butterfly-like" configuration. (d) The equilibrium configuration obtained upon electron attachment to the butterfly-like configuration. (e) The corresponding cation configuration. N atoms are represented by grey spheres, Si atoms by large light grey spheres, H atoms by small light grey spheres and C atoms by black spheres.

We note in passing that the dynamics depends on the binding configuration as well as on the molecule and the surface. A cluster model calculation indicates that pyridine adsorbs onto Si also in a "butterfly"-like configuration, where both the N atom and the C atom at the *para*-position bind to the silicon dimer atoms, Fig. 10(c). In this (less stable by *ca.* 0.2 eV) configuration, the orbitals are no longer delocalized over the nuclear backbone. Consequently both removal and addition of an electron give rise to ionic states of markedly distorted equilibria as compared to the neutral. Fig. 10(d) and (e) show the equilibrium structures of the resultant cation and anion, respectively. Examination of the neutral state energy gradients near the ionic states equilibria suggests that "butterfly"-C_5H_6N/Si would suffer resonance mediated desorption, analogous to benzene/Si.

A second simple "chemical handle" on the desorption probability utilizes the sensitivity of DIET to the reduced mass factors characterizing the dynamics. Isotopic substitution has been a major tool in DIET studies of small molecules.[17,30,31,34,47] Provided the substitution modifies significantly the reduced mass(es) corresponding to the mode(s) along which the neutral and ionic equilibria are displaced, a large isotope effect is expected;[47] the lighter isotope travels further in the excited (ionic) state during a given lifetime and is further displaced from the ground-state equilibrium upon relaxation. For large organic systems isotopic substitution would typically have too small an effect on the reduced masses that are relevant to the desorption dynamics to produce a significant isotope sensitivity of the yield. Experience in the field of gas phase organic chemistry suggests, however, that functional group substitutions that produce no or only minor changes in bonding may serve the same purpose.

To conclude this section we consider briefly the effects of dimethyl substitution of benzene at the *para*-positions, to give *p*-xylene. A *p*-xylene molecule can bind to a Si surface dimer in two configurations. In the "bridged" configuration, the two carbons carrying the methyl groups also bind with the Si dimer (*i.e.* C3 and C6 in Fig. 1), while in the "staggered" configuration a pair of carbons not attached to the methyl groups are bound to the dimer (C1 and C4 in Fig. 1). A cluster model calculation (Section III A) estimates the binding energy of the bridged configuration to be 0.43 eV, whereas the staggered configuration has a binding energy of 0.87 eV. We therefore focus on the latter. One expects the potential energy of staggered *p*-xylene/Si to be similar to that of benzene/Si since the methyl groups are remote from the surface. Numerically we find a neutral state equilibrium geometry with $\alpha_{eq} = 31°$, $z_{eq} = 2.43$ Å [as compared to $\alpha_{eq} = 32°$, $z_{eq} = 2.42$ Å for benzene/Si], an ionic state equilibrium geometry with $\alpha_{eq}^+ \approx 23°$, $z_{eq}^+ \approx 2.42$ Å [as compared to $\alpha_{eq}^+ = 20°$, $z_{eq}^+ = 2.43$ Å for benzene], a very similar binding energy and essentially the same potential gradients at $\mathbf{Q}_{eq}$ and at $\mathbf{Q}_{eq}^+$ as in the benzene/Si system.[45] In order to focus entirely on the mass effect we compute the desorption dynamics of a fictitious model, using the benzene/Si potential energy surfaces with the kinetic energy of staggered *p*-xylene/Si. The reduced mass corresponding to the ring mode, μ_X(xylene/Si) = 19.6 u differs rather little from that of benzene/Si, μ_X(benzene/Si) = 17.3 u. The reduced mass corresponding to the desorption mode, μ_Z(xylene/Si) = 106 u, is significantly larger than that of the benzene/Si system, μ_Z(benzene/Si) = 78 u.

The dashed curve in Fig. 8(a) shows the residence-time-dependent desorption probability of staggered *p*-xylene/Si. At short times, $\tau_R \lesssim 30$, motion in the ionic state is predominantly along the ring-bending mode. Due to the small difference in μ_X between benzene/Si and staggered *p*-xylene/Si, a relatively small difference between the desorption probabilities of the two species is observed. The dotted and dot-dashed curves of Fig. 7 show, respectively, the expectation values of X and Z in a staggered *p*-xylene/Si wavepacket evolving on the neutral surface subsequent to $\tau_R = 30$ fs residence in the ionic state. The inset compares snapshots of a benzene/Si and a *p*-xylene/Si wavepacket at the neutralization instance, $t = 0$. While the difference in $\langle X \rangle(t = 0)$ and $\langle Z \rangle(t = 0)$ between the two species is hardly discerned, the wavepacket contour shows that the lighter species experiences a steeper region of the potential energy surface. As τ_R increases, coupling between X and Z in the ionic potential energy surface gives rise to increasing deviation of the wavepacket motion in the ionic state from quasi-separable and the difference in μ_Z between the two species results in an increasing difference in P_{des}. At $\tau_R = 60$ fs the benzene/Si wavepacket center reverses its direction in the ionic state and hence the $P_{des}(\tau_R)$ curves cross, with the heavier species desorption probability exceeding that of the lighter one for $60 \lesssim \tau_R/\text{fs} \lesssim 120$, up until reflection of the benzene/Si wavepacket at the inner turning point.

Irrespective of the nature of the motion in the ionic state, the heavier adsorbate desorbs more slowly. Comparison of the dashed and dotted curves of Fig. 7 illustrates the accumulating difference between the expectation values of Z in the two wavepackets with increasing time. In the absence of vibrational relaxation, the (asymptotic) desorption probability does not probe a mass difference that affects only the neutral state motion. In practice, however, a sufficiently large mass may render the desorption sufficiently slow for vibrational relaxation to dominate. Clearly, more massive functional groups than the CH_3 would more efficiently stabilize the adsorbed state.

V. Conclusions

Our goal in the work discussed above has been to study the dynamics of STM-induced, resonance-mediated bond-breaking in complex, surface adsorbed organic molecules. After outlining the theory in Section II and briefly describing our numerical and experimental tools in Section III, we focussed on a prototypical organic-molecule/semiconductor system, namely a benzene ring covalently attached to a Si(100) surface. When subject to tunneling current benzene/Si undergoes a striking resonance phenomenon. As the sample bias is scanned in the -2 V to -2.5 V range, a sharp threshold is observed over which the desorption yield rises from below 5×10^{-10} to $ca.$ 5×10^{-7}, Fig. 2. While the overall bias dependence of the measured yield follows faithfully the functional form predicted by eqn. (1)–(5), the efficiency of the process and its sharp threshold are intriguing. *Ab initio* structural calculations and quantum mechanical wavepacket simulation describe the desorption event in terms of a transient excitation of a cationic resonance whose equilibrium is displaced with respect to that of the neutral, predominantly along a collective ring-bending mode. The large yield and the sharp threshold are traced to the combination of a relatively long lifetime with efficient mode coupling and favourable neutral state energetics.

The results of the benzene/Si study were extended within a qualitative framework to provide a more general understanding of the behaviour of organic-molecule/silicon systems subject to tunneling current, with possible implications in the fields of molecular electronics and nanoscale lithography. Organic molecules that retain π-bonding upon adsorption and exhibit resonance enhanced conductance are found in several cases to undergo resonant desorption, analogous to benzene/Si, while similarly bound molecules that lack π-function are stable. Our model suggests, however, several "handles" on the desorption yield that could serve to stabilize π-function containing organic adsorbates.

Clearly, much remains to be accomplished in future research. It would be of interest to examine experimentally several of the theoretical predictions of Section IV, including the behaviour of systems where aromaticity is conserved upon adsorption and the mass effect. Ongoing theoretical work extends the model to study the feasibility of STM-control of other surface reactions, including multichannel processes. Of specific interest is the use of a STM-triggered ion to induce an (otherwise energetically forbidden) surface isomerization reaction. A variant on that theme is the use of an ionic state to shuttle an adsorbed system between two different adsorbate bonding configurations. An example under current experimental and theoretical investigation is STM-conversion of benzene/Si from its stable configuration at room temperature, a tight-bridge geometry in which four Si–C bonds are formed, into the (metastable at 293 K) butterfly configuration of Fig. 1.

Acknowledgements

This work was supported in part by the Natural Sciences and Engineering Research Council of Canada.

References

1 For a review see, Ph. Avouris, *Acc. Chem. Res.*, 1995, **28**, 95.
2 For a review see, W. Ho, *Acc. Chem. Res.*, 1998, **31**, 567.
3 (*a*) B. C. Stipe, M. A. Rezaei and W. Ho, *Science*, 1998, **280**, 1732; (*b*) L. J. Lauhon and W. Ho, *Phys. Rev. B*, 1999, **60**, R8525; (*c*) L. J. Lauhon and W. Ho, *J. Phys. Chem. A*, 2000, **104**, 2463.
4 (*a*) G. Meyer, L. Bartels and K.-H. Rieder, *Superlattices Microstruct.*, 1999, **25**, 463; (*b*) L. Bartels, G. Meyer and K.-H. Rieder, *Appl. Phys. Lett.*, 1997, **71**, 213.

5   B. C. Stipe, M. A. Rezaei and W. Ho, *Phys. Rev. Lett.*, 1998, **81**, 1263.
6   B. J. McIntyre, M. Salmeron and G. A. Somorjai, *Science*, 1994, **265**, 1415.
7   (a) D. M. Eigler and E. Schweizer, *Nature*, 1990, **344**, 524; (b) M. F. Crommie, C. P. Lutz and D. M. Eigler, *Science*, 1993, **262**, 218; (c) L. J. Whitman, J. A. Stroscio, R. A. Dragoset and R. Cellota, *Science*, 1991, **251**, 1206.
8   B. C. Stipe, M. A. Rezaei, W. Ho, S. Gao, M. Persson and B. I. Lundqvist, *Phys. Rev. Lett.*, 1997, **78**, 4410.
9   STM-stimulated desorption techniques include excitation to a repulsive excited surface (with consequent breakup on either or both excited and ground states),[10–12] sequential vibrational heating,[10,11,13–15] electric-field-induced desorption[16] and desorption mediated by a single electronic transition to a bound state.[17–19]
10 (a) T.-C. Shen, C. Wang, G. C. Abeln, J. R. Tucker, J. W. Lyding, Ph. Avouris and R. E. Walkup, *Science*, 1995, **268**, 1590; (b) Ph. Avouris, R. E. Walkup, A. R. Rossi, H. C. Akpati, P. Nordlander, T.-C. Shen, G. C. Abeln and J. W. Lyding, *Surf. Sci.*, 1996, **363**, 368; (c) Ph. Avouris, R. E. Walkup, A. R. Rossi, T.-C. Shen, G. C. Abeln, J. R. Tucker and J. W. Lyding, *Chem. Phys. Lett.*, 1996, **257**, 148; (d) T.-C. Shen and Ph. Avouris, *Surf. Sci.*, 1997, **390**, 35; (e) E. T. Foley, A. F. Kam, J. W. Lyding and Ph. Avouris, *Phys. Rev. Lett.*, 1998, **80**, 1336.
11 M. Schwartzkopff, P. Radojkovic, M. Enachescu, E. Harmann and F. Koch, *J. Vac. Sci. Technol. B*, 1996, **14**, 1336.
12 R. S. Becker, G. S. Higashi, Y. J. Chabal and A. J. Becker, *Phys. Rev. Lett.*, 1990, **65**, 1917.
13 K. Stokbro, C. Thirstrup, M. Sakurai, U. Quaade, B. Y.-K. Hu, F. Perez-Murano and F. Grey, *Phys. Rev. Lett.*, 1998, **80**, 2618.
14 C. Thirstrup, M. Sakurai, T. Nakayama and K. Stokbro, *Surf. Sci. Lett.*, 1999, **424**, L329.
15 L. Bartels, M. Wolf, T. Klamroth, P. Saalfrank, A. Kühnle, G. Meyer and K.-H. Rieder, *Chem. Phys. Lett.*, 1999, **313**, 544.
16 M. A. Rezaei, B. C. Stipe and W. Ho, *J. Chem. Phys.*, 1999, **110**, 4891.
17 L. Bartels, G. Meyer, K.-H. Rieder, D. Velic, E. Knoesel, A. Hotzel, M. Wolf and G. Ertl, *Phys. Rev. Lett.*, 1998, **80**, 2004.
18 S. N. Patitsas, G. P. Lopinski, O. Hul'ko, D. J. Moffatt and R. A. Wolkow, *Surf. Sci. Lett.*, in press.
19 S. Alavi, R. Rousseau, S. N. Patitsas, G. P. Lopinski, R. A. Wolkow and T. Seideman, *Phys. Rev. Lett.*, submitted.
20 P. H. Lu, J. C. Polanyi and D. Rogers, *J. Chem. Phys.*, 1999, **111**, 9905.
21 J. B. Giorgi, R. Kühnemuth and J. C. Polanyi, *J. Chem. Phys.*, 1999, **110**, 598, and references therein.
22 H. J. Lee and W. Ho, *Science*, 1999, **286**, 1719.
23 (a) J. W. Lyding, K. Hess, G. C. Abeln, D. S. Thompson, J. S. Moore, M. C. Hersam, E. T. Foley, J. Lee, Z. Chen, S. T. Hwang, H. Choi, Ph. Avouris and I. C. Kizilyalli, *Appl. Surf. Sci.*, 1998, **130**, 221; (b) C. Syrykh, J. P. Nys, B. Legrand and D. Stiévenard, *J. Appl. Phys.*, 1999, **85**, 3887; (c) D. P. Adams, T. M. Mayer and B. S. Swartzentruber, *J. Vac. Sci. Technol. B*, 1996, **14**, 1642.
24 W. Tian, S. Datta, S. Hong, R. Reifenberger, J. I. Henderson and C. P. Kubiak, *J. Chem. Phys.*, 1998, **109**, 2874.
25 T. Seideman, *J. Chem. Phys.*, 1997, **106**, 417.
26 Significant progress has been made during the past few years in the calculation of ground state potential energy surface of adsorbate/metal systems, [see, *e.g.*, A. Gross, M. Scheffler, M. J. Mehl and D. A. Papaconstantopoulos, *Phys. Rev. Lett.*, 1999, **82**, 1209] but comparable quality calculations for excited states and lifetimes are currently impossible [T. Klamroth and P. Saalfrank, *Surf. Sci.*, 1998, **410**, 21]
27 (a) R. Cohen, N. Zenou, D. Cahen and S. Yitchaik, *Chem. Phys. Lett.*, 1997, **279**, 270; (b) R. J. Hamers, J. S. Hovis, S. Lee, H. B. Liu and J. Shan, *J. Phys. Chem. B*, 1997, **101**, 1489; (c) J. S. Hovis and R. J. Hamers, *J. Phys. Chem. B*, 1997, **101**, 9581; (d) R. A. Wolkow, *Annu. Rev. Phys. Chem.*, 1999, **50**, 413; (e) S. R. Kasi, M. Liehr, P. A. Thiry, H. Dallaporta and M. Offenberg, *Appl. Phys. Lett.*, 1991, **59**, 108; (f) M. J. Bozack, P. A. Taylor, W. J. Choyke and J. T. Yates, *Surf. Sci.*, 1986, **177**, 933; (g) J. Yoshinobu, H. Tsuda, M. Onichi and M. Nishijima, *J. Chem. Phys.*, 1987, **87**, 7332.
28 J. W. Gadzuk and C. W. Clark, *J. Chem. Phys.*, 1989, **91**, 3174.
29 J. W. Gadzuk, *Surf. Sci.*, 1995, **342**, 345.
30 For reviews see, J. W. Gadzuk, *Annu. Rev. Phys. Chem.*, 1988, **39**, 395; J. W. Gadzuk, in *Laser Spectroscopy and Photochemistry on Metal Surfaces*, ed. H.-L. Dai and W. Ho, World Scientific, Singapore, 1995; J. W. Gadzuk in *Femtosecond Chemistry*, ed. J. Manz and L. Wöste, VCH, Weinheim, 1995.
31 A recent review is given in H. Guo, P. Saalfrank and T. Seideman, *Prog. Surf. Sci.*, 1999, **62**, 239.
32 (a) P. Saalfrank, R. Baer and R. Kosloff, *Chem. Phys. Lett.*, 1994, **230**, 463; (b) P. Saalfrank and R. Kosloff, *J. Chem. Phys.*, 1996, **105**, 2441.
33 S. M. Harris, S. Holloway and G. R. Darling, *J. Chem. Phys.*, 1995, **102**, 8235; see S. M. Harris and S. Holloway, *Chem. Phys. Lett.*, 1995, **443**, 393 for application to vibrational excitation by electron beam impact.
34 For studies of photon-simulated DIET of ammonia from Cu see, for instance, (a) T. Hertel, M. Wolf and G. Ertl, *J. Chem. Phys.*, 1995, **102**, 3414; (b) H. Guo and T. Seideman, *J. Chem. Phys.*, 1995, **103**, 9062; (c) P. Saalfrank, S. Holloway and G. R. Darling, *J. Chem. Phys.* 1995, **103**, 6720; (d) W. Nessler, K.-H. Bornscheuer, T. Hertel and E. Hasselbrink, *Chem. Phys.*, 1996, **205**, 205; (e) L. Liu, H. Guo and T. Seideman, *J.*

Chem. Phys., 1996, **104**, 8757; (*f*) J. Manz, P. Saalfrank and B. Schmidt, *J. Chem. Soc., Faraday Trans.*, 1997, **93**, 957.
35 B. Boudaïffa, P. Cloutier, D. Hunting, M. A. Huels and L. Sanche, *Science*, 2000, **287**, 1658.
36 A. D. Bass and L. Sanche, *Radiat. Environ. Biophys.*, 1998, **37**, 243.
37 (*a*) B. N. J. Persson and A. Baratoff, *Phys. Rev. Lett.*, 1987, **59**, 339; (*b*) G. P. Salam, M. Persson and R. E. Palmer, *Phys. Rev. B*, 1994, **49**, 10655; (*c*) V. Mujica, M. Kemp, A. Roitberg and M. Ratner, *J. Chem. Phys.*, 1996, **104**, 7296; (*d*) M. A. Gata and P. R. Antoniewicz, *Phys. Rev. A*, 1993, **47**, 13797; (*e*) K. Stokbro, B. Y.-K. Hu, C. Thirstrup and X. C. Xie, *Phys. Rev. B*, 1998, **58**, 8038.
38 S. Alavi, R. Rousseau and T. Seideman, *J. Chem. Phys.*, in press.
39 L. J. Lauhon and W. Ho, *Surf. Sci.*, 2000, **451**, 219.
40 (*a*) G. Kresse and J. Hafner, *Phys. Rev. B*, 1993, **47**, 55; (*b*) G. Kresse and J. Hafner, *Phys. Rev. B*, 1994, **49**, 14251; (*c*) G. Kresse and J. Furthmüller, *Comput. Mater. Sci.*, 1995, **6**, 15; (*d*) G. Kresse and J. Furthmüller, *Phys. Rev. B*, 1996, **54**, 11169.
41 J. P. Perdew and Y. Wang, *Phys. Rev. B*, 1991, **45**, 13244.
42 D. Vanderbelt, *Phys. Rev. B*, 1990, **41**, 7892, as taken from the database provided with the VASP program.[40]
43 A. D. Becke, *J. Chem. Phys.*, 1993, **98**, 5648.
44 GAUSSIAN 98, Revision A.3, M. J. Frisch., G. W. Trucks, H. B. Schlegel, G. E. Scuseria, M. A. Robb, J. R. Cheeseman, V. G. Zakrzewski, J. A. Montgomery, Jr., R. E. Stratmann, J. C. Burant, S. Dapprich, J. M. Millam, A. D. Daniels, K. N. Kudin, M. C. Strain, O. Farkas, J. Tomasi, V. Barone, M. Cossi, R. Cammi, B. Mennucci, C. Pomelli, C. Adamo S. Clifford, J. Ochterski, G. A. Petersson, P. Y. Ayala, Q. Cui, K. Morokuma, D. K. Malick, A. D. Rabuck, K. Raghavachari, J. B. Foresman, J. Cioslowski, J. V. Ortiz, B. B. Stefanov, G. Liu, A. Liashenko, P. Piskorz, I. Komaromi, R. Gomperts, R. L. Martin, D. J. Fox, T. Keith, M. A. Al-Laham, C. Y. Peng, A. Nanayakkara, C. Gonzalez M. Challacombe, P. M. W. Gill, B. Johnson, W. Chen, M. W. Wong, J. L. Andres, C. Gonzalez, M. Head-Gordon, E. S. Replogle and J. A. Pople, Gaussian Inc., Pittsburgh, PA, 1998.
45 The cluster model used in Section IV agrees with the surface calculation to within 0.01 Å in bond lengths and 0.09 eV in energetics. Comparison of relative energetics computed at the 6-31G** and 6-311G** levels established an average error of 0.05 eV in the neutral state and 0.06 eV in the ionic state potential energy.
46 P. Y. Ayala and H. B. Schlegel, *J. Chem. Phys.*, 1997, **107**, 375.
47 (*a*) D. Menzel and R. Gomer, *J. Chem. Phys.*, 1964, **41**, 3311; (*b*) P. A. Redhead, *Can. J. Phys.*, 1964, **64**, 886.
48 S. Gokhale, P. Trischberger, D. Menzel, W. Widdra, H. Dröge, H.-P. Steinrück, U. Birkenheuer, U. Gutdeutsch and N. Rösch, *J. Chem. Phys.*, 1998, **108**, 5554.
49 For experimental and theoretical studies of the equilibrium structure of neutral benzene/Si(100) (reflecting some controversy in the early literature) see, M. J. Kong, A. V. Teplyakov, J. G. Lyubovitsky and S. F. Bent, *Surf. Sci.*, 1998, **411**, 286; M. Staufer, U. Birkenheuer, T. Belling, F. Nörtemann, N. Rösch, W. Widdra, K. L. Kostov, T. Moritz and D. Menzel, *J. Chem. Phys.*, 2000, **112**, 2498; K. W. Self, R. I. Pelzel, J. H. G. Owen, C. Yan, W. Widdra and W. H. Weinberg, *J. Vac. Sci. Technol. A*, 1998, **16**, 1031; U. Birkenheuer, U. Gutdeutsch and N. Rösch, *Surf. Sci.*, 1998, **409**, 213; R. A. Wolkow, G. P. Lopinski and D. J. Moffatt, *Surf. Sci.*, 1998, **416**, L1107; B. Borovsky, M. Krueger and E. Ganz, *Phys. Rev. B*, 1998, **57**, 4269; B. I. Craig, *Surf. Sci.*, 1993, **280**, L279; H. D. Joeng, S. Ryu, Y. S. Lee and S. Kim, *Surf. Sci.*, 1995, **344**, L1226 and ref. 48.
50 The probability actually increases with decreasing temperature, indicative of temperature dependence of the resonance lifetime. We note that a temperature-dependent electronic life-time was previously inferred in ref. 14, where the temperature dependence of STM-induced desorption of H from Si(100) *via* vibrational heating was studied.
51 M. Magoga and C. Joachim, *Phys. Rev. B*, 1997, **56**, 4722.

Electronic mechanism of STM-induced diffusion of hydrogen on Si(100)

K. Stokbro, U. J. Quaade, R. Lin, C. Thirstrup† and F. Grey

Mikroelektronik Centret (MIC), Technical University of Denmark, Bldg. 345E, DK-2800 Lyngby, Denmark

Received 19th April 2000
First published as an Advance Article on the web 20th November 2000

We have observed a scanning tunneling microscopy (STM) induced lateral transfer of a single hydrogen atom on the Si(100) surface. The transfer rate of the hydrogen atom is proportional to the electron dose, indicating an electron-assisted transfer mechanism. Measurements of the relations between the transfer rate and the sample bias and temperature give further support for an electronic mechanism. The bias dependence of the transfer rate shows a peak, and from a first principles electronic structure calculation we show that the position of the peak is related to the energy of a localized surface resonance. We propose that the hydrogen transfer is related to inelastic hole scattering with this surface resonance. We develop a microscopic model for the hydrogen transfer, and using the experimental data we extract information on the resonance lifetime and the transfer yield per resonant electron. The transfer takes place by tunneling through a small excited state transfer barrier. The transfer rate is increased if the hydrogen atom before the resonant excitation is vibrationally excited, and this gives rise to an increasing transfer rate with increasing sample temperature.

1 Introduction

The ultimate lithographic limit is to control the position of single atoms. This limit was achieved first on metal surfaces where an STM was used to control the position of single Xe adatoms on a single crystal nickel surface.[1] On metal surfaces diffusion barriers are small and the adatoms are only at fixed positions at cryogenic temperatures. On semiconductor surfaces there are strong directional bonds, thus diffusion barriers are higher than on metal surfaces, and adatom positions can be fixed at room temperature. This means that much stronger forces are needed to move atoms on semiconductor surfaces, compared to metal surfaces, thus observation of STM manipulation of single atoms on semiconductor surfaces is therefore much rarer.[2–4] In previous work on atom manipulation different mechanisms have been invoked, including inter-atomic forces,[1,5] the applied electric field,[6,7] or the tunnel current.[8,9] We have observed that with an STM that it is possible to control the position of a single hydrogen atom on the Si(100) surface due to a current induced mechanism. In ref. 4 we proposed that a tunnel electron excites a surface resonance, and that the hydrogen atom crosses a reduced barrier in the excited state. In this paper we present

† Present address: VirTech A/S, Kuldyssen 10, DK-2630 Taastrup, Denmark.

DOI: 10.1039/b003179h

new experimental data and theoretical evidence that this is the correct interpretation of the experiment.

2 Experimental data

The clean Si(100)-(2 × 1) surface in ultra-high vacuum (UHV) consists of rows of silicon dimers. Exposure to atomic hydrogen at a sample temperature of 400 °C yields a monohydride surface, with one hydrogen atom saturating the dangling bond of each silicon dimer atom. Single hydrogen atoms can be removed by applying a sufficiently high sample bias either positive[10] or negative,[11] between the tip and the surface. Fig. 1A shows an STM image of the monohydrate surface where a single hydrogen atom has been removed. The remaining hydrogen atom on the dimer is observed to switch sites when scanned at a negative sample bias below −1.5 V[4,12] (Fig. 1B and C). The images are scanned from bottom to top, and for each scan there are two images, one is recorded when the tip moves from left to right along a scan line, and the other image is recorded when the tip moves back from right to left. Switching events are detected by a lateral shift of the dangling bond position between two successive line scans (Fig. 1B and C). In 60% of the switching events the position of the tip at switching can be clearly detected as a discontinuity at a single pixel of a scanline (arrow in Fig. 1B). In the remaining 40% the switching takes place when the tip is outside the region where the dangling bond gives a contrast, and the exact tip position during the switching event cannot be determined. For the events where we can detect the tip position, we always find that the hydrogen atom moves away from the tip. This result is independent of the angle between the scan direction and the dimer rows. We find no correlation between the direction of scanning (forward or reverse) and the tip position during switching. Thus, switching only depends on the tip position and not on the scan direction.

We have measured the number of switching events per scan for a given bias and tunnel current. Fig. 2 shows results obtained by averaging over ten scans of a 40 × 40 Å^2 area scanned at 10^3 Å s^{-1}. The experimental data indicate that the number of switching events, n, is proportional to the

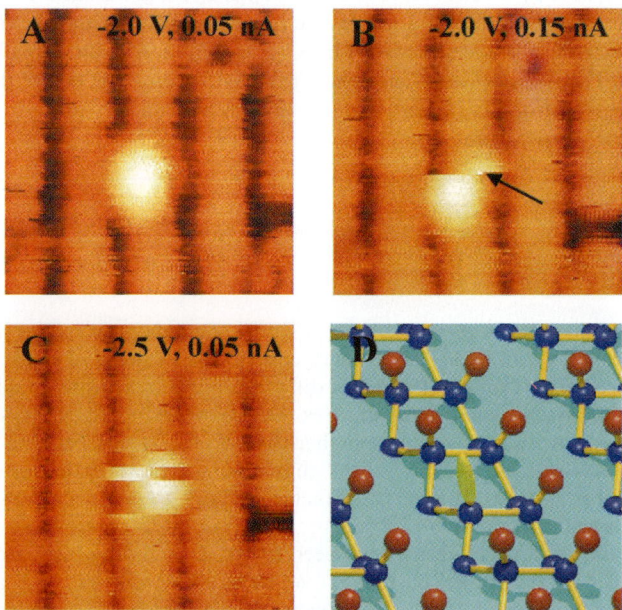

Fig. 1 A, STM image of a dangling bond (bright) on the monohydrate Si(100)-H(2 × 1) surface. B, The tip is scanning from right to left and from bottom to top. When the tip is at the position indicated by the arrow the dangling bond is switching to the right. C, At low biases the dangling bond switch sides more frequently. In this scan it switches six times. D, Geometry of the Si(100)-H(2 × 1) surface with a single dangling bond. Hydrogen atoms are red and silicon atoms blue.

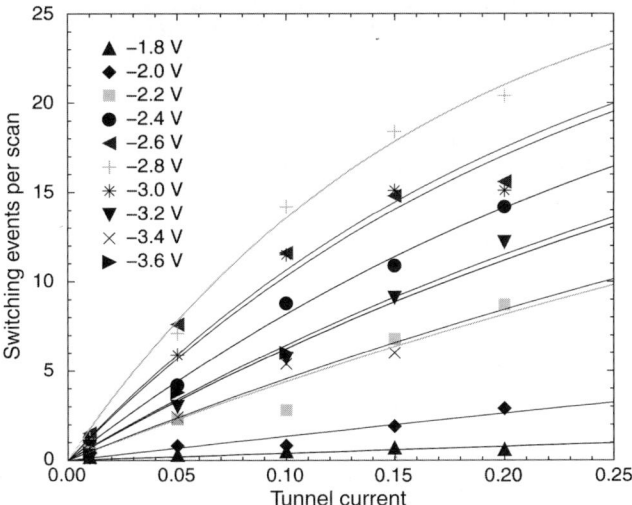

Fig. 2 The number of switching events per scan as a function of the tunnel current for 10 different biases.

dose. Assuming an exponential waiting time distribution, we have[13]

$$n = N(1 - e^{-pIt_{int}}), \qquad (1)$$

where N is the number of scanlines across the dangling bond, t_{int} is the interaction time between the hydrogen atom and the STM tip in one scanline, I is the tunnel current, and the parameter p determines the switching probability per tunnel electron. The interaction time is given by $t_{int} = 2l/v$, where $l = 5$ Å is the width of the region where the tip induces switching (the factor 2 is because each line is scanned in both directions), and v is the scan speed. Using eqn. (1) we have for each bias fitted the value of p to the data of Fig. 2, and the result is shown in Fig. 3. A peak is clearly seen at -2.7 V. Above -1.5 V, the dangling bond can be scanned for several hours

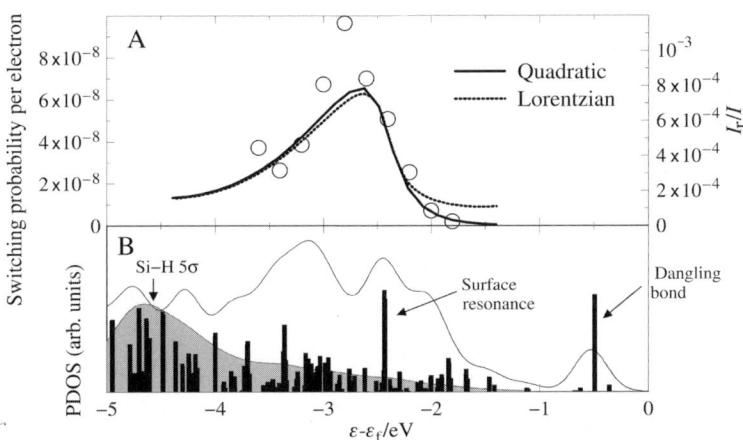

Fig. 3 A, The switching probability per tunneling electron as a function of bias voltage obtained by a least squares fit of eqn. (1) to the data in Fig. 2. The dashed bold curve shows the theoretically predicted probability assuming a Lorentzian line-shape of the resonance, and the solid curve shows the result for a line-shape corresponding to an initial quadratic decay of the resonance. The fraction of electrons from the surface resonance I_r/I is shown on the right hand axis. B, The continuous line shows the density of states in the surface region of the $c(4 \times 4)$ slab, and the bars show the overlap of each eigenstate with the occupied molecular orbitals of the unit formed by the silicon dimer and the lone hydrogen atom. The main features are peaks at -0.5 eV (dangling bond), -2.4 eV (surface resonance) and -4.7 eV (H–Si bond). The shaded area shows the density of states projected on the Si–H 5σ orbital.

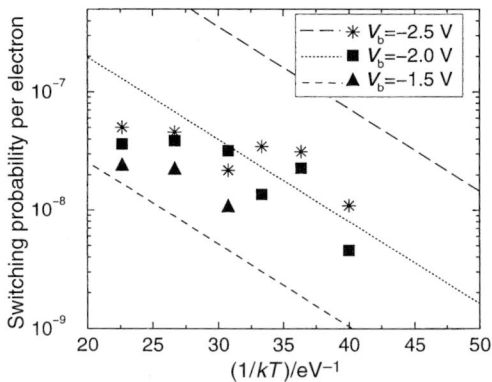

Fig. 4 Arrhenius plot of the switching probability per tunnel electron as a function of the sample temperature measured for 3 different biases. For $T < 390$ K we do not observe any switching events for a sample bias of -1.5 V. We relate the oscillations in the switching probability to fluctuations in tip geometry. The lines show theoretical predictions from the model described in the text.

without switching. Below -3.4 V, the switching probability increases again, however, at these biases desorption of hydrogen also becomes significant. The measurements shown in Fig. 2 were all obtained with the same STM tip. We have found that the value of p depends on the properties of the tip used for the experiment, however, the bias dependence is qualitatively the same and the peak position in Fig. 3 is reproducible within ± 0.25 V for different tips. At positive sample bias we see very few switching events for tunnel currents and biases below the desorption regime and there is no indication of a peak. We have also performed measurements on p-type silicon and in this case we observed very few switching events at negative sample bias and there is no indication of a peak.

Fig. 4 shows an Arrhenius plot of the measured switching probability per tunnel electron for varying sample temperature. The switching probability increases with temperature for all 3 biases. The temperature dependence shows an activated behaviour, and from a least squares fit to the data we find an activation barrier of $\sim 0.10(4)$ eV. Previously, we have investigated the relation between the number of switching events per scan and the scan speed and found $n \propto t_{int}$ in agreement with eqn. (1).[12] The switching probability on a deuterium-passivated surface was found to have the same voltage dependence as a H-passivated surface except that the ratio of the switching rates for deuterium compared to hydrogen is 0.14 ± 0.1.[4]

3 Theory

To help determine the microscopic mechanism underlying the switching phenomena, we have investigated the system using first principles electronic structure calculations. Our approach is based on density functional theory[14] within the generalized gradient approximation[15] and the electronic wavefunctions are expanded in a plane-wave basis set. The hydrogen and silicon atoms are described by ultra-soft pseudopotentials similar to those used in the study of ref. 16. These are highly accurate pseudopotentials which give absolute convergence of the total energy for a plane-wave cutoff of 20 Ry, and this is the cutoff we have used in this study.

3.1 Geometry of a dangling bond defect

We first study the geometry of a dangling bond defect on the Si(100) surface. To describe this system we have used a 6 layer Si(2 × 1) slab, and each silicon atom in the last layer is fixed in a bulk position and passivated by 2 hydrogen atoms. For the Brillouin zone integration we have used a (4 × 8) Monkhorst–Pack net, however, using only a (2 × 4) net gives almost identical results. The geometry of the dangling bond defect was optimized until forces on the atoms were less than 0.03 eV Å^{-1} and in Fig. 5A we show the resulting geometry. The changes in atomic positions relative to that of the ideal Si(100)-H(2 × 1) reconstruction[16] is less than 1%. To find the transition state for the transfer of the hydrogen atom to the other side we have placed the H atom

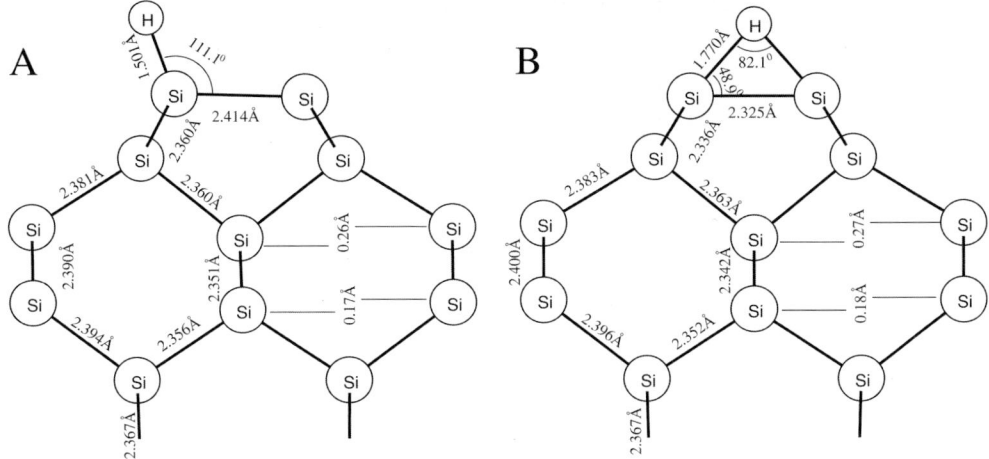

Fig. 5 A, Geometry of a single dangling bond defect on the Si(100)-H(2 × 1) surface. B, Transition state for the switching event.

in the symmetric position between the two silicon atoms. The geometry was optimized keeping the mirror symmetry, and this resulted in the geometry of Fig. 5B. The energy of this geometry is 1.32 eV higher than geometry A. The positions of the silicon atoms are very similar in geometry A and B. Keeping the silicon atoms fixed in the positions of geometry A we have calculated the reaction path for the hydrogen switching, and the result is shown in Fig. 6. With the Si atoms in frozen positions the barrier is 1.38 eV, *i.e.* only slightly larger.

From frozen phonon calculation we have determined the H frequencies in geometry A, B, and on the Si(100)-H(2 × 1) surface without defects. The results are shown in Table 1. The frequency of the hydrogen-bending mode perpendicular to the reaction path does not change along the reaction path and it is similar to the value for the Si(100)-H(2 × 1) surface.[16] The frequency of the Si–H stretch mode is in geometry A similar to the value for the Si(100)-H(2 × 1) surface, while it is reduced in geometry B. The frequency of the H-bending mode parallel to the reaction path is in geometry A quite different from the value on the Si(100)-H(2 × 1) surface, and in geometry B the frequency is imaginary. Using these frequencies we find the zero point energy corrected barrier $B_H = 1.18$ eV. From transition state theory,[17] we find that the switching rate R due to thermal

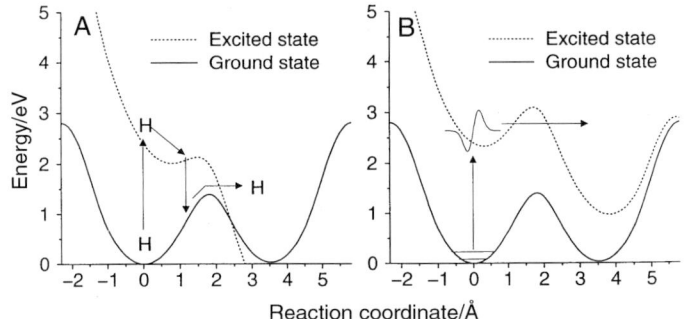

Fig. 6 The solid line shows a first principles calculation of the ground state potential, $U(x)$, for a hydrogen atom moving on a Si(100) substrate with frozen coordinates. The dotted line shows the excited state potential as obtained with the model of eqn. (6). A, $\alpha = -1$ eV Å^{-1}, this regime corresponds to conventional DIET, where the hydrogen atom can gain enough energy in the excited state to overcome the ground state barrier. B, $\alpha = -0.4$ eV Å^{-1}, which is the regime appropriate for the hydrogen transfer discussed here. In this regime the hydrogen atom tunnels through an excited state barrier. The tunnel probability is higher when the hydrogen atom is prepared in a vibrationally excited state.

Table 1 The hydrogen vibrational frequencies for the monohydrate Si(100)-H(2 × 1) surface, and the frequencies in geometry A and B of Fig. 5

	H stretch/meV	H bend$_\parallel$/meV	H bend$_\perp$/meV
H(2 × 1)	258(260[a])	78	76
Geometry A	258	177	74
Geometry B	156	125 I	74

[a] ref. 30.

diffusion over the barrier is given by

$$R = v_H e^{-B_H/kT}. \qquad (2)$$

The attempt frequency, v_H, is determined from

$$v_H = \frac{\Pi_{i=1}^3 v_i(A)}{\Pi_{i=1}^2 v_i(B)}, \qquad (3)$$

where $v_i(A)$ are the hydrogen frequencies in geometry A, and $v_i(B)$ are the real frequencies in geometry B. From the values in Table 1 we obtain $v_H = 7 \times 10^{13}$ s^{-1}. From eqn. (2) we find $R = 7$ events year^{-1} at room temperature, and $R = 1$ events s^{-1} at 430 K. Thus the switch is thermally stable up to at least 400 K.

To check that the results are converged with respect to the supercell size, we have performed calculations on a dangling bond defect in a (2 × 2) cell surrounded by a passivated surface. We found negligible changes in the atomic positions of geometry A and B, and the energy of the transition state barrier was within 0.01 eV of the result of the (2 × 1) cell. We next considered the effect of an uniform electric field perpendicular to the surface on the geometry of the dangling bond defect and the transition state barrier. The calculations were performed for a (2 × 1) and a (2 × 2) cell by placing a dipole layer in the middle of the slab. The slab consists of 12 layers and the hydrogen atoms on the back side are non-passivated in order to correctly describe the charge transfer due to the electric field.[16] The results presented in Fig. 7 show that the transition state barrier is only weakly dependent on an external electric field.

3.2 Models of the hydrogen switching

We will now discuss the microscopic mechanism underlying the STM-induced hydrogen transfer. For the tunneling conditions used in the experiment, the tip is more than 6 Å away from the surface, and the chemical interaction between the tip and the hydrogen atom is therefore negligible. Furthermore, if the transfer mechanism is related to a direct chemical interaction between the tip and the hydrogen atom, the switching rate will depend on the scan direction and this is not

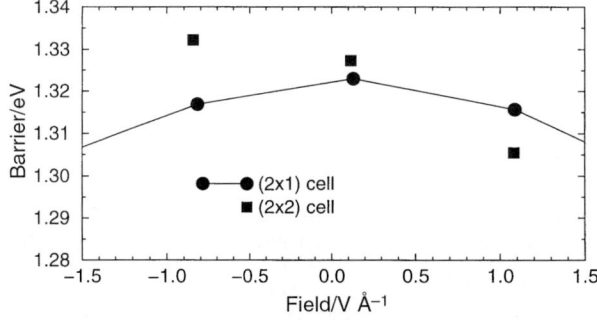

Fig. 7 Energy barrier for hydrogen switching calculated with an external electric field perpendicular to the surface. Calculations are shown for both a (2 × 1) and (2 × 2) cell.

observed. Another possible mechanism is a lowering of the diffusion barrier due to either the electric field amplitude[6] or the field gradient.[6,7] The interaction time between the tip and the hydrogen atom in Fig. 2 is $t_{int} = 0.01$ s^{-1}, and for this mechanism to give switching rates comparable to the measurements, we find from eqn. (2) that the barrier must be lowered to ~ 0.7 eV during the interaction. A barrier of this size is inconsistent with the temperature measurements of Fig. 4, which indicate an activated behaviour with a barrier around 0.1 eV. Furthermore, from Fig. 8 we see that typical electric fields are around 0.2 V Å^{-1} and Fig. 7 shows that a homogeneous field of this magnitude will have very little effect on the barrier. Our estimates of the effect of a field gradient on the barrier show that such an effect gives rise to a barrier lowering of less than 0.1 eV, in agreement with previous estimates of field-induced barrier lowering for similar systems.[2,18] We also note that none of the above mechanisms can adequately account for the sharp peak in the switching rate at -2.7 V.

The microscopic mechanism must therefore be related to an electron-induced effect, as also suggested by the experimentally observed proportionality between the switching rate and the electron dose. Barrier crossing and desorption induced by electronic transitions (DIET) is a well-documented phenomenon.[19] At negative bias, DIET involves creation of a short-lived hole state by electrons tunneling from a localized surface resonance.[11,20] Previously, we have observed that it is possible to desorb hydrogen atoms from the Si(100)-H(2 × 1) surface at negative sample bias, and shown that this is related to inelastic hole scattering with the Si–H 5σ resonance.[11,16,21] A single electron interacting with the Si–H 5σ resonance cannot lift the hydrogen atom over the desorption barrier, and the desorption is induced by multiple electron transitions (DIMET).[11] The inelastic scattering gives rise to a non-equilibrium population of the Si–H vibrational stretching levels. The non-equilibrium population can be described by an effective temperature,[22] and the solid line in Fig. 8 shows this effective temperature, T_{eff}, as a function of the sample bias. $T_{eff} = 500$ K at -2.5 V, and using this value for the temperature naively in eqn. (2) we find $R = 90$ s^{-1}. Since $Rt_{int} \sim 1$ this is the right order of magnitude for explaining the experimental data, however, several other things indicate that this is not the underlying mechanism. Firstly, the effective temperature shown in Fig. 8 was obtained with a model using the vibrational lifetime of the Si–H vibration for the ideal Si(100)-H(2 × 1) surface. On the ideal surface the vibrational relaxation is due to a decay into 3 bending modes and one silicon substrate phonon.[23] However, from Table 1 we see that the bending modes of a hydrogen atom next to a dangling bond defect are changed, and the lifetime of the Si–H stretch vibration will be substantially reduced, since it can now decay using only 2 bending modes and one silicon substrate phonon. We therefore expect that the effective temperature of a H atom next to a dangling bond defect will be substantially lower than the value shown in Fig. 8. We also note that the effective temperature describes excitations of the Si–H stretch vibration, and since this movement is perpendicular to the reaction path for switching, such an excitation will probably not lower the barrier for switching. Furthermore, vibrational heating gives rise to a power law between the switching rate and the tunnel current with a power

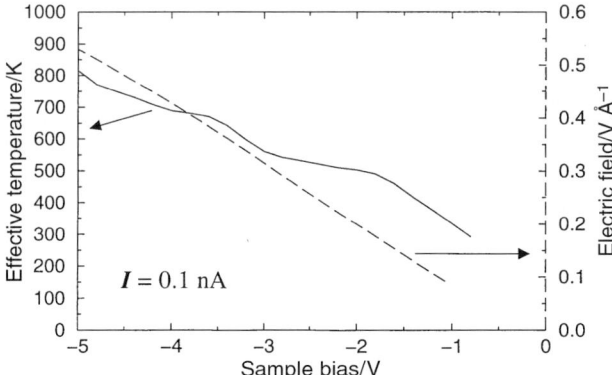

Fig. 8 The dashed line shows the electric field as a function of the sample bias for a tunnel current of 0.1 nA, the solid line shows the increased effective temperature of the Si–H stretch vibration due to vibrational heating by inelastic scattered electrons, as obtained by the model of ref. 11.

larger than 1,[11] while we observe a power of 1 in Fig. 2. Since the Si–H 5σ resonance is quite broad, the effective temperature shown in Fig. 8 has very little structure, and it cannot account for the measured peak in the bias dependence of the switching rate.

To investigate if electron scattering with another resonant level could give rise to the switching events, we have calculated the overlap of each eigenstate in the $c(4 \times 4)$ supercell with the occupied molecular orbitals of the unit formed by the silicon dimer and the lone hydrogen atom. An inelastic scattering mechanism will be most efficient when the applied bias brings the resonance in close alignment with the Fermi level of the tip, and we are therefore interested in identifying resonant levels, that have an energy close to the position of the peak in the switching rate. The result is shown in Fig. 3B. The feature at -2.4 eV is closely aligned with the tip Fermi level when the switching rate is a maximum at -2.7 V. This feature arises from an eigenstate in the slab calculation which has its main weight in the first two surface layers and is laterally localized around the silicon dimer. The state has a small but non-decaying tail into the bulk, and it is therefore a surface resonance. The surface resonance is not localized at the hydrogen atom, however, the electrostatic attraction between a hole localized in the surface resonance and the negative charge on the dangling bond will depend on the position of the hydrogen atom. The potential energy surface of the hydrogen atom will therefore be modified when a hole is localized in the resonance. Previously, we have calculated the spatial probability distribution for an electron tunneling into the resonance, and shown that the distribution is in good agreement with the measured spatial distribution of the tip position when the switching occurs.[4]

The surface resonance is nearly orthogonal to all bulk eigenstates, and the excited state lifetime, τ, will be relatively long. During this excitation the H atom will propagate on the excited state potential energy surface $U^*(x)$, and it may overcome the barrier either in the excited state or in the ground state using kinetic energy gained in the excited state. First principles calculations of $U^*(x)$ and τ are technically difficult.[24] Instead, we develop a microscopic model with τ and η_H as adjustable parameters, where η_H is the switching probability per resonance excitation. The switching probability per tunneling electron, p, is given by

$$p = \eta_H I_r / I, \qquad (4)$$

where I_r is the tunnel current from the surface resonance. This current is given by $I_r(d, V_b, \tau) = D_r(d, V_b) N_r(V_b, \tau)$, Where D_r is the tunnel rate from the surface resonance, and N_r is a weight factor which gives the fraction of the surface resonance that can tunnel into unoccupied tip states. The tunnel rate is calculated using the Tersoff–Hamann approximation, $D_r \propto |\psi^r(d+R)|^2$, where ψ^r is the resonance wavefunction, d the tip sample distance, and R is the radius of the tip.[25,26] The weight of the surface resonance above the tip Fermi level is determined from

$$N_r(V_b, \tau) = \int_{V_b}^{\infty} P_r\left(\frac{\varepsilon - \varepsilon_r}{\Delta}\right) \frac{d\varepsilon}{\Delta}, \qquad (5)$$

where P_r is the resonance line-shape, $\varepsilon_r = -2.4$ eV is the resonance energy, and Δ the resonance half width at half maximum (HWHM). The tip sample distance, d, is determined from a calculation of the total tunnel current I.

By fitting the above expression for p to the data in Fig. 3A, we obtain $\Delta = 0.13$ eV and $\eta_H = 8 \times 10^{-5}$ for a Lorentzian line shape, $P_r(x) = [1/\pi(1+x^2)]$ (dotted line). From Δ we determine the resonance lifetime $\tau = \hbar/2\Delta = 2.5$ fs. We see that the low energy tail is not well described by this model. The low energy tail is related to the decay of the resonance in the short time limit ($t < 1$ fs). A Lorentzian tail corresponds to an exponential decay law. However, in the short time limit the decay will be affected by relaxations of other electronic states which screen the hole, and this is better described by an initial quadratic decay of the resonance.[27,28] For such a decay law one obtains $P_r(x) \propto \Pi_{i=0}^{\infty}(1 + x^2/(1+4i)^2)^{-1}$, and the tail of this resonance is in better agreement with the experimental data (solid line). The fitted value of τ is 5 times longer than hole excitation of the Si–H 5σ state.[11] This is because the surface resonance is almost an eigenstate. We note that τ determines the slope of the peak near -2.4 V, but the width of the peak is an intrinsic feature related to the bias dependence of D_r, and therefore insensitive to the value of τ.

From the extracted resonance lifetime and the switching probability per resonance excitation, we will now try to make a simple model of the excited state potential $U^*(x)$. We will use the model of ref. 20

$$U^*(x) = U(x) - \varepsilon_r - \alpha(x - x_0), \tag{6}$$

where $U(x)$ is the ground state potential, $\varepsilon_r = -2.4$ eV is the resonance excitation energy, x_0 is the hydrogen equilibrium position, and $\alpha(x - x_0)$ is the change in the resonance energy when the hydrogen atom is displaced (the negative signs are because the hole resonance is emptied within an excitation). Since the mass of the hydrogen atom is one order of magnitude smaller than a silicon atom, the silicon atoms have no time to respond to a displacement of the hydrogen atom, and the ground state potential energy surface of the hydrogen atom, $U(x)$, is therefore calculated with fixed silicon positions.

In conventional DIET the hydrogen atom can overcome the barrier when it has propagated in the excited state for a time t long enough such that on return to the ground state, the kinetic energy gained in the excited state plus the potential energy at return is enough to overcome the barrier. This mechanism is illustrated in Fig. 6A, and corresponds to the regime $\alpha > 0.7$ eV Å^{-1}. For this mechanism the switching probability per resonance excitation is given by $\eta_H \sim e^{-t/\tau}$, where t is the time the hydrogen atom has to propagate in the excited state to overcome the barrier. To obtain switching probabilities comparable to the experiment we need $t = 10\tau$. Our estimates based on classical propagation on the excited state, show that the H atom will cross the barrier much faster, thus with this mechanism we would get a far too high switching probability. Furthermore, a deuterium atom will need to propagate for a time $\sim \sqrt{2}t$ on the excited state, and using $t = 10\tau$ this would give a reduction of the switching rate by 50, while only a reduction of 7 is measured.

Instead we propose that we are in the regime $\alpha < 0.7$ eV Å^{-1}. In this regime there is a barrier for switching to the other side. We propose that the hydrogen atom tunnels through this barrier in the excited state. The probability for overcoming the barrier will depend strongly on the vibrational state of the hydrogen atom before the excitation, and we can write

$$\eta_H = \sum_n \eta_H(n) e^{-n\hbar\omega/kT}. \tag{7}$$

In this equation $e^{-n\hbar\omega/kT}$ is the occupation of the nth level of the hydrogen parallel bending mode ($\hbar\omega = 0.16$ eV), and $\eta_H(n)$ is the switching probability per resonance excitation when the hydrogen atom initially is in the nth vibrational level. Since $e^{-2\hbar\omega/kT} \ll \eta_H$, only the occupation of $n = 0$ and 1 will give significant contributions. If the $n = 0$ term dominates, the switching probability will not depend on the sample temperature, which contradicts the experimental results shown in Fig. 5. Instead we propose that the $n = 1$ term dominates, thus

$$\eta_H \sim \eta_H(1) e^{-\hbar\omega/kT}, \tag{8}$$

where $\hbar\omega = 0.16$ eV, and from the value of η_H we determine $\eta_H(1) = 0.05$. This temperature dependence is in good agreement with the experimental data, as shown in Fig. 5.

For a deuterium atom the switching rate per resonance excitation is $\eta_D \sim 1 \times 10^{-5}$. The lower switching rate of the deuterium atom arises both from a difference in the vibrational levels ($\hbar\omega_D = 0.11$ eV), and differences in the switching probabilities per resonance excitation, $\eta_D(n) \neq \eta_H(n)$. In this case it is first for $n = 3$ that we have $e^{-3\hbar\omega_D/kT} \ll \eta_D$, and now there can be a significant contribution from occupations of both the $n = 0$, 1 and 2 levels. The $n = 2$ vibrational state of deuterium has an energy $2.5 \, \hbar\omega/\sqrt{2} \sim 0.28$ eV, which is similar to the energy of the $n = 1$ state of the hydrogen atom (1.5 $\hbar\omega \sim 0.24$ eV), thus we expect $\eta_H(1) \sim \eta_D(2)$. However, the occupation of this level is $e^{-2\hbar\omega/\sqrt{2}kT}/e^{-\hbar\omega/kT} \sim 0.07$ times lower than the $n = 1$ hydrogen state, and we propose that this is the origin of the lower switching rate of the deuterium atom.

4 Conclusions

By a combination of experiment and first principles theory, we have determined the microscopic mechanism for STM-induced transfer of a single hydrogen atom between two identical, stable sites on a silicon dimer. The mechanism involves excitation of a hole resonance on the silicon surface, which changes the potential energy surface of the hydrogen atom. The excited state potential still has a barrier, but if the hydrogen atom initially is in a vibrationally excited state it can tunnel through the barrier.

While controlled lateral manipulation of atoms is now well established at low temperatures[8,29] on surfaces where relatively weak tip–atom interactions can overcome small diffusion barriers, this approach is not easily extended to higher temperatures or surfaces with strongly chemisorbed species. Therefore, the use of a short-lived excited state to briefly lower a diffusion barrier represents a promising new route to controlled lateral manipulation of chemisorbed atoms and molecules.

5 Acknowledgements

This work was supported by the Danish Ministries of Industry and Research and the use of national computer resources was supported by the Danish Research Councils.

References

1. D. M. Eigler and E. Schweizer, *Nature*, 1990, **344**, 524.
2. J. J. Boland, *Science*, 1993, **262**, 1703.
3. Y. W. Mo, *Science*, 1993, **261**, 886.
4. U. Quaade, K. Stokbro, C. Thirstrup and F. Grey, *Surf. Sci.*, 1998, **415**, L1037.
5. G. Dujardin, A. Mayne, O. Robert, F. Rose, C. Joachim and H. Tang, *Phys. Rev. Lett.*, 1998, **80**, 3085.
6. J. A. Stroscio and D. M. Eigler, *Science*, 1991, **254**, 1319.
7. L. J. Whitman, J. A. Stroscio, R. A. Dragoset and R. J. Celotta, *Science*, 1991, **251**, 1206.
8. D. M. Eigler, C. P. Lutz and W. E. Rudge, *Nature*, 1991, **352**, 600.
9. T.-C. Shen, C. Wang, G. C. Abeln, J. R. Tucker, J. W. Lyding, Ph. Avouris and R. E. Walkup, *Science*, 1995, **268**, 1590.
10. J. W. Lyding, T. C. Shen, J. S. Hubacek, J. R. Tucker and G. C. Abeln, *Appl. Phys. Lett.*, 1994, **64**, 2010.
11. K. Stokbro, C. Thirstrup, M. Sakurai, U. Quaade, Ben Yu-Kuang Hu, F. Perez-Murano and F. Grey, *Phys. Rev. Lett.*, 1998, **80**, 2618.
12. F. Grey, C. Thirstrup and H. Busch, in *Large Clusters of Atoms and Molecules*, ed. T. P. Martin, Kluwer, Dordrecht, 1996, p. 463.
13. We assume that p has the same value in all scanlines, and neglect that the hydrogen atom can switch twice in one scanline. A more detailed description of the experiment without these approximations will be presented elsewhere (U. Quaade *et al.*, *Nanotechnology*, submitted).
14. P. Hohenberg and W. Kohn, *Phys. Rev. B*, 1964, **136**, 864.
15. J. P. Perdew, J. A. Chevary, S. H. Vosko, K. A. Jackson, M. R. Pederson, D. J. Singh and C. Fiolhais, *Phys. Rev. B*, 1992, **46**, 6671.
16. K. Stokbro, *Surf. Sci.*, 1999, **429**, 327.
17. P. Hänggi, P. Talkner and M. Borkovec, *Rev. Mod. Phys.*, 1990, **62**, 251.
18. B. S. Swartzentruber, A. P. Smith and H. Jónsson, *Phys. Rev. Lett.*, 1996, **77**, 2518.
19. D. Menzel, *Nuclear Instrum. Methods Phys. Res.*, 1995, **101**, 1.
20. G. P. Salam, M. Persson and R. E. Palmer, *Phys. Rev. B*, 1994, **49**, 10655.
21. C. Thirstrup, M. Sakurai, T. Nakayama and K. Stokbro, *Surf. Sci.*, 1999, **424**, L329.
22. R. E. Walkup, D. M. Newns and Ph. Avouris, in *Atomic and Nanometer Scale Modification of Materials*, ed. Ph. Avouris, Kluwer, Dordrecht, 1993, p. 97.
23. P. Guyot-Sionnest, P. H. Lin and E. M. Hiller, *J. Chem. Phys.*, 1995, **102**, 4269.
24. F. Mauri and R. Car, *Phys. Rev. Lett.*, 1995, **75**, 3166.
25. J. Tersoff and D. R. Hamann, *Phys. Rev. B*, 1985, **31**, 805.
26. K. Stokbro, U. Quaade and F. Grey, *Appl. Phys. A*, 1998, **66**, S907.
27. M. L. Goldberger and K. M. Watson, *Phys. Rev. B*, 1964, **136**, 1473.
28. R. Sadeghi and R. T. Skodje, *Phys. Rev. A*, 1995, **52**, 1996.
29. L. Bartels, G. Meyer and K.-H. Rieder, *Phys. Rev. Lett.*, 1997, **79**, 697.
30. Y. J. Chabal and K. Raghavachari, *Phys. Rev. Lett.*, 1984, **53**, 282.

Inelastic interactions of tunnel electrons with surfaces

Andrew J. Mayne, Franck Rose and Gérald Dujardin

Laboratoire de Photophysique Moléculaire, Bât. 210, Université de Paris-Sud, 91405 Orsay, France

Received 22nd May 2000
First published as an Advance Article on the web 22nd August 2000

Inelastic interactions of electrons emitted from the tip of a scanning tunnelling microscope (STM) are used to desorb individual hydrogen atoms from a Ge(111) surface. It is observed that the inelastic interactions depend not only on the electron energy and the current intensity but also on the electron emission regime of the STM tip. Quite surprisingly, it is found that tunnel electrons interact inelastically much less efficiently than field emitted electrons even though the electrons are in resonance with the Ge–H unoccuppied orbital.

Introduction

Electronic excitations and the ensuing dynamics of atoms, molecules and solid-state systems are of great interest in many areas of physics.[1] Among many phenomena, inelastic tunnelling is probably the most puzzling. So far, inelastic electron tunnelling has been much studied in quantum-well structures.[2–5] In these mesoscopic systems, attention has been focussed on the energy dissipation and the vibrational spectroscopy through inelastic electron interactions with phonons and localized vibrational modes.[1] More recently, inelastic electron tunnelling has been recognized as one of the most important issues in nanoscale physics. Indeed it plays a crucial rôle in the manipulation of adsorbed atoms and molecules on surfaces with the STM[6] or with lasers.[7,8] By using tunnel electrons issued from the tip of the STM, controlled manipulations such as rotation,[9,10] dissociation,[11,12] desorption[6,8,13–18] and association[19] of individual atoms or molecules have been performed. Following laser excitation, hot electrons produced in the substrate can tunnel through the adsorbate–substrate barrier and induce desorption[8] or reaction[20] of the adsorbed atoms or molecules. More generally, inelastic electron tunnelling is considered to be of crucial interest for future atomic-scale devices. Indeed, operating these devices in the electron tunnel regime rather than in the usual ballistic or diffusive regimes opens up many new opportunities for very high speed, low energy dissipation and quantum compatible information processing systems.

Studying inelastic electron tunnelling effects in STM and laser manipulations of adsorbed atoms or molecules on surfaces or in atomic-scale devices is markedly different from studies in quantum-well structures. In the quantum-well problems, the tunnel barriers are modelled as square potential barriers through which the transport of electrons can be relatively easily calculated.[21] For adsorbed atoms or molecules on surfaces, potential barriers are much more difficult to model since the atomic-scale structure of the whole system has to be considered. Another important difference is that inelastic electron effects do not significantly modify the quantum-well structures whereas, on the atomic-scale, inelastic effects can strongly modify the local structure. In the latter case, additional complications may arise from coupling between the inelastic electron tunnelling and the dynamics of the system.

DOI: 10.1039/b004092o

Despite their importance in atomic-scale physics, inelastic electron tunnelling processes are still very poorly understood. Two main problems deserve particular attention; the rôle of resonances and the rôle of the electron source and collector. In order to explain the relatively efficient bond breaking of adsorbates on surfaces using either the STM[15] or *via* laser excitation,[8] many authors have suggested that inelastic electron tunnelling through resonant unoccupied electronic states of the adsorbed atoms or molecules should be the dominant process. Although these resonant unoccupied states are known to exist from other high-resolution electron energy loss spectroscopy (HREELS), near-edge X-ray absorption fine structure (NEXAFS), inverse photoemission or two-photon photoemission experiments, there is no clear experimental evidence of these resonances in any inelastic electron tunnelling experiment. It is quite surprising that the yield of the inelastic process recorded as a function of the excitation energy has never revealed the presence of resonant tunnelling. So far, it has been considered that the only valid parameters when injecting electrons into adsorbed atoms or molecules on a surface were the excitation energy and the current intensity. Whether electrons are injected as tunnel electrons through a barrier or in vacuum as ballistic electrons has not been explicitly taken into account. In other words, it was assumed that the targeted adsorbed atoms or molecules can be somewhat decoupled from the source of the electrons.

In this paper, we wish to investigate these two important questions, *i.e.* the rôle of resonances and the rôle of the electron source in inelastic electron tunnelling. For that we have studied the desorption of individual hydrogen atoms from the Ge(111)-c(2 × 8) surface with the STM tip. For many reasons which will be explained later, this is the ideal system. Previous results obtained by the STM desorption of hydrogen from Si(100) surfaces[15] were more ambiguous due to the inherent complications of both the Si(100) system and the method used, thus making the analysis difficult. Here we can experiment with individual H atoms, each H atom being isolated on the Ge(111) surface whereas on the Si(100), the full layer of hydrogen may introduce additional interactions between the H atoms. The intrinsic properties of H on Ge(111) are also more favourable. In particular, the energy of the Ge–H bond is less than the σ^*(Ge–H) resonance energy which is contrary to the H–Si(100) case. The method used here to derive the desorption yield as a function of the various parameters (excitation energy, current intensity) is based on the statistical measurement of the excitation time. This will be shown to be a more accurate method than that used on H–Si(100) which is based on the statistical analysis of the extraction over many sites when the STM tip is moved across the surface.[15] From these results, we will show that electrons issued from the STM tip interact inelastically with an H atom adsorbed on the Ge(111) in a very specific manner when they are tunnel electrons as compared to field emitted electrons in vacuum. It follows that, in the tunnel regime, a fully coherent description of the STM tip, the adsorbate and the surface is required to understand the inelastic interactions.

Experimental

The STM studies have been carried out using a STM operating at room temperature in UHV at a pressure of 2×10^{-11} Torr. The STM images presented were taken with a tunnel current of 1 nA and a sample bias of -1 V or $+1$ V for the filled and empty states, respectively. The Ge(111)-c(2 × 8) surface was prepared following a similar procedure to that used in early experiments.[22] Essentially this involves a slow warming of the Ge sample to 640 °C (1 h) followed by a 500 eV Ar ion bombardment of the hot Ge surface at 640 °C (30 min). The sample is annealed for 1 h at the same temperature then cooled slowly to room temperature (1 h). This method produces reproducibly clean surfaces with large 2 × 8 domains and <1% surface defects. The sample is then exposed to hydrogen in the presence of a tungsten filament heated to 1700 °C and located 5 cm from the sample. Exposures of 1 to 10 Langmuirs (L) were sufficient. Clean STM tips are made from polycrystalline tungsten which had been electrochemically etched *ex vacuo* and then heated by electron bombardment in vacuum.

A typical surface is shown in Fig. 1(A and B—left-hand side) where the double rows of Ge adatoms can be clearly seen. To see the hydrogen, it is easier to look at the filled state image in Fig. 1(A and B—centre). Here, the restatoms and adatoms can be seen. The hydrogen atoms manifest themselves by inducing bright sites which have a triangular or square form as can be seen in Fig. 1A and B, respectively. From the surface structure diagrams for each site, shown on the

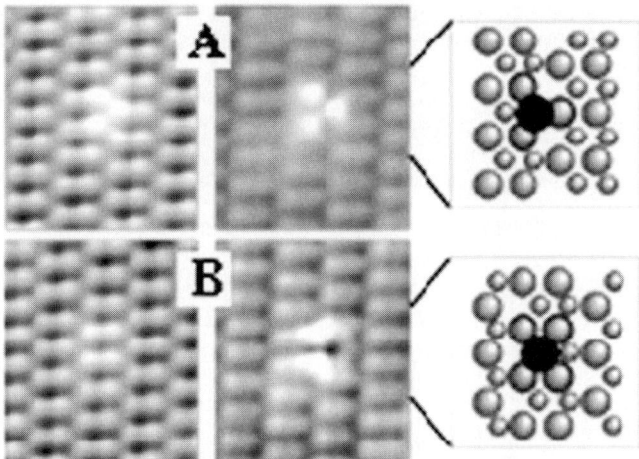

Fig. 1 A typical Ge(111)-c(2 × 8) surface (50 × 50 Å). A, An adsorbed hydrogen surrounded by 3 adatoms (triangle site). On the left-hand side is the empty state image taken at +1 V and 1 nA, in the centre is the filled state image (−1 V and 1 nA) where the higher electron density on the 3 adatoms surrounding the hydrogen can be seen. The surface structure diagram on the right-hand side shows the adsorption site of the hydrogen and its surroundings. B, the corresponding images and diagram for an adsorbed hydrogen surrounded by 4 adatoms (square site).

right-hand side of the figure, it is possible to determine that the hydrogen is adsorbed on a restatom and that the surrounding 3 or 4 adatoms have a higher electron density, thus showing up as bright atoms in the STM images. This arises because the hydrogen adsorbed on the restatom blocks the charge transfer that occurs naturally from the adatoms to the restatoms.[23]

Determining that each site was in fact a single H atom was not obvious. Many adsorption experiments were carried out by exposing the clean surface to atomic hydrogen, molecular hydrogen or water and counter experiments with nitrogen and argon. By a process of elimination the conclusion that the triangle and square sites were indeed single adsorbed hydrogens was reached. The final element of proof was derived from the manipulation experiments described in the next section.

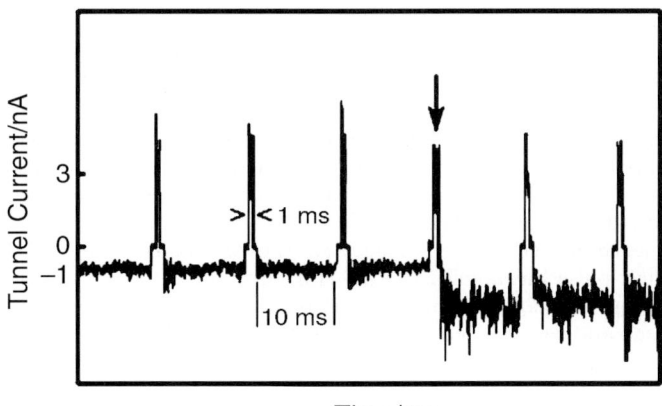

Fig. 2 A section of the pulse applied to an adsorbed hydrogen. The "write" pulse is 1 ms, the "read" pulse 10 ms, by applying a voltage of +6 V, the current required to remove the hydrogen during the "write" pulse is 3 nA. The "read" pulse corresponds to the initial conditions of −1 V and 1 nA.

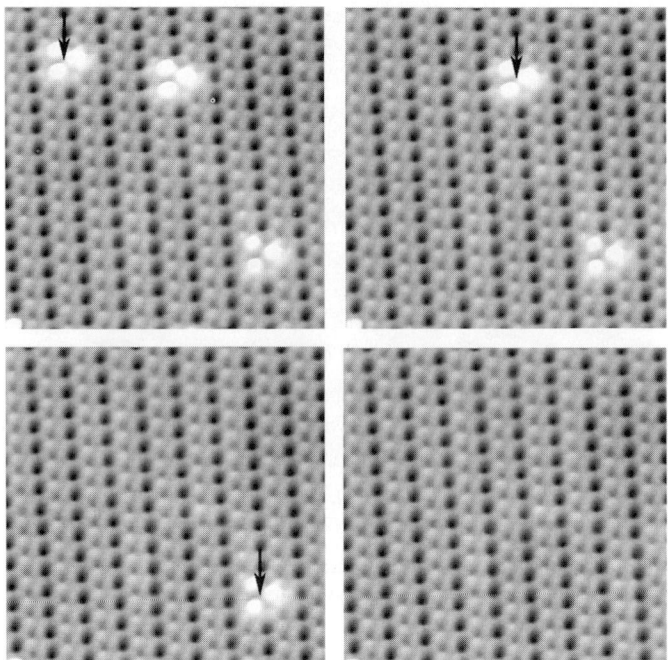

Fig. 3 A sequence of 4 consecutive STM topograph images (100 × 100 Å). The imaging conditions were −1 V and 1 nA. The first contains 3 hydrogen atoms (triangular sites) and a defect. A single pulse was applied to each H in turn (indicated by the arrow) and the extraction which results can be seen in the subsequent image.

Manipulation method

To remove a single hydrogen, the following method is used; the STM tip is placed above the atom under computer control, the feedback loop is opened, the tip is retracted a fixed distance (2–20 Å) from the surface and a positive sample bias (2–10 V) applied for a short time period (200 ms) during which the tunnel current is measured. The applied bias is then removed, the tip returned to its initial z position, the initial bias reapplied and the feedback restored.

A change in the current during the pulse, indicating that the hydrogen atom had been removed, was difficult to detect with the tip–sample distance and bias voltage used during the excitation. So, without restoring the feedback, after the "write" pulse, a "read" pulse was made by applying the initial voltage bias and the current measured, permitting the removal of the H atom to be clearly detected. By applying a sequence of a hundred or so pulses (duration 0.1 to 2 ms), each followed by a measure of the current at the image bias, before re-establishing the feedback loop, it was possible to deduce the current required and the time at which the change occurred. A section of such a sequence of pulses is shown in Fig. 2 where the removal of the hydrogen occurs during the 4th pulse. The surface is then imaged to verify that the extraction of the hydrogen has taken place and the process repeated on another H atom.

In Fig. 3, a sequence of 4 consecutive images are shown. The first contains 3 hydrogen atoms and a defect. A single pulse was applied to each H in turn (indicated by the arrow) and the result is seen in the subsequent image. As one can see, one by one the hydrogen atoms are removed leaving finally a clean and perfectly reconstructed surface.

Results and discussion

Many single manipulation events were carried out as a function of applied sample bias. The current could be varied by retracting the tip by different distances. For each pulse, the current, I, was measured and the time, τ, at which the removal occurred. From this a distribution of the number of events as a function of τ was obtained. If this distribution can be fitted with an exponential decay having a characteristic time, $\bar{\tau}$, the desorption yield, Y, in atoms per electron is $Y = (1/I\bar{\tau})$.

However, the experiment is not quite so simple and so the reasoning used above is not entirely correct. In fact, whatever the precision of the tip position for the manipulation, one must consider that the tip is not always directly over the H atom thus affecting the magnitude of the inelastic coupling between the tip and the Ge–H bond. This results in a distribution function of τ which is not strictly exponential. By studying in detail this function in correlation with the statistics of the positioning of the tip, one can deduce the minimum value of $\bar{\tau}$ when the tip is exactly centred above the hydrogen.[24] This leads to the dependence of the yield, Y, on V_s as shown in Fig. 4. For all investigated values of the bias voltage, V_s, the desorption yield, Y, was found to be independent of the current, I, in the range 0.2–6 nA. The first observation that one can make is that the yield increases dramatically above about 4.5 V. This coincides with the known work function of the Ge(111) surface at 4.8 eV.[25]

At this point it is necessary to discuss the known electronic processes associated with the Ge–H bond. It has been shown by UV photoemission[26] and by EELS[27] experiments that the binding of the H atoms to the Ge(111) surface gives rise to an occupied σ and unoccupied σ^* orbitals located 5 eV below and 3.5 eV above the Fermi level, respectively. This electronic configuration is significantly simpler than in the cases of Si(100)[28] or C(100)[29] where the existence of surface dimers result in the formation of several σ and π orbitals upon adsorption of hydrogen. From this simple electronic scheme, one can anticipate two favourable Franck–Condon electronic excitation processes. The first is the resonant electron attachment onto the unoccupied antibonding σ^*(Ge–H) orbital. Under the STM tip, this resonance at 3.5 eV might be slightly shifted in energy due to the electric field, that is, the Stark effect.[30] The second process is the $\sigma \to \sigma^*$ electronic excitation which should have an energy of about 8.5 eV. These two electronic processes can lead to the possible desorption of a hydrogen atom through a MGR (Menzel–Gomer–Redhead) type desorption model[31,32] since the σ^* orbital is believed to be antibonding, and the dissociation energy of the Ge–H bond is known to be about 1.5 eV,[33] i.e. much below the excitation energy of both

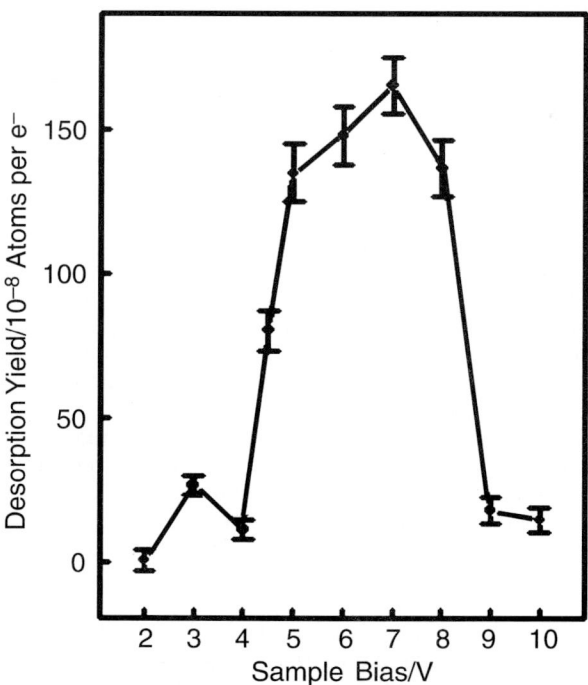

Fig. 4 Desorption yield, Y, in atoms per electron, as a function of applied bias voltage, V_s, during the pulses.

electronic processes. There does not appear to exist any known electronic process having a resonance or threshold energy in the range 4–5 eV where the sharp increase in the desorption yield is observed (Fig. 4). Therefore, we conclude that this increase is due to the transition between the tunnel regime ($V_s < 4.5$ eV) and the field emission regime ($V_s > 4.5$ eV). It is somewhat surprising that, in the tunnel regime, the inelastic coupling of the electrons is much less efficient than in the field emission regime since, in the former case, the inelastic tunnelling can be resonant or quasi-resonant with the σ^* orbital. In fact this must be related to the very different nature of tunnel and field emitted electrons. Tunnel electrons result from coupling, through evanescent waves in the tunnel barrier, between the electronic states of the STM tip close to its Fermi level with the surface states and adsorbate orbitals. In the field emission regime, electrons emitted from the STM tip through the field effect are in the vacuum when they interact with the surface and the adsorbate. The relatively inefficient tunnel regime may be explained by a weak coupling through the tunnel barrier between the STM tip and the inelastic channel.

Another interesting observation is that, in the tunnel regime, the desorption yield seems to have a maximum value at $V_s = 3$ V (see Fig. 4). Despite the uncertainties in the yield measurement and the limited number of points in the yield curve, it appears that the yield at $V_s = 4$ V is significantly less than expected from the extrapolation of the variation in yield between 2 and 3 V. We note that this maximum in the desorption yield at 3 V fits well with the energy of the σ^* resonance at 3.5 eV. A further feature of the yield curve in Fig. 4 is the strong decrease in the yield for sample bias voltages higher than 8 V. This is believed to be due to enlargement of the area of the surface irradiated by the electrons when increasing the bias voltage. In order to set the electron current to similar values for each voltage, the tip needs to be retracted further from the surface as the voltage is increased up to 10 V. This spreading out effect of the electron beam was also observed during the manipulations when, from time to time, two hydrogen atoms separated by a distance as large as 32 Å could be desorbed during the same excitation pulse.

Comparison between Ge(111)-c(2 × 8) : H and Si(100)-2 × 1 : H

A comparison between these two systems is important as it highlights the fundamental problems which need to be addressed. First, the two surfaces do not have the same reconstruction; the germanium surface is composed of adatoms and restatoms arranged in two identical but intermeshed arrays, whereas the silicon surface comprises parallel rows of dimers. Thus the two surfaces may have differing influences on the adsorbed hydrogen atoms. The experimental conditions are not quite the same either. The germanium surface is exposed to a small amount of hydrogen so that only a few of the surface dangling bond sites are occupied and also it is only the restatom sites that are occupied. In contrast, the silicon surface is completely covered, that is, all the surface dangling bonds are saturated with hydrogen.

The hydrogen bond is intrinsic to each system; the Ge–H bond energy is 1.5 eV[33] and that of Si–H is 3.5 eV.[15] This means that, in our experiments, the electrons always have enough energy so that one electron is sufficient to break the bond. As a consequence, it was observed that the yield, Y, was independent of the current for all voltages. The experiments on silicon are more difficult to interpret. Multiple vibrational excitation due to inelastic scattering with the Si–H σ^* resonance has been suggested in order to explain the desorption[15,17] over a range 2–5 eV that encompasses the Si–H bond energy of 3.5 eV. As a result the current followed a power law varying as I^n where $n = 2$–5 for voltages less than 5 V. In the H–Si(100) system, the σ^*(Si–H) resonance has about the same energy (2–3 eV) as in the H–Ge(111) case. However, the resonant effect could not be observed in the Y–V curve since in this 2–3.5 eV range, Y exhibits very strong variations with V and I. Surprisingly, no clear transition between the tunnel and field emission regimes has been reported for H–Si(100).[15] This may be due to the fact that the Si–H bond energy of 3.5 eV is not far from the value of the Si(100) work function (4.7 eV). The transition may have been hidden by a more complicated effect.

A further problem derives from the fact that the method of extracting the adsorbed hydrogens is different. The germanium surface is sparsely covered with hydrogen. This means that by placing the tip above each hydrogen, one can extract the hydrogens one by one without affecting the neighbours (which are often more than 30 Å away). Thus individual extraction times are obtained from which the probability distribution is constructed and the desorption yield, Y, determined. In the experiments on Si(100)-2 × 1 : H, the surface is completely covered by hydrogen, a line several hundred Å long is traced with the STM tip during which the hydrogens are irradiated with electrons and the number of extracted hydrogens are counted afterwards. Thus the electron dose is automatically averaged over all sites. This may introduce additional uncertainties in the yield measurements since the relative contribution of each site to the averaging may depend on the bias voltage and the current.

A related problem arises when the dynamics of the extraction are considered. For the Ge–H system, it is clear that one hydrogen is removed at a time. This is so in the Si–H case when the current is low (0.01 nA) but if the electron current is high, it is not clear what is happening; do single hydrogens leave or pairs, and if pairs then do they leave in a concerted or consecutive manner? In other words, does a newly formed dangling bond influence the ease with which the surrounding hydrogens could be extracted?

Conclusions

The desorption of individual hydrogen atoms from a Ge(111) surface with the tip of the STM has been found to be an ideal system for studying the inelastic transport of electrons through a single atom. The method used here and the inherent simplicity of the H–Ge(111) system has enabled us to demonstrate that the electrons emitted in the field emission regime which are away from the σ^*(Ge–H) resonance interact inelastically with the H atom much more efficiently than the resonant tunnel electrons. This demonstrates that the properties of the adsorbate orbitals are not sufficient to permit a clear understanding of the inelastic electron transport through a single adsorbed atom. The way in which the electrons are emitted from the STM tip, that is to say either as field emitted or tunnel electrons, seems to play a critical rôle.

Acknowledgements

We wish to thank C. Joachim for very helpful and stimulating discussions, the European TMR network "Manipulation of individual atoms and molecules" and the European IST-FET "Bottom-up-Nanomachines" (BUN) programmes for financial support.

References

1 J. W. Gadzuk, *Phys. Rev. B*, 1991, **44**, 13466.
2 N. S. Wingreen, K. W. Jacobsen and J. W. Wilkins, *Phys. Rev. Lett.*, 1988, **61**, 1396; N. S. Wingreen, K. W. Jacobsen and J. W. Wilkins, *Phys. Rev. B*, 1989, **40**, 11834.
3 M. Jonson, *Phys. Rev. B*, 1989, **39**, 5924.
4 B. Y. Gefland, S. Schmitt-Rink and A. F. J. Levi, *Phys. Rev. Lett.*, 1989, **62**, 1683.
5 J. M. Lopez-Custillo, C. Tannous and J. P. Jay-Gerin, *Phys. Rev. A*, 1990, **41**, 2273.
6 D. M. Eigler, C. P. Lutz and W. E. Rudge, *Nature*, 1991, **352**, 600.
7 S. A. Buntin, L. J. Richter, R. R. Cavanagh and D. S. King, *Phys. Rev. Lett.*, 1988, **61**, 1321.
8 L. Bartels, G. Mayer, K.-H. Reider, D. Velic, E. Knoesel, A. Hotzel, M. Wolf and G. Ertl, *Phys. Rev. Lett.*, 1998, **80**, 2004.
9 B. C. Stipe, M. A. Rezaei and W. Ho, *Science*, 1998, **279**, 1907.
10 J. K. Gimzewski, C. Joachim, R. R. Schlittler, V. Langlais, H. Tang and I. Johannsen, *Science*, 1998, **281**, 531.
11 G. Dujardin, R. E. Walkup and Ph. Avouris, *Science*, 1992, **255**, 1232.
12 B. C. Stipe, M. A. Rezaei, W. Ho, S. Gao, M. Persson and B. I. Lundqvist, *Phys. Rev. Lett.*, 1997, **78**, 4410.
13 R. S. Becker, G. S. Higashi, Y. J. Chabal and A. J. Becker, *Phys. Rev. Lett.*, 1990, **65**, 1917.
14 J. W. Lyding, T.-C. Shen, J. S. Hubacek, J. R. Tucker and G. C. Abeln, *Appl. Phys. Lett.*, 1994, **64**, 2010.
15 T.-C. Shen, C. Wang, G. C. Abeln, J. R. Tucker, J. W. Lyding, Ph. Avouris and R. E. Walkup, *Science*, 1995, **268**, 1590.
16 Ph. Avouris, *Surf. Sci.*, 1996, **363**, 368.
17 K. Stokbro, C. Thirstrup, M. Sakurai, U. J. Quaade, B. Y.-K. Hu, F. Perez-Murano and F. Grey, *Phys. Rev. Lett.*, 1998, **80**, 2618.
18 C. Thirstrup, M. Sakurai, T. Nakayama and M. Aono, *Surf. Sci.*, 1998, **411**, 203.
19 H. J. Lee and W. Ho, *Science*, 1999, **286**, 1719.
20 M. Bonn, S. Funk, Ch. Hess, D. N. Denzler, C. Stampfl, M. Scheffler, M. Wolf and G. Ertl, *Science*, 1999, **285**, 1042.

21 T. C. L. G. Sollner, W. D. Goodhue, P. E. Tannenwald, C. D. Parker and D. D. Peck, *Appl. Phys. Lett.*, 1983, **43**, 588.
22 R. S. Becker, J. A. Golovchenko and B. S. Swartzentruber, *Nature*, 1987, **325**, 419.
23 T. Klistner and J. Nelson, *Phys. Rev. Lett.*, 1991, **67**, 3800.
24 G. Dujardin, F. Rose and A. J. Mayne, to be published.
25 F. Rose, A. J. Mayne, L. Hellner and G. Dujardin, to be published.
26 R. D. Bringans and H. Höchst, *Phys. Rev. B*, 1982, **25**, 1081.
27 L. Surnev and M. Tikhov, *Surf. Sci.*, 1984, **138**, 40.
28 S. Ciraci, R. Butz, E. M. Oellig and H. Wagner, *Phys. Rev. B*, 1984, **30**, 711.
29 F. Maier, R. Grauper, M. Hollering, L. Hammer, J. Ristein and L. Ley, to be published.
30 Ph. Avouris, *Surf. Sci.*, 1996, **363**, 368.
31 D. Menzel and R. Gomer, *J. Chem. Phys.*, 1964, **41**, 3311.
32 P. A. Redhead, *J. Chem. Phys.*, 1964, **42**, 886.
33 J. Wintterlin and Ph. Avouris, *J. Chem. Phys.*, 1994, **100**, 687.

The initiation and characterization of single bimolecular reactions with a scanning tunneling microscope

Lincoln J. Lauhon and Wilson Ho*

Laboratory of Atomic and Solid State Physics and Cornell Center for Materials Research, Cornell University, Ithaca, New York 14853, USA. E-mail: wilsonho@ccmr.cornell.edu

Received 10th April 2000
First published as an Advance Article on the web 14th July 2000

A scanning tunneling microscope (STM) operating at 9 K in ultrahigh vacuum was used to initiate a bimolecular reaction between isolated hydrogen sulfide and dicarbon molecules on the Cu(001) surface. The reaction products ethynyl (CCH) and sulfhydryl (SH) were identified by inelastic electron tunneling spectroscopy (STM-IETS) and by sequentially removing hydrogen atoms from an H_2S molecule using energetic tunneling electrons. For comparison, the thermal diffusion and reaction of H_2S and CC at 45 K and H_2O and CC at 9 K were also observed.

Introduction

The extreme spatial confinement of energetic electrons in the tunnel junction of a STM permits chemical modification of adsorbates to be controlled with atomic precision.[1-3] Though the remarkable imaging abilities of the STM have been well-utilized in documenting such reactions, the diagnostic capabilities of the STM were greatly expanded by the demonstration of STM-IETS.[4] The chemical sensitivity provided by this technique has enabled STM-induced bond scission and bond formation events at the single bond level to be characterized in detail.[5,6] One example of the precision of current experiments was the selective removal of individual hydrogen atoms from acetylene (HCCH).[6] STM images and vibrational spectra acquired by STM-IETS were used to unambiguously identify the unimolecular reaction sequence HCCH → CCH + H and CCH → CC + H.

In the present paper, we demonstrate the rehydrogenation of dicarbon (CC) by abstraction of hydrogen from hydrogen sulfide (H_2S) to form ethynyl (CCH) and sulfhydryl (SH). The reactions were performed at the single molecule level by inducing the diffusion of single H_2S molecules near CC reaction sites using the STM. The identities of the reaction products were established through imaging, vibrational spectroscopy and the controlled step-wise abstraction of hydrogen from H_2S using tunneling electrons. The thermal diffusion and reaction of H_2S with CC at 45 K were observed for comparison, as was the thermal diffusion and reaction of H_2O and CC at 9 K. The ability of the STM to initiate single-bond scission and formation reactions in a well characterized environment, and to characterize the local bonding before and after reaction through STM-IETS, allows the excited states of molecules to be studied at the single molecule level.

Experimental methods

The homemade variable temperature STM used in these studies has been described elsewhere.[7] The STM and sample temperature can be varied between 9 and 350 K by varying the flow rate of a continuous-flow liquid-helium cryostat and employing a temperature controller to adjust a heater load. Experiments were performed at 9 K unless otherwise noted. Tungsten tips were sharpened *ex situ* by ac and dc etching, and *in situ* by self-sputtering and annealing. The Cu(001) surface was prepared by standard Ne$^+$ sputtering and annealing procedures. Acetylene and hydrogen sulfide were dosed on the Cu(001) surface by leaking high purity gases through a variable leak valve. A capillary array doser was used to direct the molecular flux towards the sample and minimize the amount of gas required to achieve the desired coverages, which were typically ~0.005 ML.

The vibrational excitations of single adsorbed species can be detected using STM-IETS.[4] To take a vibrational spectrum, the tip is positioned over the molecule and the feedback is turned off. The dc sample bias voltage is ramped over the range of vibrational peaks while a sinusoidal bias modulation is superimposed. When the energy of the tunneling electrons is sufficient to excite a tunneling-active vibrational mode of the molecule, the conductance increases due to the onset of an inelastic tunneling channel. The derivative of the conductance exhibits a peak when eV_{bias} is equal to the molecular vibrational energy. The conductance $\sigma = dI/dV$ and its derivative $d\sigma/dV = d^2I/dV^2$ are proportional to the first and second harmonics of the tunneling current which are measured with a lock-in amplifier. I–V curves are taken to determine the proportionality constants. Background spectra taken over clean areas of the surface are used for comparison. In general, not all vibrational modes of a given adsorbate are detected by STM-IETS, and there is not yet a general rule for predicting the appearance of tunneling-active modes.

Single bond scission

The creation of dicarbon from acetylene by inducing C–H bond scission with the STM is briefly summarized below to establish the identity of the ethynyl intermediate and dicarbon end products; CCH was produced from CC in the hydrogenation reaction which is the focus of the present work. Details of HCCH dehydrogenation have been described elsewhere.[6] Single hydrogen atoms were removed from HCCH by positioning the STM tip directly above the molecule to be dissociated and raising the sample bias to ~3 V. Upon H removal, a slight increase in the tunneling current is detected by the STM electronics and the bias voltage is automatically lowered to prevent further chemical changes. To remove the H atom from CCH, the bias voltage is raised to ~2.2 V. Upon dissociation, a sharp decrease in the tunneling current is detected and the voltage pulse is terminated. The vibrational spectra of HCCH, CCH and CC obtained by STM-IETS are shown in Fig. 1. CCH has an upshifted C–H stretch energy (395 meV) relative to that of HCCH (357 meV), which is characteristic of the rehybridization which accompanies the removal

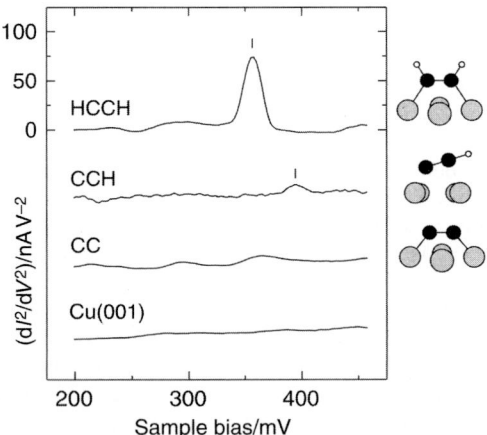

Fig. 1 STM-IETS spectra of acetylene and products of step-wise dehydrogenation. A spectrum taken over the bare copper surface is shown for comparison. The positions of the C–H stretch modes are indicated for HCCH and CCH. The rms bias modulation was 10 mV and the tunneling gap was set at 1 nA, 0.25 V. (Adapted from ref. 6.)

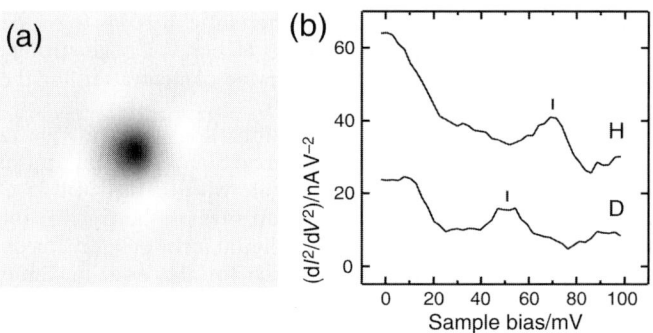

Fig. 2 (a) 15 × 15 Å² STM image of an isolated hydrogen atom taken at 0.1 nA tunneling current and 20 mV sample bias. (b) STM-IETS spectra of H and D. A background spectrum taken over the bare copper surface has been subtracted to emphasize the vibrational features. The indicated peaks are due to the excitation of the H–Cu and D–Cu stretch modes. The rms bias modulation was 7 mV and the tunneling gap was set at 0.1 nA, 50 mV. Signal averaging times were 14 and 28 min for these H and D spectra, respectively.

of one hydrogen atom. CC shows no C–H stretch mode as expected. Identification of the vibrational modes was confirmed by observing shifts in the peaks of the isotopes DCCD and CCD.

Hydrogen atoms removed from HCCH in this manner are typically found on the copper surface within 50 Å of the dissociation site. Hydrogen images as a small depression of 0.002–0.02 Å, depending on the sharpness of the tip and the tunneling conditions, and occupies the four-fold hollow site (Fig. 2(a)). These observations are in agreement with theoretical calculations of the binding energy and STM image.[8] The identification of the depression as hydrogen was confirmed by observing the Cu–H(Cu–D) stretch mode at 70(51) meV in vibrational spectra taken with the STM (Fig. 2(b)). These values are in agreement with electron energy loss spectroscopy (EELS) data.[9]

Vibrational excitation of H(D) can result in hopping between lattice sites at a rate that depends on the energy of the electrons and the tunneling current density. (The current density depends on both the current setpoint and the tip sharpness.) Similarly, excitation of C–H stretch and bending modes in HCCH and CCH, respectively, results in molecular rotation.[10,11] The transfer of vibrational energy from high to low energy modes in these molecules leads to the exploration of the local potential energy surface (PES). For hydrogen, the H–Cu stretch mode may effectively couple to a frustrated translation parallel to the surface. Under excitation with 150 meV electrons at 1 nA, the H hopping rate is 11× greater than the D hopping rate. This large difference may be used to quickly distinguish between H and D which are indistinguishable in constant current images.

Single bond formation

After dissociating a large number of acetylene molecules, the surface becomes populated with primarily dicarbon and hydrogen. The dicarbon is not mobile. The hydrogen is capable of diffusing in one of three ways: by quantum tunneling, by thermal diffusion and by electron-induced diffusion *via* vibrational excitation. We have observed all of the aforementioned phenomena. The diffusing hydrogen atoms, after long periods of observation, do not react with the CC species even when they are in close proximity. This indicates that the 200 meV diffusion barrier[12] for surface H on Cu(001) is smaller than the barrier to reaction with CC. Diffusion without reaction was observed for temperatures where quantum tunneling (9–60 K) and thermal diffusion (60–80 K) dominate. A mutual attraction between hydrogen atoms was also noted, but H_2 formation was not observed. Temperature-programmed desorption (TPD) studies indicate that atomic hydrogen desorbs recombinatively near 300 K.[9]

Hydrogen sulfide was chosen to accomplish the rehydrogenation of dicarbon for several reasons. Photodissociation of H_2S in the gas phase has been shown to produce hot hydrogen atoms.[13] Irradiation in the UV of H_2S and CO coadsorbed on Cu(111) at 68 K led to the production of a variety of products due to reactions involving hot hydrogen atoms.[14] On Cu(001) at 9

K, H_2S chemisorbs molecularly and does not diffuse thermally. It binds to copper mainly through the sulfur atom which produces a weakening of the S–H bonds.[15] The strong Cu–S interaction and lack of Cu–H interaction may give rise to a favorable configuration for the hydrogen in the S–H bond to react with another adsorbate.

In general, H_2S was not amenable to *directed* manipulation under a wide range of tunneling currents and sample biases. To encourage the desired reaction a different approach was found to be successful. By vibrationally exciting an H_2S molecule with 0.1 nA of 0.25 eV tunneling electrons, it was made to hop randomly between adsorption sites on the Cu(001) surface. The motion of the molecule was followed by the STM tip using a digital form of lateral feedback.[16] Since H_2S images as a protrusion, the tip was directed to search for the local maximum in height at a constant current. Excitation continued until the H_2S arrived, by a random walk, in the vicinity of a CC reaction site. At this point, the molecule tracking routine was terminated and a constant current image taken. The results of such a process are shown in Fig. 3. The bright H_2S of Fig. 3(a) has diffused and become less bright in Fig. 3(b), and the dark CC of Fig. 3(a) has become less dark in Fig. 3(b) after the encounter with H_2S. These changes in molecular contrast, corresponding to changes in tip height, are due to changes in the local density of states (LDOS) upon chemical transformation. A clearer picture of the reaction products was obtained by separating the bright fragment from the dark using the aforementioned excitation and tracking method on the bright fragment (Fig. 3(c) and (d)).

The darker species was identified as CCH. The streaky appearance of the image on the perimeter of the depression is due to fast rotation of the ethynyl under the imaging conditions. With a sharp tip, the upshifted C–H stretch mode characteristic of CCH could be observed by STM-IETS. The same reaction was also induced with D_2S as an additional test of the identity of the darker reaction product. In practice, these vibrational modes are difficult to observe in ethynyl due to the broadband noise generated by molecular rotation events; the ethynyl rotates very

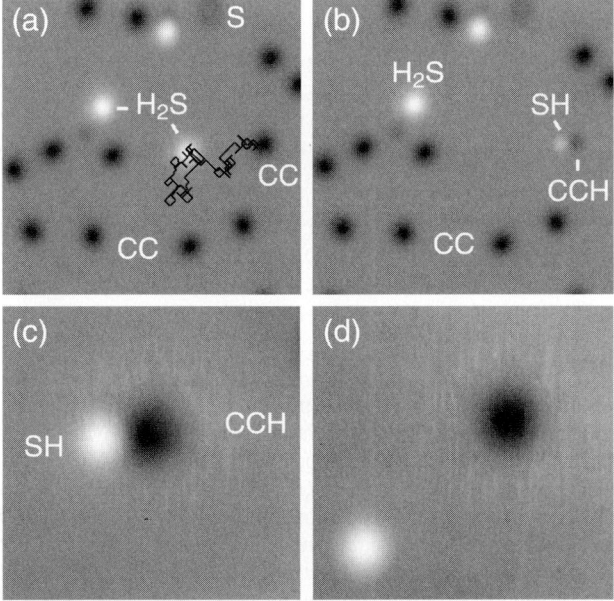

Fig. 3 (a) 100×100 Å² STM image of H_2S and CC on Cu(001) at 9 K. The image was taken at 0.1 nA and 0.1 V. By exciting the H_2S at 0.25 V, the molecule was made to hop between lattice sites in a random manner. The path of the tip following the adsorbate motion is indicated by the black line, starting near the middle of the image and ending toward the right hand side next to the CC. (b) STM image taken after the induced diffusion; the H_2S has reacted with a CC. (c) 30×30 Å² image of the newly formed SH and CCH reaction products. (d) The SH was separated from the CCH by induced diffusion to show the products more clearly. Under the imaging conditions of 0.1 nA and 0.1 V, the CCH rotates rapidly. This rotation is responsible for the streaky halo around the depression.

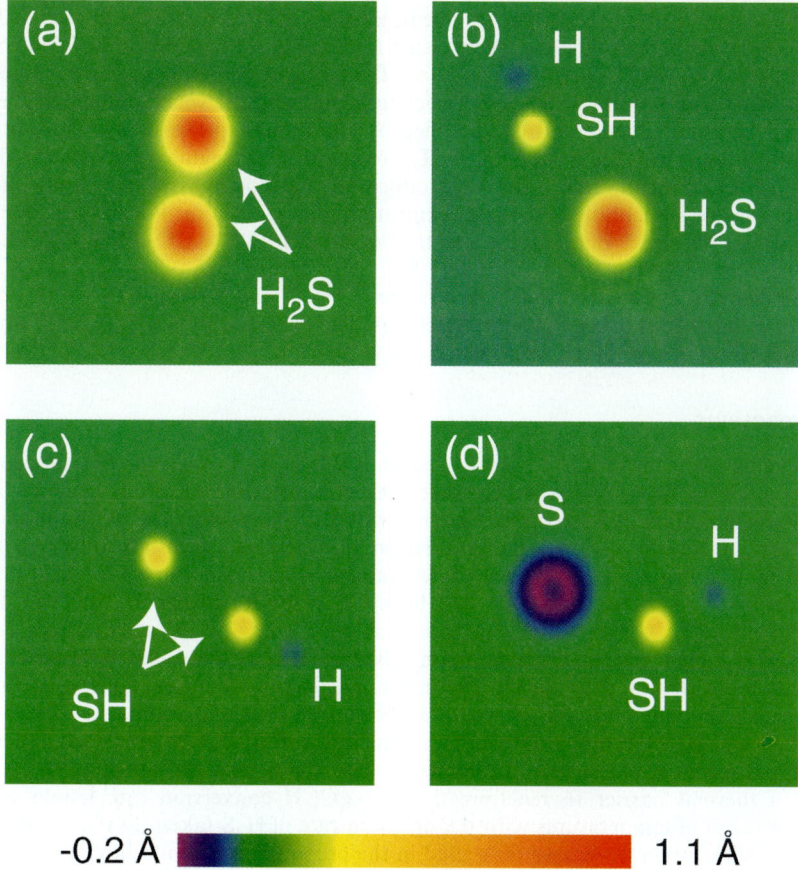

Fig. 4 Controlled hydrogen abstraction from H_2S. (a) 50×50 Å^2 STM image of two H_2S molecules. (b) The upper molecule of (a) has been dissociated to form SH and H. (c) The H_2S of (b) was dissociated to form SH and H. (d) The leftmost SH of (c) was dissociated to form S and H. Note the increased separation between the H and its parent molecule compared to that of (b) and (c). In between dissociation events, previously created atomic hydrogen was removed from the scan area. The color bar denotes the range of tip height variations in these constant current topographical images.

quickly under the tip at biases corresponding to the stretch modes. The steplike changes in the tunneling current produce noise at all frequencies within the bandwidth of the current preamplifier. In addition to the distinct stretch modes of CCH and CCD, the electron induced rotation rates are also distinguishable. As was mentioned previously, ethynyl rotates when vibrationally excited by tunneling electrons. This rotation has a characteristic onset behavior which exhibits clear isotopic differences[11] and allows CCH and CCD to be distinguished by a measurement of their respective rotation rates.

STM-IETS was performed on the bright fragments formed from CC + D_2S and CC + H_2S reactions in an attempt to identify the species expected to be SH. The species hopped from beneath the tip under excitation with electrons of energies comparable to the S–H and S–D stretch modes, preventing spectra from being obtained in this region. As was the case with hydrogen, vibrational excitation of a perpendicular mode led to parallel motion. Vibrational spectra were taken from 0–130 meV, an energy region which includes S–H and S–D bending modes.[17] Though characteristic differences between SH, SD, and background spectra were noted, no features were observed that could be definitively assigned to molecular vibrations. More sophisticated deconvolution schemes than the simple background subtraction currently employed may result in the identification of additional vibrational structure.

To unambiguously identify the bright fragment, a single hydrogen atom was removed from an isolated H_2S molecule using tunneling electrons of 1.5 eV (Fig. 4). The method of control was the same as that used in the HCCH dehydrogenation; the moment of dissociation was indicated by a drop in tunneling current, upon which the dissociating voltage pulse was terminated. This resulted in a Cu-bound H atom, which could be identified by vibrational spectroscopy, and a bright fragment identical to that produced by the CC + H_2S reaction. This fragment could be further dissociated using 1.5 eV tunneling electrons to produce an additional hydrogen atom and a dark species which was not susceptible to further chemical modification (Fig. 4(c) and (d)). We therefore conclude that the observed bimolecular reaction is in fact H_2S + CC → SH + CCH, and the electron-induced unimolecular reactions are H_2S → SH + H and SH → S + H. The CCH and H may be identified by their vibrational spectra, and the S and SH species by the controlled abstraction of H from H_2S using tunneling electrons. The experiments were repeated with D_2S to provide further confirmation of the reaction sequence.

Thermal reactions

Because tunneling electrons of 0.25 eV were used to enable H_2S to overcome the diffusion barrier, it was not possible to determine whether these energetic tunneling electrons were involved in a simultaneous surmounting of a barrier to reaction with CC. To explore the nature of the reaction further, H_2S was dosed at various temperatures in the presence of surface dicarbon to permit thermal diffusion to bring the two species into contact. After dosing at 45 K, SH species were observed next to CCH molecules which had been CC prior to dosing. In addition, H_2S were found next to intact HCCH molecules to which they were weakly attracted. These two observations suggest a high transient mobility upon adsorption. H_2S molecules dislodged from their attractive sites were observed to diffuse thermally. Due to weak interactions with the tip, the absolute thermal hopping rate of H_2S was not determined, but was estimated to be 0.5–5 s^{-1}. Several examples of H_2S thermally diffusing and reacting with CC were observed, indicating that any reaction barrier must be less than or comparable to the barrier to diffusion. To establish the existence of a thermal barrier to reaction, the CC → CCH conversion rate would need to be studied over a range of temperatures with the diffusion rate of H_2S taken into account. To obtain reasonable statistics, larger coverages than used in the present study would be desirable. This type of measurement was demonstrated for the oxidation of CO by atomic oxygen on the Pt(111) surface.[18]

Analogous to the case of H_2S at 45 K, H_2O thermally diffuses and reacts with CC at 9 K. The product OH may be separated from CCH, and further dissociated to produce O and H. Details of the experiments with H_2O will be published elsewhere. This observation supports the conclusion that there is little barrier to reaction of H_2S and CC. In the gas phase, the first bond dissociation energy of H_2O is ∼1.8 eV greater than that of H_2S, and H_2S decomposes more readily than H_2O on the Cu(001) surface.[15,17] Because the amount of thermal energy available at 9 K to drive the CC + H_2O reaction is insignificant, and because S–H bond cleavage should be even more facile, it is likely that the CC + H_2S reaction proceeds with a negligible barrier.

Non-reacting species

The hydrogenation of CCH and HCCH with the potential hydrogen donating species H_2S, SH, and H was not observed using STM-induced diffusion, nor was SH reactive towards CC. The failure of SH to react with CC could be of geometrical origin, in that S–H may adsorb vertically on four-fold hollow sites, or of energetic origin, in that S–H bond strength of SH could be larger than that of H_2S on Cu(001). These probable differences in bond energies and orientations are reflected in two experimental observations. First, the rate of SH dissociation at 1.5 V and 0.1 nA is smaller than that of H_2S under the same conditions. Second, the transient diffusion length of the 'hot' hydrogen atom removed from SH is, on average, much larger than that of the hydrogen atom removed from H_2S. The first hydrogen atom removed travels, on average, 7 Å from the parent H_2S with a narrow distribution of distances; the standard deviation is 1.5 Å. In contrast, the hydrogen atom removed from SH was rarely found within 10 Å of the parent molecule, was often observed 20–50 Å away, and occasionally was not observed at all. The initial trajectory of the abstracted hydrogen atom, which may differ for H_2S and SH, could be responsible for the differences in product distribution. Polanyi and coworkers have studied such processes for species desorbing upon photodissociation. The STM presents an additional complication in that reaction products may collide with the tip prior to surface readsorption.

Direct excitation of surface hydrogen with the same voltage pulse used for dissociation, but of even greater duration, did not result in the large diffusion lengths associated with H removed from SH, indicating that the observed diffusion is associated with energy gained during dissociation. Quantitative evaluation of energy exchange between the hot hydrogen and the lattice is necessary to relate diffusion lengths to the reaction energetics. Careful calculation of the bonding geometries and bond dissociation energies will also assist in the analysis of these experimental observations.

Summary

Vibrational excitations induced by tunneling electrons were used to bring reactants into close proximity and to analyze the reaction products. Higher energy excitation of H_2S resulted in hydrogen abstraction to produce SH, allowing the bimolecular reaction products of H_2S and CC to be identified as SH and CCH. An interesting subject of future research is the relative probabilities of diffusion and dissociation of H_2S under varying excitation energies and rates. The branching ratios depend on binding energy, diffusion and rotation barriers, the redistribution of excited state energy, both vibrational and electronic, and energy exchange with the surface. The advantage of using the STM in such studies is that the initial excitation is atomically localized and the de-excitation pathways, including rotation, diffusion, desorption, and reaction, can be clearly visualized, distinguished, and in some cases quantitatively measured.

References

1. Ph. Avouris, *Acc. Chem. Res.*, 1995, **28**, 95.
2. W. Ho, *Acc. Chem. Res.*, 1998, **31**, 567.
3. J. K. Gimzewski and C. Joachim, *Science*, 1999, **83**, 1683.
4. B. C. Stipe, M. A. Rezaei and W. Ho, *Science*, 1998, **280**, 1732.
5. H. J. Lee and W. Ho, *Science*, 1999, **286**, 1719.
6. L. J. Lauhon and W. Ho, *Phys. Rev. Lett.*, 2000, **84**, 1527.
7. B. C. Stipe, M. A. Rezaei and W. Ho, *Rev. Sci. Instrum.*, 1999, **70**, 137.
8. N. Lorente, F. Olsson and M. Persson, to be published.
9. I. Chorkendorff and P. B. Rasmussen, *Surf. Sci.*, 1991, **248**, 35.
10. B. C. Stipe, M. A. Rezaei and W. Ho, *Phys. Rev. Lett.*, 1998, **81**, 1263.
11. L. J. Lauhon and W. Ho, *Surf. Sci.*, 2000, **451**, 219.
12. L. J. Lauhon and W. Ho, to be published.
13. W. G. Hawkins and P. L. Houston, *J. Chem. Phys.*, 1980, **73**, 297.
14. D. V. Chakarov and W. Ho, *J. Chem. Phys.*, 1991, **94**, 4075.
15. J. A. Rodriguez, *Surf. Sci.*, 1990, **234**, 421.
16. L. J. Lauhon and W. Ho, *J. Chem. Phys.*, 1999, **111**, 5633.
17. K. T. Leung, X. S. Zhang and D. A. Shirley, *J. Phys. Chem.*, 1989, **93**, 6164.
18. J. Wintterlin, S. Völkening, T. V. W. Janssens, T. Zambelli and G. Ertl, *Science*, 1997, **278**, 1931.

General Discussion

Prof. Aeschlimann opened the discussion of Prof. Wolf's paper: (1) In Fig. 4 of your paper, does the decay rate of state A represent the decay of the excited state population including all refilling processes (*e.g.* decay of state B), or does it show the inelastic lifetime of an excited electron in state A?

(2) It is hard to believe that the symmetry of the excited state A does not change from a very localised state at 1 ML to a free electron state at 5 ML. Can you comment on this?

Prof. Wolf responded: With respect to your first question regarding the time resolved 2PPE measurements: The decay times, τ_{decay} for peak A should be associated with the inelastic lifetime of state A, whereas τ_{rise} relates to the rate of building up the population from higher lying states. However, our time resolution is not sufficient to fully resolve these complicated dynamics, but the actual trends of the lifetime as a function of coverage remain unchanged for different assumptions in the analysis.

The answer to the second question, regarding the polarization dependence, is that we observe the same polarization dependence (and hence symmetry of state A) at all coverages. This is the same polarization dependence as found for *e.g.* image potential states. The first C_6F_6 layer has an ordered (presumably highly symmetric) structure and exhibits a (3 × 3) LEED pattern. With increasing coverage we observe the development to a quasi delocalised free-electron-like state in the direction parallel to the surface, which obeys the same symmetry as an image potential state.

Dr Gauyacq asked: What is known about excited states and resonances in free gas phase C_6F_6 molecules? Can this give some suggestions about the state playing a role in your experiments? How does the symmetry link with the polarization dependence in the experiment?

Prof. Wolf responded: As mentioned before, we observe for a monolayer of C_6F_6/Cu(111) a (3 × 3) LEED structure, which suggests that the molecules lie flat at the surface, presumably in high symmetry sites. Our polarization measurements indicate σ symmetry, so we believe that peak A should be assigned to a σ* orbital, which has antibonding character with respect to the C–F bond. The latter is consistent with our observation of photochemical decomposition by UV light. The symmetry of the states in the gas phase has been discussed in the paper, however, different assignments exist for the LUMO of free C_6F_6 in the literature.

Prof. Winter asked: What is the growth mode in your experiments, and how did you control growth? Can you comment on the orientation of the C_6F_6 molecules in the film?

Prof. Wolf responded: We have determined our C_6F_6 coverage by thermal desorption spectroscopy, which shows clearly separated desorption peaks up to 2 layers (see also ref. 1 here). While at fractional coverages we observe in our 2PPE spectra a double peak structure, we get a single peak upon completion of each layer. Our interpretation is therefore, that at fractional coverages we have co-existing islands with different layer thickness as is typical for layer by layer growth. Similar findings using 2PPE spectroscopy are made for the layer by layer growth of Ag/Pd(111) (see ref. 2).

1 Vondrate and Zhu, *J. Chem. Phys.*, 1999, **103**, 3349.
2 R. Fisher *et al.*, *Phys. Rev. Lett.*, 1993, **70**, 654.

Dr Shluger said: Could you please explain in more detail why you are not happy with the dielectric continuum model for interpretation of your experiments? Since your system is very anisotropic and contains one or two layers of molecules, is it justified to use one dielectric constant to describe its dielectric properties? Would application of a dielectric tensor or at least two different values parallel and perpendicular to the surface improve agreement of your model with experiment? It would perhaps be interesting to try this model and see which values of dielectric constants could fit the experiment.

Prof. Wolf responded: We have used the known dielectric constant for the condensed (liquid) phase, which is isotropic. The exact value for one monolayer C_6F_6 adsorbed on Cu(111) is not known. We believe that fitting a dielectric tensor with more free parameters is not appropriate based on our limited knowledge of the orientation and structure of the adlayer.

Prof. Kleyn said: Can you give an estimate of how efficient substrate electron induced photochemistry is for the system C_6F_6/Cu(111)?

Prof. Wolf responded: For the C_6F_6/Cu(111) system we did observe a slow photochemical decomposition for excitation with UV pulses, which results in a gradual increase in the work function due to accumulation of fragments (presumably fluorine) at the surface. For the 2PPE experiments presented here we tried to keep this effect as small as possible and stopped measurement when the work function changes more than 20–30 meV. Further tests showed that there is a pronounced wavelength dependence of the dissociation cross section: In fact we cannot induce photochemistry with visible light. The cross section is small in the monolayer regime and increases for thicker layers, however we have not tried to estimate a real number. Because we accumulate the reaction products at the surface, this system is not ideal to study this reaction with time-resolved 2PPE spectroscopy.

Prof. Wolf opened the discussion of Prof. Matsumoto's paper: In your discussion of the fast α component of your TOF spectra of Xe/O/Si you suggest as a possible interpretation a "translational rainbow" effect. However, you observe the α TOF component only at high excitation energies ($h\nu = 3.5$ eV) which may be indicative of another mechanism, namely the excitation of two different electronic states and potential energy surfaces. Could you give us a further discussion of the origin of the α component?

Prof. Matsumoto responded: As you pointed out, the wavelength dependence indicates that the origins of the two velocity components in the TOF distribution are different. The reason we quoted the possibility of a translational rainbow effect is due to the narrow velocity distribution of the α component around 0.8 eV. The Antoniewicz model with a constant quenching rate of the excited state cannot explain such a sharp velocity distribution, unless the lifetime of the excited state is long enough such that the adsorbate can sample the inner turning point of the excited state potential energy curve. However, we prefer another explanation for this. Namely, it is rather natural to think that the quenching rate (the lifetime of the excited state) depends on the distance between Xe and the substrate; the quenching rate increases with decrease in the distance. This makes it possible for the adsorbates that gain a certain kinetic energy in the excited state to have a larger quenching rate.

Dr Shluger said: Could you please explain in more detail what is the thickness and structure of the oxide film on the Si surface? Do you know whether it has a developed band gap comparable to that in bulk silica? Is it possible that your laser light excites excitons in this film, similar to excitons in amorphous silica?

Prof. Matsumoto responded: It has been well established that oxidation saturates at 1 ML by annealing the surface covered by N_2O.[1] Thus, we do not think that the surface has a developed band gap comparable to that in bulk silica. However, as you pointed out, excitonic states may exist in the energy range much smaller than the band gap of the bulk silica.[2] These excitonic states, if any, can be involved in the nonthermal Xe desorption.

1   H. Kato, K. Sawabe and Y. Matsumoto, *Surf. Sci.*, 1996, **351**, 43.
2   J. Song, R. M. VanGinhoven, L. R. Corrales and H. Jónsson, *Faraday Discuss.*, 2000, **117**, 303.

Dr Lee commented: You suggested that electron acceptor oxygen inserted into the dimer bond of Si(100) surface was necessary to detect the fast component α ($\langle E \rangle = 0.85$ eV) of Xe photodesorption with $h\nu \geqslant 3$ eV in Fig. 2 in the paper. Perhaps it is worth trying an experiment similar to photodesorption of Xe from a 'chlorinated' Si(100) surface? There may be a fast component with a different kinetic energy and also different threshold photon energy from that of the Xe/O/Si(100) system.

Prof. Matsumoto responded: Because chlorine also has large electron negativity, the similar electron-transfer state proposed in the paper may exist. Thus, it may be worth trying the similar experiments as you suggest.

Prof. Kleyn said: I remember that the electron affinity of the CH_3 radical is about 1 eV, higher than that of H^-. Could this influence the surface photochemistry?

Prof. Matsumoto responded: Interestingly the photochemistry of methane is self-quenched as the coverage of photoproducts CH_3 and H increase and eventually those adsorbates lower the work function of Pt(111) by 0.9 eV.[1] We think that this lowering of the work function is mainly caused by electron donation from CH_3 adsorbates. This self-quenching can be understood in the following way. As described in the paper, the adiabatic energy necessary for the excitation is estimated to be 7 eV [=13 eV (E_i of methane) − 6 eV (the work function of Pt(111))] if the complete electron transfer occurs in the excited state. This is still larger than 6.4 eV, but the further stabilization by an image force or chemical interactions with the surface makes it possible for excitation to occur. However, if the work function decreases, more energy is needed for the excitation. Thus, the adsorbates lowering the work function, such as CH_3 or Cs, quench the photochemistry at a fixed photon energy of 6.4 eV.

In conjunction with the photochemistry of methane, let me elaborate on the excitation mechanism further, which we believe inherent on adsorbates with spherical symmetry. We describe the photochemistry of methane on Pt(111) and Xe on O/Si(100) and discuss some common features in the two cases. In order to examine the excitation mechanism invoked, it is interesting to see what would happen in the case of Xe on Pt(111) and methane on O/Si(100). In the former case, no photochemistry is observed. This is why Xe is used for a space layer to separate the electronic systems between adsorbate and substrate on many occasions. But, this does not exclude the possibility of the excitation mechanism described in the paper. Even though a similar electron-transfer state would exist in this adsorption system, a short lifetime of the excited state at Pt(111) and the heavy mass of Xe prevent measurable photodesorption. In contrast, the preliminary measurements on methane on O/Si(100) shows that methane is desorbed with a large cross section (1×10^{-17} cm^2) by the irradiation of 6.4 eV photons, although we do not see any indication of dissociation.

1   Y. Matsumoto, Y. A. Gruzdkov, K. Watanabe and K. Sawabe, *J. Chem. Phys.*, 1996, **105**, 4775.

Prof. Hasselbrink commented: In the context of the substrate mediated excitation processes it has always been argued that this pathways dominates because it allows the large number of initial substrate excitations to compensate for the loss due to quenching. The net result is then a photodesorption cross section which is comparable to a gas phase photoexcitation cross section. Now in your case of a direct excitation this argument would lead one to expect a very small cross section. Can you comment on this and have you tried to determine excitation and desorption cross sections?

Prof. Matsumoto responded: The total cross section of the methane photochemistry is estimated to be 1.8×10^{-19} cm^2 at 6.4 eV. Since desorption, dissociation and photo-induced associative desorption between a hot hydrogen and a CH_3 adsorbate produced photochemically contribute to the total cross section, it is difficult to separate the contributions from these processes. However, we can still argue whether or not the obtained cross section is reasonable for the direct excitation by considering its isotope effect. The isotope effect in desorption/dissociation cross sections can be given by,[1]

$$\frac{\sigma_1}{\sigma_2} = \left(\frac{1}{p_1}\right)^{\sqrt{(m_2/m_1)}-1}$$

where σ_1 and σ_2 are the desorption/dissociation cross sections for the isotopic labelled adsorbates with mass m_1 and m_2, respectively, and p_1 is the probability, i.e., a ratio between σ_1 and a primary excitation cross section σ_{ex}. The cross section of CD$_4$ at 6.4 eV is 1.3×10^{-19} cm^2. There are two limiting cases we can argue. If the photochemistry is mostly dominated by desorption, m_1 and m_2 would be taken as 16 (CH$_4$) and 20 (CD$_4$), respectively. The probability turns out to be 0.06. On the other hand, if dominated by dissociation, the reduced mass of H–CH$_3$ and D–CD$_3$ would be more appropriate for m_1 and m_2, respectively, since the motion along the C–H(D) coordinate is relevant to the dissociation. Then, $p_1 = 0.43$. In reality the probabilities for desorption and dissociation should fall in this range. These considerations limit the range of σ_{ex}: 4.2×10^{-19} cm^2 and 3.0×10^{-18} cm^2. The absorption cross section of gaseous methane: $\sim 2 \times 10^{-18}$ cm^2 at 8.9 eV (near onset) and 2×10^{-17} cm^2 at 10.3 eV (peak).[2] Since the excited state responsible for the absorption band is expected to be shifted to lower energy by interactions with the substrate, the estimated excitation cross sections are not physically unreasonable.

1 X.-Y. Zhu, *Annu. Rev. Phys. Chem.*, 1994, **45**, 113.
2 L. C. Lee and C. C. Chiang, *J. Chem. Phys.*, 1983, **78**, 688.

Prof. Wolf said: In your measurement on CH$_4$/Pt(111) you observed a unique polarization dependence, which clearly showed that the photochemistry originates from a direct excitation mechanism. What is known about the excitation mechanism for Xe/O/Si? Could it be that it is a substrate mediated process, for example induced by hot holes?

Prof. Matsumoto responded: Since Xe is weakly bound at the surface, the HOMO of Xe would be located at ~ 7 eV below the Fermi level. The threshold photon energy of the α component is ~ 3.5 eV. Thus, the absorption of photons with this energy cannot make hot holes energetically sufficient to transfer to the HOMO.

We mainly regard the α component in the TOF distribution of Xe as a plausible candidate for the 'direct' excitation as in the case of methane on Pt(111). However, unfortunately it is difficult to obtain conclusive evidence experimentally to support our proposal, such as the polarization and incident angle dependence of the cross section undertaken in the photochemistry of methane. This is due to the fact that there are two velocity components apparently originating in the different excited states in the TOF distribution and their angular distributions are quite different to each other. In particular, the sharp angular distribution of the α component and a fixed configuration of the laser incident beam and the line of sight of the quadruple mass spectrometer do not allow us to measure the TOF distributions with p- and s-polarization at the incident angle of $\sim 70°$ where a significant deviation from the substrate excitation is expected. Therefore, we must say at this stage that the excitation mechanism described in the paper is still a proposed model for understanding the Xe photodesorption and more experimental and theoretical work is needed.

Prof. Hasselbrink said: Whereas in Prof. Matsumoto's paper the excitation mechanism is addressed, in the paper by Prof. Wolf as well as in the paper presented by Dr Gauyacq the fate of the electron in the intermediate resonance state is addressed. In order to make a connection it would be worthwhile to discuss the excitation mechanism for the states studied with 2-photon photoemission. How does the electron get into these long-lived states?

Prof. Aeschlimann commented: We have to distinguish between: (1) a substrate mediated direct excitation, in which a bulk electron from the substrate is directly excited into the resonance state (this process must be separated from an intra-molecular direct excitation) and (2) a substrate mediated indirect excitation, in which there is first an optical excitation into a substrate bulk state, followed by a scattering (or tunnelling) process into the resonance state.

If the indirect excitation is more favorable than the direct excitation, we can conclude that the tunnelling barrier is rather low. Hence we do not expect a long resonance lifetime because in general a high electron transfer yield also results in a high back transfer yield (*i.e.* if the electron comes into the resonance easily, it also comes out easily).

Indeed, polarization dependent measurements for the long lived Cs hybrid state have shown that the resonance is mainly populated by substrate mediated direct excitations. This is also the case for the well investigated $2\pi^*$ state of CO on Cu(111). Only the σ state in the same system may be populated by an indirect excitation (see ref. 1).[1]

1 M. Wolf, *Phys. Rev. B*, 1999, **59**, 5926.

Prof. Wolf said: Based on our current knowledge the predominant excitation mechanism in two-photon-photoemission (2PPE) involves direct excitation by the first photon from an occupied initial state to an unoccupied intermediate state, from where the electron is subsequently excited by the second photon to a free electron state (inverse LEED state) in the vacuum. This has been demonstrated using an extended polarization analysis for image potential states on Cu(111) as well for CO/Cu(111) (see ref. 1). The CO/Cu(111) system (see. Fig. 1) is of particular relevance for comparison with the excitation mechanism in surface photochemistry, because both direct and indirect (substrate mediated) excitation processes are observed within the same system. From our polarization analysis we find that all the peaks in the spectra of CO/Cu(111) above $E - E_{Fermi} \geqslant$ 1.7 eV are populated *via* a direct excitation process, while the σ-state at $E - E_{Fermi} = 1.5$ eV is populated by an indirect process where the photon is first absorbed in the substrate and subsequently a secondary electron populates the σ-state. The latter mechanism is believed to be predominant in most photoinduced reactions at metals and semiconductors (see ref. 2).

At first sight the dominant excitation mechanisms in 2PPE and surface photochemistry appear to be completely different. However a closer look at the growing literature on time-resolved 2PPE studies suggests the following picture: It is now well established that the lifetimes of image potential states and of excited states of adsorbates are governed by the overlap of that part of their wavefunction, which penetrates into the substrate, with the electronic state in the bulk. The larger this overlap the stronger the coupling to electron hole pair excitation in the substrate and hence the shorter the excited state lifetime. On the other hand electron transfer from the substrate to the adsorbate will also depend on this coupling and is enhanced for stronger coupling. A mechanism which explains in a consistent manner the apparently different findings for 2PPE and surface photochemistry (*i.e.* direct *vs.* substrate mediated excitation) is illustrated in Fig. 2. In the first excitation step in 2PPE an electron is excited in a direct (optical) transition to the tail of the wavefunction inside the substrate whereas in substrate mediated excitation an electron hole pair is

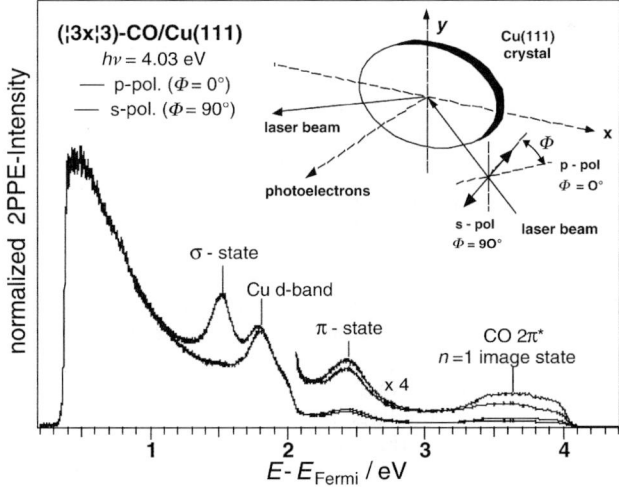

Fig. 1 Polarization dependence in 2PPE spectroscopy of CO/Cu(111). The inset shows the experimental geometry. The analysis on the dependence of the photoemission signal as a function of polarization angle Φ allows one to discriminate between direct and indirect excitation mechanisms in 2PPE (adapted from ref. 1).

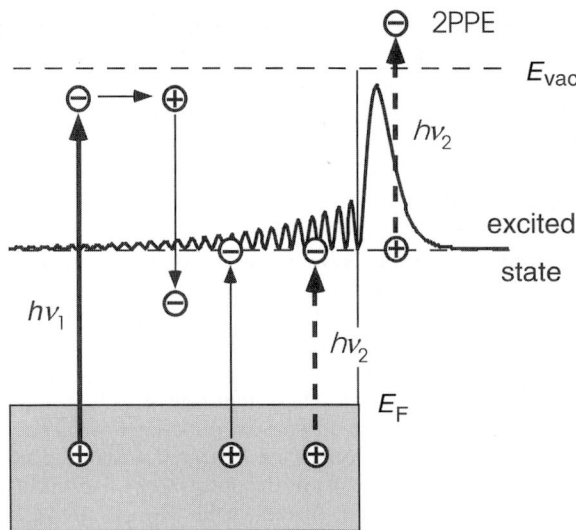

Fig. 2 Excitation mechanisms in two-photon-photoemission (2PPE) and surface photochemistry. In 2PPE the photon energy ($h\nu_2$, dashed arrows) is most often smaller compared to surface photochemistry (photon energy ($h\nu_1$ solid lines)). The excited state may be either populated by an indirect, substrate mediated process or by a direct electronic transition.

created inside the substrate. Note that in surface photochemistry the photon energy $h\nu_1$ is often in the UV range (exceeding the substrate work function) while in 2PPE the photon energy $h\nu_2$ is often considerably lower (and actually must be below the work function to avoid one-photon-photoemission). In surface photochemistry the relaxation of the primary electron into electron hole pairs may thus generate *secondary* electrons with energies which are sufficient to reach the unoccupied excited state of the adsorbate.

We propose that these secondary electrons *directly* populate the unoccupied state of the adsorbate by an electronic transition to the tail of the excited state wavefunction inside the substrate (see Fig. 2). However, this transition is now induced by the screened coulomb interaction in the electron–electron scattering event and not by the light field as in the case of the direct excitation mechanism. Note, that hot electron transport followed by electron attachment to the adsorbate is not invoked in this scenario. Under this aspect the proposed mechanism appears analogous to a one-step picture of photoemission spectroscopy compared to a three-step scenario (see ref. 3). This interpretation may bridge the apparently contradictory findings for the excitation process in 2PPE and substrate mediated surface photochemistry. For CO/Cu(111) we find exactly, that substrate mediated excitation dominates for the energetically lower lying σ-state, which can be easily reached by secondary electrons which result predominantly from excitations of Cu d-electrons.

1 M. Wolf, A. Hotzel, E. Knoesel and D. Velic, *Phys. Rev. B*, 1999, **59**, 5926.
2 F. Zimmermann and W. Ho, *Surf. Sci. Rep.*, 1995, **22**, 127.
3 S. Hüfner, *Photoelectron Spectroscopy*, Springer, Berlin, 1995.

Dr Gauyacq said: Prof. Hasselbrink's question was about the possible influence of the way the excited state is prepared on the observed lifetime. From a theoretical point of view, a wavepacket progagation calculation like the one we performed produces the energy and lifetime of the resonances within the considered problem. Since these are long lived, they should dominate the long time evolution and so correspond to the large time delay. Now, on the other hand, the short time evolution *a priori* depends on the way the excited state is prepared. If the preparation does not exactly produce the resonance state, it can prepare a wavepacket containing components other than the resonance and so the early time evolution can be modified and exhibit a complex time behaviour. As an example, in wavepacket propagation (see *e.g.* ref. 1), if the propagation starts with a wavepacket equal to an unperturbed atomic state in front of a Cu(111) surface, there are clearly two decay regimes. At short times, the wavepacket decays as on a free electron metal surface and at long times, the wavepacket has had enough time to probe the presence of the Cu(111) band gap and the wavepacket decays with a much longer lifetime. The latter is the one associated with the resonance state and the short time behaviour is a consequence of the preparation of the initial state.

1   A. G. Borisov, A. K. Kazansky and J.-P. Gauyacq, *Phys. Rev. B*, 1999, **59**, 10935.

Dr Saalfrank said: In connection with Prof. Wolf's talk I'd like to mention that we recently carried out[1] simulations of the energy- and time-resolved 2PPE spectra for image states on Cu(100), as reported by Höfer *et al.*[2] For that purpose we use a realistic one-electron potential,[3] a mapped Fourier method to calculate the corresponding eigenstates, and an open-system density matrix description. The latter accounts for the pump and probe fields, energy and phase relaxation, and energy-resolution of the final states. No "scaling laws" are assumed, but lifetimes and dipole matrix elements, for example, are calculated from the wavefunctions derived from the one-electron potential.[3]

In this way we obtain quite remarkable agreement between theory and experiment for energy-resolved 2PPE spectra at various delay times between pump and probe pulse, and also for time-resolved 2PPE spectra.

1   T. Klamroth, P. Saalfrank and U. Höfer, to be published.
2   U. Höfer, I. L. Shumay, Ch. Reuß, U. Thomann, W. Wallauer and Th. Fauster, *Science* 1997, **277**, 1480.
3   E. V. Chulkov, V. M. Siilkin and P. M. Echenique, *Surf. Sci.*, 1999, **437**, 330.

Prof. Kosloff said: What is the dephasing model? A Poisson model would seem suggestive.

A Gaussian dephasing model is based on the assumption of a fast *i.e.* delta correlated bath. Since the electronic processes under discussion are very fast it is hard to imagine an even faster bath. In the Poisson model the physical assumption is of rare uncorrelated events. The requirement is that the duration of these events is much smaller than the average time duration between these events. Elastic scattering of free electrons could be a good candidate for such a Poisson dephasing model. Practically, a Gaussian model gives good results, and is a limit of a Poisson process as well as other stochastic processes.[1]

1   R. Kosloff, *Physica A*, 1982, **110**, 346.

Dr Saalfrank responded: In our simulation of 2PPE spectra, (pure) dephasing was modelled by a term

$$\dot{\rho}_{mn} = -\frac{1}{T^*_{2,mn}} \rho_{mn}$$

for the off-diagonal elements of the density matrix. This is also what you get in a Gaussian dephasing model. However, except for CO-covered surfaces[1] the effects of pure dephasing on the time- and energy-resolved 2PPE spectra for image states at Cu(100) are comparatively small; hence the actual distinction between Gaussian and Poisson models will have almost no practical consequences in this case.

1   Ch. Reuß, I. L. Shumay, U. Thomann, M. Kutschera, M. Weinelt, Th. Fauster and U. Höfer, *Phys. Rev. Lett.*, 1999, **82**, 153.

Prof. Palmer opened the discussion of Prof. Seideman's, Prof. Stokbro's and Dr. Dujardin's papers: It is interesting to compare the two papers. Both of them invoke the idea of electron capture into a molecular resonance state, but in the first case the experimental signature (desorption of benzene from Si(100)) shows a threshold behaviour as a function of applied voltage, while in the second case (diffusion of hydrogen on the same surface) a characteristic resonance profile is observed (such as those found in resonance scattering of electrons by molecules on surfaces[1]). Would the authors please comment on the difference between their results?

1   R. E. Palmer and P. J. Rous, *Rev. Mod. Phys.*, 1992, **64**, 383.

Prof. Stokbro responded: We see this resonance profile both in the case of H diffusion and H desorption. The origin of this resonance profile has been described in detail in our papers.[1,2] I will briefly explain the main points here. We model the experiment by calculating both the tunnel current, I, and inelastic fraction, I_r, as a function of the bias, V_b. The calculation is based on our model in ref. 3 and in this model we include the effect of the electric field in the tunnel gap, E, and the tip–surface distance, h. The initial increase in I_r/I as function of V_b is straightforward to understand and has the same origin in Prof. Seideman's and my model, it is the decrease at high bias that is different in the two models. To understand the decrease at high bias in my model it is important to notice that in the experiment I is kept constant, and this results in an increasing h as a function of V_b. When h increases it will favour electrons that tunnel at the top of the tunnel barrier, which for negative sample bias means electronic states at the sample Fermi level. Thus with increasing negative V_b, the tunnel barrier for the resonant electrons will increase relative to electrons at the sample Fermi level, and this is the effect that gives rise to a decreasing fraction of resonant electrons.

1 U. Quaade, K. Stokbro, C. Thirstrup and F. Grey, *Surf. Sci.*, 1998, **415**, L1037.
2 K. Stokbro, C. Thirstrup, M. Sakurai, U. Quaade, Ben Yu-Kuang Hu, F. Perez-Murano and F. Grey, *Phys. Rev. Lett.*, 1998, **80**, 2618.
3 K. Stokbro, U. Quaade and F. Grey, *Appl. Phys. A*, 1998, **66**, S907.

Dr Dujardin commented: What you said to explain the observation of a resonance in hydrogen on Si(100) could also be valid in the case of benzene on Si(100). Therefore it does not explain the difference observed in the two systems.

Prof. Stokbro responded: As I mentioned previously we observe a peak in both the H diffusion data and in the H desorption data, so we are quite confident about the validity of our results. The exact shape of the peak depends on how large a fraction of the current goes through the resonance level relative to the rest of the current. When this fraction increases, the peak becomes less pronounced. Thus without doing a calculation it is very difficult for me to comment on whether the benzene data disagrees with a resonant process.

Prof. Seideman said: The electronic mechanism underlying the H/Si dynamics studied in Prof. Stokbro's paper[1] and the benzene/Si dynamics studied in our paper[2] is the same and the formalisms, in as far as the electronic problem is concerned, are equivalent. The difference is in the nuclear dynamics, in the fact that a surface- (rather than adsorbate-) derived resonance is involved, and in experimental details. In particular, the coupling of the substrate–adsorbate system to the tip, implicit in the Γ_t in eqn. (3)–(5) of ref. 2, depends on the sample bias–voltage in the experiment of ref. 1. Hence the voltage dependence of the observable reflects the competition between a threshold-like function, which contains the physics of the resonance-mediated process as in refs. 2–5, and a function that decreases with $|V_b|$. The reasons for which in the experiment of ref. 1 V_b-dependence of Γ_t is observed while in ref. 2 it is not, are that in the former the voltage range probed is much larger, the tip–surface distance is smaller, and a large (dominating) nonresonant component to the conductance is possible. In addition, the linear scale of Fig. 3a of ref. 1 accentuates the post-threshold regime whereas the semilog scale used in ref. 2 and in previous studies[3–5] accentuates the threshold regime and smoothes out small variations in the plateau regime. On the ordinate scale of Fig. 2 of ref. 2, which illustrates four orders of magnitude change of the yield in the threshold regime, a drop of the yield by a factor of 4–5 at large $|V_b|$, as seen in Fig. 3 of ref. 1, is unobservable and insignificant.

1 K. Stokbro, U. J. Quaade, R. Lin, C. Thirstrup and F. Grey, *Faraday Discuss.*, 2000, **117**, 231.
2 S. Alavi, R. Rousseau, G. P. Lopinski, R. A. Wolkow and T. Seideman, *Faraday Discuss.*, 2000, **117**, 213.
3 E. T. Foley, A. F. Kam, J. W. Lyding and Ph. Avouris, *Phys. Rev. Lett.*, 1998, **80**, 1336.
4 B. C. Stripe, M. A. Rezaei and W. Ho, *Phys. Rev. Lett.*, 1997, **79**, 4397.
5 L. J. Lauhon and W. Ho, *Surf. Sci.*, 2000, **451**, 219.

Prof. Rocca said: I notice that the scale in the relevant figures reporting the desorption yield and the switching probability is different, being linear in the paper of Stokbro *et al.* and logarithmic in the paper of Alavi *et al.* Moreover, in the former paper the switching probability decreases by a factor between 3 and 4 when the bias voltage is increased from -3 to 4 eV, while in the experiment reported by Alavi *et al.* the experimental points are all between -2 and -2.8 eV. For larger negative biases only a lower bound is given for the desorption yield. From experiment we therefore cannot tell that the dependence of the yield with bias voltage is different at large negative voltages.

Prof. Seideman commented: I fully agree with Prof. Rocca. It is in this sense that I meant that the difference between the two problems is in the experimental details and not in the essential physics.

Prof. Stokbro said: This is a very nice observation, and I fully agree with Prof. Rocca. Maybe the experimentalists did not pay much attention to this bias regime, and it would be interesting with some additional data points.

Prof. Palmer said: It seems to me that what Prof. Stokbro is trying to say is that there is a distinction between two different mechanisms. (i) In one mechanism, the electron tunnelling between tip and surface loses energy by exciting a particular excitation between two levels in the adsorbate/surface system. This excitation drives the dynamical process. It can only occur when the energy of the electron, *i.e.* the bias voltage, is high enough for the electron to be able to lose the necessary amount of energy. In this case a threshold behaviour is expected. People sometimes call the excitation "resonant", which may well be confusing some people! (ii) In the second mechanism, the tunnelling electron tunnels into a resonant state (*e.g.* an unoccupied molecular orbital) of the adsorbate/substrate system, to form something like a negative (or positive) ion resonance state. The formation of the resonance drives the dynamics. In this case the cross section as a function of bias is expected to show some kind of bell shaped resonance profile, at least if the centre of the resonance is located in the voltage range under consideration. (This distinction also applies in electron scattering; see ref. 1.)

Prof. Stokbro believes that his result is an example of (ii), *i.e.* resonance formation, while Prof. Seideman's result may be an example of (i), *i.e.* an energy loss (threshold) process. I think that Prof. Seideman may be saying, by contrast, that in her case the cross section is the product of two factors, one of which has a resonance shape—also arising from capture of an electron (or more specifically hole) to form a resonance—and the other is a desorption probability; the product of the two resembles a threshold phenomenon.

1 R. E. Palmer, *Prog. Surf. Sci.*, 1992, **41**, 51.

Prof. Stokbro said: This is exactly what I tried to state, thank you, Prof. Palmer, for this nice formulation.

Prof. Seideman responded: Let me disagree with Prof. Palmer. The electronic mechanism is the same in the H/Si[1] and the benzene/Si[2] problems. The nuclear dynamics, the nature of the resonance and the details of the experiment differ and this gives rise to the difference in observation.

This is readily seen from
eqn. (1)–(5) of ref. 2. Eqn. (1) gives the rate of a general STM-induced, resonance-mediated reaction and is formally exact. Eqn. (2)–(5) introduce a sequence of approximations but remain general and the qualitative physics they describe applies also to the experiment of ref. 1. Thus, it is only subject to the conditions of low temperature and energy-independence of the tip and substrate densities of states that the electronic transmission factor, W_{exc} in eqn. (2)–(4) of ref. 2, reduces to a simple sigmoidal form,

$$W_{exc}(\varepsilon, V_b, \tau) \sim \frac{\Gamma_t \Gamma_s}{\pi \hbar \Gamma} \left[\tan^{-1}\left(\frac{E_F - \varepsilon_r}{\Gamma/2}\right) - \tan^{-1}\left(\frac{E_F + \varepsilon + eV_b - \varepsilon_r}{\Gamma/2}\right) \right],$$

but its *qualitative* shape is of the threshold type irrespective of these assumptions. The reason is that the STM is not a narrow band source; at sample bias voltages beyond the resonance energy a range of electron energies contributes that encompasses the resonance lineshape.

As Prof. Palmer notes, the reaction rate, w, in eqn. (2) and (5), includes an energy integral over the product of the electronic factor W_{exc}, and a function that describes the dependence of the nuclear dynamics on the electron energy loss. Only if the nuclear factor is a sharply-peaked function of ε, the voltage dependence of W_{exc}, is duplicated in the observable.

This, however, applies also to the problem studied in ref. 1 [where P_{des} of eqn. (2.5) would be replaced by an appropriate reaction probability]. The main difference between the problems is that in the H/Si, study[1] Γ_t decreases with increasing $|V_b|$ leading to a decrease of the yield at V_b below the resonance position (which should not be mistaken for a resonance feature). In the benzene/Si study[2] V_b-dependence of Γ_t is negligible over the range of relevant voltages and hence a threshold feature is observed. Please see also my reply to Dr Dujardin's question earlier.

1 K. Stokbro, U. J. Quaade, R. Lin, C. Thirstrup and F. Grey, *Faraday Discuss.*, 2000, **117**, 231.
2 S. Alavi, R. Rousseau, G. P. Lopinski, R. A. Wolkow and T. Seideman, *Faraday Discuss.*, 2000, **117**, 213.

Prof. Polanyi commented: Prof. Stokbro has reported a rather efficient STM-tip-induced lateral-transfer of H on Si(100). The proposed model, with which I have no quarrel, is that the H-atom transfer takes place over a reduced barrier on an excited potential-surface.

Can one be confident, however, as he claims, that the H tunnels through this barrier, rather than passing over it? The transition-state theory of reaction rates can probably explain the observed isotope effect in going from H to D, without invoking any tunnelling. Quantum tunnelling is most likely to be important, moreover, only if the rate is very low, that is to say, the process is being examined in its threshold region.

Prof. Stokbro responded: What we are confident about is that the H-atom transfer takes place over a reduced barrier on an excited potential-surface. Since we have not calculated the excited state potential and made wavepacket propagation of H on the excited state potential, the details of the transfer on the excited state potential is speculative. We agree with Prof. Polanyi that the isotope effect could also be explained from ordinary transition state theory, in fact, this was how we explained the isotope effect in our first paper.[1] To be more conclusive on which model is the correct one, we plan to perform wavepacket propagation studies.

1 U. Quaade, K. Stokbro, C. Thirstrup and F. Grey, *Surf. Sci.*, 1998, **415**, L1037.

Dr Shluger said: I would like to ask Prof. Seideman and Prof. Stokbro to comment as to what extent the effects of desorption and diffusion, which they describe, can be assisted or affected by STM tips? In atomic force microscopy, displacements of molecules at atoms by tips are very common. Also adsorption of surface atoms on tips and subsequent desorption have been observed in many experiments. To what extent does adhesion between tips and adsorbed atoms and molecules affect the interpretation of experiments reported here? A related question concerns the character of the tip-induced electric field, which is important in these experiments. Is a homogeneous field model adequate?

Prof. Seideman responded: The first part of Dr Shluger's question is clearly related to Prof. Polanyi's comment in the earlier discussion of Dr Saalfrank's paper. The probabilities of transfer of adsorbates to the tip and of re-adsorption to a remote site on the surface depend on the system and the experimental details, primarily on the tip–surface distance.

In the classic "Eigler switch" experiment,[1] where the tip–surface distance varied in the 3.8–5.5 Å range, the Xe-atom center-of-mass motion was subject to a double-well potential with a variable barrier height (see Fig. 2 of ref. 2) and hence controllable and reversible transfer to and from the tip was observed. In the benzene/Si case discussed in our paper, the tip is *ca.* 7 Å from the surface. Similar tip–surface distances apply to the well-studied CO/Cu[3] and H/Si[4] desorption problems. In these studies the potential energy surface for the nuclear motion is a multidimensional function of coordinates that has many local minima, one describing a molecule adsorbed onto the tip, which are, however, separated by large regions where the potential along the desorption coordinate is constant (corresponding to a desorption continuum, see Fig. 4a of ref. 5). A probability that a molecule would hop to the tip or re-adsorb onto the surface certainly exists (with hopping to the tip being more probable with the measurement method of ref. 3 than with the method of ref. 4 and 5). For the benzene/Si case we found no evidence for (although we did not exclude) adsorption to the tip but re-adsorption to the surface events were observed. Numerically both types of events are treated as desorption, since our potential energy surface is confined to the region of one local minimum. Consistent with the calculation, our measurement counts re-adsorbed molecules and molecules that may have hopped to the tip as desorbed.

Regarding the second part of your question, the STM electric field in our study is ca. 0.3 V Å^{-1}, well below the threshold for field-induced desorption.[6] Such fields suffice to induce Stark-shifts but the resulting variation in the potential energy surfaces is below 0.1 eV, comparable to the accuracy of our potential energy surfaces. Field-gradients are negligible at a distance of 7 Å from the center of the tip. Thus, the STM-tip affects the dynamics essentially only through the dependence of the tip density of states on its structure and chemical composition.

1 D. M. Eigler, C. P. Lutz and W. E. Rudge, *Nature*, 1991, **352**, 600.
2 Ph. Avouris, *Acc. Chem. Res.*, 1995, **28**, 95.
3 L. Bartels, G. Meyer, K.-H. Rieder, D. Velic, E. Knoesel, A. Hotzel, M. Wolf and G. Ertl, *Phys. Rev. Lett.*, 1998, **80**, 2004.
4 E. T. Foley, A. F. Kam, J. W. Lyding and Ph. Avouris, *Phys. Rev. Lett.*, 1998, **80**, 1336.
5 S. Alavi, R. Rousseau, G. P. Lopinski, R. A. Wolkow and T. Seideman, *Faraday Discuss.*, 2000, **117**, 213.
6 M. A. Rezaei, B. C. Stipe and W. Ho, *J. Chem. Phys.*, 1999, **110**, 4891.

Prof. Stokbro responded: We always neglect the effect of the tip on the H desorption barrier through either the electric field or a direct chemical interaction, since our estimates show that the effects are very small. The tip is always in the tunnel regime, thus it is more than 5 Å away from the H atom, at this distance chemical interactions are much less than 1 meV. I have done some modelling of the electric field between the tip and the surface (see ref. 1), and for all the different tip geometries considered, I found that the electric field is quite constant within a radius of 5–10 Å from the center of the tip. Thus, tip effects are not important and a homogeneous field is adequate.

1 K. Stokbro, *Surf. Sci.*, 1999, **429**, 327.

Prof. Groß said: I am surprised about the small difference in the energetics obtained by the slab and the cluster calculations, in particular regarding the fact that two very different functionals had been used in both approaches. For example, in a similar system, H_2/Si(100), differences of up to 1 eV were found between these two methods.[1–3] For how many configurations did you compare slab and cluster calculations? Do you have an explanation for the surprisingly good agreement between both methods?

1 M. R. Radeke and E. A. Carter, *Annu. Rev. Phys. Chem.*, 1997, **48**, 243.
2 A. Groß and M. Scheffler, *Surf. Sci. Rep.*, 1998, **32**, 291.
3 E. Penev, P. Kratzer and M. Scheffler, *J. Chem. Phys.*, 1999, **110**, 3986.

Prof. Seideman responded: Let me first note that several of our results are consistent with the findings of Penev *et al.* (ref. 3 in your comment; the review articles ref. 1 and 2 are in qualitative accord with ref. 3, and consider the same system, although by contrast to ref. 3, they do not focus on comparing the slab and cluster models) and second, point out that these authors stress that their results pertain to the H_2/Si(100) dissociative-adsorption and associative-desorption; they do not suggest transferability of the conclusions to other surface reactions.

Thus, in accord with ref. 3, we find that the cluster model overestimates the binding energy and (as is evident from comparison of Fig. 1 and 3 of our discussion paper[4]) underestimates the buckling of the bare silicon dimers. The discrepancy of 0.09 eV in the binding energy between the slab and cluster models quoted in ref. 4 corresponds to the B3LYP functional but a similar discrepancy (0.14 eV) is obtained when the GGA functional, employed in our slab calculations, is used within the cluster approximation. In ref. 4 we used the B3LYP rather than the GGA functional since, as discussed in ref. 1 of your comment, it tends to predict more reliable barrier energies. Moreover, the relative error in the binding energy predicted by our Si_{12} cluster model, ca. 10% or 15%, is not far from the relative error found for H_2/Si (25% with the smallest cluster considered, Si_9 and 10%–20% with a Si_{15} cluster, see Table III of ref. 3) where the desorption energy is more than twice that of benzene/Si. (Clearly, such error in energetics has significant effects on the H_2/Si *dynamics* but negligible effect on the benzene/Si dynamics.)

By contrast to the H_2/Si case, however, the C_6H_6/Si structure predicted by the slab model is well-reproduced by the cluster approximation (a quantitative comparison, omitted from ref. 4, is given in ref. 5). In addition, the discrepancy in energetics between the slab and cluster models is smaller than that found for H_2/Si^3 and the results are less sensitive to the functional. Bond lengths computed within a cluster model with the GGA functional are identical to those predicted by the B3LYP cluster model (and to those predicted by the slab model) to within 0.01 Å. The binding energy is *ca.* 0.05 eV larger with the GGA than with the B3LYP functional. Importantly, the transition state energy is only 0.02 eV lower with the GGA than with the B3LYP functional.

Both differences between the H_2/Si and C_6H_6/Si systems are expected. With regard to the first feature, we note that Penev *et al.*[3] analyze in detail the reason for the failure of the Si_9 cluster to describe the energetics and trace it to features specific to the nature of the H_2–Si(100) interaction, see p. 3992 of ref. 3. With regard to the second we recall that, while different functionals are known to furnish different results, the variation with respect to functional is strongly system dependent. It may be expected to be particularly severe for H_2/Si since for small molecules the self interaction of electrons plays a much more important role than for large systems, see ref. 6. Although, to our knowledge, the system-size dependence of the functional-sensitivity has not been investigated in detail, it appears that the results found for H_2/Si are not directly transferable to large organic adsorbates.

1. M. R. Radeke and E. A. Carter, *Annu. Rev. Phys. Chem.*, 1997, **48**, 243.
2. A. Gross, *Surf. Sci. Rep.*, 1998, **32**, 291.
3. E. Penev, P. Kratzer and M. Scheffler, *J. Chem. Phys.*, 1999, **110**, 3986.
4. S. Alavi, R. Rousseau, G. P. Lopinski, R. A. Wolkow and T. Seideman, *Faraday Discuss.*, 2000, **117**, 213.
5. S. Alavi, R. Rousseau and T. Seideman, *J. Chem. Phys.*, 2000, **113**, 4412.
6. See, for example, R. G. Parr and W. Yang, *Density-Functional Theory of Atoms and Molecules*, Oxford University Press, Oxford, 1989.

Prof. Gadzuk said: As a follow up to Prof. Seideman's discussion on the use of excited states to facilitate the transfer of atoms or molecules between a surface and an STM tip, recall the "Eigler-Switch",[1] which provided the theoretical problem of the year about five years ago. The switch, which is schematically shown in Fig. 3, depends upon a current-driven, reversible transfer of an adsorbed Xe atom between the two stable minima of the double well potential. In almost all the existing theories[2] it has been argued that a tunnelling electron might temporarily reside in a

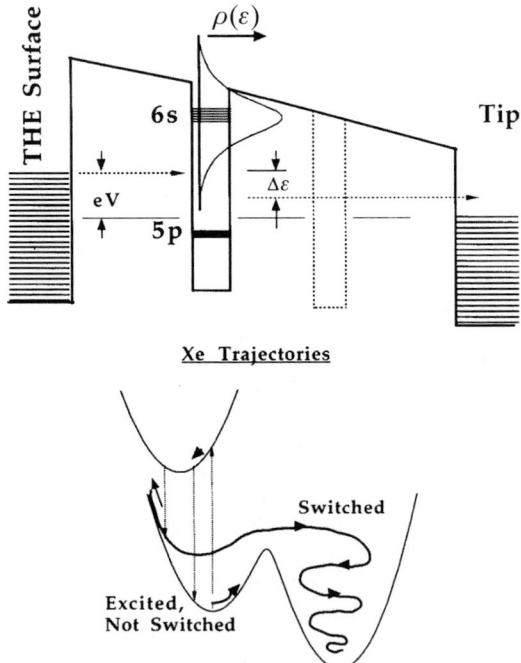

Fig. 3 STM/Xe-atom switch.

localised state on Xe, as weighted by the tail of the Xe 6s electron local density of states at the Fermi level (even though it is a bit of a stretch to say that an electron with energy ~ 5 eV off a ~ 1 eV wide resonance can still be involved in a resonance process!), converting the Xe into a Xe$^-$ temporary negative ion. Upon occupation, an additional (assumed linear) Xe$^-$/surface interaction takes place (which when added to the original $\sim$ harmonic potential, yields the displaced oscillator shown in the figure. The Xe$^-$ moves inward a bit, the electron departs, and the Xe is returned to the ground state double well potential slightly vibrationally excited. This sequence is repeated a few more times until the Xe finally accumulates enough energy to pass over the activation barrier. As a consequence of the fact that several electron hits were required to gain the requisite barrier-crossing-energy, the transfer rate is proportional to I^n, where I is the current passing through the switch and $n \sim 3$–4 is roughly related to the number of hits for switching to occur. Evidently, this can be regarded as another example of an excited electronic state (Xe$^-$/surface/STM) used to facilitate atomic rearrangement/reaction on the ground state potential energy surface.

1 D. M. Eigler, C. P. Lutz and W. E. Rudge, *Nature*, 1991, **352**, 600.
2 N. Lorente and M. Persson, *Faraday Discuss.*, 2000, **117**, 277.

Prof. Persson commented: Our rationale for invoking an electron–vibration coupling through the tail of 6s resonance of the Xe atom at the Fermi level in ref. 1 was the work by Eigler and coworkers.[2] They found from a comparison results of an atom-on-jellium model and observed STM images that the tunnelling from the tip through the atom to the surface was dominated by this tail.

1 S. Gao, M. Persson and B. I. Lundqvist, *Solid State Commun.*, 1992, **84**, 271.
2 D. M. Eigler, P. S. Weiss, E. K. Schweisser and N. D. Lang, *Phys. Rev. Lett.*, 1991, **66**, 1189.

Dr Dujardin commented: I would like to say a few more words on this problem of resonances. This is somewhat related to the problem of the "toy models" that we discussed earlier. When people try to model a system, they make severe simplifications. For example, in the case of hydrogen on Si(100) people have assumed that there exists a single σ (Si–H) electronic state involved in the STM desorption process. But in fact, if you look at the electron photoemission spectrum, you realise that there are at least four σ and π electron states related to the adsorption of H on Si(100). I believe that it is the existence of many coupled inelastic channels which prevent the observation of clean resonance effects in the desorption yield of H from Si(100).

Prof. Stokbro responded: I do not understand what you mean, which π states? The H 1s orbital only couples with states with σ symmetry, so how can you invoke π states in the inelastic coupling to the H atom? We find that the simple model with a single resonant level is in excellent agreement with the experimental data, so I would call the phenomenon we observe a clean resonance effect.

Prof. Kummel asked: What would be the yield at constant current density *vs.* voltage? This would isolate the influence of voltage from the e$^-$ irradiation area. The paper had stated that the tip–surface distance was increased at higher V to keep constant I and this increased the area of e$^-$ irradiation by the tip.

Dr Dujardin responded: To correctly calculate the yield at constant current density on the surface is a difficult task. However a simple calculation can be made. Assume a tip sample distance at 2 and 10 V of 10 and 30 Å, respectively. Assuming that the 'cone' of electrons always has the same shape then we can calculate the relative surface areas. Normalising gives a reduction in the yield by a factor 2 at 4 V, 4 at 6 V and 9 at 10 V. This causes the maximum yield to decrease but the overall shape remains unaffected. The increase between 4 and 5 V is still clearly visible.

Prof. Stokbro said: I do not find it fair to say that there is no evidence for resonant processes in STM induced desorption of H from Si(100). The Lyding–Avouris team has published six papers on the subject and our group has published five papers. These publications cover studies of the current, bias, isotope and temperature dependence of the desorption rate, and all these data are in excellent agreement with first principles models of resonant processes.

Dr Dujardin responded: We believe that some of the conclusions reached in the previous studies on STM induced desorption of H from Si(100) are very questionable. As explained in our paper, the method previously used has a serious fault that probably leads to values of the desorption yield being underestimated by more than a factor 100. The method consists of moving the STM tip along a line across the surface and in counting the number of desorbed hydrogens afterwards. The implicit assumption is that the interaction of the tip with hydrogen is always the same whatever its position. Our results show that this cannot be true. The method that we used not only enables a measure of the desorption yield but also the localisation of the inelastic interaction. We have shown that the inelastic interaction can be very localised down to 0.2 Å. This means that the efficiency of the electrons strongly depends on the tip position above the surface. It is this factor that has been omitted in previous Si(100)–H studies. This effect depends on the bias surface voltage and may affect the measurement of the yield by a factor of more than 100. If despite these errors, the experimental desorption yield could be well fitted by first principle model calculations, this is because these models contain many adjustable parameters. One central assumption of these models is that the tunnel electrons go through a *single* inelastic resonant channel, namely the Si–H $6\sigma^*$ unoccupied and 5σ occupied orbitals for positive and negative bias surface voltage respectively. This assumption could not be experimentally verified and has no real grounds for it to be considered correct since it is well known from valence band photoemission studies that there are many other close Si–H orbitals which can provide other inelastic channels. We believe that several inelastic channels contribute to the desorption of hydrogen on the Si(100)–H surface. Therefore we feel it is fair to say that the evidence for resonant processes in STM induced desorption of H from Si(100) remains inconclusive.

Dr Shluger commented: I would like to ask Dr Dujardin to explain in more detail what he means by the difference between tunnel electrons and field emitted electrons? In particular, I am very interested in his comment that by using different regimes of electron emission one can control inelastic surface processes and reactions. Could you please comment on how this control should be feasible to achieve?

Dr Dujardin responded: The wavefunctions of tunnel and field emitted electrons are very different in nature. They are evanescent waves and free electron waves respectively. Therefore it is perfectly conceivable that they couple with a specific inelastic channel in very different ways and that this difference can be used to achieve a degree of control over a particular process or reaction. We don't think that this must be always the case, it may depend on the inelastic channel. However we believe that it explains the strong increase in the H desorption yield from Ge(111) around 4.5 eV.

Prof. Stokbro said: I do not understand your notion of non-tunnelling electrons. The currents involved are in the nA range, so clearly all electrons have to cross a tunnel barrier even in the field emission regime.

Dr Dujardin responded: Tunnel electrons are found when the bias voltage is lower than the surface work function, *i.e.* when the energy of electrons emitted by the STM tip is always below the vacuum level of both the tip and the surface. When the bias voltage is higher than the surface work function electrons emitted by the STM tip are in the vacuum when they reach the surface and are in that sense ballistic.

Prof. Polanyi opened the discussion of Dr Lauhon's paper: Lauhon and Ho's very elegant experiments open the way to studies of bimolecular exchange reactions at surfaces, molecule-by-molecule.

The intention in the related field of 'surface-aligned photochemistry,' SAP, with which we have been involved, is to hold two reagents in preferred alignment at preferred positions on surfaces and then to induce a recoiling fragment from one molecule to react at a controlled impact-parameter (miss-distance) with an adjacent molecule. For SAP the reaction can occur out of the surface-plane, often ejecting the reaction product.[1]

In the case that Dr Lauhon has described, in which atomic-H recoils *in the plane of the surface*, there appear to be marvellous possibilities for 'sticky billiards' (reaction) with well-located co-adsorbates that give rise to scattered products attached to the surface. One thinks of the analogous case of the co-planar scattering of elementary particles in a cloud-chamber. Would the authors care to comment?

1 For a review see J. Polanyi and H. Rieley, *Photochemistry* in *The Adsorbed State Dynamics of Gas–Surface Interactions*, ed. C. T. Rettner and M. N. R. Ashfold, Royal Society of Chemistry, London, 1991, ch. 8, p. 329; or J. Polanyi and Y. Zeiri, *Dynamics of Adsorbate Photochemistry* in *Laser Spectroscopy and Photochemistry on Metal Surfaces*, Advanced Series in Physical Chemistry, 5, ed. H.-L. Dai and W. Ho, World Scientific Publishing Co. Pte, Ltd., Singapore, 1995, ch. 26, p. 1241.

Dr Lauhon responded: We have observed reactions between hot hydrogen atoms, produced by single-molecule dissociation, and nearby molecules. To unambiguously determine the pathway of these bimolecular reactions, we work at very low coverages. Such events are rare at low coverages, so one experimental challenge is simply to produce enough events to determine the relative probabilities of the possible reactions. Higher coverages can be fruitfully explored, however, once, the relevant molecular species are identified.

Two areas of exploration come to mind in the pursuit of surface-aligned tunnelling electron induced chemistry. The first is the selection of suitable reactants, that is, reactants whose products remain on the surface to participate in secondary reactions. The second is the identification of the reaction products. While STM images are central to adsorbate recognition, one can also use the dynamic behavior of single molecules to characterize and identify adsorbates. Inelastic electron tunnelling spectroscopy (IETS)[1] is one means of characterizing a reaction product. Controlled dissociation is another characterization tool also mentioned in the paper under discussion; if one has hydrogenated a molecule with a hot hydrogen atom, one might also reverse the process using tunnelling electrons. Isotopic substitution is quite useful in this context as shifts are observed in STM-IETS spectra, dissociation thresholds, and rotation and diffusion barriers.[2]

The ability of the STM to study thermal and tunnelling electron-induced diffusion should also prove useful in characterizing the energy dissipation of hot reaction products between their creation and eventual encounter with the targeted species.

1 B. C. Stipe, M. A. Rezaei and W. Ho, *Science*, 1998, **280**, 1732.
2 L. J. Lauhon and W. Ho, *Phys. Rev. Lett.*, 2000, **84**, 1527 and references therein.

Dr Kroes said: (1) You said that vibrationally exciting H_2S on the surface enhances the rate of diffusion of H_2S over the surface. Is this in a mechanism in which H_2S gives up its vibrational energy to jump over a barrier to diffusion or in a vibrationally adiabatic mechanism? (2) Would calculations or experiments on adiabatic dissociative chemisorption of H_2S be of any use to the interpretation of your experiments?

Dr Lauhon responded: (1) I do not think that we have performed experiments that can distinguish between these two cases for H_2S, though we have for atomic hydrogen. My general picture of the process of diffusion induced by inelastic electron excitation is that transitions between adjacent potential minima occur *via* vibrational excited states; as the molecular wavefunction becomes more delocalized, the potential barrier is reduced. The transition rate is then determined by a competition between exploration of the potential energy surface and vibrational relaxation. This is also my picture of dissociation induced by multiple vibrational excitation with the relevant coordinate being the intramolecular bond as opposed to the molecule–surface bond.

If adsorption is accompanied by significant lattice distortion, the diffusion process itself may involve energy exchange between the molecule and surface. This is manifest in activation energies and attempt frequencies that deviate from their classical values.

While the STM can be used to determine thermal activation energies *via* variable temperature diffusion measurements, calculations of the static lattice barriers are not accurate enough to prescribe this as a means of discriminating between vibrationally adiabatic and non-adiabatic diffusion mechanisms.

In the case of H/Cu(001), we were able to determine that thermal diffusion above 60 K is essentially classical, that is, diffusion is dominated by transitions between vibrational excited states with energies close to the potential barrier.[1] Between 25 and 60 K, diffusion occurs *via* phonon-assisted tunnelling and reveals a smaller barrier that arises from the local lattice distortion. Studies of tunnelling induced by inelastic electron excitation are underway.

(2) The experiments and calculations you suggest would be useful if the redistribution of translational energy upon collision, resulting in diffusion or dissociation, could be related to the distribution of vibrational energy transferred to the molecule *via* inelastic electron tunnelling.

1 L. J. Lauhon and W. Ho, *Phys. Rev. Lett.*, 2000, **85**, 4566.

Dr Kolasinski said: In the case of H/Si on H/Ge it is assumed that the hydrogen desorbs as H atoms. In the cases of H_2S/Cu and HS/Cu we are told that the S–H bond dissociates, hot H atoms are released onto the surface and H adsorbs. Do we know whether hydrogen desorbs from Si or Ge as H or H_2? Can H atoms excited by an STM tip abstract a neighbouring adsorbed H atom? When an S–H bond is dissociated, what is the branching ratio of H atom adsorption compared to desorption?

Dr Lauhon responded: Though I have spoken of removing H atoms, we cannot determine the number of electrons the proton carries with it as it leaves the molecule. Dissociation of H_2S using 1.5 eV tunnelling electrons always resulted in the production of an adsorbed H atom quite close to the reaction site.

Dissociation of SH resulted in adsorbed H in the majority of cases, but the H was farther from the reaction site. Due to the possibility of H–tip collision prior to adsorption, the branching ratio may not be well-defined, *i.e.*, one does not want to mistake desorption followed by reabsorption for direct adsorption. As we learn more about the motion of hot hydrogen, however, we may find that this desorption/reabsorption mechanism is not necessary to explain the large distances observed in this case.

Prof. Stokbro added: In the case of H atom desorption from Si(100) the H atoms desorb as single H atoms. We know this since we can desorb single H atoms in a controlled way at negative sample bias, this is described in a paper which we have submitted to *Nanotechnology*. The Lyding group has also made single atom lithography at positive sample bias, see ref. 1. In the theoretical model of the desorption experiment, we assume it is a single H atom that desorbs and our model is in excellent quantitative agreement with the experimental data. There has been some discussion how the vibrational excitation can stay localised on a single H atom. I addressed this issue in a recent paper,[2] and found it is related to an anharmonic Si–H potential. When the H atom is multiple vibrational excited, there is a large anharmonic shift in the frequency, and this prohibits the H atom in exchanging energy with the surrounding H atoms.

1 K. Stokbro *et al.*, *Nanotechnology*, 2000, **11**, 70.
2 K. Stokbro, *Surf. Sci.*, 1999, **429** 327.

Dr Dujardin commented: I would like to come back to the question by Dr Kolasinski. This question suggests that the dynamics for the desorption of hydrogen may be more complicated than expected from simple bond breaking models. I would fully agree with this. For example, in the case of the completely hydrogen covered Si(100) surface, one often observes desorption of more than a single hydrogen atom, which indicates that complicated dynamical processes involving several hydrogen atoms may occur. Even in the case when we experiment with a single hydrogen atom on a clean Ge(111) surface, things are more complicated than it seems. In addition to the hydrogen desorption, we also observed the displacement of the hydrogen atom over the surface. The two processes are in competition. It follows that the dynamics must be very complicated. In the case of acetylene on Cu(100), Dr Lauhon mentioned that the dissociation of a hydrogen atom and its displacement over the surface may involve some interaction with the STM tip. Again this shows that the dynamics can be extremely complicated.

Dr Dujardin said: You wrote in your paper that, when dissociating hydrogen from acetylene, hydrogen atoms can be found within 50 Å of the dissociation site. Does it mean that hydrogen atoms can travel over the surface by such a long distance?

Dr Lauhon responded: The removed hydrogen gains enough energy during the dissociation to either travel large distances across the surface or 'bounce' off the tip and 'land' some distance from the original reaction site. Dissociation of acetylene in the field-emission regime, which involves a large tip–surface distance and substantial non-local dissociation, suggests that the hydrogen does not need to bounce from the tip to arrive on the surface. The non-local nature of this dissociation makes it more difficult to trace individual atoms to their source, however. The majority of our acetylene dissociation experiments were performed under conditions that we later discovered were not favorable to the observation of atomic hydrogen. Recent cursory experiments suggest that the hydrogen removed from isolated acetylene molecules by 3 eV electrons is 'hotter' than hydrogen removed from H_2S.

Prof. Chesters said: In the C_2–H_2S reaction could you say something more about what the optimum separation of the reactants appears to be and whether there is a preferred crystallographic direction for the reaction?

Dr Lauhon responded: The H_2S tracking was not performed with sufficient resolution to determine the direction of reactant motion immediately prior to reaction, but such resolution is possible. In a more detailed analysis of the $H_2S + C_2$ reaction products, SH was found to prefer the fourfold hollow site (FFHS) diagonal to the C_2H FFHS (along $\langle 100 \rangle$). This indicates the necessary proximity of the reactants and also the role that the substrate can play in channeling the motion of the products away from the transition state. A similar channeling is expected to occur for the reactants into the transition state.

Dr Lee asked: In your experiments, how long did it generally take for a H_2S molecule to arrive in the vicinity of a CC reaction site? In addition, I am wondering what happened to the previously created atomic hydrogen which was removed from the scan area, in Fig. 4(c) of the paper.

Dr Lauhon responded: The time required to produce a reaction depends on the number of available reaction sites as well as the induced diffusion rate. In general, we found that hopping rates of about $1\ s^{-1}$ were convenient for following the motion of H_2S. Reaction of a chosen H_2S was typically accomplished in a few minutes.

The atomic hydrogen that was removed from the scan area by vibrational excitation remained on the surface.

Prof. Sitz said: I have two related questions about the images in Fig. 4 of the paper. First, the H_2S images look spherical; is this a result of free rotation of the molecule, or are the H-atoms just invisible? Secondly, why do the HS images appear so much smaller than the images of H_2S?

Dr Lauhon replied: We have seen no evidence for or against H_2S rotation. Evidence for rotation would include abrupt changes in tip height within a particular line scan indicative of orientational change. Evidence against rotation would include, for example, an image of C_{2v} symmetry that was not due to the tip shape. If the molecular rotation rate were much faster than the response of the STM feedback, an azimuthally symmetric image would be produced, as you point out. Because the STM does not image atoms *per se*, however, an azimuthally symmetric image is not inconsistent with a fixed molecular orientation.

The different molecular electronic structure, orientation, and binding site combine to reduce the probability of tunnelling above SH compared to H_2S. Because the S atom is expected to be closest to the surface in both cases, the height difference is primarily electronic, and not geometric, in origin.

Dr Giorgi said: Following from a dynamic perspective, how do you think the reaction occurs? You have observed the first re-hydrogenation of CC by H_2S, what about the second re-hydrogenation of CCH by HS or even another H_2S molecule? Can you shed some light as to whether it is a geometrical problem or an energetic problem? (In particular comparing H_2S + CC vs. H_2S + CCH.)

Dr Lauhon responded: We think that the lack of reactivity between SH and CC is primarily geometric in origin. The H_2S, occupying top or bridge sites, can get much closer to the hollow site C_2 than can the hollow site SH. Qualitatively, the deep depression produced by C_2 in STM images indicates a large perturbation of the surface electronic structure. This perturbation may help define the transition state by dominating the rather weak interaction between the H atoms (within H_2S and SH) and the Cu surface. The C_2H, on the other hand, does not produce such a large perturbation of the substrate. Electronic structure calculations have indicated[1] that the C atom in C_2H which is not H-terminated shifts towards the hollow site. This shift, combined with the lesser surface perturbation, could make it more difficult for H containing species to achieve favorable reaction geometries with C_2H.

1 F. Olsson and M. Persson, to be published.

Mr Al-Halabi said: My question is related to the change in the molecular contrast observed in Fig. 3a and b for H_2S and HS respectively, can this be due to the different adsorption sites of the two molecules on the copper surface, *i.e.* the different adsorption sites might yield different orientations of H_2S and HS?

Dr Lauhon responded: A change in contrast is expected after removing an H from H_2S because the electronic structure of the molecule, and hence the tunnelling probability, changes. In addition, a given adsorbate may image differently on different adsorption sites if the bonding changes in such a way as to influence the electronic states available for tunnelling. The case of benzene on Pt(111) is a good example of this effect.[1,2]

1 P. S. Weiss and D. M. Eigler, *Phys. Rev. Lett.*, 1993, **71**, 78.
2 P. Sautet and M.-L. Bocquet, *Surf. Sci.*, 1994, **304**, L445.

Dr Shluger said: I want to raise a more general issue of statistics in the type of experiments described by Dr Lauhon. The situation with which we are dealing here is very different from, for example, scattering experiments which we discussed in previous sessions. Reactions between several atoms or molecules are observed individually and also by one particular researcher. I think that there is an important issue of how probability of these individual events is established, statistical errors in STM spectral measurements are calculated, and comparisons between different groups are made. This requires standardisation of imaging conditions and tip characterisation.

Dr Lauhon responded: This question points out the greatest strengths and weaknesses of STM. The condition of the sample and adsorbate can be characterized very well, often eliminating discrepancies that arise between the aforementioned scattering experiments. The condition of the tip, however, is usually not well known. The degree to which the tip condition influences the result of a particular measurement depends greatly on the measurement in question and needs to be determined empirically. In particular, as the electronic structure of the tip varies, so will the overlap with various electronic states in the sample. To the extent that tunnelling into particular states is associated with vibrational excitation, dissociation, *etc.*, the outcome of excitation will reflect the tip structure. One promising approach towards tip characterization includes functionalization with a known molecular entity.[1] There are also quantitative means to determine the fraction of electrons participating in the vibrational and electronic excitations of single molecules.[2-4]

1 H. J. Lee and W. Ho, *Science*, 1999, **286**, 1719.
2 B. C. Stipe, M. A. Rezaei and W. Ho, *Science*, 1998, **280**, 1732.
3 D. M. Eigler, C. P. Lutz and W. E. Rudge, *Nature*, 1991, **352**, 600.
4 L. Bartels, G. Meyer, K.-H. Rieder, D. Velic, E. Knoesel, A. Hotzel, M. Wolf and G. Ertl, *Phys. Rev. Lett.*, 1998, **80**, 2004.

Prof. Seideman commented: As Dr Shluger points out, the measurement (at least with the method of ref. 1 and 2) is statistical; the error bars in Fig. 2 of ref. 2 give the statistical error in the benzene/Si(100) desorption yield.

With regards to the possibility of quantitatively comparing experiments carried out in different laboratories, it is important to note that systematic errors are expected, although difficult to quantify, and are to be added to the (known) statistical error. This is clearly illustrated in the work of Adams and coworkers[3] who repeated the H/Si desorption experiment of Avouris *et al.*[1] and reported good agreement with ref. 1. The desorption yield measured in the two laboratories for the same system and conditions are compared in Fig. 5 of ref. 3. It is evident that the agreement is indeed good inasmuch as the threshold and plateau features are reproduced. Nevertheless, the yield differs by a factor of 5–10, depending on the bias.

I don't think that one can compare this type of experiment to gas–surface scattering experiments of the quality illustrated, *e.g.*, in Prof. Sitz' poster and discussed earlier. In the latter case the essential physics is simple and well-understood; the experiment aims at, and is capable of, probing the fine details of the potential energy surface. In the former case one is exploring new phenomena and aims at understanding the qualitative physics. Our theoretical modelling is also not nearly as refined as that of gas–surface scattering calculations of the type discussed by Kroes *et al.*[4]

1 E. T. Foley, A. F. Kam, J. W. Lyding and Ph. Avouris, *Phys. Rev. Lett.*, 1998, **80**, 1336.
2 S. Alavi, R. Rousseau, G. P. Lopinski, R. A. Wolkow and T. Seideman, *Faraday Discuss.*, 2000, **117**, 213.
3 D. P. Adams, T. M. Mayer and B. S. Swartzentruber, *J. Vac. Sci. Technol. B*, 1996, **14**, 1642.
4 D. A. McCormack, G.-J. Kroes, R. A. Olsen, J. A. Groeneveld, J. N. P. van Stralen, E. J. Baerends and R. C. Mowrey, *Faraday Discuss.*, 2000, **117**, 109.

Theoretical aspects of tunneling-current-induced bond excitation and breaking at surfaces

Nicolas Lorente† and Mats Persson

Department of Applied Physics, Chalmers/Göteborg University, S-41296 Göteborg, Sweden

Received 7th April 2000
First published as an Advance Article on the web 4th October 2000

We have performed a density functional study of the electronic structure, images and vibrationally inelastic tunneling in the scanning tunneling microscope and vibrational damping by excitation of electron–hole pairs of CO chemisorbed on the (111) and (100) faces of Cu. We find that the $2\pi^*$ molecular orbital of CO turns into a broad resonance with parameters that differ significantly from those suggested by inverse and two-photon photoemission measurements. The calculated vibrational damping rate for the internal stretch mode and relative changes in tunneling conductance across vibrational thresholds are in agreement with experiment. The non-adiabatic electron–vibration coupling is well described by the Newn–Anderson model for the $2\pi^*$-derived resonance whereas this model is not able to describe the non-adiabatic coupling between the tunneling electrons and the vibration. We believe that this model misses an important mechanism for vibrational excitation in tunneling that involves the change of tunneling amplitude by deformation of the tails of the one-electron wavefunctions with vibrational coordinate.

I. Introduction

The unique ability of the scanning tunneling microscope (STM) to manipulate and characterize single molecules on surfaces by inelastic electron tunneling provides new understanding and control of the surface chemical bond and reactivity.[1] The highly localized inelastic electron tunneling in space, has made it possible to desorb single atoms,[2,3] to dissociate single diatomic molecules,[4] to break single bonds in polyatomic molecules,[5] to form new bonds by transferring atoms and molecules, reversibly, to and from the tip,[6–8] induce reversible rotational motion[9,10] and perform vibrational microscopy.[11]

Most theoretical attempts to describe and understand bond excitation and breaking by inelastic electron tunneling are based on a resonance model for the adsorbate-induced electronic structure and the non-adiabatic electron–vibration coupling.[3,7,12–15] This model, that involves the coupling to a local vibrational mode, has a long history and has been applied to various electronically non-adiabatic processes such as: desorption induced by electronic transitions and driven by laser-excited hot electrons,[16] vibrational damping by excitation of electron–hole pairs,[17] and vibrationally inelastic tunneling.[18] In the resonance model, an adsorbate-induced antibonding state is temporarily occupied by a tunneling electron, resulting in a motion of the nuclear wave packet on the excited-state potential until the resonance decays, after which the wave packet is no longer

---

† Permanent address: Laboratoire Collisions Agrégats Réactivité, UMR 5589, Université Paul Sabatier, 31062 Toulouse Cedex, France.

DOI: 10.1039/b002826f

exclusively in the lowest vibrational state and can even escape by ending up in a state above the barrier. This model has been used to rationalize observed power-law dependence on the tunneling current of the transfer rate in the atomic switch,[12,13,15] the desorption rate of H on Si(100),[3] and the dissociation rate of O_2 on Pt(111)[4] and involves, typically, a number of adjustable parameters, with the notable exception of the treatment by Stokbro and coworkers[3] where the parameters were determined from density functional calculations.

The recent development of vibrational spectroscopy and microscopy by inelastic electron tunneling[11] has made it possible to study the non-adiabatic coupling between tunneling electrons and vibrations in more detail. We have recently developed a theory of vibrationally inelastic tunneling based on density functional theory (DFT) and a generalization of Tersoff–Hamann (TH) theory. Using this theory we were able to calculate STM images for C_2H_2 chemisorbed on Cu(100) and relative changes in tunneling conductance across vibrational modes and their spatial images that were in good agreement with experiment.[19] CO on Cu provides a most interesting system to study the nature of the electron–vibration coupling in tunneling-current-induced bond excitation and breaking at surfaces and in vibrational damping by electron–hole pair excitations because, for this system, the resonance state derived from the $2\pi^*$ molecular orbital of CO has been argued to play an important role in some of these processes and experimental data of these resonance state parameters, tip-induced desorption rates, vibrationally inelastic tunneling efficiencies and vibrational damping rates are available for comparison with theory.

In this paper, we have applied our theory to the calculation of STM images, relative changes in tunneling conductance across vibrational thresholds and vibrational damping rates of CO on the (111) and the (100) faces of Cu. The non-adiabatic electron–vibration interaction was treated by perturbation theory. The one-electron wavefunctions and electron–vibration couplings were calculated in a super-cell geometry using a plane-wave and pseudopotential method. The one-electron structure of the chemisorbed molecule, in particular, the $2\pi^*$-derived state, was characterized by calculating the projected density of states on molecular orbitals. The results are discussed in relation to both available experimental data and results from a resonance model.

The paper is organized as follows. We begin by giving some details of the density functional calculations in Section II A and show in Section II B, how the local density of states (LDOS) in the TH approach to STM images and the projected density of states (PDOS) on molecular orbitals were calculated using density functional calculations. In Section II C, we present expressions for relative changes in vibrational conductance based on a generalization of the TH theory to vibrationally inelastic tunnelling and for vibrational damping rates by excitation of electron–hole pairs and show how these expressions are simplified in the Newns–Anderson model. In addition, we give some details of how these expressions were calculated using the plane-wave, pseudopotential method. We present our results for the $2\pi^*$-derived resonance feature in the PDOS, STM images and vibrationally inelastic tunneling efficiencies and vibrational damping rates and discuss them in relation to experimental data in Sections III A, III B and III C, respectively. In Section III D, we discuss our results in relation to the Newns–Anderson model for the $2\pi^*$-derived states and tip-induced desorption experiments and, finally, in Section III D give a short summary and some concluding remarks.

II. Theory

A. Some details of the density functional calculations

The chemisorption and one-electron structure calculations were based on DFT and performed using the plane-wave pseudopotential code DACAPO.[20] The exchange-correlation effects were described by the PW91 version of the generalized gradient approximation (GGA).[21]

The (111) and (100) surfaces of Cu were represented by slabs in a super cell geometry. We used slabs of various sizes, which included 2×2, 3×3 and 4×4 surface unit cells and 4 to 6 layers of Cu atoms. The vacuum region of the (111) and (100) slabs contained 4 and 5 layers, respectively. The CO molecule was only adsorbed on one side of the slab and the net surface dipole moment was compensated by a dipole layer and an energy correction term.[22] The scattering of the valence electrons from the C, O and Cu ion-cores were represented by ultrasoft pseudopotentials that allowed a plane-wave cut-off energy of 25 Ry. The surface Brillouin zones were sampled using a uniform two-dimensional mesh with a constant density of k points for the various surface unit cells.[23] The geometry of the adsorbed CO molecule was optimized using a quasi-Newton method where ionic forces were obtained from the converged electron density using Hellman–Feynman theorem. The ionic forces were also used in the construction of the dynamical matrix for the calculation of vibrational modes.

B. Local and projected density of states

In this section we give a short account of the TH theory of STM images in terms of LDOS and show how the LDOS and the PDOS on molecular orbitals have been calculated using the plane-wave, pseudopotential method in a super cell geometry.

Our calculations of the STM images are based on the TH theory[24] and involve calculation of the LDOS at the Fermi level. This theory is based on an independent electron approximation, the Bardeen approximation for the tunneling matrix element, and an s-wave representation of the tip electronic structure. At zero temperature and low tip–sample bias, V, the differential conductance, dI/dV, is then simply related to the LDOS, $\rho(\mathbf{r}_0, \varepsilon)$, as

$$\frac{dI}{dV}(V) \propto \rho(\mathbf{r}_0, \varepsilon_F + eV)$$
$$= \sum_\alpha |\psi_\alpha(\mathbf{r}_0)|^2 \delta(\varepsilon_F + eV - \varepsilon_\alpha) \tag{1}$$

where $\mathbf{r}_0$ is the location of the tip apex, ε_F the Fermi energy of the sample. In the second step, we have expanded the LDOS in one-electron wavefunctions $\psi_\alpha(\mathbf{r})$ and energies ε_α. Another related quantity that provides useful information about the electronic structure is the PDOS on an orbital $\psi_a(\mathbf{r})$ defined as,

$$\rho_a(\varepsilon) = \sum_\alpha |\langle \psi_\alpha | \psi_a \rangle|^2 \delta(\varepsilon - \varepsilon_\alpha). \tag{2}$$

Note that $\rho_a(\varepsilon)$ obeys the sum rule that it should integrate up to unity.

The calculation of the LDOS and PDOS using the plane-wave and ultrasoft pseudopotential (US-PP) method in a supercell geometry is not entirely straightforward. In the US-PP calculations, the pseudo wavefunctions, $\tilde{\psi}(\mathbf{r})$ are in general labeled by a composite index $\alpha = (n, \mathbf{k}_\parallel)$ that includes the band index, n and the wavevector $\mathbf{k}_\parallel$ in the surface Brillouin zone. These wavefunctions form only an orthonormal set when introducing the so-called overlap matrix $\hat{S}$ associated with the ultrasoft pseudopotentials.[25] So we calculated the matrix element in eqn. (2) from the states $|\psi_a\rangle = \hat{S}^{1/2}|\tilde{\psi}_a\rangle$ and $|\psi_\alpha\rangle = \hat{S}^{1/2}|\tilde{\psi}_\alpha\rangle$ that we have constructed by calculating $\hat{S}^{1/2}$ and its action on the pseudo-states.[26] In the expression for the LDOS in eqn. (1), we can use directly the pseudo-wavefunctions $\tilde{\psi}_\alpha(\mathbf{r}_0)$ because $\mathbf{r}_0$ is far outside the ion cores but, as pointed out in earlier works,[27] we need to extrapolate analytically the tails of the wavefunctions to the tip region. An additional calculational aspect is that the finite set of k points and the finite slab thickness in PP calculations make the spectrum of one-electron states discrete. We handled this aspect by introducing a Gaussian broadening of the δ function in eqn. (2) as

$$\delta(\varepsilon) \to \frac{\exp\left(-\dfrac{\varepsilon^2}{\sigma^2}\right)}{\sigma\sqrt{\pi}} \tag{3}$$

Finally, note that the periodicity of the super cell makes the sums in eqn. (1) and (2) diagonal over $\mathbf{k}_\parallel$.

C. Vibrationally inelastic tunneling and vibrational damping rates

Here we give a short account of the theory of the change in tunneling conductance across a vibrational mode and excitation of electron–hole pairs by a vibrational mode. We show also how these quantities have been calculated using DFT and the US-PP method in a supercell geometry.

We have calculated the non-adiabatic effects of the adsorbate vibration on the tunneling current using a generalization of the TH expression for the tunneling current in eqn. (1) to include the electron–vibration coupling in the sample.[28] This generalization states simply that $\rho(r_0; \varepsilon)$ in eqn. (1) should be the LDOS for the electrons interacting with the adsorbate vibration. This interaction is in general weak and can be calculated using the lowest order vibration self-energy diagram in the many-body perturbation expansion of $\rho(r_0; \varepsilon)$.[29] One finds, using this approach, that this coupling results in a discontinuity, $\Delta\rho(r_0)$ of $\rho(r_0; \varepsilon + eV)$ across the threshold, $\varepsilon = \varepsilon_F + \hbar\Omega$, for a vibrational excitation with energy, $\hbar\Omega$, resulting in a discontinuous change, $\Delta\sigma$, of the differential conductance $\sigma = dI/dV$. The relative change of this tunneling conductance, $\Delta\sigma/\sigma$ across the vibrational energy is given by,

$$\eta_{tot}(r_0) = \frac{\Delta\rho(r_0)}{\rho(r_0; \varepsilon_F)}. \tag{4}$$

The discontinuity $\Delta\rho(r_0)$ has two contributions: an inelastic contribution $\Delta\rho_{inel}(r_0)$, where the vibration is excited by the tunneling electrons in the sample, and an elastic contribution $\Delta\rho_{el}(r_0)$, which takes into account the elastic effect of the vibration on the tunneling electrons in the sample. From a calculational point of view, we find it convenient to make the following decomposition of these two contributions:

$$\Delta\rho_{inel}(r_0) = \Delta\rho_P(r_0) + \Delta\rho_I(r_0) \tag{5}$$

$$\Delta\rho_{el}(r_0) = -2\Delta\rho_I(r_0) \tag{6}$$

where in the quasi-static limit, $\hbar\Omega \to 0$,

$$\Delta\rho_I(r_0) = \delta Q^2 \sum_\beta \left| \pi \sum_\alpha \psi_\alpha(r_0) \langle \psi_\alpha | v' | \psi_\beta \rangle \delta(\varepsilon_\beta - \varepsilon_\alpha) \right|^2 \delta(\varepsilon_F - \varepsilon_\beta) \tag{7}$$

and

$$\Delta\rho_P(r_0) = \sum_\alpha |\delta\psi_\alpha(r_0)|^2 \delta(\varepsilon_F - \varepsilon_\alpha). \tag{8}$$

Here the electron–vibration coupling, v', is the derivative of the effective one-electron potential, v_{eff}, with respect to the vibrational displacement, Q, and $\delta Q^2 = \hbar/(2M\Omega)$ is the mean-square amplitude of Q and M is the mass associated with Q and $\delta\psi_\alpha(r_0)$ is the first-order change in a one-electron wavefunction $\psi_\alpha(r_0)$ when making a static displacement δQ.

Our theory for the relative change in conductance, η, across a vibrational energy, based on the TH approach, assumes that the electron–vibration coupling is restricted to the sample and ignores the contribution to η from the long-range dipole interaction for electrons tunneling between the sample and the tip. This contribution, η_{dip}, can be shown to have no elastic component and is positive.[29] As shown by Persson and coworkers,[18,30] the maximum value of η_{dip} can be estimated by the following simple result,

$$\eta_{dip}^{max} = \left(\frac{\delta\mu}{ea_0}\right)^2 \tag{9}$$

where $\delta\mu$ is the root-mean-square amplitude of the dipole moment of the charge distribution and a_0 is the Bohr radius.

The calculation of the damping rate of a vibrational mode by excitations of electron–hole pairs is also based on first-order perturbation theory in the electron–vibration coupling. In the quasi-static limit, $\hbar\Omega \to 0$, the damping rate, γ_{eh}, is given by,[31]

$$\gamma_{eh} = \frac{2\pi\hbar}{M} \sum_{\alpha,\beta} |\langle \psi_\beta | v' | \psi_\alpha \rangle|^2 \delta(\varepsilon_F - \varepsilon_\beta)\delta(\varepsilon_F - \varepsilon_\alpha), \tag{10}$$

which includes spin degeneracy. Note that γ_{eh} is independent of the vibrational frequency.

In the Newns-Anderson (NA) model of a resonance level coupled to a vibration, we have shown elsewhere[28] that the results for η_{el} and η_{inel} derived from eqn. (4), (6), (8) and (7) and the result for γ_{eh} in eqn. (10) reduce to results derived earlier by Persson and Baratoff[18] for the vibrationally inelastic tunneling probability and by Persson and Persson[17] for the vibrational damping rate, respectively. In this model a single one-electron state $|\psi_a\rangle$ with energy ε_a is interacting with a continuum of one-electron states and the electron–vibration coupling is assumed to be governed by the slope of ε_a with respect to the vibrational coordinate q. For this particular electron–vibration coupling one finds in the NA model the following results for η_{inel} and γ_{eh}:

$$\eta_{\text{inel}}(\mathbf{r}_0) = \delta Q^2 \left(\frac{\partial \varepsilon_a}{\partial q}\right)^2 \frac{1}{\Delta \varepsilon_a^2 + \gamma_a^2} \frac{\rho_a(\varepsilon_F)|\psi_a(\mathbf{r}_0)|^2}{\rho(\mathbf{r}_0;\varepsilon_F)}, \quad (11)$$

and

$$\gamma_{\text{eh}} = \frac{2\pi\hbar}{M}\left(\frac{\partial \varepsilon_a}{\partial q}\right)^2 \rho_a(\varepsilon_F)^2. \quad (12)$$

Here $\Delta \varepsilon_a$ and γ_a are defined by the Green function, $g_a(\varepsilon)$, projected on $|\psi_a\rangle$ as,

$$g_a(\varepsilon_F) \equiv \frac{1}{\Delta \varepsilon_a + i\gamma_a} \quad (13)$$

and $\rho_a(\varepsilon_F) \equiv -1/\pi \, \text{Im} \, g_a(\varepsilon_F)$ is the density of states projected on $|\psi_a\rangle$. In the case of a simple well-defined resonance, $\Delta \varepsilon_a$ and γ_a are the resonance position with respect to the Fermi level and the half-width at half-maximum, respectively. The last factor in eqn. (11) takes into account that, in general, only a fraction of the electrons tunnel through the adsorbate state $|\psi_a\rangle$.

There are several aspects in the evaluation of $\eta_{\text{el}}(\mathbf{r}_0)$, $\eta_{\text{inel}}(\mathbf{r}_0)$ and γ_{eh} using the US-PP method in a supercell geometry. A few aspects are common to those in the calculation of the TH expression for the tunneling current in eqn. (1) such as the need to broaden the δ functions in eqn. (7), (8) and (10) by the Gaussian function in eqn. (3) and to extrapolate the wavefunctions to the tip region. In this case the justification for introducing such a broadening is based on the assumption that the variation of the electronic structure on the scale of the vibrational energies is small. The other new aspects involve calculation of the matrix element $\langle \psi_\alpha | v' | \psi_\beta \rangle$ in eqn. (7), (10) and the first-order change in the wavefunction, $\delta\psi_\alpha(\mathbf{r}_0)$ in eqn. (8). We follow Head-Gordon and Tully[32] and use first order perturbation theory to re-express the matrix element in terms of wavefunction overlaps and one-electron energies as,

$$\langle \psi_{\alpha'} | \delta v | \psi_\alpha \rangle = \begin{cases} 0, & k_\| \neq k'_\| \\ \delta\varepsilon_\alpha, & n' = n, \, k_\| = k'_\| \\ (\varepsilon_\alpha - \varepsilon_{\alpha'})\langle \psi_{\alpha'} | \delta\psi_\alpha \rangle, & n' \neq n, \, k_\| = k'_\| \end{cases} \quad (14)$$

where $\delta v \equiv v' \delta Q$ and $\delta \varepsilon_\alpha$ is the first-order change in ε_α. This result for the matrix element shows that both $\langle \psi_\alpha | v' | \psi_\beta \rangle$ and $\delta\psi_\alpha(\mathbf{r}_0)$ can be calculated from the two displaced configurations by $\pm \delta Q/2$ by using central differences and choosing phases of the wavefunctions such that $\langle \psi_\alpha(\delta Q) | \psi_\alpha(-\delta Q) \rangle \simeq 1$. In the wavefunction overlaps one should use, as in the calculation of the PDOS in eqn. (2), the states obtained by acting with $\hat{S}^{1/2}$ on the pseudo-states.[26]

III. Results and discussion

A. Chemisorption bond parameters and electronic structure

The first step is to determine the equilibrium geometry of the chemisorbed CO molecule on the (100) and (111) faces of Cu. We have restricted our geometry optimization to the experimentally determined configuration of CO adsorbed on the atop site perpendicular to the surface with the C atom coordinated to the Cu atom. A relaxation of perpendicular coordinates of the C and O distances and all the coordinates for the two outermost Cu layers for the (2 × 2) surfaces give C–O and C–Cu distances of 1.17 and 1.86 Å for both Cu(100) and Cu(111), which are in good agreement with the corresponding distances 1.15 and 1.9 Å for Cu(100) obtained by an analysis of low-energy electron diffraction data.[33] In the calculations of the electronic structure for the larger unit cells we have used these coordinates for the C and O atoms but on the unrelaxed slab. The calculated chemisorption energies are 0.70 and 0.77 eV for the p(2 × 2) structures on the (111) and (100) faces, respectively.

Table 1 Calculated vibrational energies for CO on a rigid Cu(100) surface. The mode patterns are schematically depicted by showing the relative phases of the motion of the molecule with the corresponding mode labels. The values within the parentheses are from experiments.

Mode	CO	$\hbar\Omega$/meV
ν_I	←→	265 (259)[a]
ν_M	←←	41 (43)[b]
ν_R	↑↓	31 (35)[b]
ν_T	↓↓	6.6 (3.5)[c]

[a] Ref. 34. [b] Ref. 35. [c] Ref. 36. All results have been obtained for the (2 × 2) surface unit cell with four layers.

For the p(2 × 2) structure of CO on the (100) surface of Cu, we have also determined the vibrational energies of the CO molecule on the rigid surface. The resulting energies of the vibrational modes, their schematic displacement fields and labels are shown in Table 1. For the two parallel modes of the CO molecule, the frustrated translational mode T and rotational mode R, we have diagonalized the dynamical matrix whereas, for the perpendicular modes, we have obtained the mode energies by calculating the energy for approximate displacement fields: a rigid displacement of the molecule for the frustrated translational mode M and a relative displacement of the C and O atoms with a fixed centre of mass for the internal stretch mode I. As shown in Table 1 the calculated mode energies are in good agreement with the experimental values, but this comparison is not entirely correct because the relatively low mass of the Cu atoms and the single coordination of the CO–Cu bind in the atop adsorption site introduce an upward energy shift for the modes M and R when the dynamic coupling to the substrate lattice is turned on.

We have analyzed the electronic structure of CO on the Cu surface using the calculated PDOS on molecular orbitals of the free molecule. In Fig. 1, we show the calculated PDOS on orbitals participating in the bonding of CO with Cu. The results for these PDOS for CO chemisorbed on Cu(100) are very similar to those for CO chemisorbed on Cu(111) and are not shown. We have also indicated the one-electron energies of the free molecular orbitals. We find that the one-electron energy, −29.3 eV, of the free 3σ orbital with respect to the vacuum level is essentially not shifted upon chemisorption, showing that the electrostatic shift of the level is negligible. The results for the PDOS show that the character of the molecular states survives, as expected for a weakly chemisorbed molecule such as CO on Cu. The largest changes occur for the HOMO orbital 5σ and the two-fold degenerate LUMO orbital 2π*. The 4σ orbital hybridizes slightly with the 5σ orbital and the 2-fold degenerate 1π orbital is largely unaffected, whereas the 5σ orbital hybridizes with the Cu d band, located in the energy region −5 to −3 eV, and overlaps with the Cu sp band and the 4σ orbital and shifts downward in energy by about 3 eV. The anti-bonding two-fold degenerate 2π* orbitals broaden dramatically into a partially filled resonance-like feature when overlapping with the sp band and hybridize with the Cu d band. The tails of the PDOS for the 2π* orbitals and 5σ orbital, below and above the Fermi-level, are consistent with the traditional Blyholder model[37] of chemisorption, involving donation from the 5σ orbital and back donation to the 2π* orbitals.

Here we are primarily interested in the nature of the 2π* resonance-like feature in the PDOS. As shown by the result for the PDOS on the 2π* orbitals for the p(4 × 4) structure of CO on Cu(111) in Fig. 2(a), the broadening is not an effect of the dispersion of the 2π* orbitals introduced by adsorbate–adsorbate interactions. Fig. 2(a) shows also that the PDOS is well converged with respect to the number of layers for Cu(111). In Fig. 2(b), we make a detailed comparison between

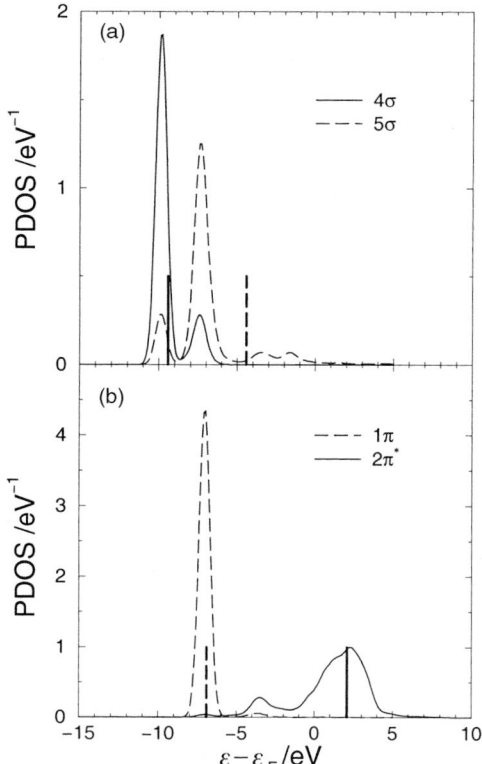

Fig. 1 Calculated PDOS of CO chemisorbed on Cu(111) in the p(2 × 2) structure. The projections are on molecular states with σ symmetry in (a) and π symmetry in (b) of the free molecule. The vertical lines are the positions of the free molecular states, slightly shifted in energy so that the energies of the 3σ states agree with the chemisorbed one. Note that both spin and orbital degeneracies are included. The PDOS have been broadened by a Gaussian with $\sigma = 0.5$ eV.

the PDOS for the 2π* orbitals for CO adsorbed on (111) and (100) faces. This comparison shows that both the half-widths, γ, at half-maximum and the positions, $\tilde{\varepsilon}_{2\pi}^*$, of the center of the peaks are similar: $\gamma_{2\pi^*} \approx 1.15$ and 1.45 eV for the (111) and (100) face, respectively, and $\tilde{\varepsilon}_{2\pi^*} - \varepsilon_F \approx 1.9$ and 2.2 eV for the (111) and (100) faces, respectively. The slightly smaller value of $\gamma_{2\pi^*}$ for the (111) surface than for the (100) surface is probably an effect of the larger bandgap on the (111) surface than on the (100) surface. Note that part of the broadening is not inherent and derives from our Gaussian broadening of the PDOS. For CO on Pd(100), Delbecq and Sautet[38] obtained similar values for 2π*-derived resonance parameters from their density functional calculations.

Before attempting to make a direct comparison of the energy and width of the 2π-derived resonance state with experiments, we have to stress that DFT can, in principle, only describe ground-state properties so that the energies of unoccupied and occupied orbitals do not have a rigorous correspondence to excitation energies that are measured in experiments. One exception is the fact that the binding energy of the highest occupied Kohn–Sham orbital for the exact exchange-correlation functional is equal to its ionization energy.[39] Our calculated values of 14.2, 11.7 and 11.6 eV for the binding energies of the 4σ-, 1π- and 5σ-derived states of CO on Cu(100) are all downshifted by about 3 eV as compared to the corresponding ionization potentials 11.5, 8.5 and 8.5 eV measured by ultraviolet photoemission spectroscopy.[40] The 2π*-derived state has been investigated experimentally by inverse photoemission spectroscopy (IPS)[41] for CO in a c(2 × 2) structure on Cu(100) and by two-photon photoemission spectroscopy (2PPE)[42] for the $\sqrt{3} \times \sqrt{3}$ structure of CO on Cu(111). In IPS one measures the energy differences of the excited states of the system in the presence of an extra electron whereas 2PPE involves a two-step process in which an electron–hole pair is created in the first step by a photon from the pump pulse and an electron is

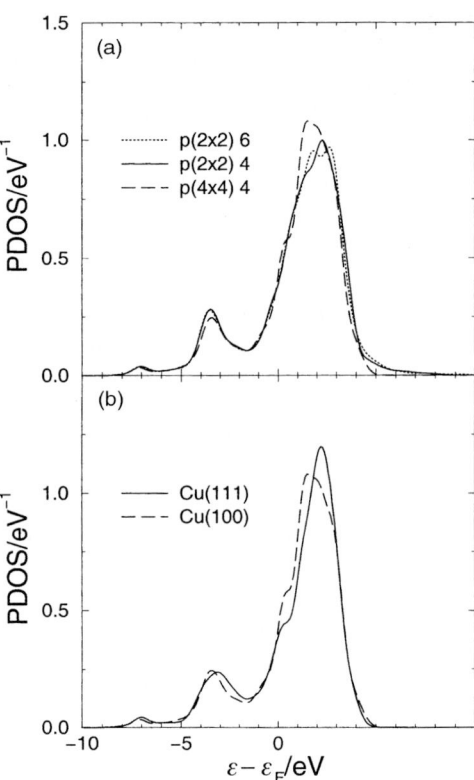

Fig. 2 Calculated PDOS on the $2\pi^*$ of CO chemisorbed on Cu(111) and Cu(100) in the p(4 × 4) structure. Same broadening as in Fig. 1 and the spin and orbital degeneracies are also included. (a) Variations of the PDOS with system size. (b) Surface dependence of the PDOS.

then emitted from this excited state by a second photon from the probe pulse. The IPS measurements identify two adsorbate-induced peaks at 2.6 and 3.6 eV above the Fermi level and the energy dependence of the cross-section suggests that the latter peak derives from a $2\pi^*$ level. In the analysis of the 2PPE measurements based on the photon polarization dependence, two peaks with π symmetry at 2.4 and 3.6 eV above the Fermi level are identified. From the sensitivity of these peaks to the adsorbate structure, the latter peak is shown to derive from the $2\pi^*$ level and its full-width at half-maximum (FWHM) is determined to be about 0.87 eV. These values for the energy and width differ significantly from the corresponding values of $\tilde{\varepsilon}_{2\pi^*} = 1.9$–$2.2$ and $2\gamma_{2\pi^*} = 2.3$–2.9 eV that we obtain for the $2\pi^*$ resonance feature in the PDOS.

B. Elastic and vibrationally inelastic tunneling in STM

Here we present our calculated topographical STM images of CO chemisorbed on the (111) and (100) faces of Cu and vibrationally inelastic tunneling signals for some selected molecular modes and their spatial images. These results are also discussed in relation to available experimental data.

We find that the calculated topographical STM images, based on the LDOS at the Fermi level, have a large lateral extension and are sensitive to the size of the surface unit cell. In Fig. 3, we show contour plots of the calculated topographical images for the (3 × 3) and (4 × 4) structures of CO on Cu(100) at a distance of about 7 Å of the tip apex from the surface plane, z_0. For both structures the width of the image of CO is about 5 Å but the corrugations are somewhat different. In the (4 × 4) structure, the image is a protrusion with a maximum corrugation of 0.35 Å whereas for the (3 × 3) structure a depression of about 0.1 Å develops on top of the protrusion and the maximum corrugation for the (3 × 3) structure is about 0.17 Å. Such a depression in the image is

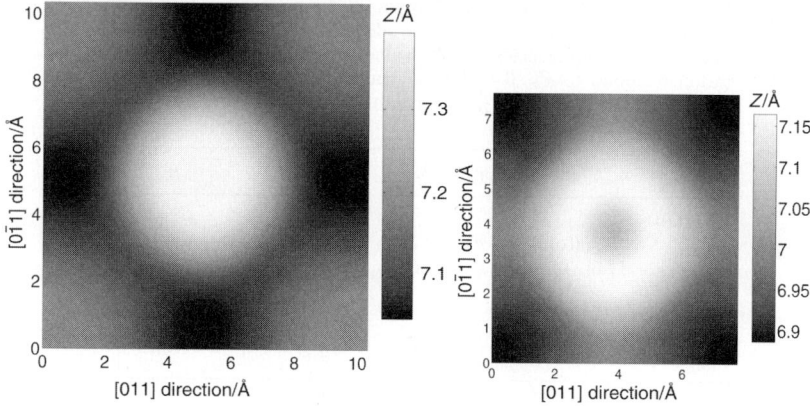

Fig. 3 Calculated topographical images of CO on Cu(100) for the p(4 × 4) and p(3 × 3) structures. The delta function has been broadened by a Gaussian with $\sigma = 0.25$ eV.

expected when the $2\pi^*$ orbitals contribute to the LDOS at the Fermi level because these orbitals have a node perpendicular to the surface. Increasing the distance z_0 from 7 to 9 Å decreases and smooths the overall corrugation. For the (4 × 4) structure the maximum corrugation decreases from 0.35 Å for $z_0 = 7$ Å to 0.15 Å for $z_0 = 9$ Å.

The sensitivity of the calculated STM image to the electronic structure is illustrated by comparing the image for CO on Cu(111), shown in Fig. 4, with the corresponding image for CO on Cu(100) in Fig. 3 at about the same average distance $z_0 \sim 7$ Å. A protrusion corresponding to the CO molecule is still found with a height of 0.08 Å. A depression is developed around the molecule, enhancing the "sombrero" shape of the image. The width of the protrusion is 4 Å (FWHM) and is about 1 Å less than the corresponding protrusion of CO on Cu(100).

Using the theory of vibrationally inelastic tunneling in Section III B, we find that the nonadiabatic coupling between the tunneling electrons and the vibration can be very efficient. In Table 2, we have summarized our results for the calculated maximum relative changes $\eta_{tot}^{(max)}$, in tunneling conductance across some vibrational modes of CO chemisorbed on Cu(100) and Cu(111) for the p(4 × 4) structure. We have included results both from short-range electron vibration coupling in the sample based on eqn. (4), (6), (7) and (8) and from long-range dipole coupling based on eqn. (9). The latter contribution is essentially zero for the R modes and negligible for the T modes while it is dominant for the strongly infrared-active I modes. This result is in sharp

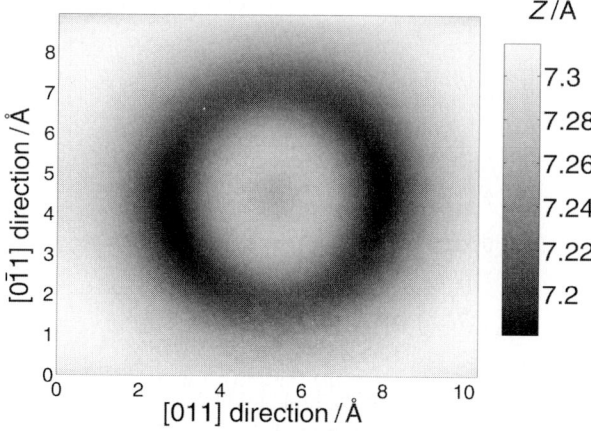

Fig. 4 Calculated topographical image of CO on Cu(111) for the p(4 × 4) structure. Same broadening function as in Fig. 3.

Table 2 Calculated relative changes in tunneling conductance, η, across some selected modes of CO adsorbed on Cu(100) or Cu(111). The calculated efficiencies are the values at maximum absolute value and correspond to $\sigma = 0.4$ eV while the values within the parentheses correspond to $\sigma = 0.25$ eV. All results have been obtained for the (4×4) surface unit cells

Mode	$\eta_{\text{tot}}^{(\text{max})}$ (%)	$\eta_{\text{inel}}^{(\text{max})}$ (%)	$\eta_{\text{el}}^{(\text{min})}$ (%)	$\eta_{\text{dip}}^{(\text{max})}$ (%)
Cu(100)				
ν_I	0.50(0.42)	0.64(0.79)	$-0.15(-0.41)$	1.3
ν_R	17.6(23.8)	19.3(25.6)	$-1.6(-2.0)$	0.0
Cu(111)				
ν_M	2.3(3.5)	2.9(4.2)	$-0.56(-0.67)$	0.08
ν_R	14.8(11.9)	17.1(15.2)	$-2.2(-5.0)$	0.0

contrast to the case of short-range coupling in the sample where the efficiencies $\eta_{\text{tot}}^{(\text{max})}$ are very large for the R modes and small for the I modes. Note that the two-fold degeneracy of the R modes makes the efficiencies a factor of two larger. The results for CO chemisorbed on the Cu(100) and Cu(111) faces show the same trends and are similar in magnitude. We find that the results for $\eta_{\text{tot}}^{(\text{max})}$ are rather weakly dependent on the broadening, giving us confidence in our calculational procedure, but the large size of the systems, slabs with a (4×4) surface unit cell and 4 layers, have prohibited us from performing an investigation of the sensitivity of the results with respect to the number of layers.

In Table 2, we have also given results for the maximum magnitudes of the positive inelastic contributions, $\eta_{\text{inel}}^{(\text{max})}$, and of the negative elastic contributions, $\eta_{\text{el}}^{(\text{min})}$, to the vibrationally inelastic tunneling efficiencies. Note that the maximum magnitudes of η_{inel} and η_{el} are not necessarily attained at the same lateral positions of the tip apex. For the R and T modes, η is dominated by the inelastic contribution whereas for the I modes $\eta_{\text{el}}^{(\text{min})}$ is comparable to $\eta_{\text{inel}}^{(\text{max})}$ but in all cases $\eta_{\text{tot}}^{(\text{max})}$ is positive. As shown by the result in eqn. (8), when the inelastic contribution η_{inel} dominates over η_{el}, the vibrationally inelastic tunneling efficiency is determined by the change of tunneling amplitude with vibrational coordinate.

We have also investigated the spatial distribution of the vibrationally inelastic efficiency. In Fig. 5, we have plotted a profile of η_{tot} for the R modes of CO on Cu(100) and Cu(111). The calculated profiles for these two surfaces are very similar and show a maximum at the molecule and a

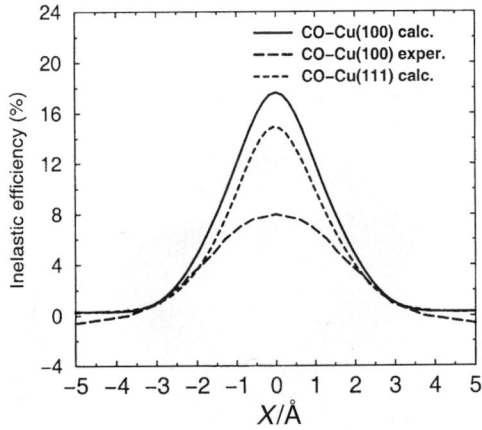

Fig. 5 Profiles of vibrational inelastic images of the frustrated rotation of CO on Cu. The (——) and (--): calculated results for the (100) and (111) faces of Cu, respectively, (—·—) experimental result for d^2I/dV taken from Lauhon and Ho,[43] scaled to the observed value of 8% for $\Delta\sigma/\sigma$. The profiles for the (111) and (100) surfaces are along the $[1\bar{1}0]$ and $[001]$ directions, respectively with the origin at the adsorption site. The delta function has been broadened by a Gaussian with $\sigma = 0.4$ eV.

FWHM of about 2.3 and 2.8 Å for the (111) and (100) faces, respectively, about 2 Å narrower than the constant LDOS image.

A general problem when comparing calculated STM images with experiments is that the images are tip dependent and that the state of the tip is not very well-characterized. In the case of the STM image of CO chemisorbed on Cu(100), Lauhon and Ho find a circular depression of about 0.2 Å with a FWHM of about 5.7 Å at a bias of 0.036 mV and a tunneling current of about 1 nA,[43] whereas for an etched tip, they find a protrusion of about 0.5 Å with a FWHM of about 2.4 Å at a bias of 0.1 V and tunneling current of 1 nA.[44] We have not attempted to calculate any absolute values of the tunneling conductances because this calculation requires certain unknown parameters that are, usually, at best educated guesses, such as the radius of curvature of the tip and its density of states. However, our average value of about 7 Å for z_0 is in the experimental range of distances[45] where TH theory is applicable.[46]

We find an overall agreement of the calculated η_{tot} with the measured $\Delta\sigma/\sigma$ for vibrational modes of CO chemisorbed on Cu(100) by Lauhon and Ho.[43] As in the experiments, we find that the R mode gives the dominant signal and that the signal from the I mode is weak. The calculated value for η for the R mode in Table 2 is about twice that of the measured value for $\Delta\sigma/\sigma$ of about 8%. In the case of the I mode, we find that η_{tot} is dominated by inelastic dipole scattering and its calculated value of 1.8% is close to the measured value of about 1.5% for $\Delta\sigma/\sigma$. As shown in Fig. 5, the calculations can also account for the observed image of the vibrationally inelastic signal. The shape of the spatial image of the calculated η_{tot} for the R mode is in good agreement with the rescaled experimental image of d^2I/dV^2, recorded by Lauhon and Ho[43] at a bias corresponding to the vibrational energy. Finally, we would like to mention that recent experiments by Moresco et al.[47] have also established that the R mode gives a dominant signal in vibrationally inelastic tunneling of CO on Cu(211).

C. Vibrational damping rates

Here we present our calculated vibrational damping rates for some selected modes of CO adsorbed on the (111) and (100) faces of Cu and discuss them in relation to experimental data.

In Table 3, we have summarized our results for damping rates, γ_{eh}, for some vibrational modes of CO on Cu by excitation of electron–hole pairs. The relatively weak dependence of the calculated γ_{eh} on the broadening and the number of layers shows that the calculated values are basically converged and gives us confidence in our calculational scheme. The dependence of the damping rates on the size of the surface unit cell is probably not significant at the present level of accuracy. We are calculating the damping rate of a collective vibrational mode at the Γ point in the surface Brillouin zone so a dependence on the surface structure might be present, at least for high coverages, but the present result indicates that our results for the p(2 × 2) structures are close to the result for an isolated molecule.

Table 3 Calculated vibrational damping rates for some selected modes of CO adsorbed on Cu(100) or Cu(111). The calculated damping rates, γ are shown as a function of the broadening σ and surface structure. The values within the parentheses correspond to a 6 layer slab and all other values are obtained for a 4 layer slab

Mode	$\gamma_{eh}/10^{12}$ s^{-1}		
	$\sigma = 0.2$ eV	$\sigma = 0.3$ eV	$\sigma = 0.4$ eV
p(2 × 2)CO–Cu(100)			
ν_I	0.53(0.57)	0.57(0.50)	0.58(0.48)
p(4 × 4)CO–Cu(100)			
ν_I	0.76	0.70	0.63
ν_R	0.47	0.52	0.52
p(4 × 4)CO–Cu(111)			
ν_M	0.14	0.11	0.11
ν_R	0.17	0.31	0.39

We find a significant variation of the γ_{eh} over the various vibrational modes and some dependence on the crystal face. The largest value of γ_{eh} is attained for the I mode whereas γ_{eh} has its smallest value for the M mode. The results for γ_{eh} of the R mode indicate that γ_{eh} is somewhat smaller on the (111) face than on the (100) face. The results for the damping rates obtained by Head-Gordon and Tully[32] using Hartree–Fock and a cluster representation of CO on Cu are similar to our results except that they find the damping rate for the M mode to be an order of magnitude smaller than our result.

We find also that our calculated γ_{eh} are in good agreement with experimental data. The most direct measurement of damping rates comes from real-time measurements using picosecond pump–probe pulses, whereas less direct measurements come from an analysis of Fano lineshapes in infrared reflection absorption spectroscopy (IRAS) but gives directly the electron–hole pair contribution. The pump–probe measurements by Morin et al.[48] of the I mode of CO in a c(2 × 2) structure on Cu(100) resulted in a value of about 0.5×10^{12} s^{-1} for the damping rate, which supports the early results obtained from infrared lineshape measurements by Ryberg.[49] This value is in excellent agreement with our calculated value of about $(0.5–0.6) \times 10^{12}$ s^{-1}. The analysis of the Fano lineshapes of the R mode of CO suggested the values 0.33×10^{12} s^{-1} and 1×10^{12} s^{-1} for γ_{eh} of the c(2 × 2) and $\sqrt{3} \times \sqrt{3}$ structure of CO on Cu(100) and Cu(111), respectively. In particular, the value for CO on Cu(100) is in good agreement with our calculated value for the R mode in Table 3. Thus this favourable comparison gives additional support for our theoretical approach to electron–vibration coupling in the excitation of electron–hole pairs in vibrational damping.

D. Some implications for tip-induced desorption

Here we discuss our results for the 2π-derived resonance state, vibrationally inelastic tunneling rates and vibrational damping rates in relation to recent work on tip-induced desorption of CO on Cu.

The observed desorption of individual CO molecules on Cu(111) by single electrons tunneling from a STM tip at a bias exceeding 2.4 V was interpreted as being induced by a transient occupation of a $2\pi^*$-derived level.[7] In their analysis of the data in terms of a 1D model for the desorption coordinate using a Menzel–Gomer–Redhead-like resonance model proposed by Gadzuk,[50] they concluded from the observed isotope effect and 2PPE measurements of the resonance width of the $2\pi^*$-derived level that the tip-induced desorption probability of about 5×10^{-9} per tunneling electron can be accounted for by a potential slope of the excited state of about 1 eV Å^{-1} and a fraction in the range 0.01–4% of the tunneling current passing through the resonance.

Our calculations of vibrationally inelastic tunneling and vibrational damping rates suggest that this picture of the mechanism for tip-induced desorption is incomplete. Our calculated values for η_{inel} give the probability for a tunneling electron with an energy above the vibrational threshhold to excite an vibration and show that the R mode is most efficiently excited with a probability as large as 10%. Note that the trend of $\eta_{inel}^{(max)}$ over various modes has no correlation with γ_{eh}. We find that the excitation probabilities $\eta_{inel}^{(max)}$ for the M and I modes cannot simply be understood in a NA model for resonance coupling through a 2π-derived level although we find that this model describes the electron–vibration coupling in the electron–hole pair damping of the I mode.

In the NA model for the vibrational damping rate, γ_{eh}, presented in Section II C, γ_{eh} is determined by the density of states, $\rho_a(\varepsilon_F)$, projected on the molecular $2\pi^*$ orbital and the slope, $\partial\varepsilon_a/\partial q$, of the average resonance energy. From our calculated PDOS on this orbital for CO in the p(2 × 2) structure on Cu(100) as a function of vibrational coordinate, we find that $\rho_a(\varepsilon) \approx 0.095$ eV^{-1} for a single $2\pi^*$ orbital and single spin, and $\partial\varepsilon_a/\partial q \approx 2.5$ eV Å^{-1} resulting in value of about 0.66×10^{12} s^{-1} for γ_{eh}. This value is very close to our calculated value in Table 3 using the full expression in eqn. (10). Thus this comparison gives strong support for the resonance model of the electron–vibration coupling for γ_{eh} of the I mode and also for the original application of this model by Persson and Persson.[17] However, in the latter application the parameters were determined indirectly from experiments.

In the case of vibrationally inelastic tunneling, the NA model presented in Section II C is not able to account for the calculated $\eta_{inel}^{(max)}$. First we make the most favourable assumption that all electrons tunnel through the $2\pi^*$-derived level, that is, the LDOS is assumed to be dominated by the contribution from the $2\pi^*$ orbital. From our calculated values of 2.5 eV Å^{-1}, 2.2 eV, and 1.45 eV for $\partial\varepsilon_a/\partial q$, $\Delta\varepsilon_a$ and γ_a for the $2\pi^*$-derived resonance-like feature of CO on Cu(100), we obtain, based on eqn. (11), a value of about 0.1% for $\eta_{inel}^{(max)}$, which is about a factor of 5 smaller than the calculated value of 0.5%. In the case of the M mode, the discrepancy is even larger than for the I mode because we find that $\partial\varepsilon_a/\partial q$ about a factor of four smaller for the M mode than for the I mode and the calculated strength is about a factor of six larger for the M mode than for the I mode. We believe that this resonance model misses an important mechanism for electron–vibration coupling in tunneling, suggested by eqn. (8), that involves the change of tunneling amplitude by deformation of tails of the one-electron wavefunctions with respect to the vibrational coordinate. More work is needed to elucidate the role of this mechanism in creating the excitations of the bond beyond the first excited vibrational state that are necessary to break the bond by single tunneling electrons.

IV. Summary and concluding remarks

We have performed a density functional study of the electronic structure, images and vibrationally inelastic tunneling in the STM, and vibrational damping rates of CO on the (111) and the (100) faces of Cu, with the objective of gaining a better understanding of the nature of the electron–vibration coupling in tunneling-current-induced bond excitation and breaking at surfaces. The calculation of the STM images, the relative changes in tunneling conductance across vibrational thresholds and their real space images were based on TH theory and its generalization to the interaction of tunneling electrons with an adsorbate vibration. The non-adiabatic electron–vibration coupling was treated by perturbation theory. The one-electron wavefunctions and electron–vibration couplings were calculated in a super-cell geometry using a plane-wave and pseudopotential method. The electronic structure of the chemisorbed molecule was characterized by calculating PDOS on molecular orbitals.

Of particular interest in this study is the character of the $2\pi^*$-derived state and its effect on the electron–vibration coupling. The unoccupied $2\pi^*$-orbital turns into a broad and partially filled resonance-like feature in the projected density of states when interacting with the sp and d bands of Cu. The energy, position and width of this feature are very similar on the (111) and (100) faces of Cu but they differ significantly from the position and width suggested by IPS and 2PPE spectroscopic measurements. The calculated vibrational damping rate for the internal stretch mode supports the NA model $2\pi^*$-derived for the electron–vibration coupling. The calculated vibrational damping rates are in good agreement with available experimental data. We find that this resonance model is not able to describe our calculated vibrational excitation probabilities in inelastic tunneling and gives too small probabilities. We believe that such a resonance model misses an important mechanism for vibrational excitation in tunneling that involves the change of tunneling amplitude by deformation of tails of the one-electron wave-functions with vibrational coordinate. The calculated relative changes in tunneling conductance across vibrational thresholds and their real space images are in agreement with available experimental data and do not show a simple correlation with calculated vibrational damping rates. Thus our results indicate that simple resonance models that involves an electron–vibration coupling only through the slope of the resonance energy do not completely describe tip-induced bond excitation and breaking.

Acknowledgements

We thank Lincoln Lauhon and Wilson Ho for sharing their experimental data with us prior to publication. We are grateful for support from the Swedish Natural Science Research Council (NFR) and the European Commission's TMR Programme, Contract #ERBFMRXCT970146, Atomic/Molecular Manipulations. The allocations of computer resources by UNICC at Chalmers and PDC in Stockholm are also acknowledged.

References

1. J. K. Gimzewski and C. Joachim, *Science*, 1999, **283**, 1683.
2. T. C. Shen, C. Wang, G. C. Abeln, J. R. Tucker, J. W. Lyding, Ph. Avouris and R. E. Walkup, *Science*, 1995, **268**, 1590.

3   K. Stokbro, C. Thirstrup, M. Sakurai, U. Quaade, Ben Yu-Kuang Hu, F. Perez-Murano and F. Grey, *Phys. Rev. Lett.*, 1998, **80**, 2618.
4   B. C. Stipe, M. A. Rezaei, W. Ho, S. Gao, M. Persson and B. I. Lundqvist, *Phys. Rev. Lett.*, 1997, **78**, 4410.
5   L. Lauhon and W. Ho, *Phys. Rev. Lett.*, 2000, **84**, 1527.
6   D. M. Eigler, C. P. Lutz and W. E. Rudge, *Nature*, 1991, **352**, 600.
7   L. Bartels, G. Meyer, K.-H. Rieder, D. Velic, E. Knoesel, A. Hotzel, M. Wolf and G. Ertl, *Phys. Rev. Lett.*, 1998, **80**, 2004.
8   H. J. Lee and W. Ho, *Science*, 1999, **286**, 1719.
9   B. C. Stipe, M. A. Rezaei and W. Ho, *Science*, 1998, **279**, 1907.
10 B. C. Stipe, M. A. Rezaei and W. Ho, *Phys. Rev. Lett.*, 1998, **81**, 1263.
11 B. C. Stipe, M. A. Rezaei and W. Ho, *Science*, 1998, **280**, 1732.
12 S. Gao, M. Persson and B. I. Lundqvist, *Solid State Commun.*, 1992, **84**, 271.
13 M. Brandbyge and P. Hedegård, *Phys. Rev. Lett.*, 1994, **72**, 2919.
14 G. P. Salam, M. Persson and R. E. Palmer, *Phys. Rev. B*, 1994, **49**, 10655.
15 S. Gao, M. Persson and B. I. Lundqvist, *Phys. Rev. B*, 1997, **55**, 4825.
16 J. W. Gadzuk, *Phys. Rev. B*, 1991, **44**, 13466.
17 B. N. J. Persson and M. Persson, *Solid State. Commun.*, 1980, **36**, 175.
18 B. N. J. Persson and A. Baratoff, *Phys. Rev. Lett.*, 1987, **59**, 339.
19 N. Lorente and M. Persson, *Phys. Rev. Lett.*, submitted.
20 We used an optimized version 1.30 of DACAPO of CAMP, DTH, Denmark.
21 J. P. Perdew, J. A. Chevary, S. H. Vosko, K. A. Jackson, M. R. Pederson, D. J. Singh and C. Fiolhais, *Phys. Rev. B*, 1992, **46**, 6671.
22 L. Bengtsson, *Phys. Rev. B*, 1999, **59**, 12301.
23 For the (2 × 2), (3 × 3) and (4 × 4) surface unit cells, we used 36, 16 and 9 k points in the surface Brillouin zone, respectively.
24 (a) J. Tersoff and D. R. Hamann, *Phys. Rev. Lett.*, 1983, **50**, 1998; (b) J. Tersoff and D. R. Hamann, *Phys. Rev. B*, 1985, **31**, 805.
25 D. Vanderbilt, *Phys. Rev. B*, 1990, **41**, 7892.
26 The calculation of $\hat{S}^{1/2}$ is described in ref. 28.
27 W. Sacks, S. Gauthier, S. Rousset, J. Klein and M. A. Esrick, *Phys. Rev. B*, 1987, **36**, 961.
28 N. Lorente, M. Persson and L. Bengtsson, to be submitted.
29 C. Caroli, P. Combescot, P. Nozieres and D. Saint-James, *J. Phys. C*, 1972, **5**, 21.
30 B. N. J. Persson and J. E. Demuth, *Solid State Commun.*, 1986, **57**, 769.
31 B. Hellsing and M. Persson, *Phys. Scr.*, 1984, **29**, 360.
32 (a) M. Head-Gordon and J. C. Tully, *Phys. Rev. B*, 1992, **46**, 1853; (b) M. Head-Gordon and J. C. Tully, *J. Chem. Phys.*, 1992, **96**, 3939.
33 S. Andersson and J. B. Pendry, *Phys. Rev. Lett.*, 1979, **44**, 363.
34 S. Andersson, *Surf. Sci.*, 1979, **89**, 477.
35 C. J. Hirschmugl, G. P. Williams and F. M. Hoffmann, *Phys. Rev. Lett.*, 1990, **65**, 480.
36 B. F. Mason, R. Caudano and B. R. Williams, *Phys. Rev. Lett.*, 1981, **47**, 1141.
37 G. Blyholder, *J. Phys. Chem.*, 1964, **68**, 2772.
38 F. Delbecq and P. Sautet, *Phys. Rev. B*, 1999, **59**, 5142.
39 J. P. Perdew, R. G. Parr, M. Levy and J. L. Balduz, *Phys. Rev. Lett.*, 1982, **49**, 1691.
40 J. L. Pascual, L. G. M. Pettersson and H. Ågren, *Phys. Rev. B*, 1997, **56**, 7716.
41 K.-D. Tsuei and P. D. Johnson, *Phys. Rev. B*, 1992, **45**, 13287.
42 M. Wolf, A. Hotzel, E. Knoesel and D. Velic, *Phys. Rev. B*, 1999, **59**, 5926.
43 L. J. Lauhon and W. Ho, *Phys. Rev. B*, 1999, **60**, R8525.
44 L. J. Lauhon, personal communication.
45 S. Heinze, S. Blügel, R. Pascal, M. Bode and R. Wiesendanger, *Phys. Rev. B*, 1998, **58**, 16432.
46 N. D. Lang, *Phys. Rev. B*, 1988, **37**, 10395.
47 F. Moresco, G. Meyer and K.-H. Rieder, *Mod. Phys. Lett. B*, 1999, **13**, 709.
48 M. Morin, N. J. Levinos and A. L. Harris, *J. Chem. Phys.*, 1992, **96**, 3950.
49 R. Ryberg, *Phys. Rev. B*, 1985, **32**, 2671.
50 J. W. Gadzuk, *Surf. Sci.*, 1995, **342**, 345.

Dynamics of charge transfer states on metal surfaces: The competition between reactivity and quenching

Ronnie Kosloff,[a] Gil Katz[a] and Yehuda Zeiri[b]

[a] *Department of Physical Chemistry and the Fritz Haber Research Center, The Hebrew University, Jerusalem 91904, Israel*
[b] *Department of Chemistry Nuclear Research Center-Negev Beer-Sheva 84190, P.O. Box 9001, Israel*

Received 17th May 2000
First published as an Advance Article on the web 13th September 2000

The dynamics of excited states of adsorbates on surfaces caused by charge transfer is studied. Both negative and positive charge transfer processes are possible. In particular we are interested in positive charge transfer from a metal surface to molecular or atomic oxygen adsorbed on the surface. Once the negatively charged oxygen on the surface loses an electron it becomes chemically activated. The ability of this species to react depends on the quenching time or back transfer. The analysis of these processes is based on a set of diabatic potential energy surfaces each representing a different charged oxygen species. The dynamics is followed by solving the multichannel time-dependent Schrödinger equation or Liouville von Neumann equation. Due to the nonadiabatic character of these reactions large isotope effects are predicted.

I. Introduction

Charged molecular and atomic species adsorbed on a metal surface are well documented phenomena. In transition metals it is energetically cheap to move electrons in and out of the d orbitals creating a partially charged molecular or atomic adsorbate. For oxygen example, two stable adsorbed molecular species, peroxide and superoxide, have been observed, differing by the amount of negative charge transferred from the metal.[1-8] The extra charge weakens the molecular bond and increases the bond distance. In addition, the binding of the molecule to the surface is strengthened due to the Coulombic attraction between the molecular charge and its image in the metal. Adsorbed atomic oxygen on a metal surface is also partially charged due to the large negative affinity of the oxygen atoms (-1.4 eV).

Adsorption of N_2 is quite different since it has a high ionization energy and negative electron affinity. Only positively charged molecular excited states of nitrogen can be stabilized. The low energy adsorbed states are the ones that bind through the π orbitals resembling the binding of CO. Atomic nitrogen creates a very strong bond with transition metals where the net charge transfer is small.

In the adsorption of CO the charge transfer to or from the metal is small. The binding energy of CO to metals is weak and is similar to the binding energy of metal carbonyl compounds. In the gas phase CO is quite an inert material. The triple bond between the carbon and oxygen is almost immune to chemical attack. The adsorption of CO on a metal surface only slightly reduces the bond energy.

These examples illustrate that molecules adsorbed on transition metal surfaces possess a variety of electronic configurations. A key question is the role of these electronic configurations in elementary catalytic processes. Insight into this question requires methods to calculate the electronic configurations, and to follow the transitions from one electronic configuration to another.

One approach is to follow the dictum of the Born–Oppenheimer approximation and to consider adiabatic dynamics on a single ground electronic state. The assumption is that the electronic structure of a dynamical encounter changes continuously from one configuration to another. The advantage of this approach is that the powerful methods of quantum chemistry, in particular density functional theory (DFT), are most effective in calculating the ground state potential surfaces.[9,10]

The other approach is a non-adiabatic one in which the nuclear dynamics is followed on more than one potential energy surface simultaneously. The topology of crossings and conical intersections between potentials determines the outcome of the chemical event. The difficulty of this approach is due to the lack of effective methods to calculate the various potentials and nonadiabatic coupling terms for extended systems on metals. There are only a few exceptions,[11–13] based on embedding a cluster of primary atoms into a continuous description of the bulk. For non-metallic surfaces, such as metal oxides, the excitation process is more localized, therefore cluster calculations are more justified and give reasonable results for excited electronic surfaces.[14,15]

Obtaining the potential surfaces and nonadiabatic coupling terms is only the first step in the task of modeling nonadiabatic chemical encounters on surfaces. The next step is to simulate the nuclear dynamics on these surfaces. The difficulty is that the cost of a full quantum representation scales exponentially with the number of degrees of freedom. For this reason current time-dependent methods for solving the Schrödinger equation are capable of representing problems with up to 6 degrees of freedom for the effective mass of hydrogen[16,17] and up to 3–4 degrees of freedom for the effective masses of nitrogen or oxygen.[18,19] This limits the application of direct quantum simulation to model problems. There is the option of employing approximate semiclassical methods for describing nuclear nonadiabatic dynamics.[20] In a careful comparison between the best semiclassical methods and full quantum calculation for multidimensional nonadiabatic transitions we found qualitative discrepancies between the two approaches.[21]

The dynamics of molecules adsorbed on metal surfaces are complicated due to the simultaneous encounter with dissipative forces from two origins: electronic and phononic. Electronic dissipative forces are caused by the interaction of the free metal electrons, or more precisely electron–hole pairs with the adsorbate. These electron–hole pairs are able to exchange a continuous amount of energy. The other source of dissipative forces is the lattice vibrations or phonons, which again form a band of energy levels. Dissipation has a strong influence on the nonadiabatic dynamics.[22] We have shown a general turnover behavior as a function of the dissipative parameters describing electronic dephasing and vibrational relaxation. For small values of these parameters the nonadiabatic rate is enhanced. Above a critical value the nonadiabatic rate decreases. In any event dissipation does not justify a reduction of the dynamical description to motion on a single adiabatic electronic surface. In this study, different quantum theories are considered to describe the short-time dissipative dynamics of an adsorbate interacting with a metal surface. The alternatives range from solving the Liouville von Neumann equation using an empirical description of the dissipation to the use of a surrogate Hamiltonian formulation cast into a wavepacket description.

II. The non-adiabatic model

In the generic non-adiabatic model the dynamics of the system is simulated by solving the time-dependent Schrödinger equation on coupled potential energy surfaces. As an example the dynamics on three potential energy surfaces and three effective coordinates is examined. The wavefunction has the form:

$$\psi = \begin{pmatrix} \psi_g(z, r, \theta) \\ \psi_m(z, r, \theta) \\ \psi_a(z, r, \theta) \end{pmatrix} \quad (2.1)$$

where z is the molecular distance from the surface, r the internuclear distance and θ the orientation angle (see Fig. 1).

Once an initial state is described, The evolution in time is generated according to the time-dependent Schrödinger equation:

$$i\hbar \frac{\partial \psi}{\partial t} = \hat{H}\psi \qquad (2.2)$$

where the Hamiltonian is described as:

$$\hat{H} = \begin{pmatrix} H_g & V_{gm} & V_{ga} \\ V_{mg} & H_m & V_{ma} \\ V_{ag} & V_{am} & H_a \end{pmatrix} \qquad (2.3)$$

and $H_l = P^2/2\mu + V_l$ where l indicates the species: g, gas phase species, m, molecular surface species and a atomic dissociated species. Within this general framework one can study the basic encounters on surfaces. Fig. 2 shows a typical topology of the charge transfer states of oxygen on a metal surface. It is clear that due to different coulombic stabilization, spin states and atomic surface bonding, the different charge transfer states cross each other.

A nonadiabatic dynamical encounter is determined first by the charged state of its initial wavefunction. Following the event in time, the system will cross from one potential to another. The topology of the crossing seams has a profound influence on the outcome, leading to propensity rules.[23] Enhancement by incident translational energy of nonadiabatic transition requires a crossing seam perpendicular to the z direction. A parallel seam will cause enhancement by vibrational excitation of the incoming molecule.

A signature of nonadiabatic dynamics is a large isotope effect. The crossing seams and conical intersections represent a breakdown of the Born–Oppenheimer approximation. As a propensity rule, when the collision energy is below the crossing seam nonadiabatic tunneling takes place and the light isotope has a larger propensity for transition. Above the crossing seam the heavy isotope which is more adiabatic has a larger propensity for nonadiabatic transition.[21,24–26]

A hidden assumption in the reduced dynamical description is that the other coordinates are frozen. The time-scale of the encounter determines if this assumption is valid. For extremely short reactions only a few degrees of freedom participate. For longer times the influence of the other

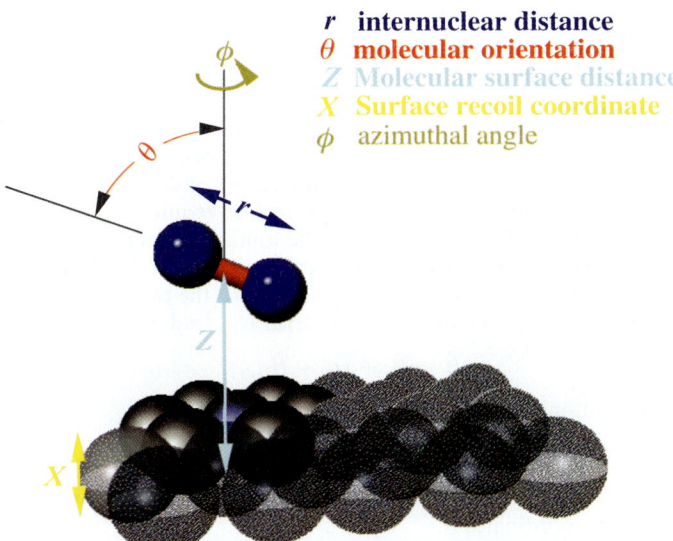

Fig. 1 The coordinate system of a molecule–surface encounter. A reduced description treats only part of the coordinates explicitly.

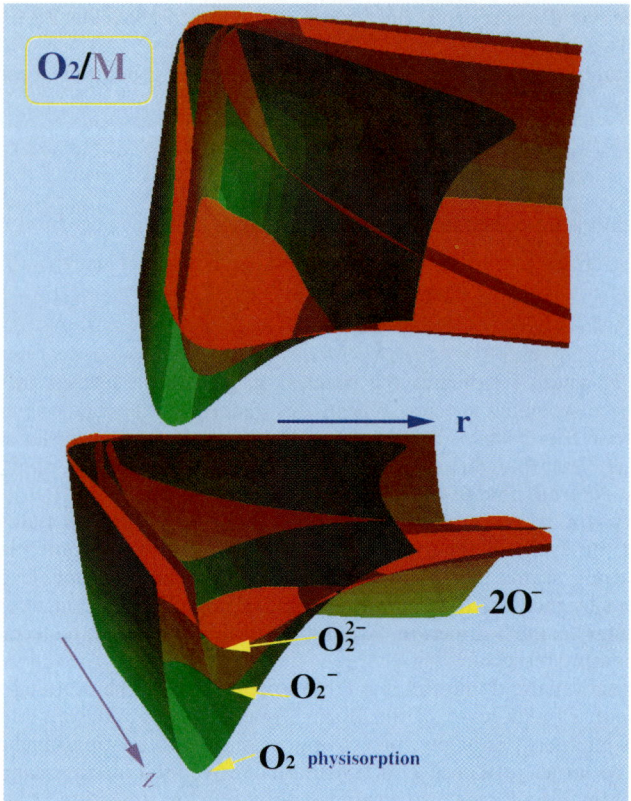

Fig. 2 Potential energy surfaces of oxygen on a metal surface showing the different charge transfer states.

degrees of freedom has to be accounted for. Due to the exponential growth of quantum computation with dimensionality one wants to avoid explicit inclusion of these degrees of freedom. The aim of a theoretical description is therefore to find an implicit description of the influence of the extra degrees of freedom.

III. Reduced dynamics

The basic idea is to distinguish between the dynamics of the primary system, the adsorbate, and that of the bulk, which consists of the electronic and lattice degrees of freedom. The focus of the study is the dynamics of the primary system which therefore requires a detailed description. The treatment of the bulk, or the bath modes, includes the minimum details required to specify their influence on the primary system. The system–bath representation is developed through the study of the total Hamiltonian. This Hamiltonian is partitioned into the primary system's bare Hamiltonian $\hat{H}_s$ the bath Hamiltonian $\hat{H}_B$ and an interaction term $\hat{H}_{int}$ leading to:

$$\hat{H} = \hat{H}_s + \hat{H}_B + \hat{H}_{int}. \tag{3.1}$$

The primary system Hamiltonian has the form:

$$\hat{H}_s = \hat{T} + V_s(\hat{R}) \tag{3.2}$$

where $\hat{T} = \hat{P}^2/2M$ is the kinetic energy of the primary system and V_s is an external potential which is a function of the system coordinates $\hat{R}$.

The idea of partitioning the system into primary and bath modes has been the key element in the quantum theory of dissipative dynamics. Starting from the work of Bloch,[27–29] reduced equations of motion for the primary system have been derived. The reduction is obtained by performing a partial trace over the bath degrees of freedom resulting in a Liouville description of the primary mode. The most well studied derivation is based on the assumption of weak coupling between the system and bath leading to a differential equation describing the system's dynamics.[30–33] In this derivation, commonly called Redfield dynamics, the influence of the bath is described by its correlation functions. This basic derivation has been supplemented by the requirement that the reduced equations of motion have the semi-group form, meaning that they preserve the complete positivity of the density operator.[34–36]

A complementary approach to dissipative dynamics is to axiomatically require a semi-group form. This leads to a general form for the reduced evolution equations.[37–39]

$$\frac{\partial \hat{\rho}}{\partial t} = -\frac{i}{\hbar}[\hat{H}, \hat{\rho}] + \mathcal{L}_D(\hat{\rho}), \qquad (3.3)$$

The generator of the dissipative dynamics, $\mathcal{L}_D$, embodies the effect of dissipation. The form of $\mathcal{L}_D$ in eqn. (3.3) describes Markovian evolution and may be cast into the Lindblad semi-group form:

$$\mathcal{L}_D(\hat{\rho}) = \hat{F}\hat{\rho}\hat{F}^\dagger - \tfrac{1}{2}\{\hat{F}^\dagger\hat{F}, \hat{\rho}\} \qquad (3.4)$$

where $\hat{F}$ is an operator defined on the Hilbert space of the primary system. These equations allow a consistent study of different dissipative models,[40,41] but require an empirical treatment when a particular system is studied.

The practical disadvantage of both the semi-group and the Redfield theories is that they are formulated in Liouville space where the state of the system is represented by a density operator. This fact squares the number of required representation points in comparison to a wavefunction description. Although powerful numerical techniques have been developed to solve the dynamics in Liouville space[32,42–46] it still is extremely taxing to treat these problems, limiting the scope of systems that can be studied.

An alternative approach, is based on constructing a surrogate finite system bath Hamiltonian.[47,48] In the limit of an infinite number of bath modes the true system's dynamics is reconstructed. The procedure constructs a surrogate Hamiltonian in which the system–bath interaction term is renormalized. To reduce further the computational effort, the bath modes are represented by the elementary two-level system (TLS). The primary system is represented by the Fourier method,[49–52] allowing a very general description. The dynamics generated by the finite surrogate Hamiltonian are able to reproduce, for a specified period of time, the true system–bath dynamics. This construction is not Markovian and therefore differs from the Redfield or semi-group treatments. The use of a finite number of degrees of freedom to represent the bath limits the length of time in which the dynamics is consistent with that of an infinite bath. The finite nature of the bath usually shows up as recurrences which eventually appear. Increasing the number of bath modes postpones the recurrence to a later time, thus the number of modes needed is determined by the time-scale of the dynamics.

The surrogate Hamiltonian approach is close in spirit to real time path integral techniques,[53,54] where a large many-body propagator is constructed and approximated. These approximations represent the bath modes as harmonic oscillators,[55–58] for which the path integration can be carried out analytically.[59]

The model of the harmonic bath has been used almost exclusively in modeling dissipative dynamics. The bath Hamiltonian $\hat{H}_B$ is then decomposed to an infinite sum of normal modes:

$$\hat{H}_B = \sum_j \varepsilon_j \hat{b}_j^\dagger \hat{b}_j. \qquad (3.5)$$

The index j is a multidimensional index describing a complex bath of Bosonic modes with energies ε_j.

Further simplification is obtained if the interaction term is a multiplication of a dimensionless geometric function $f(\hat{R})$ with a Boson mode $\hat{b}_j^\dagger$ of potential coupling strength V_j:

$$\hat{H}_{\text{int}} = f(\hat{R}) \sum_j V_j(\hat{b}_j^\dagger + \hat{b}_j). \qquad (3.6)$$

The operators $\hat{b}_j^\dagger$ and $\hat{b}_j$ are Boson type creation and annihilation operators, obeying the commutation rules: $[\hat{b}_j, \hat{b}_{j'}^\dagger] = \delta_{j,j'}\hat{I}$. This system–bath Hamiltonian already represents a drastic reduction in complexity compared to the generic system–bath entity, mainly due to the simple linear (in the bath coordinate) interaction term. The advantage of the model is that it enables an *ab initio* computational procedure. Due to the harmonic assumption classical molecular dynamics can be employed to calculate the bath correlation functions which determine the bath parameters V_j.[48]

The harmonic bath model has fundamental flaws. An important one is that rare Poisson type processes are not described properly. We speculate that Poisson processes are responsible for electronic dephasing on surfaces.[60] Their origin is elastic binary collision of electron–hole pairs with the charge distribution induced by the adsorbate. In a semi-group description the dissipative generator has the form:

$$\mathscr{L}_\mathrm{D}(\hat{\rho}) = \gamma(\mathrm{e}^{-(i/\hbar)\hat{H}_s\phi}\hat{\rho}\mathrm{e}^{+(i/\hbar)\hat{H}_s\phi} - \hat{\rho}) \qquad (3.7)$$

where γ is the collision rate and ϕ determines the phase shift per collision. Another more subtle effect is that in the harmonic bath quantum entanglement between the system and bath is truncated.

Electronic quenching is a typical example of dissipative nonadiabatic event. On metallic surfaces the process is known as the MGR model.[61–63] The process has been simulated by solving the Liouville von Neumann equation using semi-group dynamics[64–66] or by using stochastic wavepacket methods. In both descriptions the crucial parameter is the residence time on the excited electronic surface. The surrogate Hamiltonian method allows an alternative to these empirical approaches. The bath can be described as a collection of two-level systems each representing a localized electron–hole pair. The system bath coupling can be calculated directly and is the result of dipole–dipole coupling.

The influence of the surface induced dissipation can be classified into the following categories: (i) Direct quenching and excitation which is caused by free charge carriers in the metal. (ii) Electronic dephasing *i.e.* loss of phase coherence between the ground and excited surfaces. (iii) Nuclear dephasing *i.e.* loss of vibrational coherence. (iv) Nuclear relaxation decay of vibrational excitation to equilibrium. Typical time-scales of the electronic processes are in the few femtosecond range thus they are of direct relevance to the nonadiabatic encounters considered.

IV. Oxidation by hole induced processes

Oxygen negative ions are inert which is in contrast to atomic oxygen which is a very stong oxidizing agent. Molecular oxygen is a much less potent oxidizer, nevertheless the low-lying excited states of oxygen are strong oxidizers. On metallic surfaces the stable form of oxygen is negatively charged. To activate the oxygen as an oxidizing agent the amount of charge has to be reduced *i.e.* an oxygen to metal charge transfer has to take place.

Charge transfer from the oxygen to the metal can be promoted chemically. Bombarding a surface with energetic particles or with atoms which have a stong attraction to the surface will excite free charge carriers. Nienhaus *et al.*[67] have shown direct evidence for the creation of electron–hole pairs in collision of hydrogen atoms with a Cu and Ag surfaces. Lee and Rettner and Rettner and Auerbach[68,69] have suggested that these free charge carriers will promote dissociation and desorption of other adsorbed molecules.

A. Oxygen dissociation

Oxygen molecules adsorb on metals in two stable charged states peroxide and superoxide. These two states differ in the amount of negative charge, the peroxide being more negatively charged than the superoxide. The amount of excess charge on the adsorbate is located in an anti-bonding orbital, therefore the O–O bond is weakened and extended. The species is partially stabilized on the surface by an image potential, therefore the peroxide has a stronger binding and shorter equilibrium distance in the oxygen–metal coordinate compared with the superoxide. To complete the description of the O_2–metal interaction, these charged molecular states must be coupled to two additional states: one describes interaction of a neutral molecule with the substrate (physisorption) and the other leads to dissociative adsorption.

Within certain limits, the magnitude of the substrate work function can be manipulated experimentally by co-adsorption of small amounts of electron donors or acceptors.[70–72] Hence, the relative location of the potential wells in the two molecular states can be controlled.

A hole-induced process starts with the initial state of the process corresponding to an adsorbed oxygen in the most stable state, the peroxide state. The hole promotes the initial state into either the superoxide or into the physisorption states. This transition is a vertical electronic transition with no nuclear geometry change.

Starting initially in the peroxide state of the adsorbed O_2, the two possible hole-induced excitations are: Franck–Condon transitions to the physisorption state or to the superoxide state. This is described by a vertical shift of the initial nuclear wavefunction from the peroxide well to the physisorption or superoxide potentials. A third possibility is that the molecular oxygen is initially in the superoxide state. The charge transfer to the hole at the surface leads to excitation of the molecule into the physisorption state. The difference in equilibrium geometry of the peroxide and superoxide states means that the vertical transition to the physisorption potential will be to a different region and the subsequent dynamics is expected to exhibit new features. After the vertical excitation step, the dynamics is followed to final states by solving the multi-channel time dependent Schrödinger equation.[23]

The relaxation dynamics of the excited adsorbate is expected to depend on the initial state and on the amount of energy available by the hole. The maximum vertical transition that can be induced by the hole is limited by the adsorption energy of the impinging atom. Starting at the peroxide potential well the hole can induce a transition either to the physisorption state or to the superoxide state, provided that the required energy for such transitions is available. The lifetime of the hole has to be considered with respect to the dissipative processes taking place. Strong dephasing processes will increase the lifetime. Electronic quenching processes will destroy the "hole". Work in this direction is in progress.

The hole-induced desorption and dissociation play an important role in a variety of systems. Some photo-induced processes on solid surfaces exhibit similarities to the system described above. For example, the dissociation of oxygen molecules following the irradiation of the $O_2/Ni(110)$ system can be interpreted as a hole-induced desorption.[73] In this case the irradiation leads to the dissociation of adsorbed oxygen molecules. The dissociation fragments, oxygen atoms, migrate along the substrate. Eventually the excess translational energy is dissipated. The quenching rate determines the distance the fragment can move.

B. CO oxidation

Oxidation of CO in the gas phase is a slow process. In combustion conditions the reaction is carried out by free radicals for example $O(^3P)$.[74–76] This reaction is also spin forbidden. Metal surfaces can catalyze this reaction. One may ask what is the role of the metal in the mechanism of CO oxidation. In general this is quite a complex question.[77–80] It has structural aspects due to the different forms of oxygen which can reside on the surface.[81] In this study we are only interested in dynamical aspects. In the nonadiabatic framework the reactivity of oxygen either as a molecule or as an atom is related to the small gap between the ground and low-lying excited electronic states.[82,83] Preliminary electronic structure calculations we have carried out on small metal clusters show a small HOME–LUMO gap of approximately 2 eV. On the ground electronic surface the charge transfer is always from the metal to the oxygen atom.[1–4,72] The excited state in the calculation showed a back charge transfer from the oxygen to the metal. DFT calculations of spin-restricted oxygen on Pt show a small energy gap in the range 1–2 eV between the ground and excited spin-restricted oxygen forms.[71]

These excited states suggest a possible explanation for the promotion of CO oxidation on a Ru surface by producing free charge carriers by an ultrashort laser pulse.[84] Two mechanisms have been proposed to explain these observations. The first mechanism termed *electron-induced* is thought to be the result of free photo-induced electrons which occupy an anti-bonding orbital on the oxygen atom. As a result the oxygen–metal binding is weakened allowing the oxygen atom to attack the CO molecule and create CO_2. The second mechanism termed *hole-induced* is thought to be the result of a charge transfer from the oxygen atom to fill a positive hole in the metal. As a result the negatively charged oxygen becomes almost neutral, increasing its chemical reactivity substantially. This reactive oxygen will eventually lead to CO_2 formation. The reactivity of the excited oxygen is limited by the finite lifetime of this state. Quenching times are estimated to be in the femtosecond range. Increasing the lifetime will increase the oxygen oxidation power. Co-adsorption of electron acceptors may cause such an effect. A large isotope effect has been observed in the reaction[84] when the adsorbed ^{16}O is replaced with ^{18}O. A possible explanation is that, to a large extent, the excited oxygen is quenched before reacting; the lighter isotope will move further during this period, thus enhancing its chance of reacting. Other nonadiabatic explanations are possible.

C. Orientation effect in the dissociation of N_2

The favorable orientation for an activated dissociation process on a surface is parallel. The reason is that in this orientation the two new atom metal bonds are formed simultaneously. This orientation is the surface analogue of a four-center reaction.

The nonadiabatic framework is a source of a new orientation effect which is due to the different favorable orientation of the different molecular species on the surface. As an example we consider the N_2/Ru system. In the physisorption potential the molecular surface forces due to polarization of the molecule for N_2 are almost isotropic. As a result there is no preferred orientation for the physisorption state, as can be seen in Fig. 3. The well depth for this state is approximately -0.1 eV. Nitrogen, which is isoelectronic to CO, has a low-lying molecular chemisorption state where the nitrogen is perpendicular to the surface (see Fig. 3). There is experimental evidence for this state[85,86] with a well depth of -0.25 eV. The third surface is the dissociative chemisorption surface where the preferred orientation is flat.

Non-adiabatic dynamical 3-D calculations were performed on these potentials with three degrees of freedom, Z the distance from the surface, r the internuclear distance and θ the molecular orientation. The initial state was chosen to mimic a scattering encounter starting from the physisorption potential in the gas phase. At high incident kinetic energies, the direct dissociation

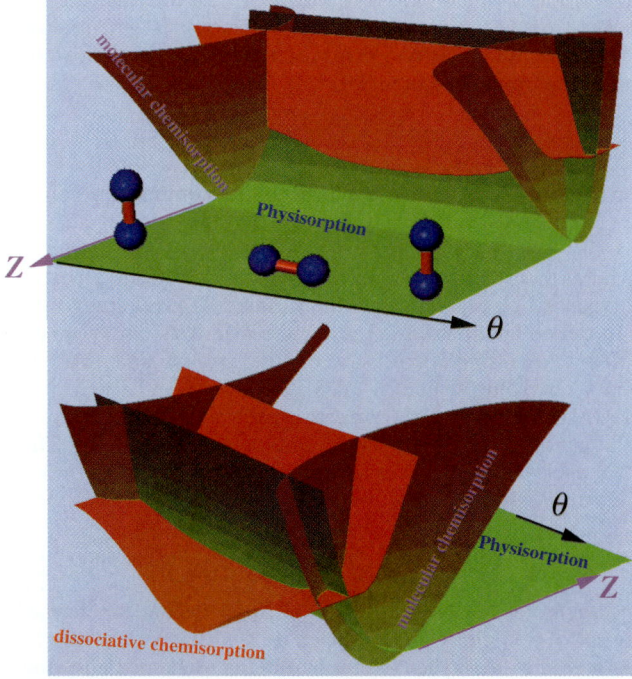

Fig. 3 Three diabatic potential energy surfaces for N_2 on Ru, representing the physisorption state, the dissociative chemisorption state and a molecular chemisorption state. The coordinates are Z, the distance of the molecule from the surface and θ, the orientation. Top: front view from the gas phase, bottom: back view.

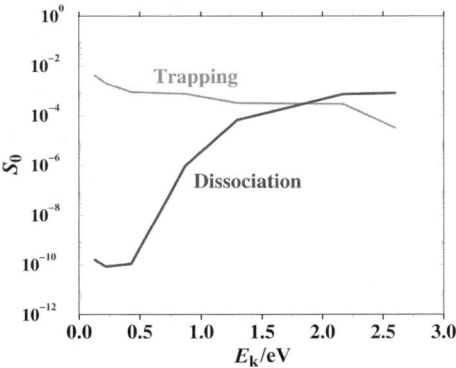

Fig. 4 Competition between dissociation and trapping for a model of N_2 on Ru.

channel was not influenced by the existence of the molecular chemisorption state, *i.e.* the dynamics was diabatic. At low incident kinetic energy an adiabatic steering force took over leading away from the favorable dissociation orientation. The competition between dissociation and trapping is shown in Fig. 4.

The overall effect of the nonreactive molecular chemisorption state on the dissociation probability is quite complex. At first one would expect that this state would always suppress the dissociation channel by diverting flux into the trapped state. But at particular energies there is a possibility of trapping in an upper adiabatic potential with favorable orientation for dissociation. The large trapping time can enhance the dissociation channel.

The orientation effect may explain the large discrepancy between the dissociation probability of the N_2/Ru system in a molecular beam experiment and a thermal experiment with comparable incident energy. The initial state of the molecular beam experiment is dominated by $j = 0$ projectiles. The thermal experiments are dominated by $j = 6-8$. For a low-energy collision the lower j states can be steered more easily than the high j states, thus they are more reactive.

V. Conclusions

Charge transfer states play important roles in surface reactions. The most relevant states are those which are easily excited *i.e.* are low in energy. The structure of the surface can alter considerably the location, charge and energetics of these excited states. For example, co-adsorption of an electron donor and oxygen will increase the charge on the oxygen and reduce its chemical reactivity. Charge transfer reactions can be promoted either chemically, by light or by a beam of electrons. On metals the indirect promotion of the reaction through free charge carriers is the dominant mechanism. The theory of these reactions lacks a reliable method to calculate excited states of adsorbates on metals. The development of such methods is a crucial step for obtaining quantitative nonadiabatic descriptions.

Reduced quantum dynamical descriptions have progressed considerably in recent years. Many new approaches have been developed that can simulate dissipative encounters. Due to the difficulty of the subject more than one approach should be compared. The theoretical challenge is to find first principle methods to calculate quenching times and dephasing times on metal surfaces.

For charge transfer reactions the quenching rate which determines the excited state lifetime is a crucial parameter. Direct measurements should be compared to first principle calculations. A more subtle issue is the role of electronic dephasing. The phenomenon can either enhance or suppress nonadiabatic transfer. The actual role in charge transfer processes on surfaces is not known. A fingerprint of these nonadiabatic reactions is a large isotope effect.

Acknowledgements

This research was supported by the German–Israel Foundation (GIF). We want to thank Ofra Citri, Micha Asscher and Faina Dobnikov for sharing unpublished results. The Fritz Haber

Research Center is supported by the Minerva Gesellschaft für die Forschung, GmbH München, FRG.

References

1. A. Raukema, D. A. Batler, F. M. A. Box and A. W. Kleyn, *Surf. Sci.*, 1996, **347**, 151.
2. L. Vattuone, M. Rocca, C. Boragno and U. Valbusa, *J. Chem. Phys.*, 1994, **101**, 726.
3. T. Greber, *Surf. Sci. Rep.*, 1997, **28**, 1.
4. P. D. Nolan, B. R. Lutz, P. L. Tanaka, J. E. Davis and C. B. Mullins, *Phys. Rev. Lett.*, 1998, **81**, 3179.
5. B. Kasemo, E. Tornqvist, J. K. Norskov and B. I. Lundqvist, *Surf. Sci.*, 1979, **89**, 554.
6. B. Kasemo, *Phys. Rev. Lett.*, 1974, **32**, 1114.
7. L. Hellberg, J. Stromquist, B. Kasemo and B. I. Lundqvist, *Phys. Rev. Lett.*, 1995, **74**, 4742.
8. A. W. Kleyn, *The Physics of Electronic and Atomic Collisions*, 1978, p. 451.
9. C. Stampfl and M. Scheffler, *Phys. Rev. Lett.*, 1997, **78**, 1500.
10. C. Stampfl and M. Scheffler, *Surf. Sci.*, 1999, **433**, 119.
11. H. Nakatsuji and H. Nakai, *Chem. Phys. Lett.*, 1990, **174**, 283.
12. H. Nakatsuji and H. Nakai, *J. Chem. Phys.*, 1993, **98**, 2423.
13. N. Govind, Y. A. Wang and E. A. Carter, *J. Chem. Phys.*, 1999, **110**, 7677.
14. X. Xu, H. Nakatsuji, X. Lu, M. Ehara, Y. Cai, N. Q. Wang and Q. E. Zhang, *Theor. Chim. Acta*, 1999, **102**, 170.
15. T. Kluner, H. J. Freund, V. Staemler and R. Kosloff, *Phys. Rev. Lett.*, 1998, **80**, 5208.
16. W. Brenig, A. Gross and U. Hofer, *Phys. Status Solidi A*, 1997, **159**, 75.
17. G. J. Kroes, *Prog. Surf. Sci.*, 1999, **60**, 85.
18. G. Katz and R. Kosloff, *J. Chem. Phys.*, 1995, **103**, 9475.
19. T. Klner, S. Thiel, V. Staemmler and H.-J. Freund, *Chem. Phys. Lett.*, 1998, **294**, 413.
20. J. C. Tully, *J. Chem. Phys.*, 1990, **93**, 1061.
21. Y. Zeiri, G. Katz, R. Kosloff, M. S. Topaler, D. G. Truhlar and J. C. Polanyi, *Chem. Phys. Lett.*, 1999, **300**, 523.
22. G. Ashkenazi, R. Kosloff and M. A. Ratner, *J. Am. Chem. Soc.*, 1999, **121**, 3386.
23. G. Katz, Y. Zeiri and R. Kosloff, *Surf. Sci.*, 1999, **425**, 1.
24. R. Kosloff and O. Citri, *Faraday Discuss. Chem. Soc.*, 1993, **96**, 175.
25. O. Citri, R. Baer and R. Kosloff, *Surf. Sci.*, 1996, **351**, 24.
26. L. Romm, O. Citri, R. Kosloff and M. Asscher, *J. Chem. Phys.*, 2000, **112**, 8221.
27. F. Bloch, *Phys. Rev.*, 1946, **70**, 460.
28. F. Bloch, *Phys. Rev.*, 1956, **102**, 104.
29. F. Bloch, *Phys. Rev.*, 1957, **105**, 1206.
30. A. G. Redfield, *Phys. Rev.*, 1955, **98**, 1787.
31. A. G. Redfield, *IBM J.*, 1957, **1**, 19.
32. W. T. Pollard and R. A. Friesner, *Adv. Chem. Phys.*, 1996, **93**, 77.
33. S. A. Egorov and J. L. Skinner, *J. Chem. Phys.*, 1996, **105**, 7047.
34. E. B. Davies, *Commun. Math. Phys.*, 1974, **39**, 91.
35. E. B. Davis, *Quantum Theory of Open Systems*, Academic Press, New York, 1976.
36. N. Van Kampen, *Stochastic Processes in Physics and Chemistry*, North-Holland, Amsterdam, 1992.
37. G. Lindblad, *Commun. Math. Phys.*, 1976, **48**, 119.
38. V. Gorini, A. Kossokowski and E. C. G. Sudarshan, *J. Math. Phys.*, 1976, **17**, 821.
39. R. Alicki and K. Lendi, *Quantum Dynamical Semigroups and Applications*, Springer-Verlag, Berlin, 1987.
40. S. Rice and R. Kosloff, *J. Phys. Chem.*, 1982, **86**, 2153.
41. R. Kosloff and M. A. Ratner, *J. Chem. Phys.*, 1982, **77**, 2841.
42. M. Berman, R. Kosloff and H. Tal-Ezer, *J. Phys. A*, 1992, **25**, 1283.
43. W. T. Pollard and R. A. Friesner, *J. Chem. Phys.*, 1994, **100**, 5054.
44. O. Kühn and V. May, *Chem. Phys. Lett.*, 1994, **225**, 511.
45. G. Ashkenazi, U. Banin, A. Bartana, R. Kosloff and S. Ruhman, *Adv. Chem. Phys.*, 1997, **100**, 229.
46. W. Huisinga, L. Pesce, R. Kosloff and P. Saalfrank, *J. Chem. Phys.*, 1999, **110**, 5538.
47. R. Baer, Y. Zeiri and R. Kosloff, *Phys. Rev. B*, 1997, **55**, 10952.
48. R. Baer, Y. Zeiri and R. Kosloff, *Surf. Sci. Lett.*, 1998, **411**, L783.
49. R. Kosloff, in *Dynamics of Molecules and Chemical Reactions*, ed. R. E. Wyatt and J. Z. Zhang, Marcel Dekker, New York, 1996.
50. R. Kosloff, *Annu. Rev. Phys. Chem.*, 1994, **45**, 145.
51. H. Tal Ezer and R. Kosloff, *J. Chem. Phys.*, 1984, **81**, 3967.
52. R. Kosloff and D. Kosloff, *J. Comput. Phys.*, 1986, **63**, 363.
53. K. Allinger, B. Karmeli and D. Chandler, *J. Chem. Phys.*, 1989, **84**, 1774.
54. K. Allinger and M. A. Ratner, *Phys. Rev. A*, 1989, **139**, 864.
55. D. E. Makarov and N. Makri, *Phys. Rev. A*, 1993, **48**, 3626.
56. N. Makri and D. E. Makarov, *J. Chem. Phys.*, 1995, **102**, 4600.

57 F. Grossmann, *J. Chem. Phys.*, 1995, **103**, 3696.
58 G. A. Voth, *Adv. Chem. Phys.*, 1996, **93**, 135.
59 R. P. Feynman, *Statistical Mechanics, a set of lectures*, W. A. Benjamin Inc., 1972.
60 U. Hofer, I. L. Shumay, C. Reuss, U. Thomann, W. Wallauer and T. Fauster, *Science*, 1997, **277**, 1480.
61 D. Menzel, *J. Electron. Spectrosc.*, 1995, **76**, 73.
62 P. R. Antoniewicz, *Phys. Rev. B*, 1980, **21**, 3811.
63 *Desorption Induced by Electronic Transitions DIET I*, ed. N. H. Tolk, M. M. Traun, J. C. Tully and T. E. Madey, Springer-Verlag, Berlin, 1983.
64 P. Saalfrank, R. Baer and R. Kosloff, *Chem. Phys. Lett.*, 1994, **230**, 463.
65 P. Saalfrank, R. Baer and R. Kosloff, *J. Chem. Phys.*, 1996, **105**, 2441.
66 P. Saalfrank, *Surf. Sci.*, 1997, **390**, 1.
67 H. Nienhaus, H. S. Bergh, B. Gergen, A. Majumdar, W. H. Weinberg and E. W. McFarland, *Phys. Rev. Lett.*, 1999, **82**, 445.
68 C. T. Rettner and L. Lee, *J. Chem. Phys.*, 1994, **101**, 10185.
69 C. T. Rettner and D. J. Auerbach, *J. Chem. Phys.*, 1996, **105**, 8842.
70 I. N. Yakovkin, V. I. Chernyi and A. G. Naumovets, *Surf. Sci.*, 2000, **442**, 81.
71 A. Kokalj, A. Lesar, M. Hodoscek and M. Causa, *J. Phys. Chem.*, 1999, **103**, 7222.
72 M. R. Salazar, C. Saravanan, J. D. Kress and A. Redondo, *Surf. Sci.*, 2000, **449**, 75.
73 Q. S. Xin and X.-Y. Zhu, *Surf. Sci.*, 1996, **347**, 346.
74 W. E. England and W. C. Ermler, *J. Chem. Phys.*, 1979, **70**, 1171.
75 S. S. Xantheas, S. T. Elbert and K. Ruedenberg, *Int. J. Quantum Chem.*, 1994, **49**, 409.
76 S. S. Xantheas and K. Ruedenberg, *Chem. Phys. Lett.*, 1990, **166**, 39.
77 C. H. F. Peden, D. W. Goodman, M. D. Weisel and F. M. Hoffman, *Surf. Sci.*, 1991, **253**, 44.
78 F. M. Hoffman, M. D. Weisel and C. H. F. Peden, *Surf. Sci.*, 1991, **253**, 59.
79 R. M. Finch, N. A. Hodge, G. J. Hutchins, A. Meagher, Q. A. Pankhurst, M. R. Siddiqui, F. E. Wagner and R. Whyman, *Phys. Chem. Chem. Phys.*, 1999, **1**, 485.
80 E. W. James, C. Song and J. W. Evans, *J. Chem. Phys.*, 1999, **111**, 6579.
81 A. Bötcher, H. Conard and H. Niehus, *Surf. Sci.*, 2000, **52**, 125.
82 D. Danovich and S. Shaik, *J. Am. Chem. Soc.*, 1997, **119**, 1773.
83 M. Filatov and S. Shaik, *J. Phys. Chem.*, 1998, **102**, 3835.
84 M. Bonn, S. Funk, Ch. Hess, D. N. Denzler, C. Stampfl, M. Scheffler, M. Wolf and G. Ertl, *Science*, 1999, **285**, 1042.
85 R. Romberg, N. Heckmair, S. P. Frigo, A. Ogurtsov, D. Menzel and P. Feulner, *Phys. Rev. Lett.*, 2000, **84**, 374.
86 R. Zehr, A. Solodukhin, B. C. Hynie, C. French and I. Harrison, *J. Phys. Chem. B*, 2000, **104**, 3094.

Self-trapped excitons at the quartz(0001) surface†

J. Song,[a,b] R. M. VanGinhoven,[a,b] L. R. Corrales[b] and H. Jónsson*[a]

[a] *Department of Chemistry 351700, University of Washington, Seattle WA 98195-1700, USA*
[b] *EMSL, Pacific Northwest National Laboratory, Richland WA 99352, USA*

Received 2nd August 2000
First published as an Advance Article on the web 6th November 2000

We have studied self-trapped excitons in α-quartz using density functional theory (DFT), both in the crystal and at the (0001) surface. The excitons are triplet excited states that distort the crystal locally. They have a long lifetime, of the order of a millisecond, and become thermally equilibrated. We have calculated the drop in the exciton energy as it approaches the surface from the interior of the crystal. In the subsurface layer of the –OH terminated (0001) surface, the energy has dropped by 0.7 eV. Another 0.4 eV drop occurs as the exciton enters the surface layer, where it breaks off an OH radical. The drop in energy can be understood from the greater ease of structural distortion at the surface. These calculations illustrate that excitons formed in the bulk could migrate out to the surface and form chemically active surface species. Molecules adsorbed at the surface could also serve as traps for the excitons and could, in principle, be induced to undergo structural or chemical transitions.

I. Introduction

Photoexcitation of oxides can lead to various interesting processes. Recently, much attention has been paid to photocatalysis, in particular on TiO_2 surfaces. There, electronic excitations created by photon absorption cause chemical reaction to occur on the oxide surface.[1] Also, metal oxide particles dispersed in aqueous solutions and exposed to gamma irradiation have been shown to induce radiolysis of water.[2,3] Clearly, studies of the electronic excitations formed in oxides due to photon absorption, the mechanism of energy transfer and subsequent chemical processes are of great interest.

Silica, SiO_2, is a particularly simple and stable oxide which could serve as a model system for studying such photoexcitation processes, although the photon energy required for excitation is too large for large-scale applications. Silica has become widely used as oxide support in model studies of metal/oxide catalysts.[4,5] Photoinduced defects can also play a role in various silica-based applications such as protective layers on electronic devices, optical fibers, and immobilization matrices for hazardous waste.[6,7]

Excitons can play an important role in the long-time evolution of silica. It has been found that triplet state, self-trapped excitons (STE) in quartz have a long lifetime, of the order of milliseconds, before recombining and giving off blue luminescence.[8–14] It has been speculated that STEs can

† Electronic Supplementary Information available. Colour versions of Fig. 1, 5 and 6 are given. See http://www.rsc.org/suppdata/fd/b0/b006289h/

DOI: 10.1039/b006289h

lead to Si–O bond breaking and degradation of the silica network. There are also indications that defects can be made mobile and/or anneal out in the presence of excitons. Previous theoretical work has shown that an STE binds to an oxygen vacancy in quartz with a 3 eV binding energy and reduces the diffusion activation barrier of the vacancy from 4 eV to less than 2 eV.[15]

In this article, an overview of results obtained from DFT calculations of STEs in α-quartz is given and new results on STEs at the quartz (0001) surface are presented. Our calculations indicate that STEs formed in the quartz crystal would tend to thermally diffuse out towards the surface, where surface –OH groups can get broken off to form OH radicals. A competing process would be the thermal activation into a different STE, which is near the singlet–triplet crossing, and subsequent non-radiative decay. We first discuss the methodology used in the calculations, and then the results obtained on STEs in the quartz crystal and, finally, at the quartz surface.

II. DFT calculations of STEs

A large system is needed to represent STEs in quartz because the self-trapping involves large displacements of atoms and substantial elastic strain. Calculations of such systems can only be handled at an approximate level. High level calculations can only be applied to small clusters with a few atoms. For the highly reliable CCSD(T) method, a single SiO_2 unit is already a large calculation. A theoretical study of STEs in quartz must, therefore, find some practical balance between the approximations in the methodology and errors due to small system size.

While DFT applies only to the ground electronic state,[16] it is possible to study the STEs in quartz because they are triplet states. With a constraint on the spin (two more electrons with spin up than spin down) in a spin-polarized DFT calculation, the lowest triplet state is the lowest energy state available. The basic theorem of DFT, the Hohenberg–Kohn theorem, holds within the triplet subspace. The calculated atomic forces can be used to relax the atomic structure to local minima on the triplet energy surface, each minimum corresponding to a triplet state exciton.[17,18]

The development of functionals for DFT has, however, mainly focused on singlet states:[19] There is little experience on how well they can be applied to higher spin states. We have carried out tests of various functionals by calculating singlet–triplet (S–T) splittings in small clusters where well established wavefunction-based methods can be carried out for comparison.[17,20] These tests indicate that the PW91 functional underestimates the S–T splitting in cases where the triplet state is delocalized. Then the semi-local description of exchange becomes problematic in the triplet state. This underestimate is most severe for the perfect crystal, where the calculated S–T splitting is 6.1 eV but the experimental estimate is 8.3 eV.[10] This is consistent with the typical underestimate of band gaps in DFT.

The B3LYP functional is more accurate as it involves exact exchange rather than just the semi-local description employed in PW91. The S–T splittings are predicted to be larger, and close to, although somewhat smaller than, those calculated by the CCSD(T) method.[17,20] The inclusion of exact exchange, however, makes it extremely costly to implement B3LYP in calculations where periodic boundary conditions are applied (and the evaluation of atomic forces at that level of theory has not yet been developed).[21] The B3LYP functional can only be applied to finite clusters at the present time. We have, therefore, developed a procedure where DFT/B3LYP cluster calculations are used to improve on the triplet state energetics of the DFT/PW91 configurations that are subject to periodic boundary conditions to eliminate surface effects. This is illustrated in Fig. 1. The atomic configuration is obtained by relaxation of the triplet state using DFT/PW91 calculations. Clusters of varying size, up to Si_8O_{25}, are then snipped out of the relaxed configuration, centered on the STE and the edge atoms (chosen to be O-atoms) are then capped with H-atoms to terminate dangling bonds. The direction of the O–H bonds is the same as the direction of the broken O–Si bonds and are chosen to have a bond length of 0.8 Å to get a similar charge on the edge O-atoms as interior O-atoms. The S–T splitting is calculated at both the PW91 and B3LYP level and the difference gives the B3LYP correction which is added to the S–T splitting calculated using PW91 and periodic boundary conditions. The correction is only important for the highly delocalized triplet states (1.2 eV for quartz and 0.6 eV for the most delocalized STE). This procedure gives an estimate of the triplet-state energy surface which appears to be in good agreement with experimental measurements.

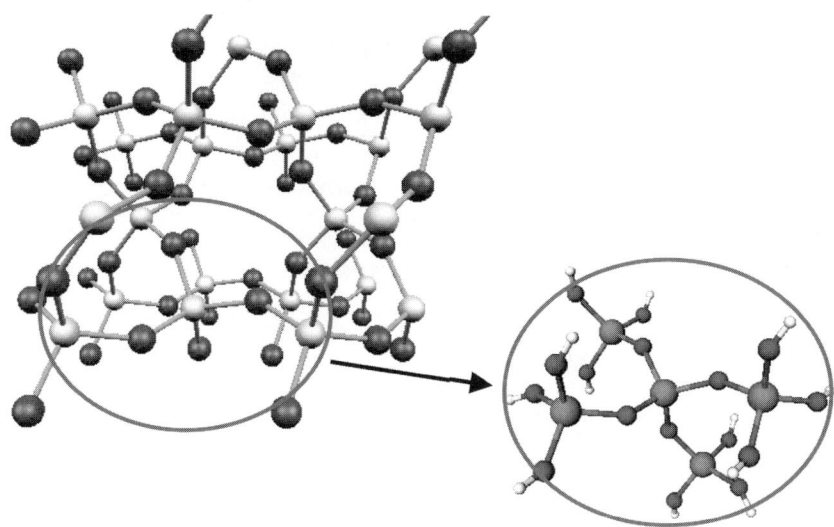

Fig. 1 The 72 atom quartz configuration used in the DFT/PW91 calculations. Periodic boundary conditions are applied. Calculations have also been carried out on clusters snipped out of the 72 atom configuration. The outermost atoms in the cluster are O atoms, which get capped with H atoms to reduce surface effects. Various wavefunction-based calculations as well as DFT calculations have been carried out for the clusters to assess the accuracy of the quasi-local PW91 functional in describing the triplet state and to estimate corrections to the DFT/PW91 results.

The plane-wave-based spin-polarized DFT calculations were carried out with the VASP code,[22] using ultrasoft pseudopotentials.[23] The energy cutoff was 29 Ry for the wavefunction and 68 Ry for the electron density. The PW91[24] exchange-correlation functional was used. The importance of using a gradient-dependent density functional rather than the local density approximation when studying silica has been demonstrated by Hamann.[25] The bulk quartz calculations were done on a 72 atom cell (eight unit cells) including just the Γ point in the k-point sampling. Additional k-points were found to have insignificant effect on both structure and singlet–triplet splittings. Two kinds of Si–O bonds are found in α-quartz. The DFT calculations predict bond lengths of 1.619 and 1.615 Å, as compared with experimental estimates of 1.612 and 1.607 Å.[26]

Low-energy electron diffraction (LEED) measurements of the (0001) surface of quartz have shown the unreconstructed (1 × 1) configuration to be stable up to 600 °C where a (3 × 1) or (1 × 3) reconstruction takes place.[27] The electronic structure of the surface has been studied by energy loss spectroscopy reflection (REELS)[28] where it was found that the surface band gap is 8.8 eV, which is very similar to the bulk band gap.[29] The surface calculations were done with a slab where the upper and lower surface were terminated with –OH groups. The surface in contact with water would be expected to be hydroxylated. A total of 84 atoms were used in the slab calculations ($Si_{20}O_{48}H_{16}$). The bottom two layers of Si and O atoms in the slab were held fixed in the perfect quartz configuration, but other atoms in the slab were allowed to move and relax to a minimum energy configuration. The capping O–H bonds in the bottom layer were pointed in the direction of the broken O–Si points, and the O–H distance was chosen to be 0.8 Å to bring the charge on the O atoms to a similar value as that of bulk O atoms.

The B3LYP calculations were carried out using Gaussian basis sets, both the 6-31G* and 6-31G* + s′ for the Si and O atoms while H atoms were represented with minimal basis. The cluster corrections were converged both with respect to cluster size and basis set.

III. Excitons in the quartz crystal

Fig. 2 shows the cluster carved out of the perfect quartz crystal configuration. An isosurface for the excess spin density in the triplet state is also shown. The distribution of the excess spin density is quite even over the whole cluster and is largest at the O-atoms. The exciton in perfect quartz is

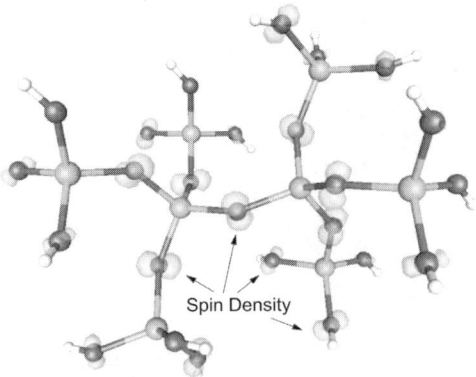

Fig. 2 A $Si_8O_{25}H_{18}$ cluster representing quartz. The 0.02 electron $Å^{-3}$ isosurface of the excess spin density of the triplet state calculated using DFT/B3LYP is shown. The excess spin density mainly resides on the O-atoms and is evenly distributed over the cluster, indicating a highly delocalized exciton.

highly delocalized. By breaking the symmetry in various ways and relaxing the system in the triplet state, we have been able to identify three different local minima corresponding to local distortions of the lattice, *i.e.* STEs.

Previously, Fisher, Hayes and Stoneham presented a model for STE in quartz obtained from unrestricted Hartree–Fock (UHF) calculations on small silica clusters (Si_5O_4 and Si_2O_7).[30] We have found very similar STE configuration in our DFT calculations, which have been carried out using significantly larger clusters, $Si_8O_{25}H_{16}$. The self-trapping mainly involves the displacement of an oxygen atom by 0.96 Å, resulting in a formally broken Si–O bond (2.5 Å). We will refer to this structure as STE-Oc. The S–T splitting calculated by the B3LYP corrected DFT is 2.8 eV, in excellent agreement with the experimentally measured luminescence, 2.6–2.8 eV[10–12] (while UHF calculations give S–T splitting of 1.6 eV). The S–T splitting predicted by the DFT calculations seems, therefore, to be quite accurate.

When one of the Si–O bonds in the 72 atom bulk crystal configuration is rotated in such a way as to formally break another Si–O bond and the system is then relaxed in the triplet state using DFT/PW91, a different STE also involving displacements of mainly O-atoms, is obtained. The structure, captured in a cluster carved out of the 72 atom configuration, is shown in Fig. 3, along with an isosurface of the excess spin density. We will refer to this as STE-Ob. It turns out to be

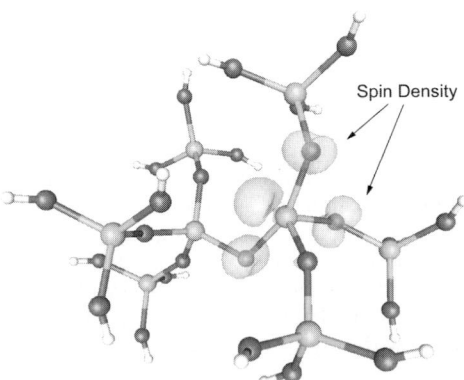

Fig. 3 One of the O-displaced self-trapped excitons, STE-Ob, in a $Si_8O_{25}H_{18}$ cluster snipped out of a 72 atom bulk configuration. The excess spin density is shown (analogous to Fig. 2) and illustrates a distribution of the hole over the three oxygen atoms while the excited electron is mainly to be found at the adjacent Si atom. Two O-atoms get displaced appreciably, by 0.6 and 0.5 Å, and the Si-atom gets displaced by 0.2 Å. The lattice relaxations extend over a large region and no bond is formally broken.

slightly lower in energy in the DFT/PW91 calculations, but slightly higher in B3LYP corrected calculations. The distortion involves roughly equally large displacements of *two* oxygen atoms (by 0.6 and 0.5 Å) bonded to the same Si atom, as well as displacements of the adjacent Si atoms (by up to 0.25 Å), in such a way that three Si–O bonds are stretched to 1.76 Å but not broken.[17,18] The singlet and triplet state energy for quartz and the STEs is shown in Fig. 4. The luminescence of this STE predicted by the B3LYP corrected calculation is 4.3 eV. The triplet state surface is very flat in this region. In fact, a minimum energy path between these two STEs has a barrier of only 0.2 eV even though large displacements are involved (one O-atom gets displaced by 0.7 Å and another by 0.6 Å. The flatness of the triplet state surface suggests that several local minima, *i.e.* STEs, may exist and that the system is quite floppy in the excited state, consistent with the large width (*ca.* 1 eV) and complex shape of the measured luminescence peak. Luminescence has also been observed at 4.0 eV in low temperature experiments (at 80 K), but was tentatively ascribed to defects or impurities.[10] Our results indicate that there may be an intrinsic STE contribution to the emission around 4.0 eV. Both the emissions centered at 2.8 and 4.0 eV seem to have more than one origin, as can be seen from the dependence of the line shape on the excitation energy.[12] The barrier to go from STE-Ob to STE-Oc is predicted to be on the order of 0.2 eV (see Fig. 4). This estimate was obtained by finding the minimum energy path between STE-Ob and STE-Oc (using the Nudged Elastic Band method,[31] using DFT/PW91, and then applying the B3LYP cluster correction. The low barrier between the STEs and the energetic preference for STE-Oc suggest that room temperature experiments would only be able to detect the 2.8 eV luminescence.

The third STE we have identified mainly involves a displacement of a Si atom. We will refer to it as STE-Si. This STE is obtained when the quartz crystal structure is distorted by displacing one oxygen atom by 0.2 Å in the direction of a Si–O bond and the system is then relaxed in the triplet state using DFT/PW91. In the end, a Si atom has moved by 0.9 Å through the plane of three of its neighboring O-atoms. STE-Si turns out to be close to a crossing of the singlet and triplet surfaces. This, we have verified by CAS-SCF calculations. The system is, therefore, expected to undergo non-radiative energy transfer to the singlet ground state if it gets trapped in STE-Si. This brings

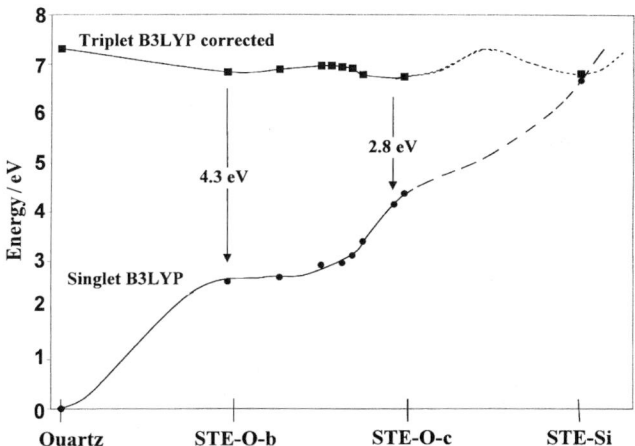

Fig. 4 A slice of the triplet and singlet energy surfaces including the perfect quartz crystal configuration and the three STEs found in bulk. The configurations are ordered with delocalization of the excess spin density increasing to the left. The configurations were obtained by relaxation of a 72 atom system subject to periodic boundary conditions using DFT/PW91. A minimum energy path was calculated between the STE-Ob and STE-Oc. Clusters were then snipped out and used to get an improved estimate of the energetics using DFT/B3LYP calculations, which include exact exchange. The correction is only significant for the more delocalized states: quartz and STE-Ob. The triplet state surface shows two local minima, the STE-Oc and STE-Ob configurations. The lower energy one has S–T splitting of 2.8 eV in close agreement with experimentally measured luminescence. An experimentally observed reduction in the intensity of the 2.8 eV luminescence with increasing temperature can be explained by the thermal activation from the STE-Oc state to the STE-Si state which is near a S–T crossing and, therefore, leads to non-radiative decay.

the system back to the perfect crystal state, in agreement with experiments which show that decay of excitons does not lead to structural changes in quartz (unlike amorphous SiO_2 where 1/100 STEs lead to Si–O bond breaking[11]). We have studied several possible paths that can take the system from STE-Oc to STE-Si, but in all cases tested to date the system preferred to make the transition through the perfect quartz configuration (which at the PW91 level is the lowest energy configuration, making it hard to calculate a minimum energy path for the transition). Our best estimate of the energy barrier for a thermal transition from STE-Oc to STE-Si is 0.5 eV (see Fig. 4).

Experiments on the temperature dependence of the luminescence intensity have indicated that the system can get thermally activated from the luminescent STE state to a new state from where the system is quenched non-radiatively.[10,14] The experimentally estimated activation barrier is 0.4 eV. The STE-Si state we have found could very likely provide this non-radiative mechanism. The predicted activation energy for the process is even in close agreement with the experimental value. Our calculations, which are summarized in Fig. 4, give a microscopic picture of the scenario deduced qualitatively from experimental measurements.[14]

IV. Excitons at the quartz (1000) surface

In order to study the behaviour of the STEs near the surface of quartz, we have carried out DFT/PW91 calculations on a quartz slab described above and shown in Fig. 5. Both sides of the slab are terminated by O-atom layers and capped with H-atoms to saturate dangling bonds. Because of the small thickness of the slab, which is limited by the large computational effort in the DFT calculations, it is only possible to represent the STE in the surface and subsurface layers in this slab.

A displacement of an O-atom in the second layer gives a STE which is quite similar to the STE-Ob in the crystal. The difference is that one of the three stretched bonds, the one pointing towards the surface, becomes longer (1.78 Å) while the other two become shorter (1.72 Å) as compared with the STE-Ob in the interior of the crystal. The Si-atom is also displaced more than

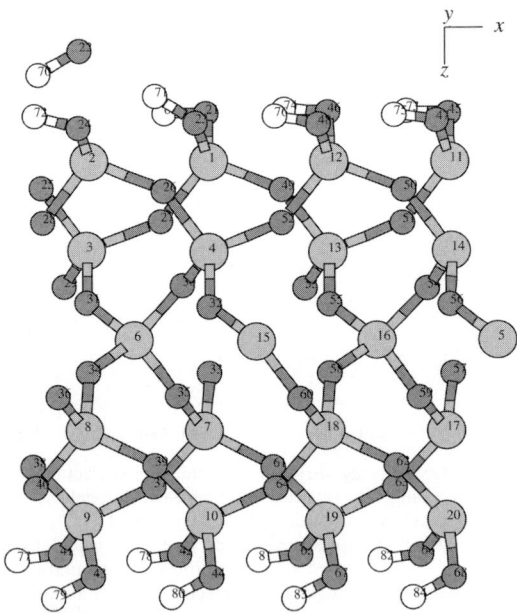

Fig. 5 A side view of the 84 atom cell ($Si_{20}O_{48}H_{16}$) used to represent a quartz slab with an –OH terminated (0001) surface. Atoms in the bottom two layers are fixed in the quartz configuration (the capping O–H bonds pointing in the direction of the broken O–Si points). Other atoms are allowed to relax in the DFT/PW91 calculation. The STE is in the surface layer and has broken an OH radical off the surface layer (top left).

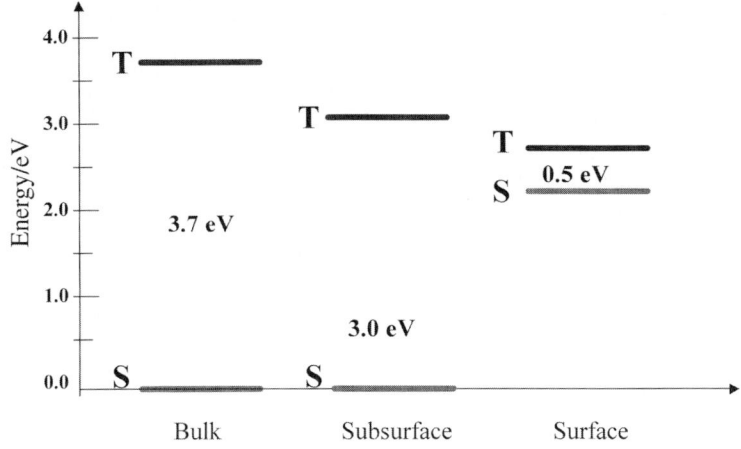

Fig. 6 The energy of the STE in bulk quartz, near and at the (0001) surface. In the subsurface layer, the STE has dropped in energy by 0.7 eV as compared with the bulk. In the surface layer, where the STE results in an OH radical being broken off, the energy drops further by 0.4 eV.

in the crystal, by 0.5 Å. The calculated S–T splitting has dropped by 0.7 eV as compared with bulk. This represents a drop in the triplet state energy and/or a rise in the singlet energy. It is difficult to compare the singlet state energy between the crystal and slab configurations, because the two are so different. We will assume here that the singlet state energy of the STE configuration is the same in the second layer as in bulk quartz. The triplet state energy is, then, lower by 0.7 eV in the subsurface layer (see Fig. 6).

In the surface layer, a similar displacement of an O-atom and subsequent structural relaxation leads to a Si–O bond rupture and spontaneous formation of an OH radical. The triplet state energy is 0.4 eV lower than in the subsurface layer and now the singlet state energy has increased significantly so the S–T splitting is only 0.5 eV. This likely represents a non-radiative channel. It is reasonable to expect this drop in the triplet state energy since the energetic cost of distortions of the lattice becomes smaller near the surface where atoms are able to move more freely. This suggests that STEs formed in bulk quartz will tend to diffuse out to the surface. Since the lifetime of STEs in quartz is on the order of 1 ms, and the diffusion barrier is estimated from our calculations to be on the order of 0.5 eV, the STEs could diffuse over a large distance at temperatures around room temperature. The transition into the STE-Si state and subsequent non-radiative decay would be a competing process, however, and in the end it may be that only a small region near the surface feeds STEs into the surface.

We have also carried out calculations for the (0001) surface with Si=O double bond reconstructed states. The Si=O terminated surface is produced from the hydroxylated surface by removing the hydrogen atoms on the top surface. To find an STE state a perturbation to the lattice structure is introduced, as in the crystal case. On the reconstructed surface having Si=O double bonds, two distinct silicon-displaced STEs are found, but the oxygen-displaced STE, STE-Ob, does not appear to exist on this particular surface. The two silicon-displaced STEs have emission energies of 1.0 and 2.5 eV. The former is more stable, is likely a non-radiative channel and the Si=O bond has stretched from 1.5 to 1.7 Å.

At this point in time, B3LYP corrections have not been applied to the surface STE configurations, but it seems clear from the comparison of the STE-Ob calculations for bulk, subsurface and surface layers that the S–T splitting decreases as the STE approaches the surface.

V. Discussion

Our calculations indicate that as the exciton approaches the surface, the energy is lowered. The increased stability of the STEs at the surface can be ascribed to the lower energetic cost of the structural distortions favored by the triplet state. The formation of an OH radical at the hydroxylated surface provides one possible mechanism for chemical processes induced by STEs. Experimental studies have, for example, demonstrated that radiolysis events can occur when metal oxide particles dispersed in aqueous solutions are exposed to gamma irradiation.[2,3] Also, one of the proposed mechanisms for photocatalysis at TiO_2 surfaces involves the production of surface hydroxyl radicals but from self-trapped holes rather than excitons.[1]

While most STEs are probably formed in the interior of the crystal, far from the surface, it is possible that thermal diffusion can bring the STEs to the surface. It is not clear what the thermal diffusivity of STEs in quartz is. If the perfect crystal is the optimal transition state for the STE hopping from one site to another, our calculations suggest an activation energy of 0.5 eV. This is likely an underestimate since the energy of the perfect crystal in the triplet state is underestimated by the DFT calculations, even at the B3LYP level. It is also possible that a lower energy path exists for the STE to hop from one site to another, bypassing the perfect crystal configuration. The perfect crystal transition state is highly delocalized and a single thermally activated diffusion hop can in principle bring the exciton a long distance away from the initial STE location.

The thermally activated transition to the STE-Si state, which is right at a crossing of the singlet and triplet surface and leads to non-radiative decay, is in competition with the diffusion of the STE. Assuming again the perfect crystal is the transition state, our calculations predict a barrier of 0.5 eV, in very good agreement with experimental measurements (0.4 eV). The relative activation energy for diffusion and transition into the non-radiative state affects, of course, strongly to what extent STEs formed in bulk can fuel surface processes. For some oxides the thermal diffusion mechanism may be active but not for others.

VI. Acknowledgements

This work was supported by the Environmental Management Science Program, Office of Environmental Management, DOE (JS), and the Division of Chemical Sciences, DOE Office of Basic Energy Sciences (LRC), and by DOE-BES grant DE-FG03-99ER45792 (RVG, HJ). The calculations were performed on a parallel IBM SP computer at the William R. Wiley Environmental Molecular Sciences Laboratory, a national scientific user facility sponsored by DOE Office of Biological and Environmental Research, located at Pacific Northwest National Laboratory.

References

1. M. R. Hoffmann, S. T. Martin, W. Choi and D. W. Bahnemann, *Chem. Rev.*, 1995, **95**, 89.
2. N. G. Petrik, A. B. Alexandrov, T. M. Orlando and A. I. Vall, *Trans. Am. Nucl. Soc.*, 1999, **81**, 101.
3. A. B. Alexandrov, A. Y. Bychkov, A. I. Vall, N. G. Petrik and V. M. Sedov, *Russ. J. Phys. Chem.*, 1991, **65**, 1604.
4. I. Zuburtikudis and H. Saltsburg, *Science*, 1992, **258**, 1337.
5. (a) X. Xu and D. W. Goodman, *J. Phys. Chem.*, 1993, **97**, 683; (b) X. Xu and D. W. Goodman, *J. Phys. Chem.*, 1993, **97**, 7711.
6. S. Rico, in *The Physics and Chemistry of SiO_2 and the Si–SiO_2 Interface*, ed. C. R. Helms and B. E. Deal, Plenum, New York, 1988, p. 75.
7. W. J. Weber *et al.*, *J. Mater. Res.*, 1998, **13**, 1434.
8. T. Tanaka, T. Eshita, K. Tanimura and N. Itoh, *Cryst. Lattice Defects Amorph. Mater.*, 1985, **11**, 221.
9. J. H. Stathis and M. Kastner, *Phys. Rev. B*, 1987, **35**, 2972.
10. C. Itoh, K. Tanimura, N. Itoh and M. Itoh, *Phys. Rev. B*, 1989, **39**, 11183.
11. N. Itoh, T. Shimizu-Iwayama and T. Fujita, *J. Non-Cryst. Solids*, 1994, **179**, 194.
12. K. S. Song and R. T. Williams, *Self-Trapped Excitons*, Springer, Berlin, 2nd edn., 1996, p. 281.
13. W. Hayes, M. J. Kane, O. Salminen, R. L. Wood and S. P. Doherty, *J. Phys. C*, 1984, **17**, 2943.
14. M. Georgiev and N. Itoh, *J. Phys.: Condens. Matter*, 1990, **2**, 10021.
15. J. Song, L. R. Corrales, G. Kresse and H. Jonsson, submitted to *Phys. Rev. Lett.*
16. (a) P. Hohenberg and W. Kohn, *Phys. Rev. B*, 1964, **136**, 864; (b) W. Kohn and L. J. Sham, *Phys. Rev. A*, 1965, **140**, 1133.
17. J. Song, H. Jónsson and L. R. Corrales, *Nucl. Instrum. Methods Phys. Res., Sect. B*, 2000, **166**, 167453.
18. L. R. Corrales, J. Song, R. M. VanGinhoven and H. Jónsson, in *NATO Advanced Studies Institute Proceedings on "Defects in Silica and Related Dielectrics"*, ed. G. Pacchioni, in press.
19. W. Kohn, A. D. Becke and R. G. Parr, *J. Phys. Chem.*, 1996, **100**, 12974.
20. R. Van Ginhoven, J. Song, M. Dupuis, K. A. Peterson, L. R. Corrales and H. Jónsson, to be published.

21 M. Stadele, M. Moukara, J. A. Majewski and P. Vogl, *Phys. Rev. B*, 1999, **59**, 10031.
22 (*a*) G. Kresse and J. Hafner, *Phys. Rev. B*, 1993, **47**, 558; (*b*) G. Kresse and J. Hafner, *Phys. Rev. B*, 1994, **49**, 14251; (*c*) G. Kresse and J. Furthmüller, *Comput. Mater. Sci.*, 1996, **6**, 16; (*d*) G. Kresse and J. Furthmüller, *Phys. Rev. B*, 1996, **55**, 11169.
23 D. Vanderbilt, *Phys. Rev. B*, 1990, **41**, 7892.
24 J. P. Perdew, in *Electronic Structure of Solids*, ed. P. Ziesche and H. Eschrig, 1991.
25 D. R. Hamann, *Phys. Rev. Lett.*, 1996, **76**, 660.
26 W. Hayes and T. J. L. Jenkin, *J. Phys. C: Solid State Phys.*, 1986, **19**, 6211.
27 F. Bart and M. Gautier, *Surf. Sci.*, 1994, **311**, L671.
28 F. Bart, M. Gautier, F. Jollet and J. P. Duraud, *Surf. Sci.*, 1994, **306**, 342.
29 T. H. DiStefano and D. E. Eastman, *Solid State Commun.*, 1971, **9**, 2259.
30 A. J. Fisher, W. Hayes and A. M. Stoneham, *J. Phys.: Condens. Matter*, 1990, **2**, 6707.
31 G. Mills, H. Jónsson and G. K. Schenter, *Surf. Sci.*, 1995, **324**, 305; G. Mills, H. Jónsson and K. W. Jacobsen, in *Classical and Quantum Dynamics in Condensed Phase Simulations*, ed. B. J. Berne, G. Ciccotti and D. F. Coker, World Scientific, Singapore, 1998, ch. 16.

Abstractive chemisorption of O_2 on Al(111)

Marcello Binetti,[a] Olaf Weiße,[a] Eckart Hasselbrink,*[a] Andrew J. Komrowski[b] and Andrew C. Kummel[b]

[a] *Fachbereich Chemie, Universität Essen, Germany*
[b] *Department of Chemistry and Biochemistry, University of California, San Diego, USA*

Received 18th May 2000
First published as an Advance Article on the web 2nd October 2000

We present experimental evidence that abstraction is a common mechanism ($\sim 50\%$) in the dissociative chemisorption of oxygen on Al(111) at a translational energy of 0.5 eV. As a result of this mechanism, individual isolated O-atoms are observed in scanning tunneling microscopy (STM). At this translational energy ordinary dissociative chemisorption processes also occur, resulting in pairs of adatoms. The ejected O-atoms originating from the abstraction reaction are detected in the gas phase using laser spectrometry. Together, these observations provide compelling evidence for the abstraction mechanism.

Introduction

Our understanding of the dynamics of dissociative chemisorption events has made rapid progress since the 1993 Faraday Discussion 96 on "*Dynamics at the Gas/Solid Interface*". The H_2/Cu-system has served as a testing ground to probe our progressing capabilities to accurately measure and theoretically predict these processes. An intriguing detailed understanding of the complex dynamics has been obtained in recent years. Quite in contrast, the O_2/Al-system continues to present the most striking puzzle. Oxygen atoms form a stronger bond to aluminium than to any other metal surface. However, several independent measurements have indicated that the sticking coefficient for thermal O_2 on room-temperature Al(111) is small ($<10^{-2}$). A recent molecular beam study by Österlund *et al.*[1] showed a strong translational energy dependence of the sticking probability, with the latter increasing to 0.8 at 0.5 eV. A molecularly chemisorbed state has not been observed for the bare surface. This is in agreement with the observation by Österlund *et al.* that the initial sticking probability is independent of surface temperature, indicating the absence of a precursor-mediated process. Apparently, dissociative adsorption of O_2 is a direct and activated process, although there is no obvious simple quantum chemical explanation for this oddity. In contrast, Zhukov *et al.* invoked a molecular precursor mechanism when interpreting their X-ray photoelectron spectroscopy (XPS) and high-resolution electron energy loss spectroscopy (HREELS) results on Al(111) oxidation kinetics.[2]

Brune *et al.*[3,4] studied the initial stages of aluminium oxidation using a room-temperature STM. At very low coverages, they found isolated O adatoms statistically separated by a distance >80 Å without any appearance of a pairwise origin. At higher coverages ($\theta = 0.03$ ML) the oxygen atoms form 1×1 islands. In a subsequent step, the oxide is formed. This study ruled out that dissociative chemisorption and oxidation are concerted processes, or that dissociation only occurs at step edges, which would have been obvious explanations for the low sticking probability. Brune *et al.* interpreted the observation of isolated atoms as a consequence of *hot* atom motion, with the energy originating from the exothermic chemisorption process.

DOI: 10.1039/b004006l

This interpretation has been the subject of an intense debate. Engdahl and Wahnström modeled the rate at which a hot atom would lose its energy due to electronic friction.[5] Wahnström et al.[6] have calculated the mobility of a hot atom on a corrugated surface such as the Al surface and concluded that the direction of motion will be rapidly randomized such that a final separation exceeding 20 Å is unlikely to result. As an alternative interpretation, the existence of an abstraction channel, favored by non-adiabatic effects, has been frequently discussed.[1,7] Since the binding energy of a single O-atom to the Al(111) surface (5.9 eV)[6] is larger than the binding energy of O_2 (5.12 eV),[8] such a process is thermodynamically allowed. Moreover, the analysis of the STM images has recently been debated.[9]

The O_2/Al(111) system has recently been addressed by a number of first-principles calculations using the density functional formalism.[10-13] However, these calculations predict a barrier free route to chemisorption, which is in stark contrast to the experimental findings. It had been speculated that the low sticking coefficient is due to an insufficient probability of the triplet to singlet conversion process, which is necessary to allow chemisorption, due to the light mass of the Al atom. Unfortunately, none of the new sophisticated spin-polarized calculations have remedied this prevailing puzzle.

In this paper, we present experimental evidence that at intermediate translational energies a predominant fraction of the chemisorption events follow an abstraction mechanism. Evidence was obtained by (i) exposing an Al(111) sample to a 0.5 eV molecular beam of oxygen molecules and collecting STM images of chemisorbed oxygen atoms, and (ii) detecting oxygen atoms in the scattered flux created by the abstraction reaction.

Experimental

The results reported in this paper were obtained using two UHV systems. The molecular beam studies using resonance-enhanced multiphoton ionization (REMPI) for O-atom detection were carried out using a standard apparatus which has been described in detail elsewhere.[14] For the purpose of this study the molecular beam produced by a pulsed valve (General Valve, Type-9) has been chopped using a homebuilt Fizeau-type chopper rotating at about 200 Hz. A fast aperture shutter (Uniblitz, LS6) is used to pick individual gas pulses at a repetition rate of 10 Hz. The resulting gas pulses have a length of about 100 µs and a peak flux of 10^{15} molecules $cm^{-2} s^{-1}$. The molecular beam has a diameter of 9 mm when it hits the sample surface at a 45° angle of incidence. In order to increase the translational energy, a short ceramic tube has been attached to the pulsed valve, which is heated to 500 K. A beam energy of 0.45 eV results if the nozzle is backed by a 10% O_2 in He mixture.

Assuming a 0.5 sticking coefficient, under these conditions it takes 6000 gas pulses until the sample is covered with 0.1 ML of oxygen. Experiments have been carried out ensuring that this coverage is never exceeded.

A Nd:YAG-laser (Spectra Physics GCR-190) pumped dye laser (Sirah Precision-Scan-G) is used to produce 452 nm light (10 ns pulse length, 0.1 cm^{-1} band width) which is frequency doubled in a β-BBO crystal yielding pulses of 2.5 mJ energy. The light is steered into the UHV-system and focused by a 25 cm lens such that the focus is located 10 mm away from the sample when measured along the direction of the surface normal. The focus is expected to have a diameter of 6 µm and a Rayleigh length of 0.5 mm. A significant background of O_2 is expected even though the focus is located outside the volume of the incoming and specularly reflected molecular beam.

O-atoms are ionized at the focus using a (2 + 1)-photon REMPI process, in which the resonant step involves the (2p) $^3P_{2,1,0}$ ground state and the (3p) $^3P_{2,1,0}$ state as intermediate.[15,16] Of the three available lines, the strongest transition (0,1,2 ← 2) at 225.66 nm was used. The resulting ions are collected by an extraction field, mass separated in a field-free time-of-flight tube and finally detected by a chevron of channel plates (Galileo 1397-1900). Due to the low signal level, individual ions events are gated with an appropriately delayed window after the laser pulse and finally counted.

The efficiency of this detection process was studied using photodissociation of NO_2 as the source for O-atoms. The resulting NO and O fragments can be detected in the same wavelength region (Fig. 1). The ion signal on the O-atom transitions has about a tenth of the strength of the

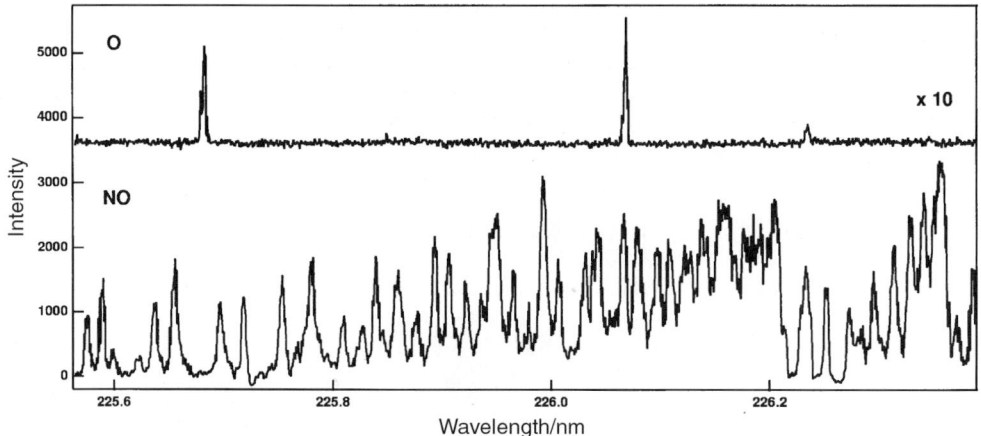

Fig. 1 Ion signal recorded for photodissociation of NO_2 using laser light of the wavelength indicated on the x-axis. Photodissociation is caused by a non-resonant 2-photon transition. The creation of ions follows in a subsequent step through a (2 + 1)- or (1 + 1)-process for O-atoms and NO-molecules, respectively. O-atoms and NO-molecules are separated by their respective flight time through a time-of-flight mass filter in the REMPI detector. The O-atom signal is expanded by a factor 10.

molecular NO signals because the latter only involves a (1 + 1)-REMPI process. Since (i) the NO and O-photofragments are produced with equal probability, (ii) the first however is distributed over about 10^2 quantum states and (iii) the NO signal shows some saturation, it can be inferred that the O-atom detection efficiency is about 10^{-2}–10^{-3}.

The Al(111) sample ($\pm 0.5°$, diameter 10 mm, purity = 99.999%) was cleaned following standard procedures using prolonged Ar^+-ion sputtering at 1 keV energy. Immediately following sputtering, the sample cleanliness was checked by Auger electron spectroscopy (AES). Sputtering cycles were repeated until the LMM Al peak (68 eV) reached its maximum. The sample was annealed to 700 K and the order of the surface was checked by low-energy electron diffraction (LEED).

The results of the O_2/Al(111) STM experiments were obtained in a separate UHV system; detailed descriptions of this system have been published.[17] The experiments were performed with an aluminium single crystal which was made by Monocrystals Company (orientation $\langle 111 \rangle \pm 1°$, dimensions: 6.4 mm diameter × 2 mm thick, purity 99.999%) and was cleaned by sputtering with Ar^+ (overnight at 1 kV, 10 µA filament; final 3.5 h at 0.5 kV and 10 µA). Immediately following sputtering, the sample was annealed once to 800 ± 20 K for 1 min. The surface was checked for cleanliness and order by Auger spectroscopy, LEED, and STM. For a new Al(111) crystal, several days or weeks of sputter and anneal cycles were required in order to obtain a surface with minimal O and C atom contaminants. The room temperature surface was dosed with a 0.5 eV O_2 molecular beam produced in a differentially pumped source chamber. The O_2 dose was controlled in order to produce surfaces with approximately 3% reacted sites (including pre-existing contaminants, see Results section). Multiple spots on the crystal's surface were dosed to ensure most of the surface contained O–Al reacted sites. The sample was then transferred from the manipulator to the STM stage. The STM data presented here were taken at room temperature in constant current mode. We were able to obtain maximum resolution for scan areas on clean, flat terraces.

Results

A typical image of the dosed surface is presented in Fig. 2. The image has not been corrected for drift, however, a planar background subtraction was performed. The surface substrate atoms are not visible, but several adsorbate features are resolved. Carbon atoms can be identified as large, faint, depressed perturbations in the surface density of states, possibly due to their occupation of a sub-surface site. Oxygen atoms are clearly visible as protrusions located at isolated sites or within

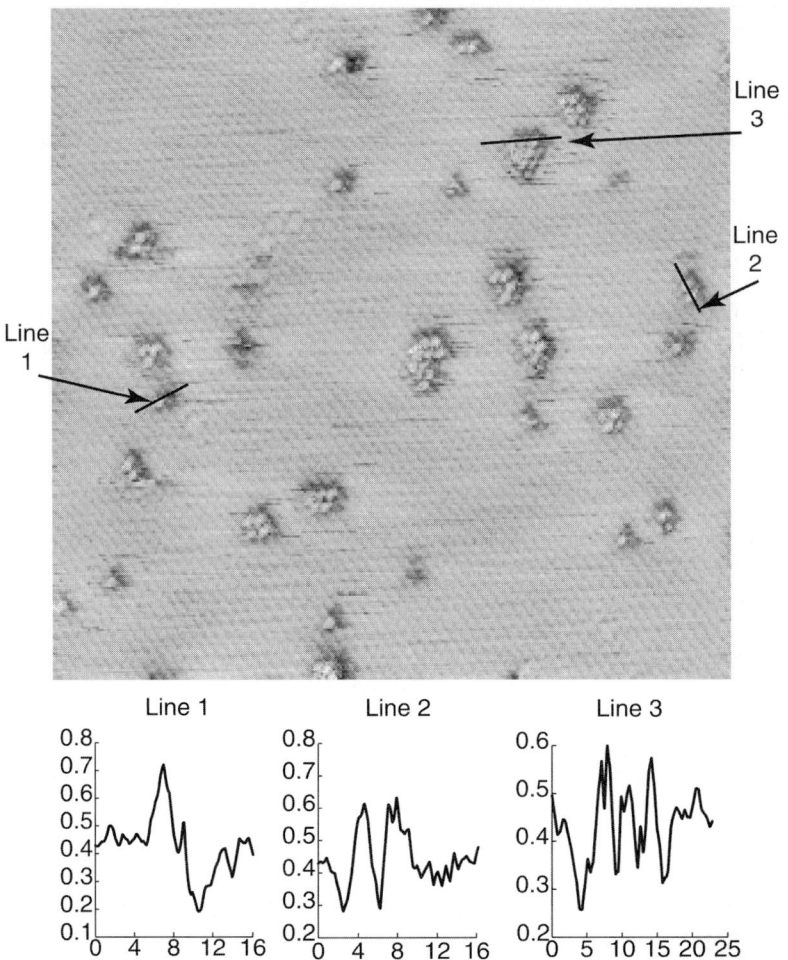

Fig. 2 Al(111) dosed with 0.5 eV O_2, constant current mode image. Tunneling conditions: $I_{tunnel} = 90$ nA, $V_{sample} = -0.2$ V. Image size 200 Å × 200 Å. Scan direction: fast, left to right; slow, bottom to top. Three line traces are drawn through O–Al reacted sites and plotted: line 1, a single oxygen adatom site, line 2, a paired site, line 3, a large oxygen island.

oxygen adsorbate islands. The line traces drawn and plotted in Fig. 2 indicate the number of O-atoms in the different O–Al reacted sites. Large oxygen islands (≥4 adatoms) clearly form 1 × 1 ordered features. The presence of carbon atoms and large oxygen islands results from our inability to obtain a clean surface with less than 1% adsorbates. Any oxygen which remains on the surface after sputtering will diffuse and form islands of O–Al during annealing. We have observed these large oxygen islands on the undosed surface. The isolated oxygen adatoms and smaller oxygen adatom islands (≤4 adatoms) are only observed on dosed surfaces. We have analyzed a set of STM images for this reaction by cataloging each oxygen feature while also recording each individually resolved O adatom. By this method, an absolute oxygen coverage of 0.030 ML can be determined for this image set. Fig. 3 shows the size distributions of adsorbates recorded in our STM images. At an O_2 incident energy of 0.5 eV, 48% of the features are single atoms, 38% atom-pairs and 13% are of larger size. Thus, 46% of the adsorbed oxygen atoms are contained in paired sites with atoms at neighboring adsorption sites, while 29% are in the form of isolated single atoms.

Fig. 4 shows the time-of-arrival spectrum recorded by the ion detector when a beam of O_2 is directed towards the clean Al(111) surface. This signal has been accumulated during the first 500

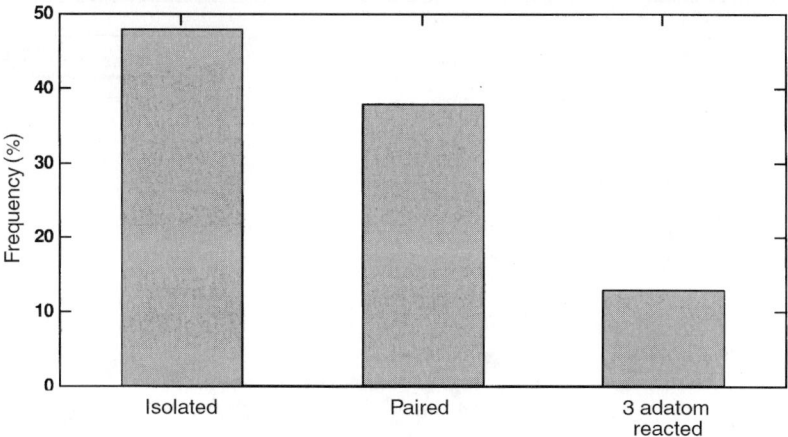

Fig. 3 For the reaction of 0.5 eV O$_2$ with Al(111), histogram showing the frequency with which isolated O-atoms, paired and three adatom reacted sites are found.

gas pulses. This data has been recorded at a somewhat shorter distance (6 mm) to the crystal than typical in order to emphasize the effect. The laser is tuned to the O-atom 0,1,2 ← 2 transition at 225.66 nm. The time-resolved trace shows two features centered at 2.5 and 3.1 µs time-of-arrival, respectively. By detuning the laser wavelength, it was verified that the creation of both signals

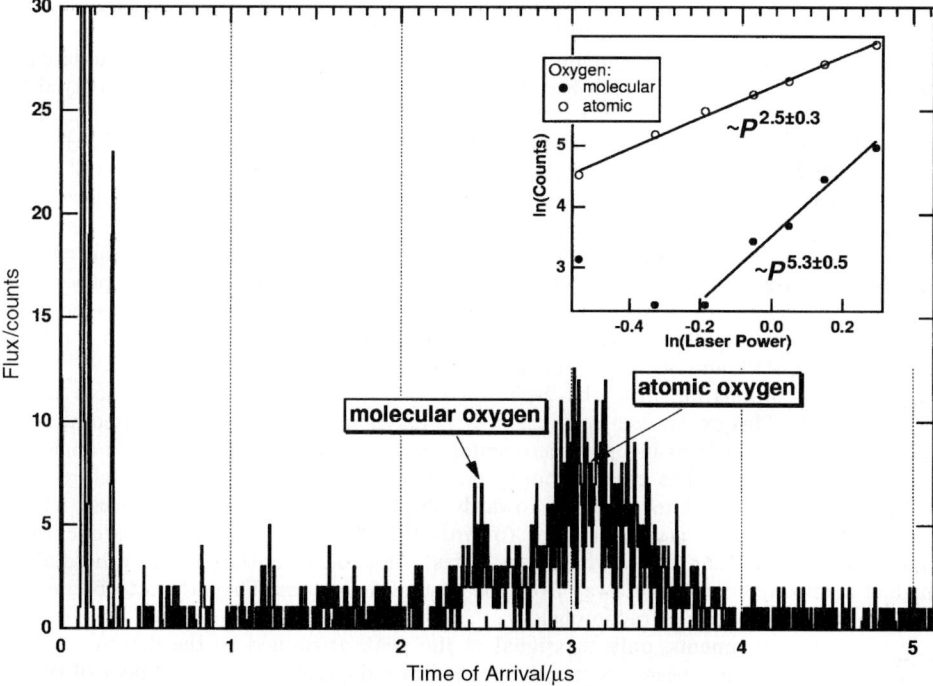

Fig. 4 Time-of-arrival spectrum at the ion detector for the laser set to 225.6 nm and an energy of 0.6 mJ pulse^{-1}. The spectrum has been accumulated during the first 500 gas pulses which scatter from a freshly prepared Al(111) sample. The laser focus has been placed 6 mm away from the surface. The spectrum exhibits O-atom signals centered at 2.5 and 3.1 µs, respectively. The strong signals at short times are due to electronic noise and scattered light. The inset shows the power dependence of the signals in a double-logarithmic graph.

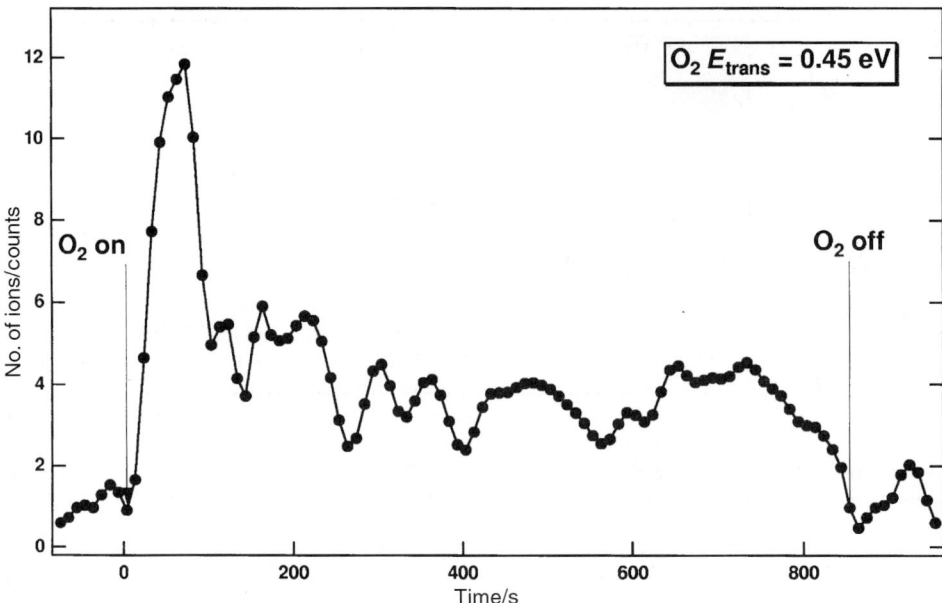

Fig. 5 Temporal evolution of the O-atom yield. At $t = 0$ s the beam shutter was opened. At $t = 840$ s the shutter was closed again. Each data point represents the number of ions counted for 100 laser pulses.

involves REMPI of oxygen atoms. Experiments at different exposure times and different placements of the laser focus with respect to the crystal lead to different weights of the two signals. In order to clarify their respective origin, time-of-arrival spectra have been recorded for different laser fluences. The area attributed to each feature has been integrated. The result is displayed in the inset of Fig. 4 in a double logarithmic graph. The first peak in the time-of-arrival spectrum scales with P^n with $n = 5.3 \pm 0.5$ dependence, whereas the second scales with P^n with $n = 2.5 \pm 0.3$. This finding is consistent with interpreting the first peak as being due to photodissociation of O_2 in the focus of the laser beam, and the second as being due to detection of atomic oxygen. Photodissociation of O_2 by a non-resonant 2-photon process and subsequent ionization of one O atom by a (2 + 1)-photon REMPI is a 5-photon process. This 5-photon dependence is well reflected by the data. The ionization of O-atoms, which have been formed in the abstraction process at the surface, is a (2 + 1)-photon process, which corresponds well with the power dependence of the later signal. The experimentally determined exponent is somewhat less than 3 which is a frequent observation in REMPI, and due to partial saturation of the ionization step.

Oxygen atoms originating from the 2-photon dissociation by 226 nm light have an excess energy of 2.95 eV. This excess energy appears as translational energy of the released atoms. The polarization of the laser light in our experimental setup was aligned under an odd angle due to experimental constraints. The corresponding momentum of the atoms was therefore directed neither perpendicular, nor in the direction towards the time-of-flight tube. Nevertheless, half of the nascent atoms had a velocity component towards the time-of-flight tube. Consequently, they arrived at earlier times than atoms which were at rest prior to the ionization. The other half of the atoms had a velocity component away from the time-of-flight tube. They are not observed, most likely since they escaped the extraction field.

In the further experiments, only the signal of the peak attributed to the detection of atoms formed by abstraction has been integrated. Fig. 5 shows the temporal development of the signal with progressing exposure of the Al(111) sample to an O_2 beam with 0.45 eV translational energy. Each data point represents the accumulated counts from 100 laser shots. At time $t = 0$ s the beam shutter is opened. A rapid rise of the signal is observed. The rise time is real since the detection system does not induce any temporal averaging beyond that connected with the accumulation of one data point. Hence, we conclude that the yield of oxygen atoms first increases with exposure.

After reaching a maximum, a rapid fall off to half the peak value is observed. Subsequently a slow decrease of the signal with time follows. At $t = 840$ s the beam shutter is closed again. The signal settles to its zero level. We expect at $t = 840$ s an O-atom coverage of about 0.2 ML. The oscillatory fine structure of the signal is not reproducible, but rather an experimental artifact.

Discussion

The STM data show the coexistence of single and paired sites. The simultaneous observation of singles and pairs unambiguously establishes that individual atoms can be seen and differentiated from pairs in STM. In a previous STM study of thermal (0.025 eV) O_2 reacted with Al(111), only isolated O adatoms were found which were separated by >80 Å at low surface coverages.[4] We attribute this difference to the higher translational energy (0.5 eV) of the O_2 molecules used in this study. The STM results are compatible with the concept that the single O-atom sites prevail at low translational energy, while the likelihood of *normal* dissociative chemisorption resulting in pairs increases as the O_2 incident energy is raised.

Laser spectrometric detection has given direct evidence of O-atoms which are ejected into the gas phase when a beam of O_2 is scattered from the surface. These O-atoms complement those single O-atoms on the Al surface observed in STM. The REMPI signal has the expected dependence on laser power, ruling out that the signal is due to photodissociation of ambient O_2 molecules followed by detection of the fragment atoms within the same laser pulse. It was demonstrated that such a scenario occurs and can be discriminated due to its power dependence from the signal arising from atoms created by the abstraction reaction.

Together, these observations provide compelling evidence that the abstraction mechanism is a realistic explanation for the observation of single chemisorbed O-atoms in STM. There remains no experimental reason to invoke hot atom motion as its origin. The latter suggestion has also been regarded as unlikely based on the results from theoretical modeling.[6]

We propose that the single isolated sites are due to abstractive chemisorption while the paired sites result from dissociative chemisorption. This explanation is consistent with the results reported from theoretical and experimental studies of halogens reacting with semiconductor[18,19] and metal surfaces.[20] Furthermore, Österlund et al.[1] reported a strong translational energy dependence of the sticking probability for O_2/Al(111); this demonstrates that the reaction is an activated process. For example, they report the absolute sticking coefficient of thermal O_2 on Al(111) to be ~ 0.01, whereas that of 0.5 eV O_2 is ~ 0.8. These sticking probability results are also consistent with the abstraction/dissociation hypothesis: thermal O_2 may rarely access ($S_0 = 0.01$) a portion of the multidimensional potential energy surface which leads to abstraction. The normal dissociative chemisorption pathway is completely closed at these energies. Higher incident energy O_2 may access regions of the complex potential energy surface unavailable to thermal O_2; therefore high incident energy O_2 may chemisorb by abstractive and/or dissociative chemisorption.

The available data are consistent with the assumption that, at energies below about 0.4 eV, the abstractive mechanism dominates, whereas at higher energies the ordinary dissociative chemisorption scenario is more significant, such that the sticking coefficient can finally reach 0.9. Work is in progress to verify this effect experimentally.

Dissociative chemisorption of O_2 is generally thought to take place by means of two electron transfer processes occurring in rapid succession.[21] We speculate that some detail in the potential energy surface prevents the molecule from orienting itself in a favorable side-on geometry. The transfer of the first electron may stabilize an upright geometry. If two electrons are transferred to the molecule in the upright geometry a very unstable situation may arise in which Coulomb forces accelerate the terminal O-atom such that it is ejected. The outcome of this scenario is an abstraction reaction. It will take further theoretical work, possibly including non-adiabatic effects, to gain a thorough theoretical understanding of this process.

Our laser spectroscopy results indicate that the initial yield of abstracted atoms slowly increases with surface coverage. This is consistent with the observation that the work function of Al(111) initially decreases.[22] With the decrease of the work function electron transfer processes to incoming O_2 molecules become energetically more favorable and occur over larger distances.[23] Consequently, the likelihood of the abstraction reaction increases with coverage.

We acknowledge fruitful discussions with Bengt Kasemo, Igor Zoric, Gerhard Ertl, Joost Wintterlin, Aart Kleyn, Alan Luntz and Bengt Lundqvist. The STM study was funded by the National Science Foundation (Grant no. CHE-9700546) and the Air Force Office of Scientific Research (Grants no. F49620-98-1-0312 and F49620-96-1-0026).

References

1. L. Österlund, I. Zoric and B. Kasemo, *Phys. Rev. B*, 1997, **55**, 15 452.
2. V. Zhukov, I. Popova and J. T. Yates, Jr., *Surf. Sci.*, 1999, **441**, 251.
3. H. Brune, J. Wintterlin, R. J. Behm and G. Ertl, *Phys. Rev. Lett.*, 1992, **68**, 624.
4. H. Brune, J. Wintterlin, J. Trost, G. Ertl, J. Wiechers and R. J. Behm, *J. Chem. Phys.*, 1993, **99**, 2128.
5. C. Engdahl and G. Wahnström, *Surf. Sci.*, 1994, **312**, 429.
6. G. Wahnström, A. B. Lee and J. Strömquist, *J. Chem. Phys.*, 1996, **105**, 326.
7. J. Strömquist, L. Hellberg, B. Kasemo and B. I. Lundqvist, *Surf. Sci.*, 1996, **352/4**, 435.
8. *Handbook of Chemistry and Physics*, ed. D. R. Linde, CRC Press, Boca Raton, FL, 75th edn., 1995, pp. 9–55.
9. G. Leonardelli, E. C. D. Gruber, M. Schmid and P. Varga, *Verhandl, DPG*, 2000, **35**, 736.
10. J. Jacobsen, B. Hammer, K. W. Jacobsen and J. K. Nørskov, *Phys. Rev. B*, 1995, **55**, 14 954.
11. T. Sasaki and T. Ohno, *Phys. Rev. B*, 1999, **60**, 7824.
12. K. Honkala and K. Laasonen, *Phys. Rev. Lett.*, 2000, **84**, 705.
13. Y. Yourshahyan, B. Razaznejad and B. I. Lundqvist, to be published.
14. T. Engel, *J. Chem. Phys.*, 1978, **69**, 373.
15. J. E. M. Goldsmith, *J. Chem. Phys.*, 1983, **78**, 1610.
16. (a) D. J. Bamford, L. E. Jusinski and W. K. Bishel, *Phys. Rev. B*, 1986, **34**, 185; (b) D. J. Bamford, M. J. Dyer and W. K. Bishel, *Phys. Rev. A*, 1987, **36**, 3497.
17. C. Yan, J. A. Jensen and A. C. Kummel, *J. Chem. Phys.*, 1995, **102**, 3381.
18. L. E. Carter, S. Khodabandeh, P. C. Weakliem and E. A. Carter, *J. Chem. Phys.*, 1994, **100**, 2277.
19. J. A. Jensen, C. Yan and A. C. Kummel, *Science*, 1995, **267**, 493.
20. K. A. Pettus and A. C. Kummel, to be published.
21. G. Katz, Y. Zeiri and R. Kosloff, *Surf. Sci.*, 1999, **425**, 1.
22. S. Utaze, S. Lacombe, L. Guillemot, V. A. Esaulov and M. Canepa, *Surf. Sci. Lett.*, 1998, **414**, L938.
23. T. Greber, *Surf. Sci. Rep.*, 1997, **28**, 1.

Chemical selectivity in the remote abstractive chemisorption of ICl on Al(111)

Kharissia A. Pettus,[a] Peter R. Taylor[b] and Andrew C. Kummel*[a]

[a] *Department of Chemistry and Biochemistry, University of California, San Diego, Mailcode 0358, 9500 Gilman Drive, La Jolla, CA 92093-0358, USA*
[b] *San Diego Supercomputer Center, University of California, San Diego, Mailcode 0505, 9500 Gilman Drive, La Jolla, CA 92093-0505, USA*

Received 6th April 2000
First published as an Advance Article on the web 2nd October 2000

The interaction of ICl and Al(111) involves remote dissociation in its chemisorption process. In remote dissociation, an electron harpoons from an Al(111) surface to an ICl gas molecule to initiate the chemisorption process. We have determined that ICl can chemisorb onto Al(111) by non-activated direct chemisorption, and the sticking probability of this direct channel is 0.65 ± 0.03. Furthermore, low energy ICl molecules that do not undergo remote dissociation can chemisorb onto Al(111) by precursor-mediated chemisorption. Not only is the interaction of ICl and Al(111) reactive, it is chemically selective. Studies with Auger spectroscopy reveal that the ratio of chlorine atoms to iodine atoms on the Al(111) is 0.32 ± 0.10 at low (0.042 ± 0.002) surface coverage. Time-of-flight mass spectrometry studies also show that chlorine atoms are the only species scattered from the surface after ICl interacts with Al(111). These results indicate that iodine-selective abstraction, in which the iodine atom of ICl chemisorbs to the aluminium surface while the chlorine atom is ejected into the gas phase, is the dominant mechanism in this reaction. Iodine-end first collisions are more reactive than chlorine-end first collisions because the lowest unoccupied molecular orbital (LUMO) of ICl is primarily composed of iodine atomic orbitals, and it is the LUMO that interacts with the harpooning electron from the aluminium.

Introduction

The reaction of iodine monochloride (ICl) and Al(111) involves remote dissociation, abstraction and chemical selectivity. Few experiments have investigated the interactions of halogens and aluminium surfaces.[1-8] Theoretical works show that halogen adsorption onto low work function surfaces like aluminium occurs by surface harpooning and remote dissociation.[9-16] When a halogen molecule approaches a low work function surface, the halogen molecule's electron affinity increases in magnitude due to the surface's ability to stabilize the negative halogen ion by its positive image charge. If the halogen molecule's effective electron affinity becomes greater in magnitude than the surface work function, the surface ejects a harpooning electron to the incoming molecule (surface harpooning). Subsequently, the halogen ion rapidly dissociates into one ion and one neutral fragment because harpooning places the negative ion on the repulsive portion of the potential above the dissociation energy (remote dissociation). The resulting ion fragment chemisorbs to the surface while the neutral fragment can later bond to the surface or can be ejected into

DOI: 10.1039/b002771p

the gas phase. A halogen molecule that does not receive an ejected electron may physisorb onto the aluminium surface and later dissociatively chemisorb.[9-16]

Studies have been done on systems similar to halogen/Al(111) systems which support the theory of surface harpooning and remote dissociation. Experiments with halogen gases and alkali metal gases and surfaces reveal that these interactions yield negative atomic halogen ions (exoions), exoelectrons and photons.[17-20] The escape of negative ions into the vacuum indicates that the dissociation of the halogen molecules occurs remotely from the surface, because if a negative ion is close enough to the surface, the ion will be drawn toward the surface by the surface's positive image charge.[17-19] However, exoelectron, exoion and photon emission are minor channels, especially on adsorbate free surfaces and may not reflect the dynamics of the majority of gas–surface collisions.[17-19] Theoretical studies with Cl_2 and potassium have also determined that surface harpooning and remote dissociation occur when a halogen molecule interacts with low work function surfaces. In addition, these studies reveal that it is possible that abstraction (the chemisorption of one atom while the other neutral atom is ejected back into the vacuum) occurs in these halogen and low work function systems.[21,22]

Former investigations of homonuclear halogen/low work function atom and surface reactions reveal that it is highly probable that surface harpooning, remote dissociation, and abstraction occur in the reaction of ICl and Al(111). These investigations, however, do not give insight into possible chemical selectivity in interhalogen/surface reactions. For this chemical intuition, analogies between the gas phase ICl/K reaction and the gas/surface ICl/Al(111) reaction are useful. Reactions between gaseous alkali metal atoms (potassium) and ICl show that alkali chloride is the principal product and alkali iodide is the minor product.[23-25] One possible reason for the dominance of the alkali chloride over the alkali iodide is the accessibility of ICl^- ground state, $^2\Sigma^+$, which leads to the non-adiabatic formation of a negative chlorine ion and an iodine neutral. The chlorine ion will bond to the metal atom while the iodine neutral leaves the site of reaction. If the ICl molecule was to have sufficient translational energy (>1 eV), then the ICl^- excited states are accessible and an iodine ion and a chlorine neutral may form.[23-25] Therefore, the ratio of alkali chloride and alkali iodide may change with increasing incident energy of ICl.

Stereo-selective experiments with ICl and K reveal that KCl is preferentially formed when K first interacts with the iodine end of ICl.[23-25] Furthermore, the iodine end is the more reactive end in the gas-phase ICl and K experiments. These results can be explained by orbital asymmetry.[23-26] When an alkali metal atom and an ICl molecule approach each other, the alkali atom preferentially sends an electron to the iodine atom of ICl because the LUMO of ICl is predominantly on the iodine atom (asymmetric molecular bonding). The additional electron does not remain on the iodine atom; instead, the electron migrates to the more electronegative atom, chlorine. The ICl^- molecule dissociates in the field of the positively charged alkali atom, and alkali chloride is then the major product.[23-25]

The iodine end of ICl is the more reactive site in many gas-phase and gas/surface reactions.[26-30] This is surprising since chlorine is more electronegative than iodine and chlorine usually forms stronger bonds. Hydrogen iodide and silicon iodide are the major products in the gas reaction of H and ICl and the gas/surface reaction of ICl and Si(111)-7×7, respectively, even though the formation of hydrogen chloride and silicon chloride are the more exothermic pathways.[26-30] These examples and the example of ICl and potassium indicate that reactions with ICl are governed by stereochemistry rather than energetics.

In this paper, we determine the reactivity and chemical selectivity in the gas/surface reaction of ICl/Al(111) and interpret the results based on conclusions drawn from the ICl and alkali metal gas reactions.

Experimental

The results of the chemisorption of ICl onto Al(111) were obtained under ultrahigh vacuum conditions. Detailed descriptions of the ultrahigh vacuum chamber, the molecular beam system, and the laser system have been published.[30-32] Experiments were performed with Monocrystals Company aluminium single crystals (orientation of crystals is $\langle 111 \rangle \pm 1$, dimensions of crystals are 10 mm in diameter $\times$ 2 mm thick and the purity of the crystals is 99.999 + %). The aluminium single crystals were cleaned by sputtering with 0.5–3 kV Ar^+ ions at normal incidence and were ordered by annealing at 673 K for 2–5 min. Surface cleanliness and order were determined by low-energy electron diffraction (LEED) and Auger spectroscopy.

The reactivity, dynamics and chemical selectivity of the ICl and Al(111) reaction were investigated. Sticking probability measurements were obtained for ICl/Al(111) using the reflection technique. These experiments were done using aluminium surface temperatures (T_{Al}) of 103, 298 and 573 K, and ICl seeded in various noble gases. In order to determine the chemical selectivity in the Al(111) and ICl reaction, clean Al(111) surfaces (T_{Al} = 103 K and 298 K) dosed with ICl seeded in helium or argon were examined with Auger spectroscopy. Electron stimulated desorption of chlorine and iodine atoms on the Al(111) was minimized by defocusing the Auger electron beam. Furthermore, the initial halogen coverages were determined by measuring halogen Auger peak intensities obtained at equal intervals over time and extrapolating to find the initial coverages.

Nonresonant multiphoton ionization (MPI, 205 nm) and time-of-flight (TOF) spectrometry were used to probe any species scattered from Al(111) (T_{Al} = 143 K in order to suppress the etching) after the surface was dosed with ICl seeded in helium. ICl was seeded in helium instead of argon during TOF experiments because at T_{Al} = 143 K ICl seeded in argon yielded no detectable abstraction products. The detection point of the laser pulses was approximately 1–3 cm from the surface. The time delay between the molecular beam pulses and the firing of the laser was changed to detect various scattered species. A TOF tube extracted ions produced by 205 nm light ionizing scattered species.

Results

Table 1 shows the sticking probabilities in the reaction of Al(111) at several temperatures with ICl at three translational energies. Sticking probabilities could not be measured at T_{Al} 103 K surface temperature with 0.1–0.2 eV ICl due to trapping of argon and krypton onto aluminium. The sticking probability of ICl on Al(111) is unaffected by surface temperature when the aluminium surface is dosed with 1.1 eV ICl molecules. This result indicates that there is no precursor-mediated chemisorption in the 1.1 eV ICl and Al(111) reaction. However, the sticking probability is dependent on aluminium surface temperature with 0.1–0.2 eV ICl molecules indicating the presence of precursor-mediated chemisorption for low translational energy ICl molecules on Al(111). The precursor channel also accounts for the higher sticking probabilities for 0.1–0.2 eV ICl than for 1.1 eV ICl on 298 K Al(111). At T_{Al} = 573 K, the sticking probability is approximately 0.65 ± 0.03 with all translational energies of ICl. These data show that the direct mechanism for ICl/Al(111) reaction is non-activated. In short, ICl chemisorption on Al(111) occurs by non-activated direct chemisorption or by precursor-mediated chemisorption.

Not only is the interaction of ICl and Al(111) reactive, it is chemically selective. Auger spectroscopy was used to determine the ratio of chlorine atoms to iodine atoms on an Al(111) surface (T_{Al} = 298 K) that had been dosed with 1.1 and 0.2 eV ICl molecules (Fig. 1). At low surface coverage (0.05 ± 0.004) and with 0.2 and 1.1 eV ICl molecules, the ratio of chlorine atoms to iodine atoms is 0.39 ± 0.05. For every chlorine atom on the surface there are 2.5 iodine atoms on the surface. This indicates that although dissociative chemisorption occurs in this reaction, iodine-selective abstraction (iodine bonds to the aluminium surface and chlorine is ejected away from the

Table 1 Sticking probability measurements for the ICl/Al reaction at various translational energies of ICl and several surface temperatures

Incident energy of ICl molecules/eV	Sticking probability
0.1	0.88 ± 0.02 (298 K)
	0.69 ± 0.03 (573 K)
0.2	0.78 ± 0.005 (298 K)
	0.66 ± 0.05 (573 K)
1.2	0.66 ± 0.05 (103 K)
	0.60 ± 0.05 (298 K)
	0.62 ± 0.05 (573 K)

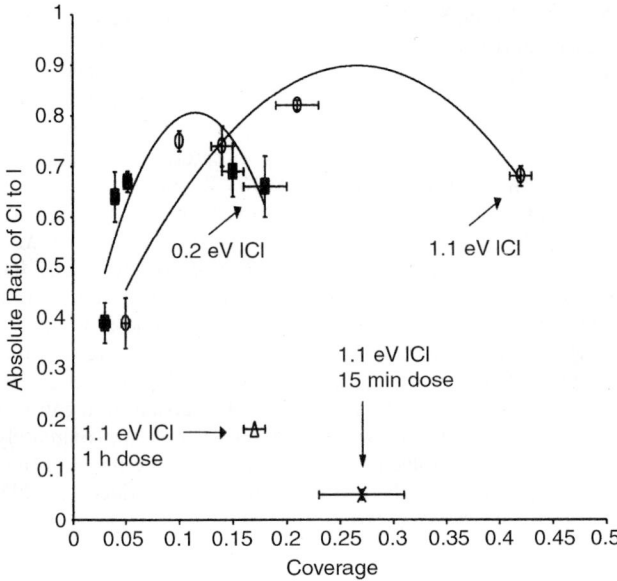

Fig. 1 Plot of the absolute ratio of chlorine atoms to iodine atoms vs. coverage on 298 K Al(111) surface ((○) ratio of Cl to I with 1.1 eV ICl molecules, and (■) ratio of Cl to I with 0.2 eV molecules) where the time of ICl dose is less than or equal to 10 min. There are two points with longer dose times and their dose times and ICl molecule energies are labeled on the graph.

surface) is the dominant mechanism when ICl reacts with Al(111).

It should also be noted in Fig. 1 that the ratio of Cl atoms to I atoms increases with increasing surface coverage. The chemical selectivity probably decreases due to extrinsic precursor-mediated chemisorption at high halogen coverage. The ratio never reaches 1 due to etching or the formation of volatile aluminium halides (probably Al_2Cl_6).[1–3] When Al(111) is dosed for a long duration (15–60 min) and with 1.1 eV ICl molecules, however, the ratio of Cl atoms to I atoms and the coverage decrease. This is the result of collision-stimulated desorption of chlorine etch products. Aluminium chloride is more volatile than aluminium iodide, so etching creates an iodine rich surface. Conversely, the Cl atom to I atom ratio does not markedly decrease at high surface coverage with exposure to 0.2 eV ICl because the 0.2 eV ICl molecules do not have enough energy to stimulate desorption of etch products.

Auger studies were repeated with 0.2 and 1.1 eV ICl on 103 K Al(111) in order to determine the ratio of chlorine atoms to iodine atoms in the absence of etching. Fig. 2 shows the results of the study. It is also important to notice that at moderate and high coverages (0.2–5) the ratio of chlorine atoms to iodine atoms is approximately 1, indicating that thermal or stimulated etching does not occur at this temperature. At low surface coverage (0.042 ± 0.002) and with 0.2 and 1.1 eV ICl, the ratio of chlorine atoms to iodine atoms is 0.32 ± 0.10. Again, the dominant mechanism in the ICl/Al(111) reaction is iodine-selective abstraction.

Nonresonant MPI (205 nm) and TOF spectrometry were used to confirm the results of the Auger studies by probing any species scattered from the Al(111) (T_{Al} = 143 K) while the surface was dosed with 1.1 eV ICl molecules. The mass (ion) spectrum in Fig. 3(a) shows that 205 nm light readily ionizes ICl. The 205 nm laser light can also dissociate ICl and ionize the resulting atoms. Iodine atoms undergo 1 + 1 multiphoton ionization while chlorine atoms undergo 2 + 1 multiphoton ionization. This accounts for most of the difference in spectral peak heights of the halogen atoms. The I_2 impurity accounts for a small portion of the atomic iodine peak height, but the large ionization cross-section of I_2 causes the I_2^+ peak to be as large as the ICl^+ peak. We have determined that the contribution of I_2 to the atomic iodine peak height is only twice that of the

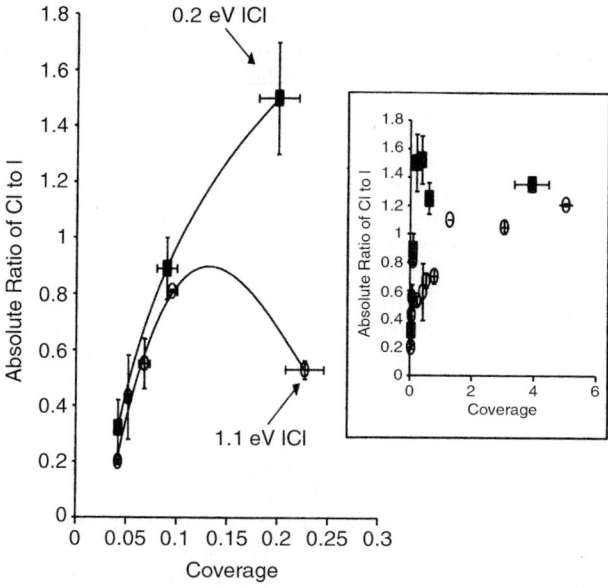

Fig. 2 Plot of the absolute ratio of chlorine atoms to iodine atoms *vs.* coverage on 103 K Al(111) surface ((○) ratio of Cl to I with 1.1 eV ICl molecules, and (■) ratio of Cl to I with 0.2 eV molecules).

molecular iodine peak height in TOF spectra. In addition, quadrupole mass spectrometry data reveal that the I_2 impurity is only 5–10% of the molecular beam.

Fig. 3(b) shows the spectrum of scattered species while 143 K Al(111) is dosed with 1.1 eV ICl molecules. At low surface coverage, the only scattered species are chlorine atoms while the peak to the left of the Cl peak is due to photoelectrons. This finding reveals that iodine atoms are chemisorbing to the surface while chlorine atoms are being ejected from the surface. Furthermore, the spectrum indicates that chlorine-selective abstraction (the chemisorption of the chlorine atom and

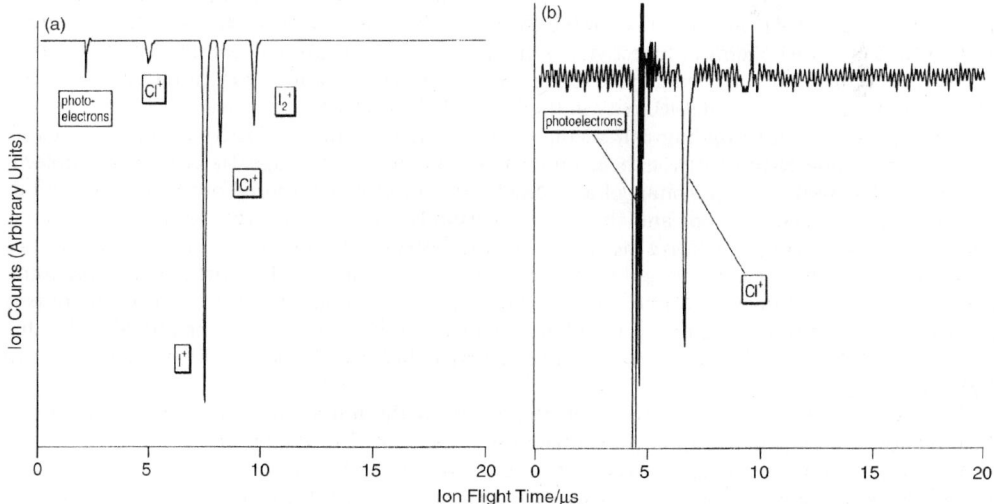

Fig. 3 (a) Ion TOF spectrum of the incident 1.1 eV ICl seeded in helium molecular beam using 205 nm light. (b) Ion TOF spectrum of scattered species while 143 K Al(111) surface is dosed with molecular beam of 1.1 eV ICl seeded in helium using 205 nm light.

the ejection of the iodine atom into the gas phase) does not occur in the ICl/Al(111) reaction at low coverage. Thus, there are two mechanisms in the direct chemisorption of ICl onto aluminium—dissociative chemisorption and iodine-selective abstraction.

Discussion

We have determined the reactivity and the chemical selectivity in the gas/surface reaction of ICl and Al(111). We found that the sticking probability for non-activated direct chemisorption in this reaction is 0.65 ± 0.03. This value is surprisingly low considering the high electron affinities of the halogens and the low work function of aluminium. Moreover, the sticking probability of 0.65 ± 0.03 exposes the inefficiency of surface harpooning in this system and the exacting conditions for ICl remote dissociation. The sticking probability is greater at low ICl translational energy where precursor-mediated chemisorption exists in the ICl/Al(111) system. This greater sticking probability at low incident translational energy of ICl is not surprising since precursor-mediated chemisorption allows a halogen molecule to be close to the Al(111) surface and to obtain an optimal orientation for chemisorption. In addition, the closer a halogen molecule is to the surface, the greater its effective electron affinity and the greater the probability for chemisorption onto the low work function surface of Al(111).[9-16]

There is chemical selectivity in the ICl/Al(111) reaction. The Auger studies show that the ratio of chlorine atoms to iodine atoms on the 103 K Al(111) and for both 0.2 and 1.1 eV ICl is 0.32 ± 0.10 at low surface coverage (0.042 ± 0.002). In addition, the TOF and MPI studies show that chlorine atoms are the only neutral species ejected from Al(111). These experiments attest to the fact that iodine-selective abstraction is the dominant mechanism, dissociative chemisorption is a minor mechanism, and chlorine-selective abstraction is non-existent in the ICl and Al(111) reaction at low coverage.

The reactivity and chemical selectivity experiments suggest that reasons for the chemical selectivity in the ICl/Al(111) cannot be solely based on gas phase potential energy curves. According to the potential energy curves for the ground and excited states of ICl$^-$, the negative chlorine ion should be the dominant ion product because the ground state, $^2\Sigma^+$, of ICl$^-$ dissociates into Cl$^-$ and I.[23-25] In contrast, our Auger and TOF studies show that iodine predominantly bonds to the surface, indicating that the iodine negative ion is formed. Furthermore, the chemical selectivity at low coverage would change with translational energy of ICl molecules if the reaction were governed solely by the gas phase potential energy curves because one of the excited states, $^2\Pi$, dissociates into I$^-$ and Cl. The excited states of ICl$^-$ should be accessible to an ICl molecule with translational energy of 1.1 eV. Therefore, 1.1 eV ICl molecules should interact with Al(111) to give a chemical selectivity different than that of 0.2 eV molecules. The Auger studies show that there is no marked change in chemical selectivity with increasing translational energy for ICl/Al(111). Indeed, the formation of Cl$^-$ being more exothermic than the formation of I$^-$ in the gas phase has little influence on the chemical selectivity in the ICl/Al(111) reaction.

Asymmetric molecular bonding can account for the high chemical selectivity and low sticking probability for non-activated direct chemisorption at low surface coverage. In asymmetric molecular bonding, the antibonding orbitals of a molecule are primarily composed of the atomic orbitals of the less electronegative atom and the bonding orbitals are primarily composed of the atomic orbitals of the more electronegative atom.[23-25] In the case of ICl (Fig. 4), the antibonding orbitals are predominantly composed of iodine atomic orbitals and the bonding orbitals are predominantly composed of chlorine atomic orbitals. It is the antibonding LUMO which would receive the harpooning electron ejected from the aluminium surface.[23-25] Therefore, the iodine-end of ICl should be the more reactive end and iodine-end first collisions should be the most reactive on Al(111).

After harpooning, the electron does not migrate from the iodine atom to the chlorine atom, which is more electronegative than the iodine atom. This is due to the fact that the magnitude of the electron affinity of an atom very close to the surface increases because of the surface's image charge.[9-16] We have done very simple calculations to determine the effective electron affinities of chlorine and iodine atoms when the ICl molecule is approaching the surface iodine-end first vs. chlorine-end first ($\Delta EA = [3.6/z]$ eV where ΔEA is the change in the electron affinity and z is the harpooning distance from the image plane).[9-16] Fig. 5 shows the effective electron affinities of

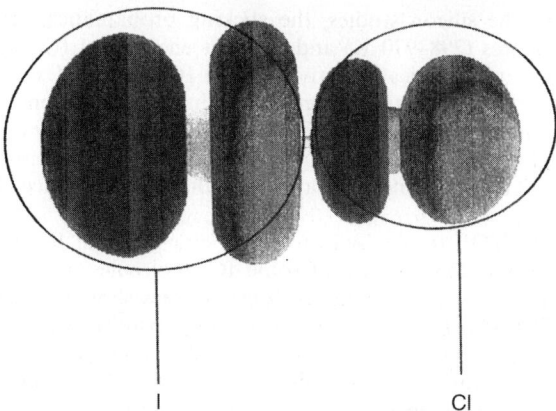

Fig. 4 Molecular orbital diagram of the chlorine and iodine atomic contributions to the ICl LUMO, which was computed by Dr Peter Taylor at the San Diego Supercomputer Center. The contributions of atomic iodine and atomic chlorine are labeled on the figure.

chlorine and iodine *vs.* the distance from Al(111) for iodine-end first and chlorine-end first collisions of ICl with the Al(111) surface. Fig. 5(b) suggests that the magnitude of effective electron affinity of iodine is greater than for chlorine in an iodine-end first ICl molecule close to the surface. Therefore, if the harpooning electron bonds to the iodine atom of an iodine-end first ICl molecule, it will remain with the iodine atom when the ICl molecule dissociates.

The chemical selectivity results coupled with the sticking probability of 0.65 ± 0.03 for non-activated direct chemisorption are consistent with a model in which surface harpooning occurs for iodine-end first ICl and side-on ICl collisions with Al(111). Furthermore, the data presented in this paper reveal that the ICl and Al(111) reaction is governed by steric effects (or that the iodine-end is the reactive site) which is similar to the results obtained in the ICl and Si (111)-7 × 7

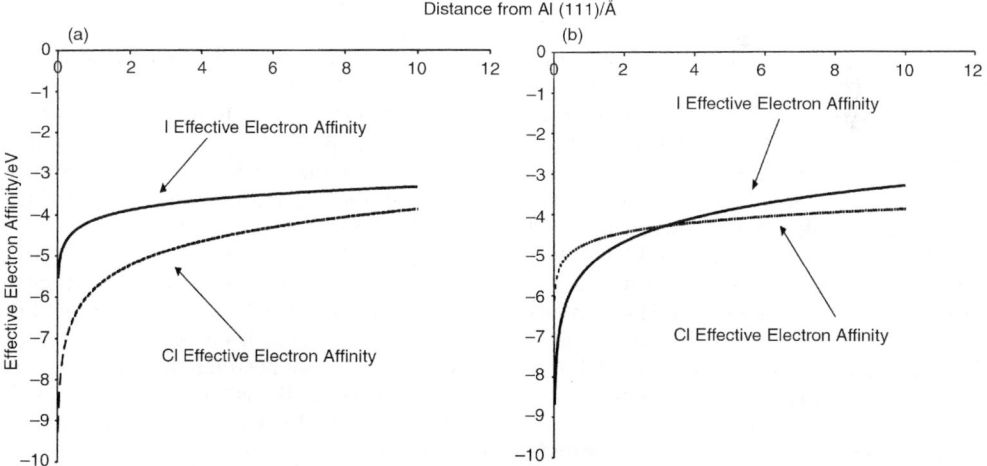

Fig. 5 (a) Plot of effective electron affinities of chlorine and iodine when an ICl molecule is approaching the surface such that the chlorine atom is closest to the surface. The change in effective electron affinity of chlorine is shown as a dashed line and the change in effective electron affinity of iodine is shown as a solid line. (b) Plot of effective electron affinities of chlorine and iodine when an ICl molecule is approaching the surface such that the iodine atom is closest to the surface. The change in effective electron affinity of chlorine is shown as a dashed line and the change in effective electron affinity of iodine is shown as a solid line.

studies.[29,30] However, in the silicon studies, the sticking probabilities were 0.89 ± 0.07 over a range of surface temperatures (298–970 K) and incident energies of ICl (0.2 and 1.2 eV).[29] The high sticking probabilities and chemical selectivity in the ICl/Si (111)-7 × 7 reaction indicate that the silicon surface rotationally steers the ICl molecules into an iodine-end configuration. This is not the case in the Al(111) and ICl reaction where the sticking probability of the direct mechanism is 0.65 ± 0.03 over a large range of surface temperatures and incident energies of ICl. Therefore, it is highly probable that few ICl molecules that approach the Al(111) surface chlorine-end first react with the surface.

Although the Auger and TOF results indicate that iodine-end first ICl and side-on ICl are the more reactive geometries, the chlorine-end of some ICl molecules may receive the harpooning electron, resulting in dissociative chemisorption. Auger studies show that dissociative chemisorption does occur in the ICl/Al(111) reaction. Unfortunately, two mechanisms lead to dissociative chemisorption in this system and these are indistinguishable by our detection methods: iodine-end first collisions in which the chlorine atoms cannot escape the surface image charge and chlorine-end first collisions in which the iodine atoms cannot escape the surface image charge.

Even if a chlorine-end first collision receives a harpooning electron, it is likely to result only in dissociation (not abstraction) due to a mass effect in the ICl/Al(111) reaction. Due to the conservation of momentum, much of the reaction energy goes to the light chlorine atom in the dissociative processes, $ICl + e^- \rightarrow I + Cl^-$ or $ICl + e^- \rightarrow I^- + Cl$. Since iodine atoms are heavy, they probably would not have a large enough kinetic energy to escape the surface if ICl dissociated into negative chlorine atoms and iodine neutrals near Al(111). In contrast, chlorine atoms are light and the ICl/Al(111) reaction provides enough energy for chlorine atoms of ICl to escape the image charge and be ejected into the vacuum.

At moderate surface coverage (≈ 0.1), the mechanism of ICl chemisorption onto Al(111) is no longer remote dissociation due to the increase in aluminium work function by the halogen adsorbates.[19,33,34] This high work function causes incoming ICl molecules to approach very close to the surface before the magnitudes of their effective electron affinities are greater than the work function of the surface. Since the ICl molecule is so close to the surface, few neutral atoms escape from the surface and both atoms of ICl bond to the surface. This idea accounts for the increase in the ratio of chlorine atoms to iodine atoms on moderately covered Al(111) surfaces. At very high coverage (0.2–5) and at low surface temperature, the ratio of chlorine atoms to iodine atoms is approximately 1 which is due to extrinsic physisorption.

Conclusions

The interaction of ICl with Al(111) was studied using the reflection technique. We have found that ICl reacts with Al(111) by non-activated direct chemisorption and the sticking probability of this mechanism is 0.65 ± 0.03. ICl with translational energies of 0.1–0.2 eV can also bond to Al(111) by precursor-mediated chemisorption. The modest sticking probability of ICl/Al(111) non-activated chemisorption is unexpected when one considers the large electron affinities of chlorine and iodine and the low work function of Al(111). However, this sticking probability is consistent with orientation-selective chemisorption of ICl molecules in the ICl/Al(111) reaction.

Auger spectroscopy, MPI, and TOF spectroscopy were used to determine the chemical selectivity and the more reactive site of ICl in the ICl/Al(111) reaction. The ratio of chlorine atoms to iodine atoms on a $T_{Al} = 103$ K Al(111) surface was determined to be 0.32 ± 0.10 by Auger and the TOF studies show that chlorine is the only neutral product scattered from the Al(111) surface. This reveals that in the interaction of ICl and Al(111) iodine-selective abstraction is the dominant mechanism. Furthermore, these studies show that the iodine end of ICl is more reactive than the chlorine end. The greater reactivity of the iodine end is because the LUMO of ICl, which interacts with the harpooning electron, is concentrated on the iodine end.

This work was funded by the National Science Foundation (Grant No. CHE-9700546 and Grant No. CHE-9700627) and the AFOSR (F49620-98-1-0312 and F49620-96-1-0026). Furthermore, the authors would like to thank Dr James McLean and Mr Andrew Komrowski for their helpful assistance in the experiments.

References

1. D. A. Danner and D. W. Hess, *J. Appl. Phys.*, 1986, **59**, 940.
2. H. F. Winters, *J. Vac. Sci. Technol. B*, 1984, **3**, 9.
3. S. Park, L. C. Rathbun and T. N. Rhodin, *J. Vac. Sci. Technol. A*, 1985, **3**, 791.
4. T. Smith, *Surf. Sci.*, 1972, **32**, 527.
5. T. Hauser, R. Greber, G. C. Xiong, B. Strehlau and O. Meyer, *Nucl. Instrum. Methods Phys. Res.*, 1993, **81**, 192.
6. J. Bormet, J. Neugebauer and M. Scheffler, *Phys. Rev. B*, 1994, **49**, 17242.
7. K. Mitsutake, J. Yamauchi, A. Sakai and M. Tsukada, *Surf. Sci.*, 1995, **324**, 106.
8. A. G. Borisov, D. Teillet-Billy and J. P. Gauyacq, *Phys. Rev. B*, 1999, **59**, 8218.
9. J. W. Gadzuk, *Comments At. Mol. Phys.*, 1985, **16**, 219.
10. T. Greber, *Surf. Sci. Reports*, 1997, **28**, 1.
11. P. A. Dowben, *CRC Crit. Rev. Solid State Mater. Sci.*, 1987, **13**, 191.
12. P. Nordlander, *Phys. Rev. B*, 1992, **46**, 2584.
13. A. R. Williams and N. D. Lang, *Surf. Sci.*, 1977, **68**, 138.
14. N. D. Lang and A. R. Williams, *Phys. Rev. B*, 1978, **18**, 616.
15. N. D. Lang and W. Kohn, *Phys. Rev. B*, 1973, **7**, 3541.
16. J. W. Gadzuk and S. Holloway, *Chem. Phys. Lett.*, 1985, **114**, 314.
17. L. D. Trowbridge and D. R. Herschbach, *J. Vac. Sci. Technol.*, 1977, **18**, 588.
18. B. Kasemo, *Surf. Sci.*, 1996, **363**, 22.
19. B. Kasemo and L. Wallden, *Surf. Sci.*, 1975, **53**, 393.
20. M. Menzinger, *Acta Phys. Pol. A*, 1988, **73**, 85.
21. L. Hellberg, J. Stromquist, B. Kasemo and B. I. Lundqvist, *Phys. Rev. Lett.*, 1995, **74**, 4742.
22. L. Hellberg, J. Stromquist, B. Kasemo and B. I. Lundqvist, *Surf. Sci.*, 1996, **352–354**, 435.
23. G. H. Kwei and D. R. Herschbach, *J. Chem. Phys.*, 1969, **51**, 1742.
24. H. J. Loesch and J. Moller, *J. Chem. Phys.*, 1992, **97**, 9016.
25. H. J. Loesch and J. Moller, *J. Phys. Chem.*, 1993, **97**, 2158.
26. R. B. Bernstein, D. R. Herschbach and R. D. Levine, *J. Phys. Chem.*, 1987, **91**, 5365.
27. J. C. Polanyi, *Chem. Scr.*, 1986, **27**, 229.
28. J. D. McDonald, P. R. LeBreton, Y. T. Lee and D. R. Herschbach, *J. Chem. Phys.*, 1972, **56**, 769.
29. Y. Liu, D. P. Masson and A. C. Kummel, *Science*, 1997, **276**, 1681.
30. K. A. Pettus, T. S. Ahmadi, E. J. Lanzendorf and A. C. Kummel, *J. Chem. Phys.*, 1999, **110**, 4641.
31. T. F. Hanisco, C. Yan and A. C. Kummel, *J. Chem. Phys.*, 1991, **97**, 1484.
32. T. F. Hanisco A. C. Kummel, *J. Phys. Chem.*, 1991, **95**, 8565.
33. E. V. Albano, *Appl. Surf. Sci.*, 1982, **14**, 183.
34. N. D. Lang, S. Holloway and J. K. Nørskov, *Surf. Sci.*, 1985, **150**, 25.

General Discussion

Prof. Stokbro opened the discussion of Prof. Persson's paper: Eqn. (1) in your paper seems to have an incorrect asymptotic behaviour. For instance, if the resonance is above the vacuum level, ψ_a is not damped, and I_{in} will increase exponentially with distance.

Prof. Persson responded: We agree that eqn. (11) has an incorrect asymptotic behaviour, in particular, if the resonance is above the vacuum level. However, in our case the resonance is below the vacuum level.

Prof. Wolf commented: In your paper you argue that the mechanism invoked for STM-induced desorption of CO/Cu(111) by transient occupation of the $2\pi^*$ derived level[1] might be incomplete. You suggest that the efficient excitation of the R-mode may be involved in the desorption process. Could you elaborate on this and how are these results consistent with the observed pronounced isotope effect?

1 L. Bartels, G. Meyer, K.-H. Rieder, D. Velic, E. Knoesel, A. Hotzel, M. Wolf and G. Ertl, *Phys. Rev. Lett.*, 1998, **80**, 2004.

Prof. Persson replied: Our calculated results and the observed efficiencies for the excitation of the frustrated rotational mode and the internal stretch mode of CO on Cu by inelastic electron tunnelling show that the former mode has a much higher excitation probability than the latter mode that involves the desorption coordinate. In addition, the favoured bonding of the C atom of CO to metal atoms implies that the transfer from the surface to the tip involves a rotation of the molecule. However, it is not clear how the excitation of this mode alone can explain the observed behaviour of the desorption rate such as the pronounced isotope effect.

Dr Gauyacq commented: To follow the preceding remark by Prof. Wolf, we can compare the present results with those presented in our paper.[1] With a theoretical approach different from that of Lorente and Persson, we found the CO($2\pi^*$) resonance on Cu(111) at higher energy, closer to the vacuum level with a width very close to the one obtained by Lorente and Persson. The two approaches then confirm the very short lifetime of the resonance and thus the absence of a stabilisation effect of the Cu(111) projected band gap in this case.

1 J.-P. Gauyacq, A. G. Borisov, G. Raşeev and A. K. Kazansky, *Faraday Discuss.*, **117**, 15.

Prof. Persson replied: In principle, the one-electron spectrum obtained in a DFT calculation does not give the correct excitation energies when adding or removing single electrons from a many-electron system. In particular, our DFT-GGA results and the experimental data suggest that we do not have a correct description of the $2\pi^*$ derived resonance of CO on Cu using this approach so I am not convinced that the agreement of our approach for the resonance lifetimes on the (100) and (111) surfaces with the ones obtained by Dr Gauyacq using a different approach confirm the absence of a stabilisation effect of the projected band gap of Cu(111).

Prof. Gadzuk said: You stated that your results require multiple resonances. I would suggest a slightly different wording which conveys a more general message. In fact what is required in order that the inelastic processes of concern here take place is a particular (but not unique!) form of electron charge distribution, hence forces, acting in the intermediate state. The system doesn't care how this charge distribution and thus perturbation Hamiltonian is formed, *i.e.* by one resonance, two resonances, or whatever. I would think that your findings for multiple resonances are a direct consequence (or limitation) of the particular forces you have associated with each resonance. In principle these forces could be produced by a single resonance, if the resonance happened to have the "right shape".

Prof. Persson replied: Our statement is that our calculated results for vibrational damping rates and inelastic efficiencies show that the electron–vibration coupling can be efficient without having a coupling through the diagonal component of the deformation potential with respect to a single resonance orbital derived from an adsorbate orbital. The prime example behind this finding is the frustrated rotational mode of CO on Cu. Simple symmetry arguments show that the corresponding vibrational coordinate does not change the energies of any adsorbate-induced resonances such as, for instance, $2\pi^*$ and one has to look at off-diagonal components of the deformation potential with respect to the adsorbate induced resonances. It is conceivable as suggested by your slightly different wording that it is possible to find a linear combination of adsorbate-induced resonances that diagonalizes the deformation potential and that one of the eigen values is dominant.

Prof. Wolf commented: Your DFT calculation yields a position of the $2\pi^*$ derived resonance with an onset close to the Fermi level. Vibrational excitation by inelastic tunnelling can occur at low bias. However, in STM-induced desorption there is a clear threshold of 2.5 eV.[1] From your calculation I would expect no threshold in contrast to the STM experiments for CO desorption. Could you please comment on this?

1 L. Bartels, G. Meyer, K.-H. Rieder, D. Velic, E. Knoesel, A. Hotzel, M. Wolf and G. Ertl, *Phys. Rev. Lett.*, 1998, **80**, 2004.

Prof. Persson responded: The conclusion of our DFT-GGA calculations of the $2\pi^*$ derived resonance is that the ground state one-electron potential in these calculations is not able to describe correctly the excited state involving one extra electron in the $2\pi^*$ derived resonance. Thus our calculated properties of the $2\pi^*$ derived resonance cannot be used to model the STM experiments of CO on Cu that involve the injection of electrons with an energy of 2–3 eV above the Fermi level.

Prof. Wolf said: I wish to note that our two-photon-photoemission experiments clearly indicate that the position of the unoccupied CO $2\pi^*$ level is located at least 3 eV above the Fermi level (see ref. 1), while your calculation places this level close to the Fermi level. There appears to be large deviations of the DFT calculation from the experiment, which may arise from the fact that DFT optimises only the ground state electron density.

1 M. Wolf, A. Hotzel, E. Knoesel and D. Velic, *Phys. Rev. B*, 1999, **59**, 5926.

Dr Shluger commented: Let us take into account the presence of a STM tip and that it may have an atom or small cluster adsorbed at the end or indeed the same molecule as on the surface, *e.g.* CO. Using your model we can imagine that the cluster or molecule on the tip can be excited by tunnelling current in the same way as those on the surface. I wonder if this could lead to some new interesting effects, which could have been already observed experimentally? In particular, resonant energy transfer from the molecule on the tip to the surface molecule. I think that tip and surface in these experiments should be treated on the same footing.

Prof. Persson answered: Your suggestion that the adsorption of an atom or small cluster on a tip can possibly give rise to some new effects in inelastic electron tunnelling is very interesting. There have been some recent experiments[1] using CO-terminated tips of inelastic electron tunnelling through adsorbed CO molecules on Cu that, for instance, indicate that there is coupling between the frustrated translational modes of CO adsorbed on the surface and on the tip.

1 F. Moresco, G. Meyer and K.-H. Rieder, *Mod. Phys. Lett. B*, 2000, **13**, 709.

Dr Darling asked: Could you elaborate on your comment that your theoretical approach is good for "certain classes of tip". Is it obvious *a priori* which STM tips fall into these classes?

Prof. Persson replied: Our experience of modelling STM images for adsorbed molecules on surfaces using DFT and the Tersoff–Hamann approximation is still rather limited. Our approach seems to give encouraging agreement with tips that give "sharp" images. The next step is to carry out a more detailed quantitative modelling of the tip, in particular, of CO terminated tips that can be fabricated experimentally in a more or less well-controlled manner.

Prof. Groß commented: I am just speculating on the reason why your CO derived levels are too low in energy according to your DFT calculations. We all know that DFT in the LDA as well as in the GGA severely underestimates band gaps. Are the deficiencies of LDA/GGA that cause the underestimation of band gaps also responsible for the apparent downshift of the CO derived excitation levels in your calculation?

Prof. Persson responded: Your suggestion that the $2\pi^*$ derived level is too low in energy in our DFT calculations is related to the underestimation of the band gaps of semiconductors and also the HOMO–LUMO gap of molecules by GGA/LDA seems plausible. A highly successful method for calculating accurate single particle spectrum of moderately correlated systems is provided by the GW approximation.[1] More recently also DFT using exact exchange have given a substantial improvement of this spectrum for molecules and semiconductors.[2] The experience from these studies is typically that the HOMO state is too high in energy in LDA/GGA because of incomplete cancellations of self-interaction effects and that the LUMO state such as $2\pi^*$ of CO is too tightly bound.

1 M. Rohlfing and S. G. Louie, *Phys. Rev. B*, 2000, **62**, 4927.
2 S. Ivanov, S. Hirata and R. J. Bartlett, *Phys. Rev. Lett.*, 1999, **83**, 5455; A. Görling, *Phys. Rev. Lett.*, 1999, **83**, 5459.

Dr Dujardin said: You calculated the spatial distribution of the inelastic process and you found that it is narrower than the distribution of the elastic process. This is what is also observed experimentally. Could you comment on the origin of this effect?

Prof. Persson replied: At this stage I do not have a definite answer but I believe that one reason for the spatial distribution of the inelastic signal being narrower than the elastic distribution is that the electron–vibration coupling is typically short ranged.

Dr Darling commented: The energy you have computed for the $2\pi^*$ level using DFT is not particularly close to the experimentally determined values. How much, if any, reliable information (general bonding features, trends, *etc.*) can we obtain about the excited states from standard DFT?

Prof. Persson replied: For semiconductors and molecules, the limitations of DFT-GGA to describe excited states are rather well understood. For instance, the HOMO–LUMO gap is typically underestimated. For adsorbates on surfaces, it would be desirable to investigate more cases to gain some more experience and understanding of the practical limitations of DFT-GGA to describe excited states for chemisorbed molecules.

Dr Kroes opened the discussion of Prof. Kosloff's paper: I have a question about the convergence of the calculations with the surrogate Hamiltonian method. How do you test convergence with respect to the number of bath modes (look for recurrences or repeat calculations with a higher number of bath modes), and can you assess in advance how many bath modes you will need for a particular problem?

Prof. Kosloff responded: Concerning the convergence surrogate Hamiltonian method with respect to the number of bath modes, one can test the convergence by repeating the calculation with an increasing number of bath modes. In principle the timescale of the process determines the number of bath modes required, for example a process with a characteristic time of τ and an energy range of E will require an energy resolution of $\varepsilon = \hbar/\tau$ and the number of bath modes will be $N \sim E/\varepsilon$. We found that for tunnelling calculations there is no characteristic timescale and a logarithmic sampling of the energy range converge the calculation.

Prof. Kummel asked: Could you further explain the hole induced dissociation of O_2^{2-} described in your paper (see section IVA)? It would seem to be in contradiction to the common notion of e^- induced dissociation of O_2 on metals via $O_2 \to O_2^- \to O_2^{2-} \to 2O$.

Do you see an activation barrier for $O_2 \to 2O$ dissociation?

Prof. Kosloff replied: In order to understand hole induced processes on metal surfaces we need some insight into the excited electronic states of oxygen on metal surfaces. The difficulty stems from the fact that the effective DFT methods which are extremely valuable for ground state properties are not suited for such a study. To address this problem Dr Dubnikova of the Hebrew Univesity, Jerusalem, performed electronic structure calculations on small metal clusters with an additional oxygen atom. What we were looking for are the charge transfer states that can be promoted by a positive hole.

To study these effects a Pt_4–O cluster as well as Pt_3–O and Pt_{19}–O were studied. The calculations was carried out with the Gaussian 98 program with the LANL2DZ basis set for Pt and cc-PCDZ for O. The ground state was optimized using u-HF and the excited states obtained in the level of uCIS. Fig. 1 shows a set of ground and excited state potentials as a function of the distance between the oxygen atom to the plane defined by the three Pt atoms. One can identify many excited states. Since we are looking for a charge transfer excitation from the oxygen to the metal we have to discriminate the excitations in the metal atoms from the charge transfer excitations. We found that the best method to identify these states is to obtain their dipole function.

Fig. 2 shows the difference between the dipole value of the ground state to the dipole value of the excited state. Excited state no. 1 is a typical metal excitation with small change in the dipole function. Most of the excitations in the cluster are of this type which would translate in the infinitely large cluster to the continuum of electron–hole pair excitations. The excited states 15 and 18 show a reversed dipole, indicating a back transfer of charge from the oxygen to the metal. The excitation energy to these states is around 1.6 eV which is in a reasonable range for hole induced processes. The binding energy of the oxygen to the cluster is 4.12 eV. An analysis of the calculations at a distance of 2.8 Å from the surface show evidence of non-adiabatic state mixing or curve crossing.

We think that these calculations reveal a universal phenomena for oxygen on most metals. Atomic oxygen sufficiently close to a metal surface will attract negative charge. The atom is not ionic, a double negative charge is suitable due to strong Coulomb repulsion. Stabilization of a double negative charge requires an ionic crystal stabilization which would mean surface recon-

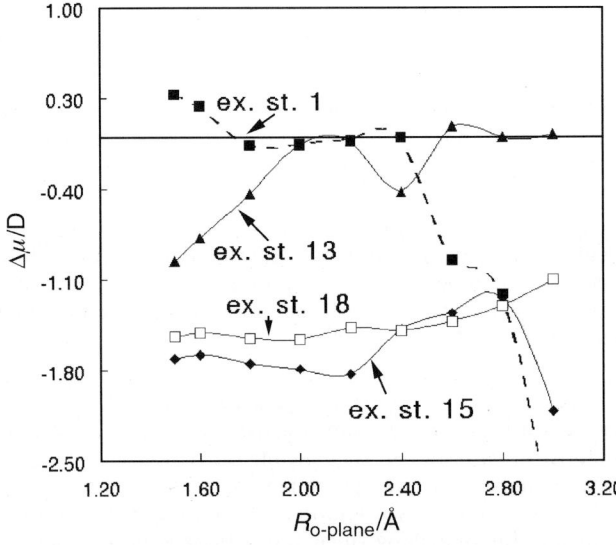

Fig. 1 Difference in the dipole moment between ground and excited state potentials.

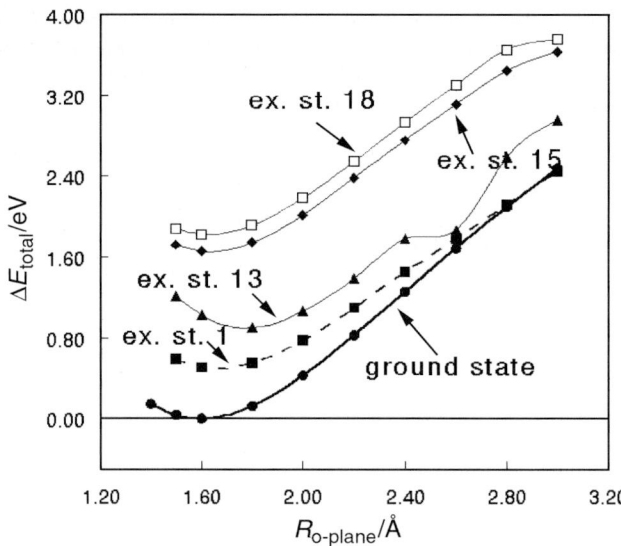

Fig. 2 Ground and excited state potential as a function of distance.

struction.[1] For this oxygen atom to become reactive it should transfer back its negative charge to the metal, a process that is induced by a positive hole.

The interaction of an oxygen molecule with a surface is more complicated. Due to its large electron affinity O_2 will also attract charge from the metal and adsorb in a negatively charged state either peroxide or superoxide which differ by the amount of negative charge transfer. This is the starting point for dissociation. One possibility is that a positive hole will attract to this species and promote the molecule to an excited state. This energy can be made available to overcome the dissociation barrier either by a series of non-adiabatic transitions to the dissociative state or by quenching back to the ground state in a MGR-like process with sufficient energy to overcome the dissociation barrier. The other possibility which is dissociation *via* an electron attachment will be hindered due to coulomb repulsion.

1   T. Somerfeld, M. K. Scheller and L. S. Cederbaum, *Chem. Phys. Lett.*, 1993, 209.

Prof. Kleyn said: We found evidence that on Ag(110) O_2 does not form the oxide at the terrace, but at steps and imperfect pairs of the surface.[1] Can you estimate the role of steps and are they needed for dissociation in your model?

1   D. A. Butler, J. B. Sanders, A. Raukema, A. W. Kleyn and J. W. M. Frenken, *Surf. Sci.*, 1997, **375**, 141.

Prof. Kosloff responded: In our model,[1] dissociation of O_2 on Ag is through a series of non-adiabatic transitions. This will take place both on terraces as well as on steps and imperfections. The possible role of such imperfections is to alter the non-adiabatic coupling elements. For example due to symmetry an oxygen molecule lying flat on the surface has a zero non-adiabatic coupling from the physisorption state to the peroxide state. An imperfection can change these coupling elements.

1   G. Katz, Y. Zeiri and R. Kosloff, *Surf. Sci.*, 1999, **425**, 1.

Dr Gauyacq commented: In the case of a charge transfer process between a molecule and a metal surface, a large number of metal electrons can be involved, for example an electron lost by the molecule can go into different metal states. Does this mean that a wavepacket approach with an expansion over electronic states like the one you presented has to include a very large number of states to describe the metal electron continuum and possible electron–hole pair excitations?

Prof. Kosloff replied: A wavepacket approach where the molecular motion is considered on a small number of potential energy surfaces can supply an adequate description only for an extremely short time. To include systematically the metal degrees of freedom we developed the surrogate Hamiltonian method. The method is still based on a wavepacket description but extra degrees of freedom are used to describe the influence of the metal. We are in the process of implementing a model including two electronic states coupled to electron–hole pairs and to phonons. By increasing the number of degrees of freedom the process can be followed for longer times. The conclusion is that the number of states is determined by the timescale of the process.

Dr Darling said: In the figures you have shown of various diabatic PESs, the crossing seams have extremely complicated shapes. Why are they so complicated—are all parts on all crossing seams important or is the dynamics dominated by certain regions of the crossings?

Prof. Kosloff responded: The topology of the crossing seams for multi-dimensional cases can be very complex. It is clear that we have to go beyond the one-dimensional Landau–Zener–Stückelberg theory. Extensive effort is now devoted to the analysis of conical intersections which is only one topological feature possible in many dimensions. The general problem of understanding multi-dimensional non-adiabatic dynamics is still open. In simulations we have found that the flux of the reaction crosses only through certain sections of the seam. When the seam is approached in a perpendicular direction a good guide is energy. At a particular collision energy the part of the crossing seam which corresponds to this energy is most important.

Dr Saalfrank commented: As a follow-on to Dr Darling's question on the complicated shape of the non-adiabatic crossing seams shown by Prof. Kosloff for various metal–O_2 systems: In the multi-dimensional case, they arise from model potentials rather than *ab initio* calculations. How robust is the shape of the non-adiabatic crossing seams on variations of the model parameters? How reliable in general are the potential and coupling parameters? How sensitive are the dynamics results on the choice of model parameters?

Prof. Kosloff replied: To address this problem I would differentiate between the long range and short range parts of the potentials. For the long range harpooning type processes the potentials are determined by molecular properties and the difference between the electron affinity of the molecule to the work function of the metal. For this it is possible to obtain quite accurate parameters. For small molecular metal distances the problem is not trivial. DFT methods can give an adequate description of ground state properties. We tried to use spin restricted DFT to get excited states but the accuracy is not good. It is possible that embedding methods such as the ones of Nakatsuji *et al.*[1] or Carter *et al.*[2] will enable *ab initio* calculations of excited states. In the meanwhile I think we have to rely on model or toy parameters. The largest uncertainty is in the non-adiabatic coupling terms.

1 H. Nakatsuji and H. Nakai, *J. Chem. Phys.*, 1993, **98**, 2423.
2 N. Govind, Y. A. Wang and E. A. Carter, *J. Chem. Phys.*, 1999, **110**, 7677.

Prof. Groß asked: If we consider CO oxidation on a Pt catalyst for a moment, do you think that electronically non-adiabatic processes are important for the chemical reaction regarding the fact that electronic excitations on transition metal surfaces are probably all quenched within 1 fs due to the strong electronic coupling?

Prof. Kosloff replied: Our first reference is the CO oxidation in gas phase. The reaction cannot take place on a single electronic surface since it requires a spin flip between reactants and products.[1] In general CO which is isoelectronic with N_2 is quite inert and does not oxidize easily.

When the reaction takes place on a metal surface new reaction pathways have to be considered. A heavy metal due to a large spin–orbit coupling will facilitate spin flips. On the other hand oxygen atoms on the metal surface are negatively charged and therefore almost inert. We think the mechanism for CO oxidation is conditioned on a back transfer of charge from the oxygen to the metal creating an oxygen radical. The lifetime of this radical is crucial in determining the yield of the reaction. Very fast quenching can completely hinder the reaction. We infer that this is an explanation to the hindering of CO oxidation on Li-precovered Pt which stabilizes the negative charge on the oxygen.² Due to fast quenching of excited states, the yield per back-transfer event will be low.

1 S. S. Xantheas and K. Rudenberg, *Int. J. Quantum Chem.*, 1994, **49**, 409.
2 I. N. Yakovkin, V. I. Chernyi and A. G. Naumovets, *Surf. Sci.*, 1999, **442**, 81.

Dr Kroes asked: In a situation where one cannot yet compute excited state PESs and has to guess them, can DFT play a role in locating the crossing seams between the ground state and excited states?

Prof. Kosloff replied: The use of DFT to obtain crossing seams is quite tricky. One can use spin restricted DFT to define a particular potential energy surface. Considering reactions on a surface of heavy metals due to spin–orbit coupling, the spin is not a conserved quantum number. Time dependent DFT can in principle define an excited potential energy surface and the non-adiabatic coupling terms. Currently I do not know of any computational study in this direction.

Dr Shluger said: Could you please comment on how nuclear motion of the surface atoms can affect the predicted adiabatic surfaces for charge transfer states? Especially in the case of comparable masses, such as O and Al, and close to the surface.

Secondly, your model seems to predict the initial stages of oxidation for Al and some other metals. Could you please comment in more detail on how, according to your calculations, the oxidation process should proceed? Is the electron transfer to O_2 a necessary step for oxygen dissociation? There has been recent *ab initio* Hartree–Fock calculations of the Al oxidation process by Prof. P. W. M. Jacobs and his group, which would be interesting to compare.

Prof. Kosloff responded: Concerning the effect of nuclear motion on the non-adiabatic process we can distinguish two cases. In the harpooning regime which is characterized by a charge transfer far from the surface the event is determined by residual electronic charge density. This electronic charge density is only slightly perturbed by nuclear motion. Close to the surface, nuclear motion can have a profound role in determining non-adiabatic transitions. We studied this effect for the dissociation of N_2 on an Fe surface.¹ For O on Al we would expect a large effect due to the comparable masses. We have not yet included this effect in the present calculations.

Concerning the second question we have not considered the transformation of the surface into an oxide. The oxidation process is very complex and requires a complete rearrangement of the surface. Our model does predict that dissociation of oxygen on a metal surface almost always starts by an electron transfer to the oxygen molecule.

1 G. Katz and R. Kosloff, *J. Chem. Phys.*, 1995, **103**, 9475.

Dr Saalfrank commented: In your paper you speculate on the role played by pure electronic dephasing processes at surfaces. You mention that Poisson processes might be more relevant here than Gaussian ones. Two questions: (1) What is the reasoning for this expectation? (2) In which way could one distinguish experimentally between Poisson or Gaussian dephasing? In particular, would line widths and shapes in absorption spectroscopy, for example, be any different?

Prof. Kosloff answered: Both Gaussian and Poisson are stochastic models which we can use to describe the dephasing process. A Gaussian is characterized by the accumulations of frequent and fast random fluctuations. We assume that these processes are fast relative to the characteristic timescale of the system. A Poisson process is characterized by a series of rare events with an exponential distribution of waiting times. The timescale of the event should be much smaller than the typical waiting time. Concerning electronic dephasing the typical timescale of the system is a few femtoseconds. It is hard to imagine any bath motion which is faster than this timescale. On the other hand an elastic scattering event can cause a large phase shift. The waiting time between events can be quite large which fits a Poisson model.

Experimentally it is quite difficult to distinguish the two processes. A transient signal in a pump–probe experiment will show for both models an exponential decay. For both cases lineshapes are Lorenzian. In order to distinguish the two cases one has to excite a broad band wavepacket with many frequency components. A Gaussian model will predict that the higher harmonics will decay quadratically faster. A sub-quadratic scaling could be an indication of a Poisson process.

Dr Darling opened the dicussion of Prof. Jónsson's paper: In all of the clusters used to fix the B3LYP correction you have more O than Si, and the O is terminated with H. What are the reasons for not terminating these clusters with Si?

Prof. Jónsson responded: The idea is to make the terminal bonds as similar to the Si–O bonds as possible. H-atoms have electronegativity that is quite similar to that of Si (1.8 *vs.* 2.1) so the bonding with O-atoms (which have electronegativity of 3.5) is going to be similar in character. We also shorten the O–H bonds to make the Mulliken charge on the outermost O atoms the same as the inner O atoms.

Dr Kroes asked: In the cluster, do you allow motion in all the degrees of freedom? Do you have confidence that these deformations will indeed occur in a 3D crystal?

Prof. Jónsson replied: The clusters are kept rigid, *i.e.* no motion is allowed at all. The configurations are entirely determined from the plane wave based DFT/PW91 calculations where periodic boundary conditions are applied. The cluster calculations are carried out only to correct the energy, not the structure.

Dr Shluger asked: Could you please explain in more detail how the surface of quartz has been constructed? Did you make the full relaxation of the slab including the lattice vectors? How strong are the surface reconstruction and rampling?

Another question concerns the way you calculate excitons in different surface layers? You conclude that excitons will tend to move towards the surface. Is this conclusion based solely on the calculated exciton formation energies, or have you also estimated barriers for exciton diffusion between different layers?

Prof. Jónsson answered: We cut the quartz crystal to make a slab with O-terminated sides. To make the hydroxylated surface (which we are primarily presenting here), we terminated both sides with H-atoms. The bottom layer is frozen and the direction of the O–H bonds chosen to be the same as the cut O–Si bonds. The top surface was then relaxed and we did not see any major reconstruction, only a slight lengthening of the surface Si–O bonds (by less than 0.04 Å).

We have not yet calculated energy barriers for the diffusion of STEs. Since the geometry in our calculations is determined by DFT/PW91 calculations and the perfect crystal has an anomalously low energy, all the transition paths calculated in the bulk case have ended up going through the perfect crystal configuration. We can therefore only place an upper bound on the energy barrier (using the B3LYP corrected energetics). The same would most likely be true for STE near the surface. We need to be able to do calculations with self-consistent energy and structure before we can address this issue. Our conclusion is, therefore, entirely based on the energy of the metastable minima for the STE, not the migration barriers.

Dr Darling said: Between the STE's involving O atom displacement there is a barrier of ~ 0.2 eV. How reasonable do you believe this estimate to be—is this barrier basically 0 eV within the limits of the computational methods?

Prof. Jónsson responded: An error bar of 0.2 eV is certainly not unreasonable for these calculations. All we can say is that the barrier is small. As in all *ab initio* calculations (except for extremely small molecules), the accuracy of the answer depends on how much the errors cancel. If one is comparing two similar configurations, with similar bonding, the energy difference between the two can be predicted with less than 0.1 eV error. But, there are also examples where the error is much larger, as large as 0.5 eV. A few years ago my group collaborated with Ken Jordan's group on a systematic comparison of various levels of DFT and wavefunction based methods. We calculated energy differences and barriers for Si–H bond breaking and Si–Si bond breaking. While the DFT/PW91 results were sometimes in error by 0.5 eV, the B3LYP results were within 0.1 eV of the best estimates (see ref. 1), Similar tests need to be performed on energy barriers in Si–O bond breaking, but they have not been undertaken yet, as far as I know.

1 P. Nachtigall, K. D. Jordan, A. P. Smith and H. Jónsson, *J. Chem. Phys.*, 1996, **104**, 148.

Dr Saalfrank commented: You mention that implementing hybrid functionals such as B3LYP in periodic electronic structure codes is difficult and uneconomic. Why is that so? Also, could you say something about the effort to run a (pure) Hartree–Fock calculation as compared to a DFT calculation, for the same periodic system?

Prof. Jónsson replied: The problem with Hartree–Fock (HF) calculations of periodic systems is the long range of the exchange interaction. There are some codes out there that can do this, but the calculations are much more tedious and time consuming than the local description of exchange in LDA or semilocal description in GGA. Only last week did I for the first time hear of a periodic HF calculation which includes analytical evaluation of atomic forces (in a presentation by Prof. Harrison at Imperial College). Even without structural relaxation, I doubt there is any B3LYP calculation out there now that includes as many atoms as we need to include in our studies of the STEs in quartz (72 atoms, without symmetry constraints). In a hybrid functional, such as B3LYP, which includes a HF calculation in addition to DFT, the effort in the HF calculation dominates. Typically, the HF calculation scales as the fourth power of the number of electrons while the DFT/GGA calculation scales as the third power.

Dr Darling addressed Prof. Kosloff: There are at present no (few?) good calculations of excited electronic states. What are the prospects for getting accurate numbers from say time-dependent DFT calculations or embedding methods, and what do you think are the likely timescales for such methods producing results?

Prof. Kosloff responded: The only adequate calculations of excited electronic surfaces of adsorbates on metals are the ones by Nakatsuji and Nakai on the O_2/Ag system.[1] The technique is a dipped embedded cluster where a small cluster containing an oxygen molecule and two or four silver atoms is used for the calculation. The method employs a CI calculation for the cluster and takes into consideration the image field due to the bulk of the metal. The orbitals are filled in such a way that a chemical equilibrium is installed between the cluster and the metal free electrons.

A promising new approach has been suggested by Govind *et al.*[2] The method employs a self consistent DFT calculation to embed a cluster. The method has not been applied yet to calculate excited states of adsorbates on metals. Discussions with Dr Kluener and Dr Carter at UCLA indicate that this will be their next step.

1 H. Nakatsuji and H. Nakai, *J. Chem. Phys.*, 1993, **98**, 2423.
2 N. Govind, Y. A. Wang and E. A. Carter, *J. Chem. Phys.*, 1999, **110**, 7677.

Prof. Jónsson added: Time-dependent DFT and embedding methods are already producing results on excited states for a given, fixed geometry. The problem is really to get the atomic forces in the excited state and be able to calculate the dynamics and/or structural relaxation of the system in the excited state. The cluster approach may have problems representing the condensed phase environment properly, and time-dependent DFT cannot be used to get the atomic force in a single excited state. We are looking at GW calculations to see if that is a viable approach. It is well established that band gaps are predicted quite accurately by GW, but systematic tests against high level, wavefunction based calculations on molecules have not been carried out yet, as far as I know.

Dr Darling commented: In your calculations you study a special type of excited state which you 'make' by constraining the electron spins. This forces your ground state into a triplet state. Are the energies for this state then as accurate as can be obtainable for the true ground state?

Prof. Jónsson replied: Excited states are, of course, generally more difficult to deal with than ground states. In principle, however, DFT applies just as well to the lowest triplet state as it applies to the lowest singlet state, but the functionals we currently have for DFT calculations have mainly been developed and tested for singlet states. We have found significant errors in the triplet state calculations using the PW91 functional, especially when the excess spin density is delocalized. More localized states seem to cause less problems. The B3LYP functional seems, however, to treat the triplet state quite well. We judge this from small cluster calculations where we have been able to carry out high level wavefunction based calculations for comparison. There can, also, be a problem with spin contamination in these calculations. That is, since the calculation is spin unrestricted, the total wavefunction may not correspond to a pure spin state. Again, our tests on small clusters indicate that this is not a serious problem in the results presented here, except possibly for the STE–Si structure which is close to where the singlet and triplet states cross and therefore calls for a multi-reference description.

Dr Darling asked: Do the experimental estimates of energies and widths of excited states from 2PPE or inverse photoemission, or determined by the voltage dependence of STM induced processes all agree well enough for us to be able to make sensible comparisons with them when we are eventually able to compute excited states accurately?

Prof. Wolf responded: In the STM experiments one must be aware of the possibility of a Stark-shift due to the bias voltage, which may hinder a direct comparison with other experimental techniques. Currently the available data base is not sufficient to draw conclusions about possible differences between 2PPE and inverse photoemission results, which is also difficult due to the poor energy resolution and/or electron induced damage in inverse photoemission. Note, that for benzene on Cu(111) we have found very good agreement with the energetic position of the LUMO between 2PPE and inverse photoemission spectroscopy (see ref. 1).

1   D. Velic, A. Hotzel, M. Wolf and G. Ertl, *J. Chem. Phys.*, 1998, **109**, 9155.

Dr Hodgson opened the discussion of Prof. Kummel's paper: It would be interesting to know the energy dependence of the abstraction process, can you extract this from the beam/STM measurement?

Prof. Kummel replied: We have at present performed the experiments at low energy with background thermal (0.025 eV) O_2 and with a 0.5 eV molecular beam of O_2 seeded in He. This is not a perfect comparison because the angle of incidence is unrestricted for the thermal gas. We will repeat the molecular beam experiment using pure O_2 (0.1 eV) and compare that to O_2 seeded in H_2 (~ 0.8 eV).

Prof. Jones asked: (1) Will abstractive chemisorption only occur for interhalogens such as ICl, or will it also occur for pure halogen?
(2) How important is the work function of the substrate. Would abstractive chemisorption be observed for ICl on a high work function surface such as Cu(111) ($\Phi = 4.9$ eV)?

Prof. Kummel responded: (1) Because the remote dissociation process is common to all the halogen molecules, it is very likely that abstraction occurs for all the lighter halogens F_2, Cl_2, and maybe Br_2. For I_2, the dynamics may be different because its vertical transition to the negative ion will provide very little momentum to push the terminal iodine away from the surface and back into the gas phase.
(2) In my opinion, for molecules such as F_2 and Cl_2 in which a vertical transition from the neutral to the negative molecular ion places the ion high up on the repulsive portion of the potential, remote abstractive chemisorption will occur even on higher work function metals such as Cu(111). The higher work function will cause the process to occur closer to the surfaces so for the heavier halogens, Br_2 and I_2, it is possible that dissociative chemisorption will dominate. For all the halogen molecules on most metal surfaces, precursor mediated chemisorption is likely to dominate at low incident translational energy. This process is more likely to result in dissociative rather than abstractive chemisorption.

Dr Kolasinski asked: How have you corrected your Auger data to account for the effects of electron stimulated desorption; the cross section for which is likely different for Cl an I?

Prof. Kummel answered: Chlorine does have a much larger electron stimulated desorption rate than iodine in our Auger spectrometer. We use a two-step process to insure this does not affect our results. (1) We lowered the electron beam energy and defocused the electron beam. (2) We recorded the intensity of the Cl, I and Al peaks as a function of time and extrapolated the Cl intensity back to zero time. We are certain this works reasonably well because at low temperature for long exposures, we get a Cl : I ratio of 1 : 1.

Prof. Groß opened the discussion of Prof. Hasselbrink's paper: I have two questions:
(1) Did you check how far apart from each other the paired oxygen atoms are as it was undertaken for O_2/Pt(111) by Wintterlin et al.?[1]
(2) How do you explain the presence of 3-atom islands? Are they due to oxygen adatom mobility like in a nucleation–growth process?

1 J. Wintterlin, R. Schuster and G. Ertl, *Phys. Rev. Lett.*, 1996, **77**, 123.

Prof. Hasselbrink responded: (1) In the pairs as well as in the larger island the oxygen atoms are separated by one lattice constant. It is well known and has been observed by Brune et al.[1] that O on Al(111) forms a 1×1 structure, *i.e.* the distance we observe in STM is consistent with an adsorption process resulting in adatoms having just that distance.

(2) We observe no significant mobility of the O-atoms in STM. We believe that the 3-atom islands are formed by dissociative adsorption of an additional molecule in the neighborhood of an adsorbed O-atom. The adsorption site of a O-atom is characterized by a high charge density. It is conceivable that the dissociative sticking probability is larger in these areas in particular when the overall sticking coefficient is small. Hence, around an O-atom an island is likely to grow and existing islands are likely to grow further. This has recently been modeled by Neuburger and Pullman.[2]

1 H. Brune, J. Wintterlin, J. Trost, G. Ertl, J. Wiechers and R. J. Behm, *J. Chem. Phys.*, 1993, **99**, 2128.
2 Neuburger and Pullman, *J. Chem. Phys.*, 2000, **113**, 1249.

Prof. Kummel added: (1) We observe that the pairs are one or two lattice spacings apart. We ruled out long range dissociation by recording in real-time the appearance of the O-atom features while dosing with background O_2. Therefore, we did not perform the same analysis as Wintterlin et al.[1]

(2) We feel the growth of islands is due to induced reactivity: the perimeter of the islands has a higher sticking probability than the rest of the surface due to a local change in the work function. This could involve either direct abstractive/dissociation chemisorption or precursor mediated dissociation through a molecular chemisorption state.

1 J. Wintterlin, R. Schuster and G. Ertl, *Phys. Rev. Lett.*, 1996, **77**, 123.

Prof. Wolf asked: Do you have an explanation for the asymmetric feature in the line scan no. 1 in Fig. 2 of the paper, associated with the monomers? How does that compare with results from other groups?

Mr Komrowski replied: There are two issues with regards to the O–Al features as they appear in high-resolution STM images. First, the asymmetry of the single protrusions (line scan no. 1), which has been noted by several research groups, is probably due to asymmetries in the tip orbitals. The shape of the tip orbitals often play a dramatic role in how surface features are imaged when the tip is in very close proximity to the surface. A more controversial issue has been the assignment of the single protrusions as O–Al monomers. A competing group has proposed that the features ascribed to O–Al monomers are actually the result of charge transfer to a substrate Al atom between two neighboring oxygen adsorbate atoms.[1] However, our assignment of the adsorbate features imaged for this study are in agreement with previous STM images of thermal O_2 on Al(111),[1] as well as theoretical calculations.[2] The line traces in Fig. 2 of our paper show that we measure a single protrusion when scanning over an O–Al monomer, a double protrusion when scanning over a pair of monomers, and three protrusions lined up in accord with the crystal lattice spacing when scanning over part of a larger adsorbate island. We have carried out additional STM experiments in attempt to confirm the assignment scheme we support: single protrusions are O–Al monomers. First, we have scanned Al(111) surfaces after dosing with a plasma-generated oxygen atom source. Significantly, we have not observed any new O–Al adsorbate features as a result of reacting the clean surface with O-atoms. Second, we have dosed clean Al(111) surfaces with thermal NO while scanning with STM. These results contain adsorbates of the same size and shape as the O–Al monomers in Fig. 2 of our paper. Moreover, we have observed paired reaction sites in STM images that can be credited to the dissociation of an NO molecule at adjacent surface sites. Each end of the paired reaction site is imaged slightly differently; consistent that one site is due to O–Al and the other N–Al.

1 M. Schmid and P. Varga, Abstracts of the 46th AVS International Symposium, Seattle, USA, SS-TuP4, 1999.
2 H. Brune, J. Wintterlin, J. Trost, G. Ertl, J. Wiechers and R. J. Behm, *J. Chem. Phys.*, 1993, **99**, 2128.
3 J. Jacobsen, B. Hammer, K. W. Jacobsen and J. K. Norskov, *Phys. Rev. B*, 1995, **52**, 14954.

Prof. Rocca commented: In order to have abstraction the O_2 molecule must adsorb with the axis normal to the surface, even if only for a very short time. Does this correspond to a molecular precursor state? If so, abstraction should not be expected for noble and transition metals where the molecular precursor state lies with the molecular axis parallel to the surface.

Prof. Hasselbrink responded: No stable molecular precursor for O_2 on Al(111) has been observed. Recent scattering experiments in my group also show no indication of a precursor. Hence, the molecule in the upright geometry may only be a scattering state. If the upright geometry is energetically favorable at large molecule surface distances compared to the flat geometry then the molecules may be funnelled into this geometry on their way in.

Abstraction is on most noble and transition metal surfaces not feasible for simple energetic reasons. The oxygen–aluminum system is peculiar in so far as the O-atom to surface bond is stronger than the intramolecular bond in O_2. Hence, the formation of a single surface bond suffices to break the molecule. In most other cases the energy from simultaneously forming two atom–surface bonds is needed to allow the breaking of the molecular bond.

Prof. Kosloff commented: Fig. 3 shows a non-adiabatic view of the O_2/Al(111) electronic potential energy surfaces. The coordinates are: z, the distance from the surface, r, the oxygen oxygen internuclear distance, $c = \sin\theta$, the angle between the orientation of the molecule and the direction perpendicular to the surface (see Fig. 1 in our paper).

Displayed is a single isosurface approximately 0.2 eV above the free gas phase oxygen molecule asymptote. The blue color represents the physisorption potential energy surface. The purple color represents the same energy value for the O_2^- (peroxide). The cyan color represent the same energy value for the O_2^{-2} (superoxide). The red color represent the same energy value for the oxygen dissociation channel.

The left hand side of the figure shows cuts through the potential at three different orientations of the molecule: Perpendicular (bottom), tilted (middle) and parallel (top).

The potential is a semi-empirical one constructed from the molecular oxygen states and polarization, the Al metal work function and image potentials. The potential form is a modification of the potential presented in ref. 1.

The dynamics of dissociation and abstraction as imagined on these electronic surfaces starts on the neutral physisorption potential (blue). Then the oxygen molecule crosses to the peroxide potential (purple) which crosses again to the superoxide (cyan). The coulombic repulsive forces on the superoxide potential will lead eventually to dissociation. In small internuclear separations there is a region of overlap between the physisorption peroxide and superoxide potential such that almost a simultaneous charge transfer event is possible. The topology of the potential crossings is consistent with the following dynamics. At low kinetic energy the dissociation will take place far from the surface. This allows the oxygen atoms to travel freely before eventually being attracted to the metal surface due to the coulomb image potential. At higher incident kinetic energy the dissociation will take place closer to the surface and the two atoms will be found close together.

1 G. Katz, Y. Zeiri and R. Kosloff, *Surf. Sci.*, 1999, **425**, 1.

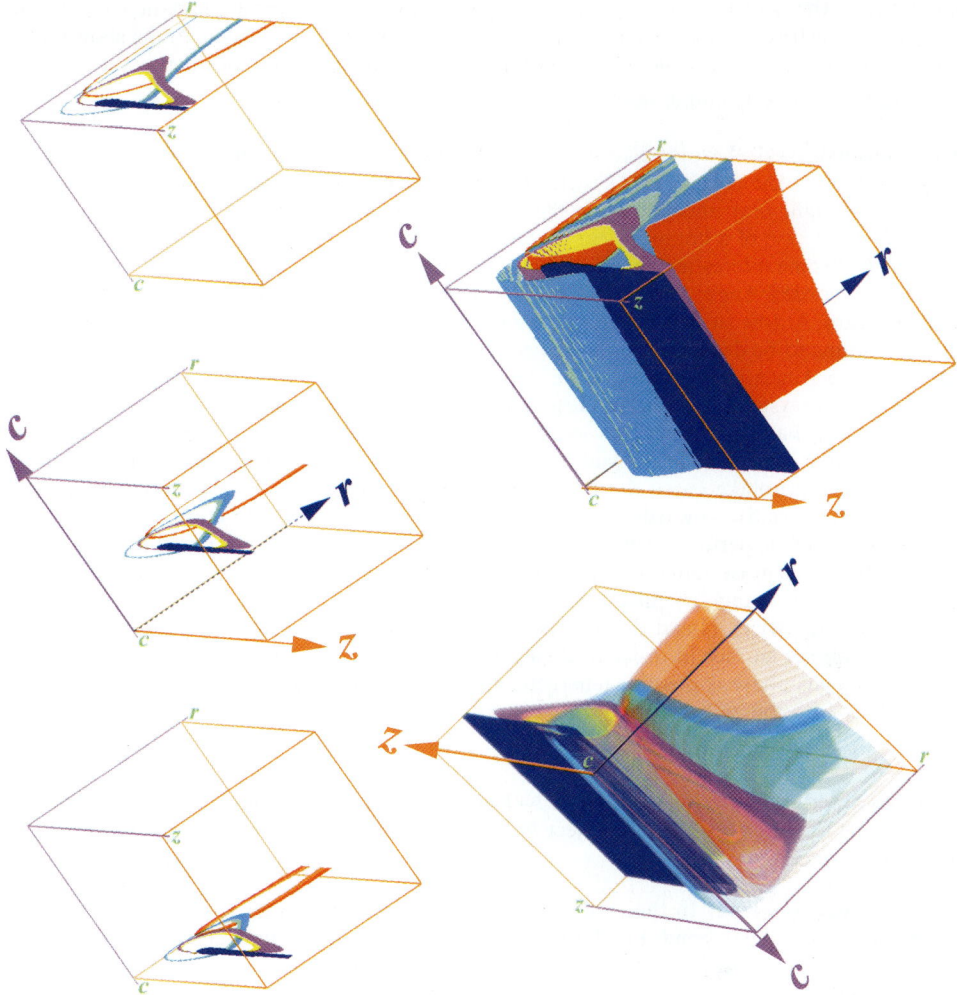

Fig. 3 Non-adiabatic view of the O_2/Al(111) electronic potential energy surfaces.

Prof. Rocca said: In response to the remark of Prof. Kosloff that Cs should decrease the sticking coefficient of O_2 by reducing the work function of the metal, I notice that the contrary (*i.e.* an increase of the dissociative sticking probability) was reported by Netzer *et al.* for the O_2/Ag(001) system.[1] The oxygen was dosed by filling the chamber with the crystal at room temperature.

1 J. A. D. Matthew, F. P. Netzer and G. Astl, *Surf. Sci.*, 1991, **259**, L757.

Prof. Kosloff replied: Considering harpooning or long range charge transfer events from metals to oxygen, different metals have a different work function. For metals with low values of workfunction the charge transfer will take place at large distances from the metal surface before more specific oxygen metal bonds are important. The direct chemical bonds between the metal and oxygen are system specific.

Prof. Kleyn asked: Do the calculations for O_2^{2-} on Al show any evidence for a 'coulombexplosion' like ejection of the O-atom?

Prof. Kosloff answered: We have not yet performed quantum dynamical calculations on the new O_2/Al potentials which include the orientation of the molecule. We intend to perform such calculations in the near future. In the past we performed quantum non-adiabatic calculations of the O_2/Al dissociation dynamics on a two-dimensional model.[1] These calculations showed a long range charge transfer process and a competition dissociation and superoxide desorption.

1 G. Katz, Y. Zeiri and R. Kosloff, *Surf. Sci.*, 1999, **425**, 1.

Mr Komrowski said: I would like to add to the discussion a comment on the experiments we have conducted on the NO/Al(111) system. Prof. Kummel and I became interested in this system because we thought that understanding the Al(111) chemistry of a heteronuclear diatomic might shed light on the case of a homonuclear diatomic, such as O_2. Instead, our results have suggested the presence of some interesting chemical dynamics in the NO/Al(111) system.

We have recorded Auger spectra of an Al(111) surface dosed with NO from a non-oriented molecular beam. At low surface coverages, we measure more oxygen on the surface than nitrogen. This result seems to be dependent upon the condition of the surface. Nevertheless, we consider the results of these Auger experiments to be definitive proof of the abstractive mechanism at low coverages for this system. A group in Goteborg which we collaborate with has also measured this effect.[1] Next, we have performed an experiment in Prof. Kleyn's laboratory in Amsterdam in which an oriented molecular beam was allowed to strike the clean surface. We measured the King and Wells sticking uptake curve during the dose. Our results show that the nitrogen end of the molecule is more reactive towards dissociation—a result which seems to contradict the results of the Auger experiment performed with the non-oriented molecular beam. Our tentative explanation is that molecule must undergo a rotation during the dissociation.

We propose a two-stage chemisorption process for the initial reaction with the clean surface. First, the molecule sticks N-end down, possibly in a molecular precursor state. Second, the molecule rotates on the surface in order to dissociate. During the dissociation process ejection of the N-atom to the gas phase is possible (energetically allowed) and may be favoured since the charge will likely stay with the more electronegative atom, oxygen.

1 I. Zoric, private communication.

Dr Kroes asked: Comparing reaction of NO on Al(111) and ICl on Al(111), why would NO flip after accepting the electron on N and eject N, as opposed to ICl where I accepts the electron and stays on the surface?

Prof. Kummel answered: In the case of ICl, a single electron transfer is sufficient to induce the breaking of the molecular bond. For NO, a single charge transfer is unlikely to result in breaking the internal bond because NO^- has a deep potential well. Instead, in a very simple picture, two charge transfers are needed to break the NO molecular bond. Therefore, NO^- will have sufficient time to flip prior to the second charge transfer process.

Prof. Kosloff commented: If the charge on an NO molecule approaching a metal surface is localized on the nitrogen end, the image force will tend to align the molecule with the nitrogen end pointing to the surface. There could be another curve crossing event taking place which will migrate the charge to the oxygen end.

Prof. Kummel replied: Our model is just a simple discrete charge transfer picture while your curve crossing model is certainly more subtle and accurate. I think we would both agree that calculations will provide a more accurate picture than our simple models.

Dr Darling asked: In your explanation of the selective abstraction reaction, is the reason for the lack of reaction when the molecule approaches the Cl-end down due to the smaller overlap between the ICl LUMO and the metal electronic wavefunctions resulting from the smaller spatial extent of the LUMO on the Cl end of the molecule?

Prof. Kummel responded: Our current hypothesis is that the chemical selectivity for I-end collision being more reactive than Cl-end first collisions is that the LUMO is concentrated on the Cl-end of the molecule. However, we fully appreciate that speculation by experimentalists upon dynamics is no substitute for high level calculations.

Prof. Sitz asked: If the reflectivity for ICl is about 0.4 for an incident energy of 1.2 eV, why is no ICl$^+$ seen in the MPI data of Fig. 3(b) of the paper?

Prof. Kummel replied: There are several reasons. (a) We are detecting near the surface normal. Therefore, diffusely scattered ICl is not detected with nearly as high efficiency as the product atoms from abstraction. (b) We are detecting at short times. Therefore, any ICl which has undergone trapping-desorption will not be detected. (c) The MPI detection process is much more sensitive to atoms than to molecules. (d) Since we want to detect the signal at very low adsorbate coverages, we cannot signal average long enough to detect the small density of inelastically scattered ICl. However, if the surface is covered with oxygen or has a high adsorbate coverage, the density of inelastically scattered ICl is high. Under these conditions, we can detect the inelastically scattered ICl by averaging our signals for many laser shots.

Dr Hodgson asked: Can you tell how much of the 0.8 eV energy left over after the O$_2$ bond breaks appears in translational energy of the O atom?

Prof. Hasselbrink replied: We have some preliminary data addressing that question. We look at the time of flight from the laser focus through the detector at low acceleration voltages. These data let us estimate that the translational energy is smaller than 0.2 eV.

Dr Kolasinski asked: Do you have any idea of what the angular distribution of ejected O atoms is?

Prof. Hasselbrink responded: We have unfortunately no data addressing this question. Due to experimental reasons, mainly to avoid a large background of scattered O$_2$ molecules, the experimental arrangement is very constrained. Secondly, the laser focus is so close to the sample and small that a variation in the angular distribution between being proportional to $\cos^1$ and $\cos^{100}$ would only result in a variation by a factor 2.5 in signal.

Dr McCoustra communicated: In your REMPI experiments, you probed only the lower of the three J components in the O(2p) ^{3}P system. Have you any evidence to suggest that this is the most populated state within the J sub-levels of this system?

Prof. Hasselbrink communicated in response: Indeed we have attempted to detect O-atoms using all three possible transitions. However, the signal on the other two transitions is very faint. Hence, we conclude that the 3p ^{3}P$_2$ is more strongly populated in the abstraction process than in photolysis of NO$_2$. This may shed some light on the orientation of electronic angular momentum and spin when the O-atom is formed.

Dr Kroes commented: In the discussion on reactions of O$_2$ the remark was made that this topic is quite different from H$_2$ + Cu where calculations would allegedly be so good that experiments would no longer be required. However, this is not the case. In a recently submitted paper[1] we compare results on scattering of $(v = 1, j = 1)$ H$_2$ from Cu(100) obtained in the group of Sitz, with wavepacket results obtained in my group, using the newest DFT PES computed in the group of Baerends. There is good agreement between the theoretical and experimental value of the survival probability $P(v = 1, j = 1 \rightarrow v' = 1, j' = 1)$.

However, the agreement is not good for rotationally inelastic scattering within $v = 1$, for $P(v = 1, j = 1 \rightarrow v' = 1, j' = 3)$ and for vibrationally inelastic scattering to high j' states, i.e., $P(v = 1, j = 1 \rightarrow v' = 0, j' = 5)$ and $P(v = 1, j = 1 \rightarrow v' = 0, j' = 7)$.

1 E. Watts, G. O. Sitz, D. A. McCormack, G. J. Kroes, R. A. Olson, J. A. Groeneveld, J. N. P. van Stralen, E. J. Baerends and R. C. Mowrey, *J. Chem. Phys.*, in the press.

Concluding remarks

Aart W. Kleyn

Leiden Institute of Chemistry, Gorlaeus Laboratories, Leiden University, PO Box 9502, 2300 RA Leiden, The Netherlands. E-mail: a.kleyn@chem.leidenuniv.nl

Received 30th October 2000
First published as an Advance Article on the web 4th December 2000

It is an honor and a heavy duty to summarize a Faraday Discussion. I will summarize the state of affairs in our field, and begin by making some connections to the outside world in several respects. Let me start by discussing the meeting and its format. Being fairly unaware of the history of Faraday Discussions, I think that although the format may not have been invented by Michael Faraday himself, he may well have been the inspiration. At times when the interest in the sciences, notably physics and chemistry, seems to be declining, one should admire Faraday even more because of his lectures aimed at the general public.[1] Currently this notion has reappeared to boost the image of our field. Faraday perhaps was lucky in his time: the more fortunate in society had time to be entertained by scientists, and in fact those scientists had a lot to offer. To a large extent our class room demonstration experiments derive from 19th century science, when experiments were more phenomenological and had more visual content. Our science is hidden behind the stainless steel walls of our vacuum chambers and can only be observed on computer displays. Nevertheless, I think we should try very hard to make some of our results palatable to future students, and perhaps even the general public by specifically aiming at bringing out important results at a much more accessible level. I feel this field offers a lot of opportunities. It may be harder to show dynamics than statics (such as structures) but is more fun, because we can look at wavepacket evolution, and classical motion.

The current format of a Faraday Discussion forms a striking contrast to a public lecture. There is hardly any lecture, and the knowledge base of the participants in the field has to be very high. Their knowledge should reflect the state of the art at that very moment, and this is facilitated by reading papers setting that state of the art before the meeting. Feynman recalls that he used to read the Physical Review. Currently this will be physically impossible, even if we allocate 168 hours per week to it. So the patterns of information uptake have changed, because it seems that the time available for uptake of scientific information is decreasing, while the amount is strongly increasing. The question if participants to a Faraday Discussion, in particular students and novices to the field, actually read all material provided would be an interesting topic for study. To have grasped the essentials of the papers is crucial because otherwise (part of) the Discussion is hardly accessible. This is true at any conference, but in particular at Faraday Discussions because the tutorial content, the presentations, is very small. Nevertheless, the number of discussion items at this Discussion per participant is very high, which is a very healthy sign. The number of students and young scientists is also high, which is essential. The average age of the participants at a conference should remain constant in time or decrease.

In our field the sophistication of theory and experiment is increasing all the time. This should be true for any field, but we, in particular, may be facing the problem that we will have 'done it all'. What will be the future: will computers compute everything we want them to and will STM's self replicate? In such a scenario what is left to us? Do we want to generate endless movies about molecules that we should watch, or do we want intuition boosters, toy models, concepts to explain to students, everything in low dimensionality? It is interesting to note that these words were used in the presentations, even in their titles. Our knowledge base and technical abilities are increasing all the time. We want to be able to understand that information in simpler terms: the fun of the field may not be only to do the experiment or the calculation, but also to make it accessible using an appropriate metaphor, to create a simple picture. I am sure that this is entirely along the lines of thinking of Michael Faraday. At this very moment, this reflection is purely academic. It was clear from this Discussion that 'excited states at surfaces' are mainly not understood.

From the future back to the past: where did this work begin? For me a good starting point are the 1930s: the flame experiments of Michael Polanyi, the father of our speaker John Polanyi. He showed in very simple experiments that the interactions between atoms and molecules could act over a larger range than their typical size, represented by the van der Waals radii.[2] As an aside, it would be nice to repeat those experiments as demonstrations, but this might be forbidden by the ever increasing number of safety officers, and the attempt might even be a danger to the image of chemistry. The large cross sections that were measured in reactions such as $Na + Cl_2 \rightarrow NaCl + Cl$ were attributed to a long distance (many Ångstroms) electron jump from the Na to the Cl_2. This jump is as the harpoon thrown to a whale with a rope on it in order to increase the capture radius of the fishing boat. The electron is the harpoon, the Coulomb force the rope. The Cl_2^- formed subsequently explodes (dissociative electron attachment). The Cl^- is attracted by the Na^+ to form a bond and the Cl radical takes off. Gas phase molecular beam experiments have refined the picture considerably in the 'alkali age' around the 1970s.[3] In these experiments the role of excited states was carefully examined. I was involved in some experiments specifically focused on the excited states, namely by detecting them as final states. In other words, we were looking at ion-pair formation. I want to show one experiment in particular. The reasons for my choice will be obvious: I had the picture in my computer when preparing my concluding remarks just before the conference dinner, they serve my purpose, they were fun to do, and are nice (for me) to discus. We observed in 1980 structure in a differential cross section for scattering of Cs from the O_2 structure that could be related to the number of vibrations that a transient O_2^- ion could make during the collision (the overall process is: $Cs + O_2 \rightarrow Cs^+ + O_2^- \rightarrow Cs + O_2$). In an indirect way processes occurring in a time of about 40 fs could be made 'visible': a demonstration of harpooning and its time dependence.[4] This and many related experiments and their theoretical analysis demonstrated that electronic excitations were extremely powerful in order to transfer energy between degrees of freedom in a molecular system in the gas phase. It is for exactly the same reason that we now have a Faraday Discussion on 'Excited states at surfaces'. Let us compare the present achievements at surfaces to those in the gas phase about 20 years ago.

In the experiment that I mentioned and in any scattering experiment there is only very limited control over the experimental conditions at the molecular level. The big difference with the current state of the art in surface dynamics is control. At first, in the scattering experiment, the impact parameter could not be set. At this Discussion we heard how atoms can be manoeuvred at the atomic level by scanning tunnelling microscopes (STM's). Reactants in a surface reaction can be made to collide at a specific place. In addition, the STM induces changes by providing an extremely strong field, or induces electronic transitions by providing an intense electron current. More control is possible: laser excitation can put molecules in specific ro-vibrational states before interaction with the surface. Finally, one can even hope to control a surface process by tailoring the shape of femtosecond laser pulses.

At this Discussion unification was an important item, albeit implicitly. The power of the STM to induce electronic transitions has made a bridge between the structure oriented STM'ers and the dynamics oriented laser and particle beam scientists. The interest of all of them in electronic transitions has even caused the ion-scatterers to join. Excited states at surfaces manifest themselves in many ways. Also the bridge between bound state oriented spectroscopists and continuum-state oriented scatterers is being formed. The unification of this methodology is going to make it possible to study what has been promised for years: chemistry at surfaces and polyatomic molecules.

This Discussion has certainly featured a number of 'evergreens': problems and topics that are discussed for perhaps decades, but never really solved. The electron–hole pair continuum is one of them. Also at this meeting, the interactions at metals are mostly being discussed in terms of discrete states. The role of the continua of the solid, electron–hole pairs and phonons, is not yet clear. Connected is the method of the dynamics calculation: fully quantum, semi-classical or fully classical descriptions of the motion of the nuclei is used in all cases. The relation of the various methods and the importance of truly quantum effects for the motion of heavy particles is still under debate. Closely connected to this point is the issue of dimensionality. Full quantum calculations for H_2 interacting with Cu and other metal surfaces or Si are feasible, as long as the degrees of freedom of the solid, points mentioned above, are neglected. The question is to what extend classical calculations are sufficient to describe the overall features of the interaction. This point also depends on the choice of the potential. In fact, calculations of molecule surface interactions can be separated into three parts, as long as the influence of electronically excited states is not (too) relevant. The calculation of the potential in a restricted phase space, the parametrization of the potential to allow a description of the full multidimensional phase space, and finally the dynamical calculation, already mentioned. The first step, the computation of the potential, has seen great leaps forward thanks to the introduction of the gradient corrections for the variations in density of the electron gas, used in DFT. Nevertheless, these calculations fail to reproduce some very qualitative phenomena observed experimentally. The oxidation of Al may serve as an example here.

In the list some younger loots on the tree of evergreens should be mentioned. 'Steering', the reorientation of an approaching molecule to follow the minimum energy path to, for instance, adsorption or dissociation is a concept widely in use. Likewise, reagent orientation is an effect often mentioned that can be examined only in very few cases. It remains high on the agenda. A topic less discussed in our present Discussion was 'transient mobility', 'transient trapping' and other forms of adsorption of molecules, that seem to maintain some of their initial (potential) energy during part of the dynamical process at the surface. Clearly it is close to the topic of 'excited states at surfaces', but as in other cases, the nature of the excited transients is yet to be fully identified. A final evergreen is H_2 on Cu surfaces. This really is the fruitfly of surface dynamics. It is the only system where in some cases there is almost quantitative agreement between the results of DFT plus quantum dynamics calculations and state of the art state-to-state scattering studies. This success would have been hard to predict during the alkali age, and many years thereafter when the H_2–metal problem was too formidable to solve theoretically. That the H_2–Cu interaction can be described very nicely using the best theoretical methods demonstrates how rapid the progress in the description of gas–surface dynamics has been.

Science needs to work on evergreens: many problems are tough and need many years for a solution. Nevertheless, it is nice if novel problems are introduced as well. This was certainly the case here. Perhaps it was the first time that organic molecules have been introduced to a dynamics meeting. It shows that we are getting brave enough to turn away from the most basic model systems. Local spectroscopy has been introduced. In some cases the STM can be used as a vibrational spectrometer. This is a specificity of study in the lateral sense that was really hard to believe until very recently. The only thing that seems to be lacking at the moment is time resolved vibrational spectroscopy at the molecular level. Near field optical microscopy in combination with femtosecond lasers may provide the future answers here. Electronically excited states at surfaces imply an extra degree of freedom in the calculations: electronically non-adiabatic effects need to be studied explicitly. Several methods have been introduced at the Discussion. A major obstacle at this stage is that the excited state surfaces cannot be computed very well, and cannot be easily introduced into DFT calculations. The excited state at surfaces will remain to challenge us for at least a decade.

In a field that is making such rapid progress towards a full understanding of interactions at the molecular level one might ask the question again: 'we will know it all, so what?'. Will we really use femtosecond lasers in the production of fine chemicals? Will we be able to have well-defined nano structures self assemble? Will we understand all relevant heterogeneous surface chemistry in the atmosphere? Will we be able to control perhaps simple chemical reactions such as steam reforming, the water gas shift reaction or methane partial oxidation sufficiently well, that we can solve some global energy problems and build a zero emission vehicle? Time will tell, but we have to realise that it is always hard for a chemist to beat 'shaking and baking'.

Coming to an end of a conference I feel one always comes to the end of a success story. Conferences always bring new ideas, introduce new knowledge or people. This seems to beat a statistical interpretation of reality, which is counter intuitive. I believe the secret is in the special environment of the conference, its total focus on science and interaction between scientists. It is successful because a group of very dedicated scientists, the organising committee, assisted by excellent professionals from the RSC, have invested an extraordinary amount of energy in the success of the Faraday Discussion on "Excited states at surfaces". I am extremely grateful for their efforts.

References

1   J. M. Thomas, *Michael Faraday and the Royal Institution*, Adam Hilger, Bristol, 1991.
2   M. Polanyi, *Atomic Reactions*, Williams & Norgate, London, 1932.
3   D. R. Herschbach, *Adv. Chem. Phys.*, 1966, **10**, 319.
4   A. W. Kleyn, J. Los and E. A. Gislason, *Phys. Rep.*, 1982, **90**, 1.

Additions and corrections

PFI-ZEKE photoelectron spectra of the methane cation and the dynamic Jahn–Teller effect

R. Signorell and F. Merkt

Faraday Discuss., 2000, **115**, 205 (DOI: 10.1039/a909272b).

On page 221, eqn. (12) gives an incorrect value for the adiabatic ionisation potential of CD_2H_2. The correct value is as listed in Table 3 of the article and the corrected form of eqn. (12) should read:

$$IP = (101\,852.3 \pm 1.4)\ cm^{-1} \qquad (12)$$

An error occurred in the late stages of production which resulted in the column headings in Table 3 being misaligned and some of the table footnotes not appearing. The correct version of Table 3 is reproduced below:

Table 3 Adiabatic ionization potentials of CH_4 and its isotopomers

	IP/cm^{-1}	$IP-IP(CH_4)/cm^{-1}$ experiment	$IP-IP(CH_4)/cm^{-1}$ ab initio	Ref.
CH_4	$101\,773 \pm 35^{a,b}$	0	0	This work
	$100\,900^{a}$	—	—	5
	$103\,078^{a}$	—	—	3
	$102\,513 \pm 161^{a}$	—	—	8
	$101\,706 \pm 81^{a}$	—	—	11
	$101\,706^{c}$	—	—	25
	$98\,803^{c}$	—	—	30
CDH_3	$101\,820 \pm 40^{a,b}$	47 ± 53	52	This work
	$101\,810.0 \pm 1.6^{d}$	37.0 ± 35	52	This work
CD_2H_2	$101\,861 \pm 25^{a,b}$	88 ± 43	102	This work
	$101\,852.3 \pm 1.4^{a,e}$	78.1 ± 35	102	This work
CD_4	$102\,210 \pm 25^{a,b}$	437 ± 43	429	This work
	$102\,197.1 \pm 1.5^{a,e}$	424.1 ± 35	429	This work
	$103\,804 \pm 161^{a}$	—	—	8
	$102\,094 \pm 121^{a}$	—	—	11

[a] Experimental value. [b] The uncertainty represents the half width at half maximum of the rotational contour of the lowest band. [c] Theoretical value. [d] Value determined from the tentative assignment of the rotationally resolved PFI-ZEKE photoelectron spectrum (see Section 4). This value corresponds to the $N = 0$, $K = 0$, $\Gamma_{evr} = A \rightarrow N^+ = 0$, $K^+ = 0$, $\Gamma_{evr}^+ = E$ transition. The influence of the electric field ionization sequence has been corrected for by a shift of $1.1\ cm^{-1}$. The uncertainty includes contributions from the calibration accuracy of both lasers, the accuracy of the field correction and the width of the observed lines. [e] Value determined from the analysis of the rotationally resolved PFI-ZEKE photoelectron spectra in Section 4.4.

On p. 221, lines 3 and 14 and p. 224, line 4, *vn* should read $\tilde{v}$.

The above errors occurred as a result of a technical problem after the proofs had been seen by the author.

The authors wish to make the following additional amendments to the paper:

On p. 226 under Conclusions, the fifth line before the end of the page should read:

By contrast, the significantly smaller gaps in the spectra of CDH_3^+ and $CD_2H_2^+$ result from the zero-point energy difference of inequivalent minima.

The first line of the caption to Fig. 7 should read:

Fig. 7. C_{3v} symmetrical potential $V(q)$ of CH_4^+ along the one-dimensional pseudorotational reaction co-ordinate q.

Direct and indirect DIET and DIMET from semiconductor and metal surfaces: What can we learn from 'toy models'?

Peter Saalfrank, Gisela Boendgen, Cécile Corriol and Tohru Nakajima

Faraday Discuss., 2000, **117**, 65 (DOI: 10.1039/b002734k).

The form of eqn. (6) appears incorrectly; the last term in this equation should read: $-\frac{1}{2}\hat{\rho}\hat{C}_k\hat{C}_k^\dagger$ rather than $-\frac{1}{2}\hat{\rho}\hat{C}_k^\dagger\hat{C}_k$. Also in the paragraph following eqn. (18), $L_{des} = 12$ should be replaced by $I_{des} = 12$.

The Royal Society of Chemistry apologises for these errors and any consequent inconvenience to authors and readers.

List of Posters

Real-time investigation of excited electronic alkali states on single crystal surfaces **M. Bauer, S. Pawlik** and **M. Aeschlimann**, *Universität Essen, Germany*

Sticking of HCl to ice surfaces at hyperthermal energies **A. Al-Halabi, A. W. Kleyn** and **G. J. Kroes**, *Leiden University, The Netherlands*

Dissociation of hydrogen halides on ice surfaces: *ab initio* study **A. Al-Halabi**, *Leiden University, The Netherlands*, **R. Bianco** and **J. T. Hynes**, *University of Colorado at Boulder, USA*

Hot ethene desorption *via* dissociative electron attachment: surface coverage, surface composition and electron irradiation effects **A. S. Y. Chan, S. Turton** and **R. G. Jones**, *University of Nottingham, UK*

Scanning probe energy loss spectroscopy: angular resolved measurements **B. J. Eves, F. Festy, K. Svensson** and **R. E. Palmer**, *University of Birmingham, UK*

Effects of electron-hole pair creation on molecule–surface interactions **D. Bird, M. Graham** and **J. Trail**, *University of Bath, UK*

Dynamics of dissociative hydrogen adsorption on an oxygen precovered stepped platinum (533) surface **B. E. Hayden, C. Mormiche** and **T. S. Nunney**, *University of Southampton, UK*

Charge transfer, dissociation dynamics and non-adiabatic events: Cl_2 on potassium **L. Hellberg, J. Strömqvist, B. Kasemo** and **B. Lundqvist**, *Chalmers University of Technology and Göteborg University, Sweden*

Ultrafast photochemistry in the VUV *via* high harmonic generation **L. Hernández-Pozos, D. Riedel, S. Baggott, K. W. Kolasinski** and **R. E. Palmer**, *University of Birmingham, UK*, **J. S. Foord**, *University of Oxford, UK*

Site poisoning during photodissociation of OCS on Ag(111) **D. Lennon**, *University of Glasgow, UK*, **R. T. Kidd** and **S. R. Meech**, *University of East Anglia, UK*

The effect of corrugation on the quantum dynamics of dissociative and diffractive scattering of H_2 from Pt(111) **E. Pijper** and **G. J. Kroes**, *Leiden University, The Netherlands*, **R. A. Olsen** and **E. J. Baerends**, *Vrije Universiteit, The Netherlands*

Interaction of D_2 with a sulfur covered Pd(100) surface **M. Rutkowski, D. Wetzig** and **H. Zacharias**, *Westfälische-Wilhelms Universität, Germany*

Surface interband transitions and surface plasmon dispersion: O/Ag(001) **L. Savio, L. Vattuone** and **M. Rocca**, *Università di Genova, Italy*, **V. De Renzi, S. Gardonió, C. Mariani** and **U. del Pennino**, *Università di Modena e Reggio Emilia, Italy*, **G. Cipriani, A. Dal Corso** and **S. Baroni**, *INFM Sezione di Trieste, Italy*

The role of internal degrees of freedom in gas–surface interaction: C_2H_4 on Ag(001) **L. Vattuone, L. Savio, U. Valbusa** and **M. Rocca**, *Università di Genova, Italy*

State-to-state scattering at a reactive surface: $H_2(v = 1, J = 1)$ from Cu(100) and Pd(111) **G. O. Sitz, M. Gostein, E. Watts, L. Shackman** and **M. Isakson**, *University of Texas at Austin, USA*

Negative ion states of chlorobenzene adsorbed on the Si(111) 7x7 surface: from resonance scattering to molecular manipulation **P. A. Sloan, K. Svensson** and **R. E. Palmer**, *University of Birmingham, UK*

6-D molecular quantum dynamics of H_2 dissociating on a Cu(100) surface **M. F. Somers, D. A. McCormack** and **G. J. Kroes**, *Leiden University, The Netherlands*, **R. A. Olsen, J. A. Groeneveld, J. N. P. van Stralen** and **E. J. Baerends**, *Free University, The Netherlands*, **R. C. Mowrey**, *Naval Research Laboratory, Washington, USA*

An energy map of surface and defect states for photo-induced surface reactions: MgO(001) surface **P. V. Sushko** and **A. L. Shluger**, *University College London, UK*

Vibrational deactivation of $HCl(v = 2)$ on MgO(100) **M. Suchan, M. Korolik, S. Hawkins, W. Kim, H. Reisler** and **C. Wittig**, *University of Southern California, USA*

Photoinduced electron transfer in thiolate self-assembled monolayers **H. Wang, P. Winget, C. J. Cramer** and **X. Zhu**, *University of Minnesota, USA*, **T. Vondrak**, *University of Minnesota, USA and University of East Anglia, UK*

Photochemistry of NO/Ag(111) system: the dynamics and origin of the high temperature site **T. Vondrak, D. J. Burke** and **S. R. Meech**, *University of East Anglia, UK*

Vibration–rotational coupling of H_2 on Cu(111) **Z. S. Wang, G. R. Darling** and **S. Holloway**, *University of Liverpool, UK*

Ultrafast electron transfer *via* excited states at surfaces **R. S. Shankar** and **F. Willig**, *Hahn-Meitner-Institut, Germany*, **V. May**, *Humboldt-Universität zu Berlin, Germany*

STM-induced desorption of H from Si(100) in the DIMET-limit: beyond the "truncated oscillator" model **I. Andrianov, G. Boendgen** and **P. Saalfrank**, *University College London, UK*

Quantum photochemistry **P. Persson**, *Uppsala University, Sweden*

List of Participants

Prof. M. Aeschlimann *Universität GH Essen, Germany*
Mr A. S. Al-Halabi *Leiden University, The Netherlands*
Dr G. M. Allcock *Royal Society of Chemistry, UK*
Prof. S. E. Andersson *Chalmers University of Technology, Sweden*
Dr I. V. Andrianov *University College London, UK*
Dr H. Arnolds *University of Cambridge, UK*
Mr S. Baggott *University of Birmingham, UK*
Prof. D. M. Bird *University of Bath, UK*
Ms G. Boendgen *University College London, UK*
Mr D. J. Burke *University of East Anglia, UK*
Prof. M. Chesters *University of Nottingham, UK*
Prof. D. C. Clary *University College London, UK*
Dr G. R. Darling *University of Liverpool, UK*
Dr G Dujardin *Université Paris-Sud, France*
Mr B. J. Eves *University of Birmingham, UK*
Mr A. J. Farebrother *University College London, UK*
Mr V. Fiorin *University of Nottingham, UK*
Prof. J. W. Gadzuk *National Institute for Standards and Technology, USA*
Dr J.-P. Gauyacq *Université Paris-Sud, France*
Dr J. B. Giorgi *University of Toronto, Canada*
Mr M. C. Graham *University of Bath, UK*
Prof. A. Groß *Technical University of Munich, Germany*
Miss C. L. Hall *Royal Society of Chemistry, UK*
Mr J. J. W. Harris *University of Cambridge, UK*
Prof. E. Hasselbrink *Universität GH Essen, Germany*
Mr R. A. Hatton *University of Nottingham, UK*
Mr L. A. Hellberg *Chalmers University of Technology, Sweden*
Dr L. Hernandez *University of Birmingham, UK*
Dr A. Hodgson *University of Liverpool, UK*
Prof. S. Holloway *University of Liverpool, UK*
Prof. R. G. Jones *University of Nottingham, UK*
Prof. H. Jónsson *University of Washington, USA*
Mr G. Katz *Hebrew University of Jerusalem, Israel*
Dr H. Kleine *University of Cambridge, UK*
Prof. A. W. Kleyn *Leiden University, The Netherlands*
Dr K. W. Kolasinski *University of Birmingham, UK*
Mr A. J. Komrowski *University of California, San Diego, USA*
Prof. R. Kosloff *Hebrew University of Jerusalem, Israel*
Dr G.-J. Kroes *Leiden University, The Netherlands*
Prof. A. C. Kummel *University of California, San Diego, USA*
Dr L. J. Lauhon *Harvard University, USA*
Dr T. G. Lee *University of Toronto, Canada*
Dr D. Lennon *University of Glasgow, UK*
Prof. Y. Matsumoto *Graduate University for Advanced Studies, Japan*
Dr A. J. Mayne *Université Paris-Sud, France*
Dr D. A. McCormack *Leiden University, The Netherlands*
Dr M. R. S. McCoustra *University of Nottingham, UK*
Dr S. R. Meech *University of East Anglia, UK*
Miss C. Mormiche *University of Southampton, UK*
Dr T. Nunney *University of Southampton, UK*
Dr R. A. Olsen *Vrije Universiteit, The Netherlands*
Mr F. Olsson *Chalmers University of Technology, Sweden*
Prof. R. E. Palmer *University of Birmingham, UK*
Mr L. M. Perdigao *University of Birmingham, UK*
Prof. M. Persson *Chalmers University of Technology, Sweden*
Mr P. Persson *Uppsala University, Sweden*

Mr E. Pijper *Leiden University, The Netherlands*
Prof. J. C. Polanyi *University of Toronto, Canada*
Dr S. Ramakrishna *Hahn-Meitner-Institut, Germany*
Miss S. Riaz *Royal Society of Chemistry, UK*
Prof. M. A. Rocca *Università di Genova, Italy*
Mr M. Rutkowski *Westfälische Wilhelms-Universität, Germany*
Dr P. E. Saalfrank *University College London, UK*
Mr H. A. Scheld *FOM — Institute AMOLF, The Netherlands*
Ms F. H. Scholes *University of Cambridge, UK*
Prof. T. Seideman *National Research Council of Canada, Canada*
Mr J. Z. Sexton *University of California, San Diego, USA*
Dr A. Shluger *University College London, UK*
Dr I. G. Shuttleworth *University of Nottingham, UK*
Prof. G. O. Sitz *University of Texas at Austin, USA*
Mr M. Skegg *University of Nottingham, UK*
Mr P. A. Sloan *University of Birmingham, UK*
Mr M. F. Somers *Leiden University, The Netherlands*
Prof. K. Stokbro *Technical University of Denmark, Denmark*
Ms M. Suchan *University of Southern California, USA*
Mr J. P. R. Symonds *University of Cambridge, UK*
Prof. A. L. Utz *Tufts University, USA*
Dr M. Volk *University of Liverpool, UK*
Dr T. Vondrak *University of East Anglia, UK*
Dr Z. Wang *University of Liverpool, UK*
Dr S. B. Wilkes *Royal Society of Chemistry, UK*
Prof. H. Winter *Humboldt-University Berlin, Germany*
Prof. M. Wolf *Fritz-Haber-Institut der Max-Planck-Gesellschaft, Germany*
Dr S. Wright *University of Liverpool, UK*

Index of Contributors*

Aeschlimann, M., 56, 58, 61, 257, 260
Alavi, S., **213**
Al-Halabi, A. S., 60, 274
Bach, C., **99**
Baerends, E. J., **109**
Binetti, M., **313**
Bird, D. M., 187, 188
Boendgen, G., **65**
Borisov, A. G., **15, 27**
Chesters, M., 273
Clary, D. C., 174, 183
Corrales, L. R., **303**
Corriol, C., **65**
Darling, G. R., **41**, 62, 63, 64, 166, 180, 332, 333, 336, 338, 339, 340, 344
Dujardin, G., 57, 168, **241**, 264, 269, 270, 272, 333
Gadzuk, J. W., **1**, 55, 59, 60, 61, 163, 168, 170, 172, 174, 188, 268, 331
Gahl, C., **191**
Gauyacq, J.-P., **15, 27**, 55, 56, 57, 58, 62, 64, 257, 262, 331, 335
Giorgi, J. B., **85**, 273
Grey, F., **231**
Groeneveld, J. A., **109**
Groß, A., 63, **99**, 167, 168, 169, 171, 172, 173, 174, 184, 267, 333, 336, 341
Hasselbrink, E., 172, 259, 260, **313**, 341, 342, 345
Hecht, T., **27**
Ho, W., **249**
Hodgson, A., **133**, 166, 175, 178, 179, 180, 189, 340, 345
Holloway, S., 60, 173
Hotzel, A., **191**
Ishioka, K., **191**
Jones, R. G., 340
Jónsson, H., **303**, 338, 339
Juurlink, L. B. F., **147**
Katz, G., **291**
Kazansky, A. K., **15, 27**
Kleyn, A. W., 55, 63, 165, 171, 258, 259, 335, 343, **347**
Kolasinski, K. W., 59, 174, 272, 340, 345
Komrowski, A. J., **313**, 341, 344
Kosloff, R., **41**, 166, 169, 171, 174, 176, 179, 263, **291**, 333, 334, 335, 336, 337, 339, 342, 343, 344
Kroes, G. J., **109**, 172, 179, 185, 186, 187, 189, 271, 333, 337, 338, 344, 345
Kummel, A. C., 269, **313, 321**, 334, 340, 341, 344, 345
Lauhon, L. J., **249**, 271, 272, 273, 274
Lee, T. G., 59, **85**, 259, 273

Lin, R., **231**
Lopinski, G. P., **213**
Lorente, N., **277**
Matsumoto, Y., 58, **203**, 258, 259, 260
Mayne, A. J., **241**
McCormack, D. A., **109**
McCoustra, M. R. S., 182, 345
Mowrey, R. C., **109**
Nakajima, T., **65**
Naumkin, F. Y., **85**
Olsen, R. A., **109**
Palmer, R. E., 61, 263, 265
Persson, M., 55, 189, 269, **277**, 331, 332, 333
Pettus, K. A., **321**
Plihal, M., **1**
Polanyi, J. C., **85**, 164, 165, 166, 167, 181, 266, 270
Quaade, U. J., **231**
Raşeev, G., **15**
Raspopov, S. A., **85**
Rocca, M. A., 264, 342, 343
Rose, F., **241**
Rousseau, R., **213**
Saalfrank, P. E., 57, **65**, 161, 163, 164, 165, 167, 173, 188, 263, 336, 337, 339
Seideman, T., 162, 185, **213**, 264, 265, 266, 267, 274
Shluger, A., 56, 58, 59, 64, 167, 179, 184, 258, 266, 270, 274, 332, 337, 338
Sitz, G. O., 186, 188, 189, 273, 345
Skelly, J. F., **133**
Smith, R. R., **147**
Song, J., **303**
Stokbro, K., 161, 162, 163, 168, 188, **231**, 264, 265, 266, 267, 269, 270, 272, 331
Taylor, P. R., **321**
Thirstrup, C., **231**
Utz, A. L., **147**, 180, 181, 182, 183, 184
VanGinhoven, R. M., **303**
van Stralen, J. N. P., **109**
Wang, J., **85**
Watanabe, K., **203**
Weiße, O., **313**
Winter, H., **27**, 55, 59, 61, 257
Wolf, M., 57, 161, 162, 165, 188, **191**, 257, 258, 260, 261, 331, 332, 340, 341
Wolkow, R. A., **213**
Wright, S., **133**
Zeiri, Y., **41, 291**
Zhong, Q., **191**

* The page numbers in **bold** type indicate papers submitted for discussions.

General Discussions of the Faraday Society/Faraday Discussions of the Chemical Society

Date	Subject	Volume
1907	Osmotic Pressure	Trans. 3
1907	Hydrates in Solution	3
1910	The Constitution of Water	6
1911	High Temperature Work	7
1912	Magnetic Properties of Alloys	8
1913	Colloids and their Viscosity	9
1913	The Corrosion of Iron and Steel	9
1913	The Passivity of Metals	9
1914	Optical Rotary Power	10
1914	The Hardening of Metals	10
1915	The Transformation of Pure Iron	11
1916	Methods and Appliances for the Attainment of High Temperatures in a Laboratory	12
1916	Refractory Materials	12
1917	Training and Work of the Chemical Engineer	13
1917	Osmotic Pressure	13
1917	Pyrometers and Pyrometry	13
1918	The Setting of Cements and Plasters	14
1918	Electric Furnaces	14
1918	Co-ordination of Scientific Publication	14
1918	The Occlusion of Gases by Metals	14
1919	The Present Position of the Theory of Ionization	15
1919	The Examination of Materials by X-Rays	15
1920	The Microscope: Its Design, Construction and Applications	16
1920	Basic Slags: Their Production and Utilization in Agriculture	16
1920	Physics and Chemistry of Colloids	16
1920	Electrodeposition and Electroplating	16
1921	Capillarity	17
1921	The Failure of Metals under Internal and Prolonged Stress	17
1921	Physico-Chemical Problems Relating to the Soil	17
1921	Catalysis with special reference to Newer Theories of Chemical Action	17
1922	Some Properties of Powders with special reference to Grading by Elutriation	18
1922	The Generation and Utilization of Cold	18
1923	Alloys Resistant to Corrosion	19
1923	The Physical Chemistry of the Photographic Process	19
1923	The Electronic Theory of Valency	19
1923	Electrode Reactions and Equilibria	19
1923	Atmospheric Corrosion. First Report	19
1924	Investigation on Oppau Ammonium Sulphate-Nitrate	20
1924	Fluxes and Slags in Metal Melting and Working	20
1924	Physical and Physico-Chemical Problems relating to Textile Fibres	20
1924	The Physical Chemistry of Igneous Rock Formation	20
1924	Base Exchange in Soils	20
1925	The Physical Chemistry of Steel-Making Processes	21
1925	Photochemical Reactions of Liquids and Gases	21
1926	Explosive Reactions in Gaseous Media	22
1926	Physical Phenomena at Interfaces, with special reference to Molecular Orientation	22
1927	Atmospheric Corrosion, Second Report	23
1927	The Theory of Strong Electrolytes	23
1927	Cohesion and Related Problems	24
1928	Homogeneous Catalysis	24
1929	Crystal Structure and Chemical Constitution	25
1929	Atmospheric Corrosion of Metals, Third Report	25
1929	Molecular Spectra and Molecular Structure	26
1930	Colloid Science Applied to Biology	26
1931	Photochemical Processes	27
1932	The Adsorption of Gases by Solids	28
1932	The Colloid Aspect of Textile Materials	29

Date	Subject	Volume
1933	Liquid Crystals and Anisotropic Melts	29
1933	Free Radicals	30
1934	Dipole Moments	30
1934	Colloidal Electrolytes	31
1935	The Structure of Metallic Coatings, Films and Surfaces	31
1935	The Phenomena of Polymerization and Condensation	32
1936	Disperse Systems in Gases: Dust, Smoke and Fog	32
1936	Structure and Molecular Forces in (a) Pure Liquids, and (b) Solutions	33
1937	The Properties and Function of Membranes, Natural and Artificial	33
1937	Reaction Kinetics	34
1938	Chemical Reactions Involving Solids	34
1938	Luminescence	35
1939	Hydrocarbon Chemistry	35
1939	The Electrical Double Layer (owing to the outbreak of the war the meeting was abandoned, but the papers were printed in the *Transactions*)	35
1940	The Hydrogen Bond	36
1941	The Oil-Water Interface	37
1941	The Mechanism and Chemical Kinetics of Organic Reactions in Liquid Systems	37
1942	The Structure and Reactions of Rubber	38
1943	Modes of Drug Action	39
1944	Molecular Weight and Molecular Weight Distribution in High Polymers (Joint Meeting with the Plastics Group, Society of Chemical Industry)	40
1945	The Application of Infra-red Spectra to Chemical Problems	41
1945	Oxidation	42
1946	Dielectrics	42 A
1946	Swelling and Shrinking	42 B
1947	Electrode Processes	Disc. 1
1947	The Labile Molecule	2
1947	Surface Chemistry (Jointly with the Sociéitéi de Chimie Physique at Bordeaux Published by Butterworths Scientific Publications Ltd	
1947	Colloidal Electrolytes and Solutions	Trans. 43
1948	The Interaction of Water and Porous Materials	Disc. 3
1948	The Physical Chemistry of Process Metallurgy	4
1949	Crystal Growth	5*
1949	Lipo-proteins	6
1949	Chromatographic Analysis	7
1950	Heterogeneous Catalysis	8
1950	Physico-chemical Properties and Behaviour of Nuclear Acids	Trans. 46
1950	Spectroscopy and Molecular Structure and Optical Methods of Investigating Cell Structure	Disc. 9
1950	Electrical Double Layer	Trans. 47
1951	Hydrocarbons	Disc. 10
1951	The Size and Shape Factor in Colloidal Systems	11
1952	Radiation Chemistry	12
1952	The Physical Chemistry of Proteins	13
1952	The Reactivity of Free Radicals	14
1953	The Equilibrium Properties of Solutions on Non-electrolytes	15
1953	The Physical Chemistry of Dyeing and Tanning	16
1954	The Study of Fast Reactions	17
1954	Coagulation and Flocculation	18
1955	Microwave and Radio-frequency Spectroscopy	19
1955	Physical Chemistry of Enzymes	20
1956	Membrane Phenomena	21
1956	Physical Chemistry of Processes at High Pressures	22
1957	Molecular Mechanism of Rate Processes in Solids	23
1957	Interactions in Ionic Solutions	24
1958	Configurations and Interactions of Macromolecules and Liquid Crystals	25
1958	Ions of the Transition Elements	26
1959	Energy Transfer with special reference to Biological Systems	27
1959	Crystal Imperfections and the Chemical Reactivity of Solids	28
1960	Oxidation-Reduction Reactions in Ionizing Solvents	29
1960	The Physical Chemistry of Aerosols	30
1961	Radiation Effects in Inorganic Solids	31
1961	The Structure and Properties of Ionic Melts	32
1962	Inelastic Collisions of Atoms and Simple Molecules	33
1962	High Resolution Nuclear Magnetic Resonance	34
1963	The Structure of Electronically Excited Species in the Gas Phase	35
1963	Fundamental Processes in Radiation Chemistry	36
1964	Chemical Reactions in the Atmosphere	37
1964	Dislocations in Solids	38
1965	The Kinetics of Proton Transfer Processes	39

Date	Subject	Volume
1965	Intermolecular Forces	40
1966	The Role of the Absorbed State in Heterogeneous Catalysis	41
1966	Colloid Stability in Aqueous and Non-aqueous Media	42
1967	The Structure and Properties of Liquids	43
1967	Molecular Dynamics of the Chemical Reactions of Gases	44
1968	Electrode Reactions of Organic Compounds	45
1968	Homogeneous Catalysis with Special Reference to Hydrogenation and Oxidation	46
1969	Bonding in Metallo-organic Compounds	47
1969	Motions in Molecular Crystals	48
1970	Polymer Solutions	49
1970	The Vitreous State	50
1971	Electrical Conduction in Organic Solids	51
1971	Surface Chemistry of Oxides	52
1972	Reactions of Small Molecules in Excited States	53
1972	The Photoelectron Spectroscopy of Molecules	54
1973	Molecular Beam Scattering	55
1973	Intermediates in Electrochemical Reactions	56
1974	Gels and Gelling Processes	57
1974	Photo-effects in Adsorbed Species	58
1975	Physical Adsorption in Condensed Phases	59
1975	Electron Spectroscopy of Solids and Surfaces	60
1976	Precipitation	61
1977	Potential Energy Surfaces	62
1977	Radiation Effects in Liquids and Solids	63
1977	Ion–Ion and Ion–Solvent Interactions	64
1978	Colloid Stability	65
1978	Structures and Motion in Molecular Liquids	66
1979	Kinetics of State Selected Species	67
1979	Organization of Macromolecules in the Condensed Phase	68
1980	Phase Transitions in Molecular Solids	69*
1980	Photoelectrochemistry	70*
1981	High Resolution Spectroscopy	71*
1981	Selectivity in Heterogeneous Catalysis	72*
1982	Van der Waals Molecules	73*
1982	Electron and Proton Transfer	74*
1983	Intramolecular Kinetics	75*
1983	Concentrated Colloidal Dispersions	76*
1984	Interfacial Kinetics in Solution	77*
1984	Radicals in Condensed Phases	78*
1985	Polymer Liquid Crystals	79*
1985	Physical Interactions and Energy Exchange at the Gas–Solid Interface	80*
1986	Lipid Vesicles and Membranes	81*
1986	Dynamics of Molecular Photofragmentation	82*
1987	Brownian Motion	83*
1987	Dynamics of Elementary Gas-phase Reactions	84*
1988	Solvation	85*
1988	Spectroscopy at Low Temperatures	86*
1989	Catalysis by Well Characterised Materials	87*
1989	Charge Transfer in Polymeric Systems	88
1990	Structure of Surfaces and Interfaces as studied using Synchrotron Radiation	89*
1990	Colloidal Dispersions	90*
1991	Structure and Dynamics of Reactive Transition States	91*
1991	The Chemistry and Physics of Small Metallic Particles	92*
1992	Structure and Activity of Enzymes	93*
1992	The Liquid/Solid Interface at High Resolution	94*
1993	Crystal Growth	95*
1993	Dynamics at the Gas/Solid Interface	96*
1994	Structure and Dynamics of Van der Waals Complexes	97*
1994	Polymers at Surfaces and Interfaces	98*
1994	Vibrational Optical Activity: From Fundamentals to Biological Applications	99*
1995	Atmospheric Chemistry: Measurements, Mechanisms and Models	100*
1995	Gels	101*
1995	Unimolecular Reaction Dynamics	102*
1996	Hydration Processes in Biological and Macromolecular Systems	103*
1996	Complex Fluids at Interfaces	104*
1996	Catalysis and Surface Science at High Resolution	105*
1997	Solid State Chemistry: New Opportunities from Computer Simulations	106*
1997	Interactions of Acoustic Waves with Thin Films and Interfaces	107*
1997	Dynamics of Electronically Excited States in Gaseous, Clusters and Condensed Media	108*
1998	Chemistry and Physics of Molecules and Grains in Space	109*
1998	Chemical Reaction Theory	110*

Date	Subject	Volume
1998	Molecular Interactions of Biomembranes	111*
1999	Physical Chemistry in the Mesoscopic Regime	112*
1999	Stereochemistry and Control in Molecular Reaction Dynamics	113*
1999	The Surface Science of Metal Oxides	114*
2000	Molecular Photoionisation	115*
2000	Bioelectrochemistry	116*

* Available for purchase, for current information on prices etc. please contact the Sales and Promotion Department, The Royal Society of Chemistry, Thomas Graham House, Science Park, Milton Road, Cambridge, UK CB4 0WF.